NONLINEAR EVOLUTION EQUATIONS: INTEGRABILITY AND SPECTRAL METHODS

Edited by

A. Degasperis, A. P. Fordy *and* M. Lakshmanan

Manchester University Press
Manchester and New York
Distributed exclusively in the USA and Canada by St. Martin's Press

Published by Manchester University Press
Oxford Road, Manchester M13 9PL, UK
and Room 400, 175 Fifth Avenue,
New York, NY 10010, USA

Distributed exclusively in the USA and Canada
by St. Martin's Press, Inc.,
175 Fifth Avenue, New York, NY 10010, USA

British Library cataloguing in publication data applied for

Library of Congress cataloging in publication data applied for

ISBN 0-7190-3273-3 *hardback*

Printed in Great Britain
by Biddles Ltd, Guildford and King's Lynn

CONTENTS

PREFACE

Soliton theory is a beautifully interdisciplinary subject which has had enormous impact, both on mathematics and the physical sciences. This book is a selection of articles presented at the workshop: "Nonlinear Evolution Equations : Integrability and Spectral Methods" held in the beautiful neoclassical XVIII century Villa Olmo, Como, Italy, 4-15 July 1988; this villa is the residence of the Centro di Cultura Scientifica "A. Volta", an active scientific organisation which kindly hosted the workshop.

Although this meeting was not 'officially' part of the NEEDS series, it certainly belonged to it in spirit. We had a broad mix of nationalities, with 82 participants from 17 countries. As well as the time allocated for the formal lectures and posters published here, there was ample free time for private discussions and collaboration.

The workshop was supported by the Italian Ministry of Public Education, University of Rome "La Sapienza", Istituto Nazionale di Fisica Nucleare, Consiglio Nazionale delle Ricerche, International Centre for Theoretical Physics (Trieste), University of Leeds, Science and Engineering Research Council (UK), Bharatidasan University, Department of Science and Technology (India) and the Indian National Science Academy. We thank all of these organisations for their generosity.

We should also like to thank the many people, both in Leeds and Rome, who helped us with typing, etc., particularly Anne Austick, Angela Di Silvestro, Marcella Mastrofini and Lynne Trundle. Last, but not least, we thank Giulio Casati, Federico Canobbio-Codelli, the Director of the Centre, and his staff (particularly Donatella Marchegiano, Manuela Troglio and Livia Marazzi) for providing the friendly and efficient environment for the workshop.

A. Degasperis, Rome.
A.P. Fordy, Leeds.
M. Lakshmanan, Tiruchirapalli.

CONTRIBUTORS

M. Ablowitz,
Program in Applied Mathematics,
Campus Box 426, Boulder,
Colorado 80309-0426, USA.

Y. Akutsu,
Institute of Physics,
Kanagawa University,
Rokkakubashi, Kanagawa-ku,
Yokohama 221, Japan.

S. Andal,
Department of Mathematics,
SRI G.V.G. Visalakshi College,
Udumalpet, Tamil Nadu,
India.

M. Antonowicz,
Institute of Theoretical
Physics, Warsaw University,
Hoza 69, 00681 Warsaw,
Poland.

C. Athorne,
Department of Mathematics,
University of Glasgow,
Glasgow G12 8QW, Scotland.

D. Atkinson,
Institute for Theoretical
Physics, Postbox 800,
9700 AV Groningen,
The Netherlands.

B.V. Baby,
International Centre for
Theoretical Physics,
Trieste, Italy.

B.M. Barbashov,
Laboratory of Theoretical
Physics, Joint Institute for
Nuclear Research, 141980,
Dubna, USSR.

A.O. Barut,
International Centre for
Theortical Physics, Trieste,
Italy, & Dept. of Physics,
University of Colorado,
CO 80309, USA.

B. Basu Mallick,
Theoretical Nuclear Physics
Division, Saha Institute of
Nuclear Physics, 92 Acharya
Prafulla Chandra Road,
Calcutta-700009, India.

G. Baumann,
Department of Mathematical
Physics, University of Ulm,
D-7900 Ulm, West Germany.

R. Beals,
Department of Mathematics,
Yale University,
Box 2155 Yale Station,
New Haven, CT 06520, USA.

G. Boffetta,
Istituto di Cosmo-Geofisica
del CNR, Corso Fiume 4,
10133 Torino, Italy.

M. Boiti,
Dipartimento di Fisica
dell' Università, and Sezione
INFN di Lecce, 73100 Lecce,
Italy.

R.K. Bullough,
Department of Mathematics,
UMIST, P.O. Box 88,
Manchester M60 1QD, England.

C.J. Camacho,
Department of Physics and
Astronomy, University of
Maryland, College Park,
MD20742, USA.

Contributors

S. Carillo,
Dipartimento di M.M.M. per le
Sc. Appl., Università di
Roma I, Italy.

R. Carroll,
Mathematics Department,
University of Illinois,
Urbana, Ill. 61801, USA.

P.J. Caudrey,
Department of Mathematics,
UMIST, P.O. Box 88,
Manchester M60 1QD, England.

B. Chakraborty,
Department of Mathematics,
Jadavpur University,
Calcutta 700032, India.

Yu-zhong Chen,
Department of Mathematics,
UMIST, P.O. Box 88,
Manchester M60 1QD, England.

Yi Cheng,
University of Science and
Technology of China, Hefei,
Anhui, China.

A.M. Chervyakov,
Laboratory of Computing
Techniques and Automation,
Joint Institute for Nuclear
Research, 141980, Dubna, USSR.

J. Cieslinski,
Warsaw University in
Białystok, Physics Institute
41 Lipowa Strett,
15-424 Białystok, Poland.

R.R. Coifman,
Department of Mathematics,
Yale University, Box 2155
Yale Station, New Haven,
CT 06520, USA.

R. Conte,
Service de physique du solid
et de resonance magnetique,
Centre d'etudes nucleaires
de Saclay, F-91191 Gif-sur-
Yvette Cedex, France.

A. Csizmazia,
30 Rue Léon Bocquet,
Saint Maur-des Fossés,
94100 France.

M. Daniel,
Department of Physics,
Bharathidasan University,
Tiruchirapalli - 620 024,
India.

A. Degasperis,
Dipartimento di Fisica,
Università di Roma I,
00185 Roma, Italy.

T. Deguchi,
Institute of Physics,
College of Arts & Sciences,
University of Tokyo, Komaba,
Meguro-ku, Tokyo 153, Japan.

S. De Lillo,
Dipartimento di Fisica,
Università di Perugia,
06100 Perugia, Italy.

C.R. Doering,
Department of Physics,
Clarkson University,
Potsdam, NY 13676, USA.

I.Ya. Dorfman,
Institute of Chemical Physics,
Academy of Science of the USSR,
ul. Kosygina 4, 117977 Moscow
GSP-1, USSR.

J.N. Elgin,
Department of Mathematics,
Imperial College,
London SW7 2BZ, England.

A.S. Fokas,
Department of Mathematics,
Clarkson University,
Potsdam, NY 13676, USA.

A.P. Fordy,
Department of Applied
Mathematical Studies & CNLS,
University of Leeds,
Leeds LS2 9JT, England.

N.C. Freeman,
School of Mathematics,
University of Newcastle upon
Tyne, Newcastle upon Tyne
NE1 7RU, England.

B. Fuchssteiner,
FB 17, Warburger Strasse 100,
Gesamthochschule,
Postfach 1621, 4790 Paderborn,
FRG.

V.S. Gerdjikov,
Institute for Nuclear Research
and Nuclear Energy,
boul. Lenin 72, 1784 Sofia,
Bulgaria.

J.D. Gibbon,
Department of Mathematics,
Imperial College,
London SW7 2BZ, England.

J. Gibbons,
Department of Mathematics,
Imperial College,
London SW7 2BZ, England.

C. Gilson,
Department of Mathematics,
University of Glasgow,
Glasgow G12 8QW, Scotland.

B. Grammaticos,
L.P.N., Université Paris VII,
Tour 24-14, 5ème étage,
2 Place Jussieu, 75251 Paris,
France.

F.A. Grünbaum,
Department of Mathematics,
University of California,
Berkeley, CA 94720, USA.

Gu Chaohao,
University of Science and
Technology of China, Hefei,
Anhiu, China.

C. Guha-Roy,
Department of Mathematics,
Jadavpur University,
Calcutta 700032, India.

H. Hasegawa,
Department of Phsyics,
Kyoto University,
Kyoto 606, Japan.

J. Hietarinta,
Dept. of Physical Sciences,
University of Turku,
20500 Turku, Finland.

D.D. Holm,
T7, Los Alamos National
Laboratory, New Mexico 87545,
USA.

M. Jaworski,
Institute of Physics, Polish
Academy of Sciences,
Al. Lotnikow 32/46,
02-668 Warsaw, Poland.

S. Kalaiselvi,
Department of Mathematics,
SRI G.V.G. Visalakshi College,
Udumalpet, Tamil Nadu, India.

P. Kaliappan,
Department of Mathematics,
Nallamuthu Gounder Mahalligam
College, Pollachi - 642 001,
Tamil Nadu, India.

V.G. Kamensky,
Landau Institute for Theoretical Physics, GSP-1, 117940,
Kosygin St. 2, Moscow V-334,
USSR.

D.J. Kaup,
Department of Mathematics,
Clarkson University,
Potsdam, NY 13676, USA.

Y. Kodama,
Department of Mathematics,
Ohio State University,
Columbus, OH 43210, USA.

K. Konno,
Department of Physics,
College of Science & Technology, Nihon University,
1-8 Kanda Surugadai,
Chiyoda-ku, Tokyo 101, Japan.

B.G. Konopelchenko,
Institute of Nuclear Physics,
630090 Novosibirsk,
USSR.

A. Kundu,
Theoretical Nuclear Physics Division, Saha Institute of Nuclear Physics, 92 Acharya Prafulla Chandra Road,
Calcutta-700009, India.

M. Lakshmanan,
Department of Physics,
Bharathidasan University,
Tiruchirapalli - 620 024,
India.

J. Léon,
Laboratoire de Physique Mathématique, U.S.T.L.,
34060 Montpellier, France.

F. Lund,
Departamento de Fisica,
Facultad de Ciencias Fisicas y Matematicas, Universidad de Chile, Casilla 487/3,
Santiago, Chile.

J.-M. Maillet,
CERN, Theory Division,
CH-1211 Genève 23,
Switzerland.

D. Maison,
Max-Planck Institut für Physik und Astrophysik, - Werner-Heisenberg-Institut für Physik -
P.O. Box 40 12 12, Munich, FRG.

S.V. Manakov,
Landau Institute for Theoretical Physics, GSP-1,
117940, Kosygin St. 2,
Moscow V-334, USSR.

I. Marshall,
Department of Applied Mathematical Studies,
University of Leeds,
Leeds LS2 9JT, England.

L. Martina,
Dipartimento di Fisica dell' Università, and Sezione INFN di Lecce, 73100 Lecce,
Italy.

V.K. Mel'nikov,
Laboratory of Theoretical Physics, Joint Institute for Nuclear Research,
141 980, Dubna, USSR.

M. Musette,
Dienst TENA, Vrije Universiteit Brussel, Pleinlaan 2,
B-1050, Brussels, Belgium.

K. Nakamura,
Fukuoka Institue of Technology, Higashi-ku,
Fukuoka 811-02, Japan.

F. Nijhoff,
Department of Mathematics & Computer Science,
Clarkson University,
Potsdam, NY 13676, USA.

J.J.C. Nimmo,
Department of Mathematics,
University of Glasgow,
Glasgow G12 8QW, Scotland.

T.F. Nonnenmacher,
Department of Mathematical Physics, University of Ulm,
D-7900 Ulm, West Germany.

M.C. Nucci,
Dipartimento di Matematica,
Università di Perugia,
06100 Perugia, Italy.

A.R. Osborne,
Istituto di Cosmo-Geofisica del CNR, Corso Fiume 4,
10133 Torino, Italy.

F. Pempinelli,
Dipartimento di Fisica,
Università di Lecce and Istituto Nazionale di Fisica Nucleare, 73100, Italy.

K. Porsezian,
Department of Physics,
Bharathidasan University,
Tiruchirapalli - 620 024,
India.

N. Rajeswari,
Department of Mathematics,
SRI G.V.G. Visalakshi College,
Udumalpet, Tamil Nadu, India.

A. Ramani,
Centre de Physique Théorique,
Ecole Polytechnique,
91128 Palaiseau, France.

A.G. Reyman,
Steklov Mathematical Institute,
Leningrad Branch, Fontanka 27,
Leningrad 191011, USSR.

P.C. Sabatier,
Laboratoire de Physique Mathématique,
34060 Montpellier Cedex,
France.

P.M. Santini,
Dipartimento di Fisica,
Università "La Sapienza", Roma and INFN, Sezione di Roma,
Italy.

S. Sarkar,
Department of Mathematics,
Jadavpur University,
Calcutta 700032, India.

C.L. Schultz,
University of Colorado at Boulder, Program in Applied Mathematics, Campus Box 426,
Boulder, Colorado 80309-0426,
USA.

M. Stirland,
Department of Mathematics,
UMIST, P.O. Box 88,
Manchester M60 1QD, England.

J. Timonen,
Department of Physics,
University of Jyväskylä,
Seminaarinkatu 15, SF-40100,
Jyväskylä, Finland.

M. Wadati,
Institute of Physics,
College of Arts & Sciences,
University of Tokyo, Komaba,
Meguro-ku, Tokyo 153, Japan.

Contributors

H. Yoshida,
National Astronomical
Observatory, Osawa, Mitaka,
Tokyo 181, Japan.

Zhou Zixiang,
Institute of Mathematics,
Fudan University,
Shanghai, China.

Part 1

INVERSE SPECTRAL METHODS AND OTHER SOLUTION TECHNIQUES

A.S. Fokas

1. SOME RECENT RESULTS IN SOLITON THEORY

In this paper I summarize some results obtained in the period of September 1987 - June 1988, regarding: (a) An initial-boundary value problem for the nonlinear Schrödinger equation, (b) the biHamiltonian formulation of the Landau-Lifshitz equation, (c) soliton cellular automata, and (d) some rigorous results on inverse scattering in multi-dimensions.

1. An initial-boundary value problem for the nonlinear Schrödinger equation

I have developed a method for studying initial boundary value problems associated with integrable nonlinear evolution equations in 1 + 1 (i.e. in one spatial and one temporal dimension). For concreteness I have considered the nonlinear Schrödinger (NLS) in the variable $q(x,t)$, x in $[0,\infty)$, with a mixed boundary condition, i.e. $q_x(0,t) + \alpha q(o,t)$ (α is an arbitrary real constant) and $q(x,0)$ are given. The following results have been obtained [1]:

(i) If $q_x(0,t) + \alpha q(0,t) = 0$ there exists an inverse scattering formalism which completely linearizes the above problem.

(ii) I $q_x(0,t) + \alpha q(0,t) = f(t)$, $f(t) \neq 0$, then $q(x,t)$ can be obtained by solving a linear integral equation uniquely defined in terms of appropriate scattering data. These data satisfy a single nonlinear integrodifferential equation uniquely defined in terms of $f(t)$. The question of linearizing this nonlinear equation remains open.

Let us consider for concreteness a Dirichlet boundary condition. Then the linear analogue of the NLS is

$$iq_t = q_{xx},\ 0 \leq x < \infty,\ q(x,0),\ q(0,t) \text{ given,} \tag{1}$$

and can be solved by the sine transform,

$$\hat{q}(k,t) = \int_0^\infty dx q(x,t) \sin kx, \quad q(x,t) = \frac{2}{\pi}\int_0^\infty dk \hat{q}(k,t)\sin kx, \tag{2}$$

where the sine data satisfy

$$\hat{q}_t = ik^2\hat{q} - ikq(0,t). \tag{3}$$

1.1 *A nonlinear analogue of the sine transform*

Consider the odd extension of the potential $q(x,t)$. Then the associated linear eigenvalue problem becomes

$$\varphi_x = ikJ\varphi + \tilde{Q}\varphi, \quad -\infty < x < \infty, \quad \tilde{Q} \doteq Q(x)H(x) - Q(-x)H(-x), \tag{4}$$

$$J \doteq \begin{pmatrix} -1 & 0 \\ 0 & 1 \end{pmatrix}, \quad \tilde{Q} = \begin{pmatrix} 0 & \tilde{q}(x) \\ \tilde{r}(x) & 0 \end{pmatrix},$$

where $H(x)$ denotes the Heaviside distribution. Let $\varphi = \Phi \exp(ikxJ)$, then (4) becomes

$$\Phi_x = ik[J,\Phi] + \tilde{Q}\Phi. \tag{5}$$

Let the matrices Ψ, Φ solve

$$\Psi = I - \int_x^\infty d\xi e^{ik(x-\xi)\hat{J}}\, \tilde{Q}\Psi, \quad \Phi = I + \int_{-\infty}^x d\xi e^{ik(x-\xi)\hat{J}}\, \hat{Q}\Phi, \tag{6}$$

where if F is an arbitrary 2x2 matrix and Y is a diagonal matrix, then $\exp(\tilde{Y})F \doteq \exp(Y)F\exp(-Y)$. Then it can be shown that:

(i) Ψ,Φ solve (5).
(ii) $\Psi = (\Psi^-,\Psi^+)$, $\Phi = (\Phi^+,\Phi^-)$, where $+(-)$ denotes analyticity in the upper(lower) half k-complex plane.
(iii) Ψ,Φ are related via

$$\Psi = \Phi e^{ikx\hat{J}} S(k), \quad S(k) \doteq \begin{pmatrix} a & \bar{b} \\ b & \bar{a} \end{pmatrix} \doteq I - \int_{-\infty}^\infty d\xi e^{-ik\xi\hat{J}}\, \tilde{Q}(\xi)\Psi(\xi,k). \tag{7}$$

(iv) $\bar{a}(k), a(k)$ are $+,-$ functions respectively.
(v) $\det S(k) = 1$.
(vi) $\Psi(-x,-k) = \Phi(x,k)$.
(vii) $S(k)S(-k) = I$, or $\bar{a}(-k) = a(k), \bar{b}(-k) = -\bar{b}(k), b(-k) = -b(k)$.
(viii) $a \to 1, \bar{a} \to 1, \bar{b} \to \frac{q(0)}{ik}, \ b \to -\frac{r(0)}{ik}$ as $k \to \infty$. (8)

Using the above and assuming that $a,\bar{a}$ have no zeros in their domains of definition it can be shown that (7a) defines a Riemann-Hilbert problem for Ψ(or Φ) in terms of the scattering data S(k). Having obtained Ψ, the potential q follows:

$$q(x) = -\frac{1}{\pi}\int_{-\infty}^{\infty} dk\frac{\bar{b}}{a}e^{-2ikx}\Psi_{1}^{-}. \tag{9}$$

It is interesting to note that the linear limit of (7b),(9) yields the sine transform (2).

1.2 *Evolution of the scattering data*

Consider the equation

$$Q_t = iQ_{xx}J + 2iqrQJ,\ Q(0,t) \doteq \begin{pmatrix} 0 & q(0,t) \\ r(0,t) & 0 \end{pmatrix},\ Q(x,0) \text{ given,} \tag{10}$$

which reduced to the NLS if $r = \pm q^*$ where * denotes complex conjugate. It is found that the t-part associated with (10) is considerably modified: Let Ψ,Q solve (6a),(10), then $\psi \doteq \Psi\exp(ikxJ)$ solves

$$\psi_t = 2ik^2[\psi,J] - (i\tilde{q}\tilde{r}J + 2k\tilde{Q} + i\tilde{Q}_xJ)\psi$$
$$- 4kH(-x)\psi\psi^{-1}(0,t,k)Q(0,t)\psi(0,t,k). \tag{11}$$

Equation (11) implies

$$S_t = -2ik^2[J,S] - 4kS\Psi^{-1}(0,t,k)Q(0,t)\Psi(0,t,k). \tag{12}$$

1.3 *A nonlinear equation for the scattering data*

The main difficulty associated with the inhomogeneous boundary value problems is the dependence of the evolution of the scattering data on quadratic products of eigenfunctions evaluated at x = 0. It seems quite remarkable that it is possible completely to eliminate these products and obtain equations involving only the scattering data. The scattering data b,$\bar{b}$ satisfy the following equations:

$$\frac{\bar{b}_t - 4ik^2\bar{b}}{4k} = -q(0,t) + \int_{-\infty}^{\infty} dk'\frac{\frac{\bar{b}(k')}{8k'} - \frac{\bar{b}(k)}{8k}}{k' - k}\frac{\partial}{\partial t}H\ln(1 - b\bar{b})(k'), \tag{13}$$

$$\frac{b_t + 4ik^2 b}{4k} = -r(0,t) + \int_{-\infty}^{\infty} dk' \frac{\frac{b(k')}{8k'} - \frac{b(k)}{8k}}{k' - k} \frac{\partial}{\partial t} H \ln(1 - b\bar{b})(k'), \tag{14}$$

where H denotes the Hilbert transform, i.e.

$$(Hf)(x) \doteq \frac{1}{\pi} \fint_{-\infty}^{\infty} d\xi \, \frac{f(\xi)}{\xi - x} . \tag{15}$$

Having obtained $b, \bar{b}$, then $a, \bar{a}$ follow by solving the Riemann-Hilbert problem

$$a\bar{a} = 1 + b\bar{b}, \; a, \bar{a} \longrightarrow 1 \text{ as } k \longrightarrow \infty. \tag{16}$$

In the case of the NLS the above equations reduce to

$$\frac{b_t + 4ik^2 b}{4k} = -r(0,t) + \int_{-\infty}^{\infty} dk' \frac{\frac{b(k')}{8k'} - \frac{b(k)}{8k}}{k' - k} \frac{\partial}{\partial t} H \ln(1 \pm |b|^2)(k'). \tag{17}$$

Remarks

(i) In the linear limit equation (17) reduces to

$$b_t + 4ik^2 b \sim -4kr(0,t).$$

This is precisely the time evolution of the sine transform (see (3)).

(ii) A two pole solution of (17) yields a solution for b which corresponds to a soliton travelling toward the origin.

Indeed, let $r = -q^*$, then $\bar{b} = -b^*$ and we find that a solution b of (17) is

$$b = 4kb_1 \left(\frac{1}{k - k_1} - \frac{1}{k - k_1^*} \right), \; b_1 = c_1 e^{-4ik_1^2 t} (k_1 + \lambda_1)(k_1 - \lambda_1^*),$$

where $k_{1_t} = 0$, $\mathrm{Im} k_1 > 0$, $\mathrm{Im}\lambda_1 > 0$. Then (6) implies $\bar{a} = \frac{(k+\lambda_1)(k-\lambda_1^*)}{(k+k_1)(k-k_1^*)}$. Letting $k_1 = r_1 e^{i\varphi}, \lambda_1 = \rho_1 e^{i\theta}$ we find:

$$\rho_1 = r_1, \; \sin^2\theta = \sin^2\varphi - 16|b|^2.$$

Thus

$$\sin\theta = \sin\varphi\tanh w, w = -(4r^2\sin 2\varphi)t - \frac{1}{2}\ln 4^3|c|^2 r^4.$$

The solution of the inverse problem then yields

$$r(x,t) = \frac{2i(k_1^* - k_1)Fe^{2ik_1 x}}{1 + |F|^2 e^{2i(k_1 - k_1^*)x}}, \quad F \doteq 8k_1^2 c_1 e^{-4ik_1^2 t},$$

$$r(-x,t) = -r(x,t).$$

(iii) We are now investigating the dynamics of the general pole solution

$$b = 4k\sum_{j=1}^{n} b_j(t)\left(\frac{1}{k - k_j} - \frac{1}{k - k_j^*}\right).$$

If the dynamics of $b_j(t)$ is linearizable, this would provide a strong argument that equation (17) is linearizable.

2 The biHamiltonian formulation of the Landau-Lifshitz equation

This problem has been studied in [2]. The Landau-Lifshitz (LL) equation describes nonlinear spin waves in an anisotropic ferromagnet. It was pointed out in [3] that is the most general magnet model admitting an r-matrix formulation. It contains the sine-Gordon (SG), nonlinear Schrödinger (NLS) and the Heisenberg model (HM) equations as particular or limiting cases. It is well known that the NLS, SG and HM equations possess **recursion operators**. A recursion operator of an equation in Hamiltonian form generates (a) a hierarchy of integrable equations, (b) a second Hamiltonian operator, and more generally a hierarchy of Poisson structures. In spite of the intense scrutiny of the LL the question of finding its recursion operators had remained open. We have recently obtained algorithmically the recursion operator of the LL equation and have also established its biHamiltonian formulation.

The LL equation is given by

$$\mathbf{S}_t = \mathbf{S} \wedge \mathbf{S}_{xx} + \mathbf{S} \wedge J\mathbf{S} \qquad (18a)$$

where

$$J = \mathrm{diag}(J_1, J_2, J_3), \ \mathbf{S} = (S_1, S_2, S_3), \ |\mathbf{S}|^2 = \mathbf{S}\cdot\mathbf{S} = 1. \qquad (18b)$$

In the above the diagonal matrix J is a measure of the anisotropy, $J_1 < J_2 < J_3$, S is an x- and t-dependent vector of unit norm in

$\mathbf{R}^3$, and $\cdot$, $\wedge$ denote the usual scalar and vector products.

The partially anisotropic HM and the HM equations correspond to $J_1 = J_2 < J_3$, and $J_1 = J_2 = J_3$ respectively. The analysis of the LL is technically more complicated than that of HM, SG and NLS. This is because the isospectral linear eigenvalue problem associated with LL involves elliptic functions [4]:

$$U_x(x,t,\lambda) = -i(\Sigma_{j=1}^{3} S_j(x,t)W_j(\lambda)\sigma_j)U(x,t,\lambda) \doteq -iLU, \tag{19a}$$

where the Pauli spin matrices are given by

$$\sigma_1 \doteq \begin{pmatrix} 0 & 1 \\ 1 & 0 \end{pmatrix}, \quad \sigma_2 \doteq \begin{pmatrix} 0 & -i \\ i & 0 \end{pmatrix}, \quad \sigma_3 \doteq \begin{pmatrix} 1 & 0 \\ 0 & -1 \end{pmatrix}, \tag{19b}$$

and

$$W_1(\lambda) \doteq \rho\frac{1}{\mathrm{sn}(\lambda,k)}, \quad W_2(\lambda) \doteq \rho\frac{\mathrm{dn}(\lambda,k)}{\mathrm{sn}(\lambda,k)}, \quad W_3(\lambda) \doteq \rho\frac{\mathrm{cn}(\lambda,k)}{\mathrm{sn}(\lambda,k)}, \tag{20a}$$

with

$$k \doteq \left(\frac{J_2 - J_1}{J_3 - J_1}\right)^{1/2}, \quad 0 < k < 1, \quad \rho \doteq \sqrt{J_3 - J_1}. \tag{20b}$$

In the isospectral problems associated with the HM, SG and NLS equations the spectral parameter λ ranges over the complex plane C. However the natural range of λ in (19) is an elliptic curve: The torus $E = \mathbf{C}/\Gamma$ where Γ is the lattice generated by $4K$ and $4iK'$, where K and K' are the complete elliptic integrals of moduli k and $k' = \sqrt{1 - k^2}$.

The method of deriving the recursion operator from an isospectral problem makes crucial use of a certain expansion in powers of the spectral parameter λ. The main difficulty we encountered in applying this method to LL stemmed from the fact that λ moves on an elliptic curve. This problem was bypassed by using the parametrization

$$\nu \doteq \frac{1}{2} W_1W_2W_3, \quad \mu \doteq W_3^2, \tag{21}$$

$$\nu^2 = \frac{1}{4}\mu(\mu + \alpha)(\mu + \beta); \quad \alpha \doteq -\frac{1}{4}(J_1 - J_3), \quad \beta \doteq -\frac{1}{4}(J_2 - J_3). \tag{22}$$

The recursion operator of the LL equation Φ_{LL} is given by

$$\Phi_{LL} = \Phi_{HM}^2 - \frac{1}{4}((4AS)\wedge(S\wedge\cdot))$$

$$- (D^{-1}\{S\cdot 4AS\wedge(S\wedge\cdot)\})S_x - (D^{-1}\{S\cdot(S\wedge\cdot)_x\})(4AS\wedge S)$$

where Φ_{HM} is the recursion operator of the HM,

$$\Phi_{HM} = -\frac{1}{2}[S\wedge D - \{D^{-1}(S\wedge S_x\cdot)\}S].$$

The Hamiltonian operators associated with the LL equation are given by

$$S\wedge \text{ and } \Phi_{LL}(S\wedge\cdot).$$

3 Soliton cellular automata

This problem has been studied in [5]-[7]. Cellular automata (CA) [8] is one of the most rapidly developing areas of mathematical physics. Significant applications of CA have occurred in computer science, engineering, physics and biology. A particular class of CA, called filter automata, has recently been proposed in [9] and [10]. This cellular automaton consists of a collection of zeros or ones, which evolve in time according to the following rule: Let

$$a^t:\ \ldots 0a_0^t\ldots a_i^t\ldots a_L^t 0\ldots,\quad L < \infty \tag{23}$$

be the state at time t, where $a_i^t = 0$ or 1 for all i and $a_0^t = 1$, $a_L^t = 1$ are the first and the last 1 respectively in a^t. Then the next state is calculated, sequentially from left to right, as follows:

$$a_i^{t+1} = \begin{cases} 1, & s \text{ even} \\ 0, & s \text{ odd or zero} \end{cases} \qquad \text{where } s \doteq \sum_{j=r}^{1} a_{i-j}^{t+1} + \sum_{j=0}^{r} a_{i+j}^{t}, \tag{24}$$

(one assumes that $a_i^{t+1} = 0$ for i far enough to the left). In the above r is a fixed integer called the **radius** and $r \geq 2$.

As an illustration of the above "parity rule" suppose that $r = 3$, then a_i^{t+1} depends only on the following (note the convention that increasing t is downwards)

$$\begin{array}{ccccccc} & & & i & i+1 & i+2 & i+3 \\ & & & \cdot & \cdot & \cdot & \cdot \\ \cdot & \cdot & \cdot & \cdot & & & \\ i-3 & i-2 & i-1 & & & & \end{array}$$

For example, if $r = 3$, a^t is given as below, a^{t+1} is as follows:

a^t: · · · 0 1 1 0 1 1 1 0 0 0 0 0 1 1 0 0 0 0 1 1 1 1 0
a^{t+1}: 0 0 1 0 1 0 1 0 0 1 0 0 1 0 0 0 0 0 0 1 1 1 1 0 0

Figure 1.

It is clear from the above that if two collections of ones are separated by a large enough number of zeros then their evolution from t to t+1 is independent of each other. Thus we can think of a given state as a collection of **particles**. The authors of [9],[10] reported the following remarkable property of particles evolving according to the above parity rule: For most of the initial configurations considered in the numerical experiments the particles emerging after interaction were identical to the original ones.

For example let r = 4 and consider the particles A = 1001111001 and B = 1111010111 separated by 6 zeros at t = 0. At t = 1 the particles begin to interact and A′ = 1101001100, B′ = 1110001111. At t = 9 the interaction ends with B′ in front of A′. We note that we always take enough bits of zeros at the end of a particle so that the particle consists of the smallest multiple of r +1. These units of r + 1 bits are called **basic strings**. For example, the particle A consists of the 3 basic strings 1001,1110,0100.

```
t=0                         10011110010000001111010111
t=1                        110100110000001110001111
t=2                      1010111001000101000011101
t=3                    1010100110010000000011011
t=4                  101101101100010001001101
t=5                110010111001010100010101
t=6              1001111101101000001110101
t=7            11011010010010011001011
t=8         1011110011010001001111001
t=9       111000111100000011010011
```

Figure 2.

Thus we see that the above filter cellular automaton is capable of modelling solitonic collisions. The mathematical and physical significance of solitons is well established and it provides additional motivation for studying the above CA.

Much progress has been made in using CA to produce remarkable patterns to to study their statistical properties. However, it has not been possible in general to follow analytically the precise evolution of a given automaton. until now this was also the case with the CA (24). Thus, the observations about solitonic collisions made in [9] are only based upon computer experiments.

We have recently developed mathematical tools needed for the analytical study of (24). The following results have been obtained:

(i) We have studied completely the evolution of single particles [5]. We have found precise conditions satisfied by the initial configuration which guarantee that a given particle is periodic. Particles are not periodic if one of the following occurs: either splitting takes place or a basic string at the beginning or at the end of the particle disappears. If the particle is periodic both its period and its displacement are expressed analytically. In particular the displacement is expressed in an elegant formula which is extensively used in the study of particle interactions.

(ii) The interaction of **simple particles** has been completely analyzed [6]. A particle is called simple if it consists of one basic string. It is quite interesting that one is able to give a complete characterization of the interaction of any two simple particles. In particular: a) every simple particle is periodic with period equal to the number of its 1's. b) If ℓ_A, ℓ_B is the number of 1's in the two simple particles A,B then B is faster than A if $\ell_B > \ell_A$. c) If $\ell_B > \ell_A$ and A,B begin interacting with A in front of B, then after $2\ell_A$ time steps, A and B will emerge as two particles with B in front of A. This interaction is **solitonic**, i.e. the particles emerging from the interaction are identical to those entering the interaction. Although the faster particle B is in front of A, the particles A and B may interact again; however eventually they will get separated with the faster particle moving in front of the slower particle. (The number of interactions can be given explicitly and depends on the relative positions of 1's in the two particles). c) If $\ell_B = \ell_A$, and the particles are close enough so that they can interact, there exist two cases. Either they will interact once and then they separate travelling independently of each other, or they will form a new periodic configuration by interacting for ever.

(iii) The results of simple particles have been extended to arbitrary particles [7].

Example

Consider the interaction of A = 1000011, B = 1110001 separated by 10 zeros. Then the theory we have developed produces the following:

t = 1	Interaction begins
t = 7	Interaction ends
t = 9	Interaction begins again
t = 17	Interaction ends
t = 18	Interaction begins again
t = 24	Interaction ends, the particles will never interact again.

The above results are verified by following numerically the interaction of A and B (wrap-around notation is used due to space limitation):

```
t=0                                                        10000110000000000111000
t=1                                                       1110000000000001100011
t=2                                                  11000010000010000110 1
t=3                                             100001100001010011011
t=4                                           11100010110110111
t=5                                      1100011110110011000 01
t=6                                 100011010110001000011
t=7                               110001100000000111
t=8                          1000111000000001100001
t=9                        1111000000001000011
t=10                  111000100001010011001
t=11             110001100010110110011
t=12        10001110010111110 0111
t=13      11110101110100 1111
t=14 111010011010001110001
```

```
t=14                                                       111010011010001110001
t=15                                                     110100010100001100011
t=16                                                101000001000001000111
t=17                                            1000011000000001111
t=18                                           1110000000001110001
t=19                                     110000100001100001001
t=20                                100001100011010010011
t=21                               11100110110100111
t=22                          110011111010001100001
t=23                     100111010100001000011
t=24                  111000100000000111
t=25             1100011000000001100001
t=26        1000111000000001000011
t=27      11110000000000000111
t=28 1110001000000000001100001
```

```
t=28                              1110001000000000001100001
t=29                         1100011000000000001000011
t=30                    1000111000000000000000111
t=31                  1111000000000000001100001
t=32             1110001000000000001000011
t=33        1100011000000000000000111
t=34   1000111000000000000000001100001
t=35 111100000000000000001000011
```

Figure 3.

4 Rigorous aspects of inverse scattering in multidimensions

The following inverse problem was considered in [11]. Given appropriate inverse data reconstruct the potential $q(x_0, x)$ associated with

$$\mu_{x_0} + \sigma \sum_{\ell=1}^{n} (J_\ell \mu_{x_\ell} + ik_\ell[J_\ell,\mu]) = q\mu, k\in\mathbf{C}^n, \sigma = \sigma_R + i\sigma_I, \sigma_I \neq 0, n > 1, \quad (25)$$

where $q(x_0,x)$ is an N x N matrix-valued off-diagonal function in $\mathbf{R}^{n+1}$, J_ℓ are constant diagonal N x N matrices (we denote the diagonal entries of J_ℓ by $J_\ell^1,\dots,J_\ell^N$ and we assume $J_\ell^i \neq J_\ell^j$ for $i \neq j$), and $\mu(x_0;x,k)$ is an N x N matrix-valued function. A special case of (25) with $\sigma = i$ and $n = 1$ has been used to linearize the Davey-Stewartson II equation [12]. It was shown **formally** in [11] that:

(i) There exists a unique solution of (25) bounded for all complex values of k and tending to I, uniformly in the other spatial variables as $|x_0| + |x_1| \to \infty$. This solution satisfies a linear integral equation.

(ii) $\frac{\partial\mu}{\partial k_p}(x_0,x,k)$ can be expressed in terms of μ and inverse data $T^{ij}(k_R,k_I,m), k_R\in\mathbf{R}^n, k_I\in\mathbf{R}^n, m\in\mathbf{R}^{n-1}$.

(iii) $T^{ij}(k_R,k_I,m)$ are appropriate scattering data if they satisfy a certain nonlinear equation (the so-called T equation).

(iv) If (iii) is satisfied then q can be reconstructed in terms of T^{ij}. Furthermore, the problem of reconstruction is reduced to solving an inverse problem for 2x2 matrices in two spatial dimensions.

Sung and the author have recently proven the above results rigorously [13]. For example, to prove (i) above introduce a new coordinate system for the ith row

$$x_0 = 2\tilde{x}_0^i, x_1 = 2J_1^i(\sigma_R\tilde{x}_0^i + \sigma_1\tilde{x}_1^i), x_\ell = \tilde{x}_\ell^i + 2J_\ell^i(\sigma_R\tilde{x}_0^i + \sigma_I\tilde{x}_1^i), \quad (26)$$

and let $\tilde{q}^{ij}$ denote q^{ij} in the new coordinates. Consider the space

$$L^\infty(\mathbf{R}^{n-1},L^1(\mathbf{R}^2))$$

$$= \{f(x_o.x_1,x''); \quad \sup_{x''\in\mathbf{R}^{n-1}} \iint |\mathbf{f}(x_0,x_1,x'')|dx_0dx_1 < \infty\}$$

with norm

$$||f||_{(1,\infty)} = \sup_{x''\in\mathbf{R}^{n-1}} \iint |f(x_0,s_1,x'')|dx_0,dx_1.$$

Assume that q has the following properties:

(A1) $q\in L^\infty(\mathbf{R}^{n+1})$.

(A2) $q \to 0$ as $|x_0| + |x_1| \to \infty$ uniformly in x''.

(A3) In each $\tilde{x}^i$ coordinate, $\tilde{q}^{ij} \in L^\infty(\mathbf{R}^{n-1}, L^1(\mathbf{R}^2))$.

(A4) $||\tilde{q}^{ij}||_{(1,\infty)} \cdot ||q||_\infty << 1, \forall i,j$.

(A5) q is continuous.

Then (i) above is true, furthermore this unique solution μ has the following properties:

(P1) μ is continuous in x_0, x.

(P2) μ depends continuously on k, uniformly in x_0, x.

(P3) $\frac{\partial\mu}{\partial k_p} \in L^\infty_{NxN}$.

(P4) $\frac{\partial\mu}{\partial k_p}$ is continuous in k, x_0, x.

(P5) $\lim_{|k_p| \to \infty} \mu^{ij} = \delta^{ij}$ for fixed x_0, x and fixed $k_1, \ldots, k_{p-1}, k_{p+1}, \ldots, k_n$.

Assuming that q and its derivatives up to order n are integrable on $\mathbf{R}^{n+1}$ and satisfy (A1),(A3), it can be shown that (ii) is true. Furthermore, it follows that the T-equation is a necessary condition on T^{ij}. By putting appropriate conditions on certain norms of T^{ij} it can also be shown that the T-equation is a sufficient condition for T^{ij} to be the scattering data of some q. Also this q is uniquely reconstructed in terms of T.

Acknowledgements

This work was partially supported by the Office of Naval Research under Grant Number N00014-88k-0447, National Science Foundation under Grant number DMS-8803471, and Air Force Office of Scientific Research under Grant Number 87-0310.

References

[1] Fokas, A.S. (1989). An Initial-Boundary Value Problem for the Nonlinear Schrödinger Equation. *Physica* **35D**, 167.

[2] Barouch, E., Fokas, A.S. & Papageorgiou, V. (1988). The BiHamiltonian Formulation of the Landau-Lifshitz Equation. *J. Math. Phys.* **29**, 2628.

[3] Fadeev, L.D. & Takhtajan, L.A. (1987). *Hamiltonian Methods in the Theory of Solitons.* Springer Verlag.

[4] Sklyanin, E.K. (1979). *On Complete Integrability of the Landau-Lifshitz Equation*, preprint, LOMI, E-3-79.

[5] Fokas, A.S., Papadopoulou, E. & Saridakis, Y. (1988). *Particles in Soliton Cellular Automata.* Clarkson University.

[6] Fokas, A.S., Papdapoulou, E., Saridakis, Y. & Ablowitz, M.J. (1988). *Interactions of Simple Particles in Soliton Cellular Automata*. Clarkson University, INS #97.

[7] Fokas, A.S., Papadopoulou, E. & Saridakis, Y. (1988). *Soliton Cellular Automata.* Clarkson University.

[8] Wolfram, S. (1986). *Theory and Applications of Cellular Automata.* World Scientific.

[9] Park, J.K., Steiglitz, K. & Thurston, W.P. (1986). *Physica D* **19**, 423.

[10] Steiglitz, K., Kamal, I. & Watson, A. (1988). *IEEE Trans. on Computers* **37**, 138.

[11] Fokas, A.S. (1988). *J. Math. Phys.* **27**, 1737.

[12] Fokas, A.S. (1983). *Phys. Rev. Lett.* **51**,

[13] Sung, L. & Fokas, A.S. (1988). *Inverse Scattering in Multidimensions*. Clarkson University, INS #98.

R. Beals and R.R. Coifman

2. THE SPECTRAL PROBLEM FOR THE DAVEY-STEWARTSON AND ISHIMORI HIERARCHIES

1 Introduction

We give a complete analysis of the inverse scattering problem for the Davey-Stewartson II and Ishimori I hierarchies, illustrating the $\bar{\partial}$-method. We emphasize that each of the formulas and identities derived has an analogue in other such contexts.

The Davey-Stewartson II (DS-II) system is

$$q_t = 2iq_{x_1x_2} - 4irq, \qquad r_{x_1x_1} + r_{x_2x_2} = r(|q|^2)_{x_1x_2}.$$

The associated linear system is

$$(u_1)_{x_1} + i(u_1)_{x_2} = 2q_1u_2, \qquad (u_2)_{x_1} - i(u_2)_{x_2} = 2q_2u_1, \qquad \text{(LS)}$$

specialized to $q_2 = \bar{q}_1$. The inverse scattering theory of LS was analysed formally by Fokas & Ablowitz [4] and rigorously by Beals & Coifman [1].

The Ishimori I system

$$\xi_t = \xi \times (\xi_{x_2x_2} - \xi_{x_1x_1}) - \rho_{x_1}\xi_{x_2} - \rho_{x_2}\xi_{x_1},$$

$$\rho_{x_1x_1} + \rho_{x_2x_2} = 2\xi \cdot (\xi_{x_1} \times \xi_{x_2}),$$

where ρ is a real scalar and ξ is a real unit 3-vector. Konolpelchenko & Matkarimov [6] analysed the associated linear system and noted that it is gauge equivalent to LS, specialized to $q_2 = -\bar{q}_1$.

Here we return to the inverse scattering theory of LS and obtain: the tangent maps to the direct and inverse scattering transformations; representations of a class of flows including the standard hierarchy; recursion formulas for the hierarchy; hamiltonian form of the flows; a new quadratic conservation law; a complete global (large potential) result for the case $q_2 = \bar{q}_1$. We proceed formally, with comments on rigorous validity at the end.

To save space and emphasize the generality of the development, we adopt the mathematician's device (or vice) of using a condensed notation. Set $x = x_1 + ix_2$ and

$$\partial_{\bar{x}} = \frac{1}{2}\left(\frac{\partial}{\partial x_1} + i\,\frac{\partial}{\partial x_2}\right),\quad \partial_x = \frac{1}{2}\left(\frac{\partial}{\partial x_1} - i\,\frac{\partial}{\partial x_2}\right).$$

We consider a second copy of $\mathbb{C}$ with variable z and operator $\partial_{\bar{z}}$, whose fundamental solution which vanishes at ∞ is the Cauchy transform

$$Cf(z) = \frac{1}{\pi}\int (z-z')^{-1} f(z')\,d\mu(z'),$$

$d\mu$ = Lebesque measure. All matrices are 2x2, complex. We write 1 for the identity matrix and

a^d = diagonal part of matrix a,

a^{off} = off-diagonal part of matrix a,

$$J = \frac{1}{2}\begin{pmatrix}-i & 0\\ 0 & i\end{pmatrix},\qquad A(x,z) = \begin{pmatrix} e^{ix\bar{z}+i\bar{x}z} & 0\\ 0 & e^{-ixz-i\overline{xz}}\end{pmatrix},$$

$A_z(\cdot) = A(\cdot, z)$.

The following are operations on matrices a or matrix-valued functions f of the variables x, z, or x and z together:

$$\mathcal{J}a = [J,\, a] = 2J\,a^{off} = -2a^{off}J,$$

$$E_z f = f^d + A_z^{-1} f^{off} = f^d + f^{off} A_{\bar{z}},$$

$$Df = \begin{bmatrix}\partial_{\bar{x}} & 0\\ 0 & \partial_x\end{bmatrix} f = \begin{bmatrix}\partial_{\bar{x}} f_{11} & \partial_{\bar{x}} f_{12}\\ \partial_x f_{21} & \partial_x f_{22}\end{bmatrix}$$

$$D_z = D + z\mathcal{J} = E_z^{-1} D E_z.$$

The fundamental solution of D_z vanishing at $|x| = \infty$ is

$$G_z = E_z^{-1} G E_z,\qquad G(x,y) = \frac{1}{\pi}\begin{pmatrix}(x-y)^{-1} & 0\\ 0 & (\bar{x}-\bar{y})^{-1}\end{pmatrix}.$$

We denote by Q both the matrix

$$Q = \begin{pmatrix} 0 & q_1 \\ q_2 & 0 \end{pmatrix},$$

and the operation of left multiplication by Q. Thus LS is: Du = Qu.

2 The direct and inverse problems for LS

With some minor changes in normalization we follow Beals & Coifman [1]. We look for a matrix Jost solution of LS:

$$\psi(x,z) = m(x,z) \begin{bmatrix} e^{ixz} & 0 \\ 0 & e^{-i\bar{x}\bar{z}} \end{bmatrix}, \quad m \to 1 \quad \text{as } |x| \to \infty \tag{1}$$

The problem for the Jost function m is $D_z m = Qm$, and the associated integral equation

$$m = 1 + G_z Q m$$

leads to the **direct formula**

$$m(\cdot,z) = (I - G_z Q)^{-1} 1 \tag{2}$$

This leads to

$$\partial_{\bar{z}} m = [\partial_{\bar{z}}, (I - G_z Q)^{-1}]1 = (I - G_z Q)^{-1}[\partial_{\bar{z}}, G_z]Q(I - G_z Q)^{-1}1$$

$$= (I - G_z Q)^{-1}[\partial_{\bar{z}}, G_z]Q_m = (I - G_z Q)^{-1} E_z^{-1} S$$

where the **scattering data** S is

$$S(z) = -\mathcal{T} \int E_z[Q(x)m(x,z)]d\mu(z). \tag{3}$$

Now an easy calculation shows that $D_z(fE_z^{-1}S) = (D_{\bar{z}}f)E_z^{-1}S$, so $G(fE_z^{-1}S) = [G_{\bar{z}}f]E_z^{-1}S$ and we obtain the $\bar{\partial}$-equation

$$\partial_{\bar{z}} m = Tm \tag{4}$$

where

$$Tf(x,z) = f(x,\bar{z})S(z)A(x,-\bar{z}). \tag{5}$$

The Riemann-Lebesque lemma implies that $m(x,z) \to 1$ as $z \to \infty$, so (4) gives the integral equation $m = 1 + CTm$ and thus the **inverse formula**

$$m = (I-CT)^{-1}1. \tag{6}$$

Now D_z commutes with T, so (6) implies

$$D_z m = [D_z, (I-CT)^{-1}]1 = (I-CT)^{-1}[D_z,C]T(I-CT)^{-1}1$$
$$= (I-CT)^{-1}[D_z,C]Tm = (I-CT)^{-1} Q(x)1 = Q(x)m$$

with $Q = [D_z,C]Tm$; thus

$$Q(x) = \mathcal{J} \int [Tm](x,z)d\mu(z). \tag{7}$$

3 **Some identities**

The operators D_z and $\partial_{\bar{z}}$ are invariant under the involutions

$$U_{\pm}f(x,z) = \begin{pmatrix} 0 & \pm 1 \\ 1 & 0 \end{pmatrix} \overline{f(x,\bar{z})} \begin{pmatrix} 0 & \pm 1 \\ 1 & 0 \end{pmatrix} \tag{8}$$

while $U_{\pm}(Qf) = \pm Q^{*}U_{\pm}$. It follows that $U_{\pm}m$ is the Jost function corresponding to $\pm Q^{*}$ and that the corresponding scattering data is

$$\pm Q^{*} \leftrightarrow \pm S(\bar{z})^{*} \tag{9}$$

Thus also $\pm Q \leftrightarrow \pm S$. We show in Section 4 that $-Q^{t} \leftrightarrow S(-\bar{z})^{t}$ where a^{t} denotes the transpose, so also $\bar{Q} \leftrightarrow -\overline{S(-z)}$.

The formulas (3), (7) and the identity

$$\mathrm{tr}(a\,\mathcal{J}(bc)) = -\mathrm{tr}(\mathcal{J}(ab)c) \quad \text{if } a = a^{\mathrm{off}}, \quad b = b^{\mathrm{off}} \tag{10}$$

imply the quadratic identity

$$\int \mathrm{tr}[Q(x)^2]d\mu(x) = \int \mathrm{tr}[S(\bar{z})S(z)]d\mu(z). \tag{11}$$

with (9) this implies

$$\int \mathrm{tr}\, Q^{*}Q = \int \mathrm{tr}\, S^{*}S \quad \text{if} \quad Q = \pm Q^{*}. \tag{12}$$

Finally, consider the diagonal part of the integral in (3):

$$\Delta(z) = \int(E_{-z}Qm)^{d} = \int[Q(x)m(x,z)]^{d}d\mu(x). \tag{13}$$

Then $\partial_{\bar{z}}\Delta = \int Q(Tm)^{\mathrm{off}}$ and (3), (7), (10) imply

$$\partial_{\bar{z}}\Delta = [\mathcal{J}S(\bar{z})]S(z) = 2JS(\bar{z})S(z). \tag{14}$$

The expression $S(\bar{z})S(z)$ which occurs in (11), (14) is invariant under the flows discussed in Section 5, so the same is true of Δ and of $\int \mathrm{tr}[Q^2]$.

4 A dual problem and the tangent maps

Set

$$\tilde{Q} = -Q^t, \quad \tilde{m} = (I - G_z\tilde{Q})^{-1}1. \tag{15}$$

We define two matrix function-valued inner products:

$$\langle f,g\rangle_1 = \int [f(x,z)^t g(x,z)]^{\mathrm{off}} d\mu(x), \tag{16}$$

$$\langle f,g\rangle_2 = \int [f(x,z)\; g(x,z)]^{\mathrm{off}} d\mu(z). \tag{17}$$

The transpose of Q with respect to $\langle\ ,\ \rangle_1$ is $-\tilde{Q}$, while a calculation shows that the transpose of D_z satisfies

$$D_z^t(fA)_z = -(D_{-\bar{z}}f)A_z.$$

Since $Qm = Q(I-G_zQ)^{-1}1 = (I - QG_z)^{-1}Q1$, formula (3) implies

$$S(z) = -\mathcal{T}\langle 1A_z, Qm\rangle_1 = \mathcal{T}\langle [\tilde{Q}(1 - G_{-\bar{z}}\tilde{Q})^{-1}1]A_z, 1\rangle_1$$

$$= \mathcal{T}\langle \tilde{Q}\tilde{m}(\cdot,-\bar{z})A_z, 1\rangle_1 \tag{18}$$

Using (18), the scattering data for $\tilde{Q}$ is

$$\tilde{S}(z) = -\mathcal{T}\langle 1A_z,\ \tilde{Q}\tilde{m}\rangle_1 = [\mathcal{T}\langle \tilde{Q}\tilde{m}, 1A_z\rangle_1]^t$$

$$= [\mathcal{T}\langle \tilde{Q}\tilde{m}A_{-\bar{z}}, 1\rangle_1]^t = S(-\bar{z})^t.$$

Similar calculations yield the tangent maps. In Newton's notation

$$(Qm)^{\cdot} = \dot{Q}m + Q[(1 - G_zQ)^{-1}1]^{\cdot}$$

$$= \dot{Q}m + Q(I - G_zQ)^{-1}\dot{Q}m = (I - QG_z)^{-1}\dot{Q}m$$

thus we obtain the **tangent map** to the direct map:

$$\dot{S} = -\mathcal{T}\langle 1A_z, (Qm)^{\cdot}\rangle_1 = -\mathcal{T}\langle \tilde{m}(\cdot,-\bar{z})A_z,\ \dot{Q}m\rangle_1 \tag{19}$$

Similarly (7) is $Q = \mathcal{T}\langle Tm, 1\rangle_2$ so

$$\dot{Q} = \mathcal{J}\langle (Tm)^{\cdot}, 1\rangle_2 = \mathcal{J}\langle (I-TC)^{-1}\dot{T}m, 1\rangle_2.$$

With respect to $\langle\ ,\ \rangle_2$, C is its own transpose and the transpose of T is $\tilde{T}$, given by $\tilde{S}$, the scattering data for $\tilde{Q}$. Thus the tangent map to the inverse map is

$$\dot{Q} = \mathcal{J}\langle \dot{T}m, \tilde{m}\rangle_2. \tag{20}$$

5 Flows

The first part of our discussion follows Beals & Coifman [2], Section 5. Suppose $\varphi = \varphi(z)$ is a diagonal matrix-valued function. Let Φ denote right multiplication by φ. Then $[\Phi,T]$ has the same form as T and we may consider the evolution $\dot{T} = [\Phi,T]$ or

$$\dot{S} = S\varphi - \hat{\varphi}S, \qquad \hat{\varphi}(z) = \varphi(\bar{x}). \tag{21}$$

The associated potential evolution is, by (20),

$$\dot{Q} = \mathcal{J}\langle \dot{T}m, \tilde{m}\rangle_2 = \mathcal{J}\langle (Tm)\varphi, \tilde{m}\rangle_2 - \mathcal{J}\langle T(m\varphi), \tilde{m}\rangle_2$$

and since $Tm = \partial_{\bar{z}} m$, $T^t\tilde{m} = \partial_{\bar{z}}\tilde{m}$, and $(\partial_{\bar{z}})^t = \partial_{\bar{z}}$, we have

$$\dot{Q} = \mathcal{J}\langle m\partial_{\bar{z}}\varphi, \tilde{m}\rangle_2. \tag{22}$$

In particular, if $\varphi(z) = z^k 1$ then the distribution $\partial_{\bar{z}}\varphi$ picks out the coefficient of z^{-k-1} in the asymptotic expansion at ∞, so

$$\dot{S} = (z^k - \bar{z}^k)S \leftrightarrow \dot{Q} = \{\text{coefficient of } z^{-k-1}$$
$$\text{in the expansion of } \mathcal{J}[m(x,z)\tilde{m}(x,-z)^t]\}. \tag{23}$$

The expansions of m and $\tilde{m}$ may be computed recursively from the equations $D_z m = Q_m$, $D_z\tilde{m} = \tilde{Q}\tilde{m}$. Note that k=2, $Q=Q^*$ gives DS-II.

The flows (21) commute and preserve $S(\bar{z})S(z)$. they are hamiltonian with respect to the (complex) symplectic form

$$\Omega_2(\dot{S}_1,\dot{S}_2) = i\int \operatorname{tr}[\dot{S}_1(z)\ \mathcal{J}\dot{S}_2(\bar{z})]d\mu(z). \tag{24}$$

In fact take Δ to be the matrix function in (13), (14) and set

$$f(S) = f_\varphi(S) = -i\int \operatorname{tr}[(\partial_{\bar{z}}\varphi)(z)\Delta(z)]d\mu(z)$$
$$= i\int \operatorname{tr}[\varphi(z)\ S(\bar{z})\ \mathcal{J}S(z)]d\mu(z) = i\int \operatorname{tr}[\mathcal{J}(S\varphi - \hat{\varphi}S)].$$

Thus with respect to the form (24), the hamiltonian vector field is

$$H_f(S) = S\varphi - \hat{\varphi}S = \dot{S}.$$

A straightforward calculation using (3), (7) shows

$$\Omega_2(\dot{S}_1,\dot{S}_2) = \Omega_1(\dot{Q}_1,\dot{Q}_2) \equiv i \int tr[\dot{Q}_1(x)\ \mathcal{T}\dot{Q}_2(x)]d\mu(x).$$

Thus the scattering map is a canonical transformation. Note that either symmetry $Q=Q^*$ or $Q=-Q^*$ makes the symplectic forms Ω_1, Ω_2 real.

6 **The Ishimori I system**

The linear system associated with I-I is

$$Du = Q(\partial_{\bar{x}} - \partial_x)u \tag{25}$$

where $Q(x) \to 0$ as $|x| \to \infty$,

$$2Q = \xi_1\sigma_1 + \xi_2\sigma_2 + \xi_3\sigma_3, \quad \xi\cdot\xi = 1$$

$$\sigma_1 = \begin{pmatrix} 0 & 1 \\ 1 & 0 \end{pmatrix}, \ \sigma_2 = \begin{pmatrix} 0 & -i \\ i & 0 \end{pmatrix}, \ \sigma_3 = \begin{pmatrix} -1 & 0 \\ 0 & -1 \end{pmatrix} = -2iJ.$$

Again we look for matrix solutions (2) and obtain

$$D_z m = Q(\partial_{\bar{x}} - \partial_x)m + izQm, \quad m \to 1 \quad \text{as } |x| \to \infty. \tag{26}$$

Arguing as above we obtain $\partial_{\bar{z}}m = Tm$ with S given by replacing Q in (3) by $Q(\partial_{\bar{x}} - \partial_x)$. This seems to contradict the development in Section 2, which appears to suggest that the solution of $\partial_{\bar{z}}m = Tm$ satisfies an equation of a different form from (26). However we cannot expect that m(x,z) in (26) has limit 1 as $z \to \infty$. If it has a limit $m_\infty(x)$ then (26) implies

$$\xi_1\sigma_1 + \xi_2\sigma_2 + \xi_3\varphi_3 = -m_\infty\sigma_3 m_\infty^{-1} \tag{27}$$

and the condition for this to be possible is that

$$-\xi\cdot\xi = \det(\Sigma\xi_j\sigma_j) = \det(-\sigma_3) = -1,$$

which "explains" the constraint above.

It is the function $\hat{m} = m_\infty^{-1}m$ which is the solution of an inverse problem (6) and therefore satisfies an equation $D_z\hat{m} = \hat{Q}\hat{m}$.

Comparison with (26) gives

$$\hat{Q} = -m_\infty^{-1} D m_\infty. \tag{28}$$

Note that (27) determines m_∞ only up to right multiplication by a diagonal matrix function; then the requirement that $\hat{Q}$ in (28) be off-diagonal (and the x-asymptotics) determine m_∞ uniquely. Thus we obtain the gauge equivalence with LS.

Note finally that the condition that ξ be real is equivalent to $Q = Q^*$ and thus to $m = U_- m$, U_- as in (8). This in turn is equivalent to $\hat{Q} = -\hat{Q}^*$ and $S(z) = S(-\bar{z})^*$. As we noted before, the symplectic forms Ω_j are real in this case. The $\varphi(z) = z^2$ flow on S gives Ishimori I.

7 Rigorous considerations

Suppose that the entries q_j of Q belong to $L^{2-\varepsilon}(\mathbb{R}^2) \cap L^{2+\varepsilon}(\mathbb{R}^2)$ for some $\varepsilon > 0$. Then $G_z Q$ is a compact map in the space of bounded continuous matrix-valued functions. If the q_j have small enough norms then $\|G_z Q\| < 1$ for all z and (2) defines m. If in addition the q_j belong to the Schwartz space $S(\mathbb{R}^2)$ then m satisfies (4) with scattering data S having entries $s_j \in S(\mathbb{R}^2)$. Conversely if the s_j belong to $S(\mathbb{R}^2)$ and have small norm in $L^{2-\varepsilon} \cap L^{2+\varepsilon}$, then (6) defines an m which satisfies $D_z m = Qm$ with $q_j \in S(\mathbb{R}^2)$. Under these conditions each of the preceding formulas can be derived rigorously.

In the DS-II specialization $Q = Q^*$ there is no obstruction to inverting $I - G_z Q$: it has trivial kernel. Similarly I-CT has trivial kernel when $S(\bar{z}) = S(z)^*$. Thus in this case the direct and inverse maps $Q \leftrightarrow S$ are nonlinear bijections. Also in this case, as we noted above, the symplectic forms Ω_j are real and the flows in Section 5 are genuinely hamiltonian. To see the injectivity, suppose $f = G_z Q f$ for some z, so $f \to 0$ as $|x| \to \infty$. Set

$$u_\pm(x) = f_{11}(x) \pm e^{-i\bar{x}z - ix\bar{z}}\, \overline{f_{21}(x)},$$

$$v_\pm(x) = e^{i\overline{xz} - ixz}\, f_{21}(x) \pm \overline{f_{22}(x)}.$$

Then $u_\pm$, $v_\pm$ each satisfy an equation of the form $\partial_{\bar{x}} u = r\bar{u}$, $r \in L^{2-\varepsilon} \cap L^{2+\varepsilon}$, $u = 0$ at $|z| = \infty$. Let $w = 0$ where $u = 0$, $w = \bar{u}/u$ where $u \neq 0$, and set $\tilde{u} = u \exp(-Crw)$. Then $\partial_{\bar{x}} \tilde{u} = 0$, $\tilde{u} = 0$ at ∞, so

$\tilde{u} \equiv 0$ by Liouville's theorem and thus each of $u_{\pm}$, $v_{\pm} = 0$. Similarly, if for some x we have a solution $g = CTg$, the same argument applies with

$$u_{\pm}(z) = g_{11}(z) \pm \overline{g_{12}(\bar{z})}, \quad v_{\pm}(z) = g_{21}(z) \pm \overline{g_{22}(\bar{z})}.$$

Acknowledgement

This research was supported by the National Science Foundation, grant DMS 8402737.

References

[1] Beals, R. & Coifman, R.R. (1985). Multidimensional inverse scattering and nonlinear partial differential equations. In *Microlocal Analysis,* Proc. Symp. Pure Math. **43**, 45. Amer. Math. Soc.: Providence.

[2] Beals, R. & Coifman, R.R. (1987). Scattering and inverse scattering for first order systems, II. *Inverse Problems* **3**, 577.

[3] Davey, A. & Stewartson, K. (1974). On three-dimensional packets of surface waves. *Proc. Roy. Soc.* **A338**, 101.

[4] Fokas, A.S. & Ablowitz, M.J. (1984). On the inverse scattering transform of multidimensional nonlinear equations related to first-order systems in the plane. *J. Math. Phys.* **25**, 2494.

[5] Ishimori, Y. (1984). Multi-vortex solutions of a two-dimensional nonlinear wave equation. *Prog. Theor. Phys.* **72**, 33.

[6] Konolpelchenko, B.G. & Matkarimov, B.T. (1989). On the inverse scattering transform for the Ishimori equation. *Phys. Letts. A* **135** , 183.

Cherie L. Schultz and Mark J. Ablowitz

3. EXACT SOLUTIONS OF NONLINEAR EVOLUTION EQUATIONS IN THE STRONG COUPLING LIMIT

We find exact closed form solutions to the strong coupling limit of two nonlinear evolution equations of physical interest, the general Davey-Stewartson equation in (n+1) dimensions and the Zakharov equation. In the case of Davey-Stewartson the Classical Inverse Scattering Transform is shown to reduce to the obtained solution via perturbation.

1 Introduction

In recent years much progress has been made in formulating the multidimensional generalization of the classical inverse scattering transform (CIST) method of solving nonlinear wave equations. However, difficulties still remain, including a complete understanding of the Hamiltonian structure and conservation laws of some of these theories. Indeed, these problems seem to be due to the inherent nonlocality present in integrable nonlinear equations in higher dimensions, and the necessity of taking into account in the CIST formalism the boundary conditions specified for such nonlocal quantities [10]. In this paper we consider two cases of nonlinear evolution equations in a special limit, the so-called "strong-coupling" limit, which preserves this feature of nonlocality, but which allows the exact, closed form solution of the initial-value problem.

2 Strong-coupling limit of the Davey-Stewartson equation and its higher dimensional generalizations

We first consider the so-called Davey-Stewartson (DS) I and II equations [2], which have been exactly solved by the CIST [3]. The DS equations are obtained as a special case of a more general system describing the evolution of packets of water waves, subject to gravity and surface tension, that travel predominantly in one direction but in which the wave amplitudes modulate slowly in the horizontal direction (see [4] for review). The DS system can also be viewed as a generalization to 2+1 dimensions of the nonlinear Schrödinger equation, and as such is a natural candidate to begin the study of multidimensional quantum systems [5].

The DS equations are given by

$$iq_t + \frac{1}{2}(\sigma_1 q_{xx} + q_{yy}) = \phi q - q^2 r$$

$$-ir_t + \frac{1}{2}(\sigma_1 r_{xx} + r_{yy}) = \phi r - r^2 q$$

$$\phi_{xx} + \frac{1}{a}\phi_{yy} = -\frac{b}{a}(qr)_{xx} \qquad (1)$$

where the physical water wave problem is obtained by taking $r = \sigma_2 q^*, \sigma_2 = \pm 1$, where q^* is the complex conjugate of q. The parameters $(\sigma_1, \sigma_2, a, b)$ are functions of two dimensionless parameters $h \equiv kh; T = k^2 T/g$, where k is the wave number in the x-direction of the underlying wave train, h the depth, T the surface tension and g gravity. The standard DS equations result from a suitable long wave limit ($kh \to 0$) of (1) in which case it is found that

DS: $\sigma_1 = \frac{-1}{a} = \sigma^2$, $b = +2/\sigma_2$, $\sigma_1, \sigma_2 = \pm 1$, $q \to q/\sigma$, and $r \to r/\sigma$.

For DSI, $\sigma^2 = +1$ and for DSII, $\sigma^2 = -1$.

The strong coupling limit of (1) is obtained by letting $q \to q/\varepsilon$, $r \to r/\varepsilon$, $t \to \varepsilon^2 t$ and letting $\varepsilon \to 0$ [1]. In this limit (1) reduces to

$$iq_t = \phi q - q^2 r$$

$$-ir_t = \phi r - r^2 q$$

$$\phi_{xx} + \frac{1}{a}\phi_{yy} = -\frac{b}{a}(qr)_{xx} \qquad (2)$$

The general solution to (2) is readily obtained. Multiplying the first equation by r, the second by q and subtracting, we find

$$\frac{\partial}{\partial t}(qr) = 0 \qquad (3)$$

Hence

$$q(x,y,t)r(x,y,t) = q(x,y,0)r(x,y,0) \equiv (qr)_0(x,y) \qquad (4)$$

We shall be primarily concerned with the boundary value problem of $(q,r) \to 0$ at $R = \sqrt{x^2 + y^2} \to \infty$. (4) implies that $\phi(x,y,t) = \phi(x,y,0) \equiv \phi_0(x,y)$. Thus the last equation in (2) can be solved by elementary methods, i.e.

$$\phi(x,y,t) = \phi_0(x,y) = \iint G(x - \xi, y - \eta) \left(-\frac{b}{a}(qr)_0(\xi,\eta)\right)_{\xi\xi} d\xi d\eta \tag{5}$$

with G(x,y) satisfying

$$G_{xx} + \sigma' G_{y'y'} = \delta(x)\delta(y') \tag{6}$$

where $y' = \sqrt{|a|}\, y$ and $\sigma' = \text{sgn}(a)$. For $a > 0$, $\sigma' = +1$ (relevant for the strong coupling limit of DSII), the usual Green's function for the Laplacian is satisfactory; i.e. $G(x,y) = \frac{1}{4\pi}\log(x^2 + y^2)$ corresponds to the boundary values $\phi \to 0$ as $r \to \infty$. However for $a < 0$, $\sigma' = -1$ (relevant for the strong coupling limit of DSI), there are other natural Green's functions which arise:

(a) $G(x,y) = \theta(x - y)\theta(x + y)$

(b) $G(x,y) = \frac{1}{4\pi i}\ln(x^2 - y^2) + \frac{1}{4}(\theta(-x - y))(\theta(x - y) - \theta(-x + y))$

(c) $G(x,y) = \frac{1}{4}(\theta(x + y) - \theta(-x - y))(\theta(x - y) - \theta(-x + y))$

Case (a) corresponds to the Riemann-Hilbert formulation of the CIST solution of (1), discussed by Fokas & Ablowitz [3] and (b) to the "$\bar{\partial}$ limit" approach of Nachman & Ablowitz [6]. Case (c) corresponds to a "symmetric" case of (1).

The general solution to (2) is written in the following form:

$$q(x,y,t) = q(x,y,0)\exp(-i\psi_0(x,y)t)$$

$$r(x,y,t) = r(x,y,0)\text{ex}(i\psi_0(x,y)t) \tag{7}$$

where

$$\psi_0(x,y) = \phi_0(x,y) - (qr)_0(x,y) \tag{8}$$

and $\phi(x,y)$ is given by (5).

The Hamiltonian structure of the more general evolution equation, (1) is only completely understood in Case (b) [10], though here the Green's function, G(x,y) is seen to have an imaginary homogeneous part, and we cannot in general let $r=q^*$. The Hamiltonian structure in Case (a) is not yet well-understood even in the strong coupling limit. It is straightforward to write down the Hamiltonian generating the equations of motion, (1) as well as its limit, (2), corresponding to (c), and we consider that finding the CIST in this "symmetric" case is an important problem.

We note that the exact analysis of (2) goes through in any number of dimensions. Consider

$$
\begin{aligned}
iq_t &= \phi q - q^2 r \\
-ir_t &= \phi r - r^2 q \\
L\phi &= -(qr)_{x_1 x_1}
\end{aligned}
\tag{9}
$$

where

$$
L = \frac{\partial^2}{\partial x_1^2} - \sum_{j=2}^{n} \sigma_j \frac{\partial^2}{\partial x_j^2} \tag{10}
$$

with $\sigma_j = \pm 1$. The general solution to (9) with suitable boundary conditions on $\phi(x_1,\ldots,x_n)$ for decaying q, r as $R \equiv \sqrt{\Sigma_{j=1}^{n} x_j^2} \to \infty$ is given by

$$
\begin{aligned}
q(x_1,x_2,\ldots,x_n,t) &= q(x_1,\ldots,x_n,0)\exp(-it\psi_0) \\
r(x_1,x_2,\ldots,x_n,t) &= r(x_1,\ldots,x_n,0)\exp(it\psi_0)
\end{aligned}
\tag{11}
$$

where

$$
\psi_0 = \phi(x_1,x_2,\ldots x_n,0) - q(x_1,\ldots,x_n,0)r(x_1,\ldots,x_n,0) \tag{12}
$$

3 Strong coupling limit from the inverse scattering transform

We can obtain the strong coupling limit of DS via a perturbation expansion of its underlying linear equations, or "Lax pair", which provides the basis for the CIST analysis. The DS equation, (1), is a result of the compatibility condition between the following two equations for the (2 x 2) matrix-valued function, μ (see [7,8])

$$
\begin{aligned}
\mu_x &= Q\mu - \sigma J\mu_y + ik\sigma(\mu J - J\mu) \\
\mu_t &= A\mu - \frac{iQ}{\sigma}\mu_y + iJ\mu_{yy} + \frac{k}{\sigma}Q_\mu - 2kJ\mu_y - ik^2(J\mu - \mu J)
\end{aligned}
\tag{13}
$$

where

$$
J = \begin{pmatrix} J_1 & 0 \\ 0 & J_2 \end{pmatrix} = \begin{pmatrix} 1 & 0 \\ 0 & -1 \end{pmatrix}, \quad Q = \begin{pmatrix} 0 & q \\ r & 0 \end{pmatrix},
$$

and A is also a (2 x 2) matrix-valued function whose components satisfy

$$(\frac{\partial}{\partial x} + \sigma\frac{\partial}{\partial y})A_{11} = -\frac{i}{2\sigma^2}(\frac{\partial}{\partial x} - \sigma\frac{\partial}{\partial y})(qr)$$

$$(\frac{\partial}{\partial x} - \sigma\frac{\partial}{\partial y})A_{22} = \frac{i}{2\sigma^2}(\frac{\partial}{\partial x} + \sigma\frac{\partial}{\partial y})(qr)$$

$$A_{12} = \frac{i}{2\sigma^2}(q_x - \sigma q_y)$$

$$A_{21} = \frac{-1}{2\sigma^2}(r_x + \sigma r_y) \tag{14}$$

with the mean field ϕ related to the A's by

$$\phi = i(A_{11} - A_{22}) + qr \tag{15}$$

In order to obtain the strong coupling limit of DS from (13) we first make the following transformations, where ε is a small parameter

$$t = \varepsilon^2\tilde{t},\ Q = \tilde{Q}/\varepsilon,\ A = \frac{\tilde{A}^d}{\varepsilon^2} + \frac{\tilde{A}^0}{\varepsilon} \text{ and } k = \tilde{k}/\varepsilon^2$$

where $\tilde{t},\tilde{Q},\tilde{A}^d,\tilde{A}^0,\tilde{k}$ will be considered as order one quantities and $\tilde{A}^d,\tilde{A}^0$ are the diagonal and off-diagonal part of A respectively. Call

$$\tilde{\mu}(x,y,t,\tilde{k}) = [d(x,y,t,k) + \varepsilon f(x,y,t,k)] \tag{16}$$

where d has only diagonal elements and f only off-diagonal elements. We assume that q,r are given initially as smoothly varying functions decaying rapidly to zero at infinity. Expanding d,f as $d = d^{(0)} + \varepsilon^2 d^{(1)} + \dots$ and $f = f^{(0)} + \varepsilon^2 f^{(1)} + \dots$ we obtain a sequence of equations from which the following results can be deduced. The first of equations (13) gives, to leading order,

$$f^{(0)} = \frac{1}{2ik\sigma}JQd^{(0)} \tag{17}$$

with $d^{(0)}$ satisfying

$$(\frac{\partial}{\partial x} + \sigma J\frac{\partial}{\partial y})d^{(0)} = -\frac{1}{2ik\sigma}JQ^2d^{(0)} \tag{18}$$

This can be integrated in the general case to find

$$d_i^{(0)} = C_i(t)\exp(\frac{i}{2k\sigma}J_i\Phi_i) \tag{19}$$

where $C_i(t)$ is an arbitrary function of time and

$$\Phi_i(x,y,t) = \iint \hat{G}_i(x - x',y - y')(qr)(x',y',t)dx'dy' \tag{20}$$

with

$$\left(\frac{\partial}{\partial x} + \sigma J_i \frac{\partial}{\partial y}\right)\hat{G}_i = \delta(x)\delta(y) \tag{21}$$

In the cases of interest we have

$$\hat{G}_i(x,y) = \frac{1}{2\pi(x + iJ_i y)}$$

for DSII and

(a) $\hat{G}_i(x,y) = J_i\delta(x + J_i y)\theta(J_i x)$

(b) $\hat{G}_i(x,y) = \int\frac{dl}{2\pi}e^{ilJ_i(x+J_i y)}[\theta(x)\theta(-J_i l) - \theta(-x)\theta(J_i l)]$

(c) $\hat{G}_i(x,y) = \frac{1}{2}\delta(x + J_i y)(\theta(x) - \theta(-x))$

for DSI.

Similarly, at leading order the time evolution for d gives

$$d_t^{(0)} = A^d d^{(0)} + \frac{k}{\sigma}(d_x^0 - \sigma J dy^{(0)}) \tag{22}$$

Substitution of (19) into (22) using (14) yields the result:

$$\frac{\partial}{\partial t}(qr) = 0 \tag{23}$$

At order ε^2, making use of (17), we find

$$Qd^{(1)} - 2ik\sigma J f^{(1)} = f_x^{(0)} + \sigma J f_y^{(0)} \tag{24}$$

and

$$f_t = A^d f + A^0 d - \frac{i}{\sigma}Qd_y - 2kJ f_y \tag{25}$$

From these second order results, along with (23) and (14) we obtain

$$q_t = q(A_{11} - A_{22})$$

$$r_t = -r(A_{11} - A_{22}) \tag{26}$$

Thus, taking into account (23), (14) and (15), we have obtained the DS equations in the strong coupling limit via a perturbation expansion of the underlying linear equations (13).

4 Limit of the Zakharov equation

We consider as a second case the one-dimensional Zakharov equation, used to model electron waves in a turbulent plasma [9].

In dimensionless units, these equations are

$$(i\partial_t + \frac{3}{2}\partial_x^2 - \delta n + i\gamma)E = 0$$

$$(\partial_t^2 - c_s^2\partial_x^2)\delta n = c_s^2\partial_x^2|E|^2 \qquad (27)$$

Here E is the Langmuir wavefield envelope, δn is the low-frequency deviation of the plasma density from its average value, γ is the rate of damping (forcing), and c_s^2 is an ion acoustic speed. In the appropriate limit this reduces to the system of equations

$$(i\partial_t - \delta n + i\gamma)E = 0$$

$$(-i\partial t - \delta n - i\gamma)E^* = 0 \qquad (28)$$

and

$$(\partial c_t^2 - c_s^2\partial_x^2)\delta n = c_s^2\partial_x^2|E|^2 \qquad (29)$$

Multiplying the first of equations (28) by E^* and the second by E and subtracting, one finds

$$\partial_t|E|^2 = -2\gamma|E|^2 \qquad (30)$$

from which we obtain

$$|E(x,t)|^2 = |E_0|^2e^{-2\gamma t} \qquad (31)$$

where $E(x,0) \equiv E_0(x)$. We also define $\partial_x^2|E_0|^2 \equiv c_s^2F(x)$, where the initial conditions determine the function, F. Then (29) becomes

$$(\partial_t^2 - c_s^2\partial_x^2)\delta n = F \qquad (32)$$

which can be integrated, giving

$$\delta n(x,t) = \int\int G_R(x - x',t - t')F(x',t')\, dx'\, dt' \qquad (33)$$

where G_R is the retarded Green's function. Since δn is known, equation (28) can be solved for E, with the result that

$$E(x,t) = E_0(x)e^{-\gamma t - i\int_0^t \delta n(x,t')dt'} \qquad (34)$$

Acknowledgements

We very much appreciate illuminating discussions with A.S. Fokas. This work was partially supported by the National Science

Foundation under Grant No. DMS-8803471, the Office of Naval Research under Grant No. N00014-88-K-0447 and the Air Force Office of Scientific Research under Grant No. AFOSR-88-0073.

References

[1] Ablowitz, M.J. & Schultz, C.L. (1989). *Stud. Appl. Math.* **80**, 229.

[2] Davey, A & Stewartson, K. (1974). *Proc. Roy. Soc. 1* **A388**, 191.

[3] Fokas, A.S. & Ablowitz, M.J. (1984). *J. Math. Phys.* **25**, 2492. Also see Fokas, A.S. & Ablowitz, M.J. (1983). In *Nonlinear Phenomena*, Wolf, K.B. (ed.). Lecture Notes in Physics, Vol. 189, Springer-Verlag: Berlin, and references therein.

[4] Ablowitz, M.J. & Segur, H. (1979). *J. Fluid Mech.* **92**, 691.

[5] Schultz, C.L., Ablowitz, M.J. & Bar Yaacov, D. (1987). *Phys. Rev. Lett.* **59(25)**, 2825.

[6] Nachman, A.I. & Ablowitz, M.J. (1984). *Stud. Appl. Math.* **71**, 251.

[7] Ablowitz, M.J. & Haberman, R. (1975). *Phys. Rev. Lett.* **35**, 1185.

[8] Ablowitz, M.J. & Segur, H. (1981). Solitons and the Inverse Scattering Transform. *SIAM Studies in Applied Mathematics No. 4.* SIAM: Philadelphia.

[9] Zakharov, V.E. (1972). *Sov. Phys. JETP* **35**, 908.

[10] Schultz, C.L. & Ablowitz, M.J. (1989). *Phys. Lett.* **135**, 433.

4. RATIONAL REFLECTION COEFFICIENTS

The study of the inverse Schrödinger problem, in the case that the reflection coefficient is a meromorphic function of the momentum, was greatly advanced by the work of Kay & Moses [1,2]. Except for some specific examples, the results were limited to the case that the potential vanishes on a half-line. When both the left and the right reflection coefficients are known (or at least their poles in the upper half plane), the generalization to an uncut potential is straightforward. Sabatier [3,4] has recently given an algorithm that facilitates the computation of the potential in this case. When one is provided only with the right reflection coefficient, however, the determination of the potential, from the Marchenko equation, presents difficulties for negative x, which can be overcome by solving partial differential equations [5,6]. An alternative procedure has been given by Calogero & Degasperis [7], in which one first deduces the transmission coefficient, and thence the left reflection coefficient, after which one solves two separate problems, with two separate reflection coefficients, one for positive and one for negative x.

Here we sketch the method of solving the problem, given both the left and the right reflection coefficients, when there are also bound states - a refinement that was often missing from earlier treatments. We deduce the general form of the potential, and we give a specific example. Details of the more difficult problem, in which only one reflection coefficient is known, will be published later [8].

If the real potential, V(x), satisfies the integrability condition

$$\int_{-\infty}^{\infty} dx[1 + |x|]|V(x)| < \infty, \tag{1}$$

then two Jost solutions, $f_\pm(k,x)$, of the Schrödinger equation,

$$\left\{- \left[\frac{d}{dx}\right] + V(x)\right\} f_\pm(k,x) = k^2 f_\pm(k,x) \ , \tag{2}$$

exist, such that, respectively,

$$f_+(k,x) - e^{ikx} \xrightarrow[x\to\infty]{} 0, \tag{3a}$$

$$f_-(k,x) - e^{-ikx} \xrightarrow[x\to-\infty]{} 0. \tag{3b}$$

The right and left transmission and reflection coefficients are defined by the equalities

$$f_+(-k,x) + R_+(k)f_+(k,x) = T_-(k)f_-(k,x), \tag{4a}$$

$$f_-(-k,x) + R_-(k)f_-(k,x) = T_+(k)f_+(k,x). \tag{4b}$$

Eq. (4a) describes flux of unit intensity, incident from the right, which is partially transmitted to the left, with transmission coefficient T_-, and partially reflected back to the right, with reflection coefficient R_+. Eq. (4b) describes the symmetric situation, in which unit flux from the left is transmitted with coefficient T_+, and reflected with coefficient R_-.

By taking Wronskians of both sides of (4a) with respect to $f_\pm(k,x)$, we find

$$R_+(k) = -\frac{W[f_+(-k,.),f_-(k,.)]}{W[f_+(k,.),f_-(k,.)]}, \tag{5a}$$

$$T_-(k) = -\frac{2ik}{W[f_+(k,.),f_-(k,.)]}, \tag{5b}$$

where we have used the fact that $W[f_+(k,.),f_+(-k,.)] = -2ik$. The x-dependence of the Jost functions has been suppressed, since the Wronskians themselves are x-independent. Similarly, taking Wronskians of (4b) with respect to $f_\pm(k,x)$, we obtain

$$R_-(k) = -\frac{W[f_+(k,.),f_-(-k,.)]}{W[f_+(k,.),f_-(k,.)]}, \tag{5c}$$

$$T_+(k) = -\frac{2ik}{W[f_+(k,.),f_-(k,.)]}, \tag{5d}$$

It follows that $T_+(k) = T_-(k)$, so we shall drop the suffix on the transmission coefficient in what follows. Eqs. (5) imply also that

$$R_-(k) = -\frac{T(k)}{T(-k)} R_+(-k). \tag{6}$$

The Schrödinger equation (2), subject to (1), has in general a finite number of real, square integrable solutions, $\psi_n(x)$ (bound states), corresponding to $k = ip_n$, $p_n \geq 0$, $n = 1,2,\ldots,N$ (N may be zero):

$$\left\{-\left[\frac{d}{dx}\right]^2 + V(x)\right\}\psi_n(x) = -p_n^2\psi_n(x), \tag{7}$$

with

$$\int_{-\infty}^{\infty} dx[\psi_n(x)]^2 = 1. \tag{8}$$

The asymptotic behaviour is

$$\psi_n(x) \begin{cases} \sim c_n \; e^{-p_n x}, \quad x\to\infty, & (9a) \\ \sim \bar{c}_n \; e^{p_n x}, \quad x\to-\infty, & (9b) \end{cases}$$

where we neglect the possibility $p_n = 0$ (zero-energy bound state), for simplicity.

We define the positive spectral transform by

$$M_+(x) = \sum_{n=1}^{N} c_n^2 e^{-p_n x} + \frac{1}{2\pi}\int_{-\infty}^{\infty} dk e^{ikx} R_+(k), \tag{10}$$

and the positive Marchenko kernel, $K_+(x,y)$, as the solution of

$$K_+(x,y) + M_+(x + y) + \int_x^{\infty} dz\, K_+(x,z)M_+(z + y) = 0, \tag{11}$$

for $y > x$. The potential is related to K_+ by

$$V(x) = -\,2 \lim_{\varepsilon\downarrow 0} \frac{d}{dx} K_+(x,\, x + \varepsilon). \tag{12}$$

Since R_- and T (as well as R_+ and the bound states), can be calculated from V, it follows that, when the inverse method works, the whole problem is determined, once M_+ has been specified.

A corresponding situation obtains, for negative x, if one starts with the negative spectral transform,

$$M_-(x) = \sum_{n=1}^{N} \bar{c}_n^{\,2} e^{p_n x} + \frac{1}{2\pi}\int_{-\infty}^{\infty} dk\, R_-(k)e^{-ikx}. \tag{13}$$

The negative Marchenko kernel solves

$$K_-(x,y) + M_-(x + y) + \int_{-\infty}^{x} dz\, K_-(x,z)M_-(z + y) = 0, \tag{14}$$

for $y < x$. The potential is given by

$$V(x) = 2 \lim_{\varepsilon\downarrow 0} \frac{d}{dx} K_-(x, x - \varepsilon). \tag{15}$$

To summarize: V is determined by the set $\{R_+, p_n, c_n^2\}$ **or** the set $\{R_-, p_n, \bar{c}_n^2\}$. Clearly, one set may not be specified independently of the other.

We shall consider the case in which $R_+(k)$ is meromorphic in the entire k-plane. this can only happen if the Jost solutions are meromorphic, which means that $T(k)$ and $R_-(k)$ are also meromorphic. Consider first the case $x > 0$, and close the k-contour in (10) in the upper half plane. One finds

$$M_+(x) = \sum_i a_i e^{-b_i x}, \tag{16}$$

where ib_i runs over the positions of the poles of $R_+(k)$, in the upper half of the k-plane, and the bound-state positions, ip_n. The coefficients, a_i, come from the residues of the poles, and from the bound-state weights, c_n^2. The sum in (16) contains a finite number of terms. The positive Marchenko equation, (11), can be written

$$K_+(x,y) = -\sum_i a_i \left\{ e^{-b_i(x+y)} + \int_x^\infty dz K_+(x,z) e^{-b_i(z+y)} \right\}, \tag{17}$$

from which it is clear that K_+ must have the form

$$K_+(x,y) = -\sum_i a_i f_i(x) e^{-b_i y}. \tag{18}$$

Substitution of this into (17), and evaluation of the z-integral, yields the algebraic relation

$$\sum_j \Delta_{ij}(x)\, f_j(x) = e^{-b_i x}, \tag{19}$$

where

$$\Delta_{ij}(x) = \delta_{ij} + \frac{a_j}{b_i + b_j} e^{-(b_i+b_j)x}. \tag{20}$$

Solution for the f's and hence for K_+, follows on inversion of the matrix Δ. From the standard expression for the inverse of a matrix as the matrix of cofactors, divided by the determinant of the matrix, it is clear that K_+ has the form of a "rational function of exponentials": to be precise, for $x > 0$,

$$\lim_{\varepsilon \downarrow 0} K_+(x, x+\varepsilon) = \frac{\sum_i \alpha_i e^{-\beta_i x}}{\sum_i \gamma_i e^{-\delta_i x}}, \tag{21}$$

where the α_i, β_i, γ_i, δ_i are determined by the a_i, b_i, as implied by the above matrix inversion procedure.

To tackle $x < 0$, we consider most expeditiously the negative Marchenko equation, closing the contour in (13) in the upper half k-plane. Now we find

$$M_-(x) = \sum_i e_i e^{d_i x}, \tag{22}$$

where id_i runs over the positions of the poles of $R_-(k)$ in the upper half of the k-plane, and the bound state positions, ip_n. The coefficients, e_i, come from the residues of the poles, and the weights, $\bar{c}_n^{-2}$. The negative Marchenko equation, (14), becomes

$$K_-(x,y) = -\sum_i e_i \left\{ e^{d_i(x+y)} + \int_{-\infty}^{x} dz K_-(x,z) e^{d_i(z+y)} \right\}. \tag{23}$$

Hence we can write

$$K_-(x,y) = -\sum_i e_i g_i(x) e^{d_i y}, \tag{24}$$

where

$$\sum_j \Gamma_{ij}(x) g_j(x) = e^{d_i x}, \tag{25}$$

with

$$\Gamma_{ij}(x) = \delta_{ij} + \frac{e_j}{d_i + d_j} e^{(d_i+d_j)}. \tag{26}$$

Inversion of the matrix, Γ, yields the g_i, and hence K_-, that clearly has a form analogous to (21), which we may write, for $x < 0$,

$$\lim_{\varepsilon\downarrow 0} K_-(x,x-\varepsilon) = \frac{\sum_i \bar{\alpha}_i e^{\bar{\beta}_i x}}{\sum_i \bar{\gamma}_i e^{\bar{\delta}_i x}}, \tag{27}$$

where $\bar{\alpha}_i$, $\bar{\beta}_i$, $\bar{\gamma}_i$, $\bar{\delta}_i$ are functions of the c_i, d_i.

The potential itself is given by (12) and (15), and since the expressions (21) or (27) retain their generic form after differentiation, it follows that the generic form of the potential, in the case that the reflection coefficients are meromorphic, is

$$V(x) = e^{2\mu_- x} \frac{\rho_0^- + \Sigma_i \rho_i^- \exp(\zeta_i^- x)}{1 + \Sigma_i \sigma_i^- \exp(\eta_i^- x)} \theta(-x) + C\delta(x) +$$

$$e^{-2\mu_+ x} \frac{\rho_0^+ + \Sigma_i \rho_i^+ \exp(-\zeta_i^+ x)}{1 + \Sigma_i \sigma_i^+ \exp(-\eta_i^+ x)} \theta(x) \tag{28}$$

Here $\zeta_i^\pm$ and $\eta_i^\pm$ are all positive, and $\mu_\pm$ is the imaginary part of the pole of $R_\pm(k)$, in the upper half of the k-plane, that is closest to the real axis, or the minimum bound-state position, say p_1, whichever is the smaller.

We have found the general form of the potential that corresponds to meromorphic reflection coefficients, which are $O\left(\frac{1}{k}\right)$ as $k \to \infty$. In the case that the pole positions and residues for $R_+(k)$ and for $R_-(k)$, in the upper half plane, are prescribed (as well as the bound-state quantities, p_n, c_n^2, c_n^{-2}) the above would be a constructive method of finding the potentials. Often, however, one would know $R_+(k)$ (together with the bound-state data) only. Then, although (28) is still true, it is harder to find the expression for $V(x)$ when $x < 0$. One has in this case to close the contour in (10) in the lower half plane, thus picking up contributions from the poles of $R_+(k)$ in Im $k < 0$. The equation to be solved is still (11), but since $M_+(z + y)$ has a discontinuity at $z + y = 0$, the Ansatz (18) no longer works. One has in fact a set of simultaneous partial differential equations. Some simple cases have been worked out [6,8]. For example, if

$$R_+(k) = \frac{i\alpha}{k-i\alpha}, \tag{29}$$

and there is a bound-state at $p_1 = \alpha$, with weight $c_1 = \sqrt{\gamma}$, then

$$V(x) = -2\alpha^2 \operatorname{sech}^2(\alpha x + d)\theta(-x) - 2\alpha\delta(x) + 2\alpha^2 \operatorname{cosech}^2(\alpha x + c)\theta(x), \tag{30}$$

where

$$c = \frac{1}{2} \log \frac{2\alpha}{\alpha-\gamma}, \tag{31}$$

and

$$d = \frac{1}{2} \log \frac{\alpha-\gamma}{2\gamma}. \tag{32}$$

This result may be obtained by solving the positive Marchenko equation (11) for positive and for negative x [6], or by first obtaining the transmission coefficient from a dispersion relation [7,8], and thence $R_{-}(k)$ from (6). The potential for $x < 0$ is then obtained from the negative Marchenko equation (14).

Acknowledgements

I would like to acknowledge discussions with P.C. Sabatier on, and with A. Degasperis above, the Lago di Como.

References

[1] Kay, I. & Moses, H.E. (1957). The determination of the scattering potential from the spectral measure function. Part IV. «Pathological» scattering problems in one dimension. *Supplemento al novo cimento* **5**, 230.
[2] Kay, I. (1960). The inverse scattering problem when the reflection coefficient is a rational function. *Communications on pure and applied mathematics* **XIII**, 371.
[3] Sabatier, P.C. (1983). Rational reflection coefficients and inverse scattering on the line. *Nuovo cimento* **78B**, 235.
[4] Sabatier, P.C. (1983). Rational reflection coefficients in one-dimensional inverse scattering problems and applications. *Conference on inverse scattering: theory and application,* p. 75, Bee Bednar, J. et al. (eds.). SIAM: Philadelphia.
[5] Atkinson, D. & Geschiere, J. (1986). Inverse scattering: an illicit reflection coefficient. *Inverse problems* **2**, L51.
[6] Atkinson, D. (1987). Meromorphic reflection coefficient. *RCP 264 Rencontre 1987*, to be published.
[7] Calogero, F. & Degasperis, A. (1982). *Spectral transform and solitons.* North Holland: Amsterdam.
[8] Atkinson, D. (1988). Rational reflections and the Marchenko problem. In preparation.

P.J. Caudrey

5. TWO-DIMENSIONAL SPECTRAL TRANSFORMS

A fairly general two-dimensional spectral problem is examined and appropriate spectral data are defined. A method of solving the inverse problem is described. The application to nonlinear evolution equations is outlined and the familiar results obtained for the Kadomtsev-Petviashvili and Davey-Stewartson equations.

1 Introduction

Since Gardner et al. discovered the IST (inverse spectral transform) as a method of solving nonlinear evolution equations (2) in 1967 it has been greatly developed. The results for nonlinear equations in 1+1 dimensions (i.e. on spatial and one temporal dimension) are by now classical (for example see [3,4]) but it is only a few years ago that the IST for equations in 2+1 and n+1 dimensions began to be understood [5]). In 1+1 dimensions the inverse problem is essentially a classical Riemann-Hilbert problem. However, for 2+1 dimensions it is found to lie outside R-H theory. A powerful method for solving the inverse problem for some 2+1 dimensional equations is the D-bar method which was originally used by Beals & Coifman [6].

The IST can also be used for equations involving integral and difference operators. Cases of discrete equations in 1+1 dimensions have been studied by many writers [7-11].

We will consider a large class of non-linear partial differential equations or systems of equations in 2+1 dimensions. All equations or systems of equations in this class can be written in the following form [12]

$$\mathbf{U}_t(x,y,t,\frac{\partial}{\partial y}) - \mathbf{V}_x(x,y,t,\frac{\partial}{\partial y}) + [\mathbf{U}(x,y,t,\frac{\partial}{\partial y}),\ \mathbf{V}(x,y,t,\frac{\partial}{\partial y})] = \mathbf{0} \tag{1}$$

where the **Lax pair U** and **V** are NxN matrix functions of x, y, t and the differential operator $\frac{\partial}{\partial y}$. Equation (1) is the compatibility condition for the **spectral equation**

$$\frac{\partial}{\partial x} \underline{\psi} = \mathbf{U}(x,y,t,\frac{\partial}{\partial y}) \underline{\psi} \tag{2}$$

and the **auxiliary spectral equation**

$$\frac{\partial}{\partial t} \underline{\psi} = \mathbf{V}(x,y,t,\frac{\partial}{\partial y}) \underline{\psi}. \tag{3}$$

The non-linear partial differential equation (1) can now be solved in three stages as follows:-

(i) Certain **Jost solutions** of (2) are found for an initial time $t = t_0$ and certain pieces of information known as the **spectral data S** are extracted from these solutions.

(ii) The time evolutions of the spectral data S are deduced from (3) and hence these data can be found for a later time t.

(iii) The **inverse spectral problem**, that of finding $\mathbf{U}(x,y,t,\frac{\partial}{\partial y})$ from S is solved for this later time.

The amazing fact is that each of these three stages is carried out by solving a **linear** problem.

The rest of the paper is arranged as follows. In Section (2) the spectral data S in the case when

$$U(x,y,t,(\frac{\partial}{\partial y}) = \mathbf{A}(\frac{\partial}{\partial y}) + \mathbf{B}(x,y) \tag{4}$$

where $B(x,y) \to 0$ as $x^2 + y^2 \to \infty$ is derived. The time evolutions of **S** are found in Section (3) and the method of solution of the inverse spectral problem is described in Section (4). In Section (5) and Section (6) two important examples, the Kadomtsev-Petviashvili equation and the Davey-Stewartson equations are considered.

2 The direct spectral problem

We take as our spectral equation

$$\frac{\partial}{\partial x} \underline{\psi} = \{\mathbf{A}(\frac{\partial}{\partial y}) + \mathbf{B}(x,y)\} \underline{\psi} \tag{5}$$

where $\underline{\psi}$ is an N-element column vector and $\mathbf{A}(\frac{\partial}{\partial y})$ and $\mathbf{B}(x,y)$ are NxN matrices. The elements of **A** are functions of the differential operator $\frac{\partial}{\partial y}$ and

$$\mathbf{B}(x,y) \to 0 \text{ as } x^2 + y^2 \to \infty. \tag{6}$$

We introduce the eigenvalues and the eigenvectors of $A(\zeta)$,

$$\mathbf{A}(\zeta) \underline{v}_i(\zeta) = \lambda_i(\zeta) \underline{v}_i(\zeta) \tag{7}$$

together with the reciprocal (row) eigenvectors $\underline{\tilde{v}}_i(\zeta)$

$$\underline{\tilde{v}}_i(\zeta)\ \underline{v}_j(\zeta) = \delta_{ij}. \tag{8}$$

We assume that these quantities are regular throughout the cut complex ζ-plane and that as ζ crosses a cut all that happens is that the suffixes $i = 1,2,\ldots,N$ are permuted.

The Jost functions $\underline{\phi}_i(x,y,\zeta)$ are defined to be solutions of (5) such that

$$\underline{\phi}_i(x,y,\zeta) = \exp\{\lambda_i(\zeta)x + \zeta y\}\ \underline{\Phi}_i(x,y,\zeta) \tag{9}$$

where the $\underline{\Phi}_i(x,y,\zeta)$ satisfy the boundary conditions

$$\underline{\Phi}_i(x,y,\zeta) \to \underline{v}_i(\zeta) \text{ as } x \to -\infty \tag{10}$$

and remain bounded for $-\infty < x < \infty$. Finding these is equivalent to solving the matrix Fredholm equations

$$\underline{\Phi}_i(x,y,\zeta) = \int_{-\infty}^{\infty}\int_{-\infty}^{\infty} \mathbf{K}_i(x,x',y,y',\zeta)\ \underline{\Phi}_i(x',y',\zeta)\ dx'\ dy' + \underline{v}_i(\zeta) \tag{11}$$

where the kernels are given by

$$\mathbf{K}_i(x,x',y,y',\zeta) = \frac{1}{2\pi}\sum_{j=1}^{N}\int_{-\infty}^{\infty} \exp\{(\lambda_j(\zeta+ia)-\lambda_i(\zeta))(x-x') + ia(y-y')\}$$
$$\times\ \{\theta(x-x') - \theta(\mathcal{Re}(\lambda_j(\zeta+ia)-\lambda_i(\zeta)))\} \tag{12}$$
$$\times\ \underline{v}_j(\zeta+ia)\ \underline{\tilde{v}}_j(\zeta+ia)\ da\ \mathbf{B}(x',y')$$

with

$$\theta(\alpha) = \begin{cases} 1 \text{ if } \alpha \geq 0 \\ 0 \text{ if } \alpha < 0 \end{cases}.$$

In general $\underline{\phi}_i(x,y,\ )$ and $\underline{\Phi}_i(x,y,\zeta)$ are not regular functions of ζ because the kernels are not regular. In fact

$$\frac{\partial}{\partial\bar{\zeta}} \mathbf{K}_i(x,x',y,y',\zeta)$$

$$= -\frac{1}{4\pi}\sum_{j=1}^{N}\int_{-\infty}^{\infty} \exp\{(\lambda_j(\zeta+ia)-\lambda_i(\zeta))(x-x') + ia(y-y')\}$$

$$\times\ \overline{\frac{\partial}{\partial\zeta}(\lambda_j(\zeta+ia)-\lambda_i(\zeta))}\ \delta(\mathcal{Re}(\lambda_j(\zeta+ia)-\lambda_i(\zeta)\})$$

$$\times\ \underline{v}_j(\zeta+ia)\ \tilde{\underline{v}}_j(\zeta+ia)\ da\ \mathbf{B}(x',y') \tag{13}$$

where a bar $^{-}$ over an expression denotes its complex conjugate. From this it follows that providing the homogeneous Fredholm equation has only the trivial solution

$$\frac{\partial}{\partial\bar{\zeta}} \underline{\phi}_i(x,y,\zeta) = \sum_{j=1}^{N}\int_{-\infty}^{\infty} \underline{\phi}_j(x,y,\zeta+ia)\ R_{ij}(a,\zeta)\ da. \tag{14}$$

where

$$R_{ij}(a,\zeta) = -\frac{1}{4\pi}\int_{-\infty}^{\infty}\int_{-\infty}^{\infty} \exp\{-\lambda_j(\zeta+ia)x' - (\zeta+ia)y'\}$$

$$\times\ \tilde{\underline{v}}_j(\zeta+ia)\ \mathbf{B}(x',y')\ \underline{\phi}_i(x',y',\zeta)\ dx'\ dy'$$

$$\times\ \overline{\frac{\partial}{\partial\zeta}(\lambda_j(\zeta+ia)-\lambda_i(\zeta))}\ \delta(\mathcal{Re}(\lambda_j(\zeta+ia)-\lambda_i(\zeta)\}). \tag{15}$$

The $R_{ij}(a,\zeta)$ form part of the spectral data.

A complication arises if for some value of ζ

$$\mathcal{Re}(\lambda_j(\zeta+ia)-\lambda_i(\zeta)\} = 0 \tag{16}$$

for all (real) a. In this case $\underline{\phi}_i(x,y,\zeta)$ has a finite (jump) singularity. We label the two sides of the singularity + and - and denote the limit of any function as the singularity is approached from either side by replacing ζ by ζ^+ or ζ^- as appropriate. After some rather tedious manipulation it can be shown that

$$\underline{\phi}_i(x,y,\zeta^+) - \underline{\phi}_i(x,y,\zeta^-) = \sum_{k=1}^{N}\int_{-\infty}^{\infty} Q_{ik}(b,\zeta)\ \underline{\phi}_k(x,y,\zeta^-+ib)\ db. \tag{17}$$

where $Q_{ik}(b,\zeta)$ satisfies

$$\sum_{k=1}^{N}\int_{-\infty}^{\infty} Q_{ik}(b,\zeta)\, P_{kil}(\zeta^{-}+ib,\zeta^{+},\zeta+ic)\, db$$

$$= - P_{iil}(\zeta^{-},\zeta^{+},\zeta+ic) - Q_{il}(c,\zeta). \qquad (18)$$

with

$$P_{ijk}(\zeta^{-},\zeta^{+}+ia,\zeta+ib) = \frac{1}{2\pi}\int_{-\infty}^{\infty}\int_{-\infty}^{\infty} \exp(-\lambda_k(\zeta+ib)x' - (\zeta+ib)y')$$

$$\times \{\theta(\mathcal{R}e(\lambda_k(\zeta^{+}+ib)-\lambda_j(\zeta^{+}+ia)\}) - \theta(\mathcal{R}e(\lambda_k(\zeta^{-}ib)-\lambda_i(\zeta^{-})\})\}$$

$$\times \underline{\tilde{v}}_k(\zeta+ib)\, \mathbf{B}(x',y')\underline{\phi}_i(x',y',\zeta^{-})\, dx'\, dy'. \qquad (19)$$

The quantities $Q_{ik}(b,\zeta)$ belong in the set of spectral data.

The question of what happens when the homogeneous Fredholm equation has non-trivial solutions is as yet unanswered in the general case. However, in one important special case the answer is known. This is when the non-trivial solutions occur for values of ζ lying in a region of the ζ-plane for which the only integer value of j and the only real value of a such that

$$\mathcal{R}e\{\lambda_j(\zeta+ia)-\lambda_i(\zeta)\} = 0 \qquad (20)$$

are $j = i$ and $a = 0$. In this case $\mathbf{K}_i(x,x',y,y',\zeta)$ is a regular function of ζ throughout the region and the Jost function is analytic. For simplicity we assume that the poles are simple. In the neighbourhood of a pole, $\zeta_i^{(k)}$, $k = 1,2,\ldots,n_i$ we can put

$$\underline{\phi}_i(x,y,\zeta) = \underline{\phi}_i^{(k)}(x,y)/(\zeta-\zeta_i^{(k)}) + \underline{\tilde{\phi}}_i^{(k)}(x,y,\zeta). \qquad (21)$$

Then it can be shown that there exist constants $\gamma_i^{(k)}$ such that

$$\underline{\tilde{\phi}}_i^{(k)}(x,y,\zeta_i^{(k)}) = \gamma_i^{(k)}\, \underline{\phi}_i^{(k)}(x,y). \qquad (22)$$

These constants also belong to the spectral data.

The complete set of spectral data is

$$S = \Big\{ R_{ij}(a,\zeta),\ i,j = 1,2,\ldots,N,\ a \in \mathbb{R}, \zeta \in \mathbb{C};$$

$$Q_{ik}(b,\zeta),\ i,k = 1,2,\ldots,N,\ b \in \mathbb{R},$$

ζ runs along the finite singularities of $\underline{\phi}_i(x,y,\zeta)$;

$$\left.\zeta_i^{(k)}, \gamma_i^{(k)}, \; k = 1,2,\ldots,n_i, \; i = 1,2,\ldots,N \right\}. \tag{23}$$

Finally in this section the solution of the inverse problem requires that

$$\underline{\Phi}_i(x,y,\zeta) - \underline{v}_i(\zeta) \to \underline{0} \quad \text{as} \quad \zeta \to \infty. \tag{24}$$

Whether this condition is satisfied or not depends on the individual case. So far it has proved to be true in most cases of practical interest. It can usually be deduced from (11) either with or without the assistance of the Riemann-Lebesque lemma.

3 The time dependence of the spectral data

The time dependence of the wave function $\underline{\psi}$ is assumed to be governed by the auxiliary spectral equation,

$$\underline{\psi}_t = \mathbf{V}(x,y,t,\frac{\partial}{\partial y})\,\underline{\psi} \tag{25}$$

where

$$\mathbf{V}(x,y,t,\frac{\partial}{\partial y}) \to \mathbf{L}(\frac{\partial}{\partial y}) \tag{26}$$

as $x^2 + y^2 \to \infty$ in the same way as $\mathbf{U}(x,y,t,\frac{\partial}{\partial y}) \to \mathbf{A}(\frac{\partial}{\partial y})$. From (1) it follows that $\mathbf{A}(\zeta)$ and $\mathbf{L}(\zeta)$ commute so that they have common eigenvectors,

$$\mathbf{L}(\zeta)\,\underline{v}_i(\zeta) = \mu_i(\zeta)\,\underline{v}_i(\zeta). \tag{27}$$

This is satisfied by $\exp\{\mu_i(\zeta)t\}\,\underline{\phi}_i(x,y,t,\zeta)$ and hence

$$\frac{\partial}{\partial t}\underline{\phi}_i(x,y,t,\zeta) = \{\mathbf{V}_i(x,y,t,\frac{\partial}{\partial y}) - \mu_i(\zeta)\,\mathbf{I}\}\,\underline{\phi}_i(x,y,t,\zeta). \tag{28}$$

The evolution of the spectral data follows,

$$\frac{\partial}{\partial t} R_{ij}(t,a,\zeta) = \{\mu_j(\zeta+ia)-\mu_i(\zeta)\}\,R_{ij}(t,a,\zeta), \tag{29}$$

$$\frac{\partial}{\partial t} Q_{ik}(t,b,\zeta) = \{\mu_k(\zeta+ib)-\mu_i(\zeta)\}\,Q_{ik}(t,b,\zeta), \tag{30}$$

$$\frac{d}{dt}\zeta_i^{(k)} = 0 \tag{31}$$

and

$$\frac{d}{dt}\gamma_i^{(k)}(t) = -\frac{\partial}{\partial\zeta}\mu_i(\zeta)\Big|_{\zeta=\zeta_i^{(k)}}. \tag{32}$$

4 The inverse spectral problem

This is the problem of finding $\mathbf{B}(x,y)$ from the spectal data S given by (23). Besides S we also have $\mathbf{A}(\zeta)$ and the information associated with it, viz. the location of the cuts in the ζ-plane, the permutations of the suffixes across the cuts and the locations of the curves Γ along which $\mathcal{Re}\{\lambda_j(\zeta+ia)-\lambda_i(\zeta)\} = 0$ for all (real) a.

The first step in solving this problem is to find the functions $\underline{\phi}_i(x,y,\zeta)$ which have the properties given by (9,14,17,21,22 and 24) and such that across the cuts

$$\underline{\phi}_i(s,y,\zeta^+) = \sum_{j=1}^{N} P_{ij}\,\underline{\phi}_j(x,y,\zeta^-) \tag{33}$$

where P_{ij} are the elements of the appropriate permutation matrix. Once these have been obtained the spectral equation (5) easily gives $\mathbf{B}(x,y)$.

We use the generalization of Cauchy's integral formula

$$w(\zeta') = \frac{1}{2\pi i}\iint_R \frac{\partial w/\partial\bar\zeta}{\zeta-\zeta'}\,d\zeta\wedge d\bar\zeta + \frac{1}{2\pi i}\int_C \frac{w(\zeta)}{\zeta-\zeta'}\,d\zeta \tag{34}$$

where R is a region in the complex ζ-plane whose boundary is the contour C and ζ' lies in R. Using the properties of $\underline{\phi}_i(x,y,\zeta)$ this gives

$$\begin{aligned}\underline{\phi}_i(x,y,\xi') &= \exp\{\lambda_i(\zeta)x + \zeta y\}\,\underline{v}_i(\zeta') \\ &+ \frac{1}{2\pi i}\iint_{\zeta\text{-plane}} \frac{1}{\zeta-\zeta'}\sum_{j=1}^{N}\int_{-\infty}^{\infty}\underline{\phi}_j(x,y,\zeta+ia)R_{ij}(a,\zeta)da\,d\zeta\wedge d\bar\zeta \\ &+ \frac{1}{2\pi i}\int_\Gamma \frac{1}{\zeta-\zeta'}\sum_{j=1}^{N}\int_{-\infty}^{\infty}Q_{ij}(a,\zeta)\,\underline{\phi}_j(x,y,\zeta^-+ia)\,da\,d\zeta \\ &+ \frac{1}{2\pi i}\int_{\text{cuts}}\frac{1}{\zeta-\zeta'}\sum_{j=1}^{N}(P_{ij}-\delta_{ij})\,\underline{\phi}_j(x,y,\zeta^-)\,d\zeta \\ &+ \sum_{k=1}^{n_i}\frac{1}{\zeta'-\zeta_i^{(k)}}\,\underline{\phi}_i^{(k)}(x,y)\end{aligned} \tag{35}$$

where the residues $\underline{\phi}_i^{(k)}(x,y)$ satisfy

$$\gamma_i^{(k)} \phi_i^{(k)}(x,y) = \exp\{\lambda_i(\zeta)x + \zeta y\} \underline{v}_i(\zeta_i^{(k)}$$

$$+ \frac{1}{2\pi i} \iint_{\zeta\text{-plane}} \frac{1}{\zeta-\zeta_i^{(k)}} \sum_{j=1}^{N} \int_{-\infty}^{\infty} \underline{\phi}_j(x,y,\zeta+ia)R_{ij}(a,\zeta)da\ d\zeta_\wedge d\bar{\zeta}$$

$$+ \frac{1}{2\pi i} \int_\Gamma \frac{1}{\zeta-\zeta_i^{(k)}} \sum_{j=1}^{N} \int_{-\infty}^{\infty} Q_{ij}(a,\zeta)\ \underline{\phi}_j(x,y,\bar{\zeta}^{-}+ia)\ da\ d\zeta$$

$$+ \frac{1}{2\pi i} \int_{\text{cuts}} \frac{1}{\zeta-\zeta_i^{(k)}} \sum_{j=1}^{N} (P_{ij} - \delta_{ij})\ \underline{\phi}_j(x,y,\zeta^{-})\ d\zeta$$

$$+ \sum_{\substack{l=i \\ l\neq k}}^{n_i} \frac{1}{\zeta_i^{(k)}-\zeta_i^{(l)}} \underline{\phi}_i^{(l)}(x,y). \tag{36}$$

These equations (35,36) form a set of matrix/Fredholm equations. Since they are singular there is no unresolved doubt as to the existence and uniqueness of their solution. However, in practical cases there is apparently no difficulty.

5 The Kadomtsev-Petviashvili equation

The Kadomtsev-Petviashvili equation (KP) is

$$(u_t - 6uu_x + u_{xxx})_x + 3\sigma^2 u_{yy} = 0 \tag{37}$$

where $\sigma = i$ (KPI) or $\sigma = -1$ (KPII). In their usual forms the spectral and auxiliary spectral equations are

$$- \psi_{xx} + u\psi = \sigma\psi_y \tag{38}$$

and

$$\psi_t = (- 4 \frac{\partial^3}{\partial x^3} + 3 \frac{\partial}{\partial x} u + 3u \frac{\partial}{\partial x} - 3\sigma v)\ \psi \tag{39}$$

where

$$v_x = u_y. \tag{40}$$

The algebra is simplified by interchanging the roles of x and y. Then N = 1 and

$$\mathbf{A}(\zeta) = -\zeta^2/\sigma, \tag{41}$$

$$\mathbf{B}(x,y,t) = u(x,y,t)/\sigma \tag{42}$$

and

$$\mathbf{V}(x,y,t,\zeta) = -r\zeta^3 + 6\zeta u(x,y,t) = 3\{u_x(x,y,t) - \sigma v(x,y,t)\} \tag{43}$$

Let us consider first the case $\sigma = -1$ (KPII). This gives

$$R(a,\zeta) = R(\zeta)\ \delta(a+2\mathcal{I}m\zeta) \tag{44}$$

where

$$R(\zeta) = \frac{i}{2\pi}\ \mathrm{sgn}(\mathcal{I}m\zeta) \int_{-\infty}^{\infty}\int_{-\infty}^{\infty} \exp\{-\bar{\zeta}x - \bar{\zeta}^2 y\}\ u(x,y)\ \phi(x,y,\zeta)\ dx\ dy. \tag{45}$$

There are no values of ζ such that $\mathcal{R}e\{\lambda(\zeta\_ia)-\lambda(\zeta)\} = 0$ for all a. Thus $\phi(x,y,\zeta)$ has no finite singularities.

Current belief is that the homogeneous version of the Fredholm equation has only the trivial solution for all ζ. However this has yet to be proved.

In the case when $\sigma = i$, $K(x,x',y,y',\zeta)$ is regular except on the imaginary axis $\mathcal{R}e\ \zeta = 0$. Thus in this case

$$R(a,\zeta) = 0. \tag{46}$$

Across $\mathcal{R}e\ \zeta = 0$ there is a finite sinngularity. Putting $\zeta = i\xi$ where ξ is real and labelling the sides of the imaginary axis according to the sign of $\mathcal{R}e\ \zeta$ we have that $Q(a,i\xi)$ satisfies

$$\int_{-\infty}^{\infty} Q(a,i\xi)\ P(i\xi+ia-0,i\xi+0,i\xi+ib)\ da$$
$$+ P(i\xi-0,i\xi+0,i\xi+ib) + Q(b,i\xi) = 0 \tag{47}$$

where

$$P(i\xi+ia-0,i\xi+0,i\xi+ib) = -\frac{i}{2\pi}\int_{-\infty}^{\infty}\int_{-\infty}^{\infty} \exp\{i(\xi+b)^2 y' - i(\xi+b)x'\}$$
$$\times\ \{\theta(-b) - \theta(b-a)\}\ u(x',y')\ \phi(x',y',i\xi+ia-0)\ dx'\ dy'. \tag{48}$$

Off the imaginary axis the conditions specified in Section (2) are satisfied so that $\phi(x,y,\zeta)$ is analytic and assuming that the poles are simple we can put

$$\phi(x,y,\zeta) = \phi^{(k)}(x,y)/(\zeta-\zeta^{(k)}) + \tilde{\phi}^{(k)}(x,y,\zeta) \tag{49}$$

in a neighbourhood of the pole at $\zeta = \zeta^{(k)}$, $k = 1,2,\dots,n$ where

$$\tilde{\phi}^{(k)}(x,y,\zeta^{(k)}) = \gamma^{(k)}\,\phi^{(k)}(x,y). \tag{50}$$

6 **The Davey-Stewartson equations**

The spectral equation

$$\frac{\partial}{\partial x}\underline{\psi} = \begin{pmatrix} \alpha\frac{\partial}{\partial y} & u \\ \pm\bar{u} & -\alpha\frac{\partial}{\partial y} \end{pmatrix}\underline{\psi} \tag{51}$$

and the auxiliary spectral equation

$$\frac{\partial}{\partial t}\underline{\psi} = \begin{pmatrix} i\frac{\partial^2}{\partial y^2} \pm \frac{\partial}{2\alpha^2}|u|^2 - \frac{i}{2}w + \frac{i}{2}v & \frac{i}{\alpha}u\frac{\partial}{\partial y} + \frac{i}{2\alpha^2}(u_x + \alpha u_y) \\ \pm\frac{i}{2}\bar{u}\frac{\partial}{\partial y} \mp \frac{i}{2\alpha^2}(\bar{u}_x - \alpha\bar{u}_y) & -i\frac{\partial^2}{\partial y^2} \mp \frac{i}{2\alpha^2}|u|^2 + \frac{i}{2}w + \frac{i}{2}v \end{pmatrix}\underline{\psi} \tag{52}$$

have as their compatibility conditions

$$\left.\begin{aligned} iu_t + \frac{1}{\alpha^2}(u_{xx} + \alpha^2 u_{yy}) &= \mp\frac{1}{\alpha^2}|u|^2u + wu \\ w_x + \alpha v_y &= \pm\frac{2}{\alpha^2}(|u|^2)_x \\ v_x + \alpha w_y &= 0 \end{aligned}\right\} \tag{53}$$

which are the Davey-Stewartson equations when $\alpha = i$ (DSI) or 1 (DSII). In this case $N = 2$,

$$A(\zeta) = \begin{pmatrix} \alpha\zeta & 0 \\ 0 & -\alpha\zeta \end{pmatrix}, \tag{54}$$

$$B(x,y,t) = \begin{pmatrix} 0 & u(x,y,t) \\ \pm\bar{u}(x,y,t) & 0 \end{pmatrix}, \tag{55}$$

$$L(\zeta) = \begin{pmatrix} i\zeta^2 & 0 \\ 0 & -i\zeta^2 \end{pmatrix} . \tag{56}$$

In the case $\alpha = i$ the delta functions which appear enable the integration in (14) to be easily carried out to give

$$\frac{\partial}{\partial\bar{\zeta}} \underline{\phi}_1(x,y,\zeta) = R_1(\zeta)\ \underline{\phi}_2(x,y,\bar{\zeta}) \tag{57}$$

and

$$\frac{\partial}{\partial\bar{\zeta}} \underline{\phi}_2(x,y,\zeta) = R_2(\zeta)\ \underline{\phi}_1(x,y,\bar{\zeta}) \tag{58}$$

where

$$R_1(\zeta) = \mp \frac{i}{2\pi} \int_{-\infty}^{\infty}\int_{-\infty}^{\infty} \exp(\bar{\zeta}(ix' - iy')\}\ \bar{u}(x',y') \times \phi_{1,1}(x'.y',\zeta)\ dx'\ dy' \tag{59}$$

and

$$R_2(\zeta) = \frac{i}{2\pi} \int_{-\infty}^{\infty}\int_{-\infty}^{\infty} \exp(-\bar{\zeta}(ix' - iy')\}\ \bar{u}(x',y') \times \phi_{2,2}(x',y',\zeta)\ dx'\ dy'. \tag{60}$$

There are no finite singularities and while it seems possible that there are values of ζ for which the homogeneous Fredholm equation has non-trivial solutions these will not satisfy the conditions in Section (2).

When $\alpha = 1$ the Jost functions are analytic functions of ζ except on the imaginary axis where

$$\underline{\phi}_2(x,y,i\xi+0) - \underline{\phi}_2(x,y,i\xi-0) = \int_{-\infty}^{\infty} Q_{21}(b,i\xi)\ \underline{\phi}_1(x,y,i\xi+ib-0)\ db \tag{61}$$

with

$$Q_{21}(b,i\xi) = -\frac{1}{2\pi} \int_{-\infty}^{\infty}\int_{-\infty}^{\infty} \exp\{-i(\xi+b)(x' + y')\}u(x',y') \times \Phi_{2,2}(x',y',i\xi-0)\ dx'\ dy'. \tag{62}$$

For the jump in $\underline{\phi}_1(x,y,\zeta)$ interchanging the roles of the two sides simplifies the algebra

$$\underline{\phi}_1(x,y,i\xi+0) - \underline{\phi}_1(x,y,i\xi-0) = \int_{-\infty}^{\infty} Q_{12}(b,i\xi)\ \underline{\phi}_2(x,y,i\xi+ib+0)\ db \tag{63}$$

where

$$Q_{12}(b,i\xi) = \pm \frac{1}{2\pi} \int_{-\infty}^{\infty}\int_{-\infty}^{\infty} \exp\{-i(\xi+b)(x' + y')\}\ \bar{u}(x',y')$$
$$\times\ \Phi_{1,1}(x',y',i\xi+0)\ dx'\ dy'. \tag{64}$$

References

[1] Gardner, C.S., Green, J.M., Kruskal, M.D. & Miura, R.M. (1967). *Phys. Rev. Lett.* **19**, 1095.
[2] Gardner, C.S., Green, J.M., Kruskal, M.D. & Miura, R.M. (1974). *Comm. Pure Appl. Math.* **27**, 97.
[3] Ablowitz, M.J. & Segur, H. (1981). *Solitons and the Inverse Scattering Transform.* SIAM Stud. Appl. Math.
[4] Calogero, F. & Degasperis, A. (1982). *Spectral Transform and Solitons 1.* North Holland: Amsterdam.
[5] Ablowitz, M.J. & Nachman, A.I. (1986). *Physica* **18D**, 223.
[6] Beals, R. & Coifman, R.R. (1980-82). *Scattering, Transformations Spectrales and Equations d'Evolution Nonlineare I, II et Seminaire Goulauoic-Meyer-Schwartz 1980-81, exp. 22, 1981-82, exp. 21.* Ecole Polytechnique: Palaiseau.
[7] Caudrey, P.J. (1984). *Wave Phenomena: Modern Theory and Applications,* p. 221, Rogers, C. & Moodie, T.B. (eds.). Elsevier Science Publishers B.V.: North Holland.
[8] Case, K.M. & Kac, M. (1973). *J. Math Phys.* **14**, 594.
[9] Manakov, S.V. (1974). *Zh. Eksp. Teor. Fiz.* **67**, 543.
[10] Bruschi, M., Manakov, S.V., Ragnisco, O. & Levi, D. (1980). *J. Math. Phys.* **21**, 2749.
[11] Caudrey, P.J. *Spectral Transforms,* Soliton Theory: A Survey of Results, Fordy, A.P. (ed.). Manchester University Press, in press.
[12] Caudrey, P.J. *Two-Dimensional Spectral Transforms,* Soliton Theory: A Survey of Results, Fordy, A.P. (ed.). Manchester University Press, 1990.

A. Degasperis

6. INTEGRO-DIFFERENTIAL EVOLUTION EQUATIONS ASSOCIATED WITH A RIEMANN-HILBERT SPECTRAL PROBLEM

We consider a spectral problem, which is a 2x2 matrix Riemann-Hilbert problem; the associated evolution equations, which are integral in the space-variable and differential in the time-variable, describe nonlinear wave propagation with nonpolynomial dispersion law. Here we give an operater Λ, which allows us to introduce both the class of evolution equations and the corresponding Bäcklund transformations. In addition, we solve, by linearization, the elementary Bäcklund transformations, and provide an explicit algebraic method of computing the multi-soliton solutions.

1 Introduction

The aim of this section is to introduce the problem and the notation. Here, and in the following, the discussion is confined to the presentation of the main results. Details and proofs are given in the papers quoted below. The starting point is the Riemann-Hilbert (RH) problem of determining a two-dimensional vector $\psi(x,z)$, which is analytic in the strip

$$0 < \mathrm{Im}x < \eta \quad , \quad -\infty<\mathrm{Re}x<+\infty \; , \tag{1}$$

bounded there everywhere (icnluding the point at infinity and the straight lines $\mathrm{Im}x = 0$ and $\mathrm{Im}x = \eta$), and whose boundary values

$$\psi^{(+)}(x,z) = \psi(x+i\varepsilon,z),\; \psi^{(-)}(x,z) = \psi(x+i\eta - i\varepsilon,z),\; \varepsilon\to 0^+,\; \mathrm{Im}x = 0, \tag{2}$$

are related to each other by the algebraic relation

$$\psi^{(-)}(x,z) = [z\sigma + u(x) + \mathbf{V}(x)]\psi^{(+)}(x,z), \qquad \mathrm{Im}x = 0 \tag{3}$$

Here the variable z appears in (3) as a (spectral) parameter, σ is the Pauli matrix

$$\sigma = \begin{pmatrix} 1 & 0 \\ 0 & -1 \end{pmatrix}, \tag{4}$$

$\mathbf{V}(x)$ is the off-diagonal matrix

$$\mathbf{V}(x) = \begin{pmatrix} 0 & v^{(-)}(x) \\ v^{(+)}(x) & 0 \end{pmatrix}, \tag{5}$$

and the function u(x) is related to it via the definition

$$u(x) = [1 + v^{(+)}(x)v^{(-)}(x)]^{1/2}. \tag{6}$$

We note that these formulae define, in fact a family of RH problems depending on the positive real parameter η, which is the width of the strip of analyticity (1). This remark is of interest because, in the limit $\eta \to 0$, the RH problem goes into the generalized Zakharov-Shabat first order differential equation ($f_x \equiv \partial f/\partial x$)

$$\psi_x(x,k) = [- ik\sigma + \begin{pmatrix} 0 & q^{(-)}(x) \\ q^{(+)}(x) & 0 \end{pmatrix}] \psi(x,k) , \quad -\infty < x < +\infty \tag{7}$$

In order to show this, and also for future reference, we give the equation (3) the formal version

$$E\psi(x,z) = [z\sigma + u(x) + \mathbf{V}(x)]\psi(x,z) , \quad \text{Im}x = 0, \tag{8}$$

where $\psi = \psi^{(+)}$ and E is the shift operator

$$E = \exp(i\eta\partial/\partial x); \tag{9}$$

then, it is easily seen that setting $z = \eta k$ and

$$\mathbf{V}(x) = i\eta\begin{pmatrix} 0 & q^{(-)}(x) \\ q^{(+)}(x) & 0 \end{pmatrix} \tag{10}$$

yields, as $\eta \to 0$, the differential equation (7). It should be pointed out, moreover, that the x-dependent matrix multiplying ψ in the r.h.s. of (8) is not a generic 2x2 matrix, as it depends only on the two complex functions $v^{(+)}(x)$ and $v^{(-)}(x)$, see (5) and (6), in a rather peculiar way. This structure stems from the condition that the determinant

$$\det [z\sigma + u(x) + \mathbf{V}(x)] = 1 - z^2 \tag{11}$$

be x-independent, this implying that the total index of the RH problem (3) is vanishing.

The equation (3), or, equivalently, (8), has been introduced [1,2] with the aim of investigating the associated evolution equations, with the asymptotic condition that $v^{(+)}(x)$ and $v^{(-)}(x)$ vanish, as $|x| \to \infty$, sufficiently fast. For this class of matrices $\mathbf{V}$, one may define a spectral problem, as well as a corresponding inverse problem. We do not deal here with these interesting problems, which are discussed in [3], but we merely note that, for finite η, these are similar to those associated with a differential operator, with a continuum and a discrete spectrum [4,5]. Thus, for instance, the continuum spectrum of (3) is $-1 < z < 1$, where the Jost matrix solution $\mathbf{F}(x,z)$ is defined by the asymptotic condition

$$\mathbf{F}(x,z) \to \exp[-ix\mathbf{G}(z)], \quad x\to-\infty \tag{12}$$

where $\mathbf{G}(z)$ is the diagonal matrix satisfying the equation

$$\exp[\eta\mathbf{G}(z)] = 1 + z\sigma; \tag{13}$$

on the other hand, the transition matrix $\mathbf{S}(z)$, as in the standard scattering theory, is given by the asymptotic behaviour

$$\mathbf{F}(x,z) \to \exp[-ix\mathbf{G}(z)]\mathbf{S}(z), \quad x\to+\infty. \tag{14}$$

2 The operator Λ and Bäcklund transformations

The basic problem we are concerned with here is the following. Suppose the matrix $\mathbf{V}^{(2)}(x)$ is given, and its corresponding Jost solution $\mathbf{F}^{(2)}(x,z)$ is known, we wish to find a transformation, $\mathbf{V}^{(2)} \to \mathbf{V}^{(1)}$, to a new matrix $\mathbf{V}^{(1)}$, which is induced by a linear transformation of the Jost solution, namely

$$\mathbf{F}^{(1)}(x,z) = \mathbf{T}(x,z)\mathbf{F}^{(2)}(x,z)\mathbf{N}(z), \tag{15}$$

where the x-independent matrix $\mathbf{N}$ is a normalization factor. Moreover, the transformation matrix $\mathbf{T}(x,z)$ is required to be polynomial in the spectral variable z, and to go, as $x\to\pm\infty$, to a diagonal matrix. This last condition clearly implies, together with the asymptotic behaviour (12), that

$$\lim_{x\to-\infty} \mathbf{T}(x,z) = [\mathbf{N}(z)]^{-1}. \tag{16}$$

Thus, the two corresponding RH problems of type (3), or in the equivalent form (8), are

$$\mathbf{E}\, \psi^{(j)} = [z\sigma + u^{(j)} + \mathbf{V}^{(j)}]\, \psi^{(j)}, \qquad j = 1,2, \tag{17}$$

where, with obvious notation, $u^{(j)}$ is given in terms of $\mathbf{V}^{(j)}$ by (6) and (5). Then, in analogy with the theory of differential equations [5], we refer to the linear transformation (15) as a Darboux transformation, and to its corresponding mapping of $\mathbf{V}^{(2)}$

into $\mathbf{V}^{(1)}$ as Bäcklund transformation. The way to obtain the relevant formulae, with technical differences, is similar to that which applies to the differential equation (7) [6]; therefore, the derivation of the following expression is omitted here, and given elsewhere [7]. For any two given polynomials g(z) and h(z),

$$g(z) = \sum_{m=0}^{M} g_m z^m , \quad h(z) = \sum_{m=0}^{M} h_m z^m , \tag{18}$$

we introduce the Darboux matrix, which defines the transformation (15), via the expression

$$\mathbf{T}(x,z) = g(z) + h(z)\sigma + \frac{1}{2}(1 - i\mathbb{D})\mathbf{G}_a - \frac{1}{2}(1 + i\mathcal{J})\mathbf{G}_c, \tag{19}$$

where $\mathbf{G}_a$ is the off-diagonal matrix

$$\mathbf{G}_a = h^{\#}(z,\Lambda)(\mathbf{V}^{(1)} + \mathbf{V}^{(2)}) + g^{\#}(z,\Lambda)[(\mathbf{V}^{(1)} - \mathbf{V}^{(2)})\sigma], \tag{20}$$

and $\mathbf{G}_c$ is the diagonal matrix

$$\mathbf{G}_c = (\mathbf{V}^{(1)}\mathbf{G}_a - \mathbf{G}_a \mathbf{V}^{(2)})/(u^{(1)}+u^{(2)}). \tag{21}$$

Moreover, we use the following notation and definitions:

$$h^{\#}(z,z') = [h(z) - h(z')]/(z-z') , \quad g^{\#}(z,z') = [g(z)-g(z')]/(z-z'), \tag{22}$$

$$\Lambda\cdot = \frac{i}{2}\sigma\{(u^{(1)}+u^{(2)})\mathbb{D}\cdot + \mathbf{V}^{(1)} \mathcal{J}[(\mathbf{V}^{(1)}\cdot - \cdot \mathbf{V}^{(2)})/(u^{(1)}+u^{(2)})] -$$
$$-\{\mathcal{J}[\mathbf{V}^{(1)}\cdot - \cdot\mathbf{V}^{(2)})/(u^{(1)}+u^{(2)})]\}\mathbf{V}^{(2)}\} , \tag{23}$$

$$\mathbb{D}\, f(x) = -(1/\eta)P \int_{-\infty}^{+\infty} dy \,\{\sinh [\pi(x-y)/\eta]\}^{-1} f(y), \tag{24}$$

$$\mathcal{J}\, f(x) = (1/\eta)P \int_{-\infty}^{+\infty} dy \,\mathrm{cotgh}\, [\pi(x-y)/\eta]f(y) - (1/\eta) \int_{-\infty}^{+\infty} dy\, f(y). \tag{25}$$

The Bäcklund transformation corresponding to the matrix (19), via the relation (15), is then expressed by the nonlinear integral equation

$$h(\Lambda)(\mathbf{V}^{(1)} + \mathbf{V}^{(2)}) + g(\Lambda)[(\mathbf{V}^{(1)} - \mathbf{V}^{(2)})\sigma] = 0, \tag{26}$$

which, of course, for a given matrix $\mathbf{V}^{(2)}$, should be solved to yield $\mathbf{V}^{(1)}$. The basic object in this approach is the operator Λ, see (23), whose dependence on $\mathbf{V}^{(1)}$ and $\mathbf{V}^{(2)}$, is responsible for the nonlinearity of the equation (26). Its expression (23) is given in terms of the two convolution integral operators $\mathbb{D}$ and $\mathcal{I}$, which are the inverse of each other, namely

$$\mathbb{D}\,\mathcal{I} = \mathcal{I}\,\mathbb{D} = 1. \tag{27}$$

In fact, $\mathcal{I}$ is one of the infinitely many inverses of $\mathbb{D}$, and its expression (25) has been chosen so as to satisfy the asymptotic condition

$$\lim_{x\to+\infty} \mathcal{I}\, f(x) = 0. \tag{28}$$

In this respect, it should be noticed that the present spectral theory is a nonlocal extension of the spectral transform approach based on the (local) differential equation (7), the parameter η measuring the extent of nonlocality. Thus the operators $\mathbb{D}$ and $\mathcal{I}$ are the nonlocal versions of the derivative and integral operations, respectively. This assertion is motivated by the following formulae, which are easily proved:

$$\mathbb{D} = \mathrm{tg}(\tfrac{1}{2}\eta\ \partial_x), \tag{29a}$$

$$\mathbb{D}\exp(ikx) = i\,\mathrm{tgh}(\tfrac{1}{2}\eta k)\exp(ikx), \tag{29b}$$

$$\mathbb{D} \sim \tfrac{1}{2}\eta\ \partial_x \text{ as } \eta \to 0,\ \mathbb{D} \to H \text{ as } \eta \to \infty, \tag{29c}$$

$$\mathbb{D}(fg) = f\mathbb{D}g + g\mathbb{D}f + \mathbb{D}[(\mathbb{D}f)(\mathbb{D}g)]. \tag{29d}$$

Therefore, in evolution equations describing wave propagation (see below), the nonlocal effects introduced by the operator $\mathbb{D}$ are relevant for wave-lengths which are of the order of, or smaller than, η (see (29b)). Moreover, by varying the parameter η, the operator $\mathbb{D}$ interpolates between the derivative operation and the Hilbert transform opertor H, as shown by (29c). Finally, the identity (29d) displays the generalization to $\mathbb{D}$ of the Leibnitz rule of differentiation.

The Bäcklund equation (26), with two generic polynomials g(z) and h(z), can hardly be solved by direct computation. Of course, a way of solution is given by the spectral transform method, provided one can solve the corresponding Bäcklund transformation for the spectral transform of $\mathbf{V}^{(1)}$ in terms of that of $\mathbf{V}^{(2)}$. Since we are not dealing here with the spectral method, we content ourselves with showing that, indeed, the Bäcklund transformation (26) induces a simple, algebraic relationship between the reflection coefficients on the continuum spectrum. By definition, the off-diagonal reflection matrix $\alpha(z)$ and the diagonal transmission matrix $\beta(z)$ are uniquely defined in terms of the transition matrix $\mathbf{S}(z)$, see (14), by the formulae

$$\mathbf{S}(z) = [1 + \alpha(z)]\beta^{-1}(z), \tag{30a}$$

$$\alpha(z) = \begin{pmatrix} 0 & \alpha^{(-)}(z) \\ \alpha^{(+)}(z) & 0 \end{pmatrix}, \quad \beta(z) = \begin{pmatrix} \beta^{(+)}(z) & 0 \\ 0 & \beta^{(-)}(z) \end{pmatrix}. \tag{30b}$$

On the other hand, combining the asymptotic behaviour (14) with the Darboux transformation (15), with (16) and the limit

$$\lim_{x\to+\infty} \mathbf{T}(x,z) = g(z) + h(z)\sigma, \tag{31}$$

we obtain the relation

$$\alpha^{(1)}(z) = [g(z) + h(z)\sigma]\,[g(z) - h(z)\sigma]^{-1}\alpha^{(2)}(z), \tag{32}$$

which explicitly solves the Bäcklund transformation (26) for the reflection matrix corresponding to $\mathbf{V}^{(1)}(x)$.

3 A class of explicit solutions

In analogy with what has been already done [8,9] for the differential equation (7), we introduce here the two elementary Bäcklund transformations $\mathbf{B}^{(\pm)}$, which are given by the equation (26) with the special choices

$$\mathbf{B}^{(+)}: g(z) = z + b_+ - z_+\,, \quad h(z) = z - b_+ - z_+, \tag{33a}$$

$$\mathbf{B}^{(-)}: g(z) = z + b_- - z_-\,, \quad h(z) = -z + b_- + z_-, \tag{33b}$$

where $b_\pm$ and $z_\pm$ are complex parameters. Hereafter we discuss only the Bäcklund transformation $B^{(+)}$, since the discussion of $B^{(-)}$ is completely similar; moreover, in order to simplify the notation, we set $b_+=b$, $z_+=z_o$, $\mathbf{V}^{(2)}=\mathbf{V}$ (which is supposed to be known) and

$$v^{(+)(1)}(x) = -4ib\, f(x), \tag{34}$$

which defines the unknown function $f(x)$. Then, by using (33a), the equation (26) implies, for $f(x)$, the quadratic integral equation

$$[u + v^{(-)}\mathbb{D}f]\mathbb{D}f + v^{(-)}f^2 - iz_o f + \tfrac{1}{4}v^{(+)} = 0, \tag{35}$$

where u is given by (6). We note that this equation is the nonlocal extension of the Riccati equation, which occurs for the equation (7), and to which it reduces when $\eta\to0$. It is remarkable that, even in the present case, the nonlinear equation (35) can be linearized in the standard way. Namely, if $\psi\begin{pmatrix}\psi_1\\ \psi_2\end{pmatrix}$ is a vector solution of the RH problem (3), with $z=z_o$, then the solution $f(x)$, and therefore $v^{(+)(1)}(x)$, see (34), can be expressed in terms of the ratio

$$\varphi(x) = \psi_2(x)/\psi_1(x), \tag{36}$$

of the components of the vector ψ. In fact, we find the explicit algebraic expression

$$v^{(+)(1)} = -b(v^{(-)}\varphi^2 + 2u\varphi + v^{(+)})/(v^{(-)}\varphi + u + z_o); \tag{37a}$$

for the other function $v^{(-)(1)}$, it turns out convenient to solve first for $u^{(1)}$ (see (6)), i.e.

$$u^{(1)} = u + \frac{1}{2}v^{(-)}(v^{(-)}\varphi^2 + 2z_o\varphi - v^{(+)})/(v^{(-)}\varphi + u + z_o) +$$

$$+\frac{i}{2}(v^{(+)(1)}/b)\mathbb{D}\, v^{(-)}, \tag{37b}$$

and then to obtain $v^{(-)(1)}$ by using (6), namely

$$v^{(-)(1)} = (u^{(1)2}-1)/v^{(+)(1)}. \tag{37c}$$

Similar formulae hold, of course, for the elementary Bäcklund transformation $\mathbf{B}^{(-)}$ defined by (33b), and they read

$$v^{(-)(1)} = b(v^{(+)}\varphi^2 + 2u\varphi + v^{(-)})/(v^{(+)}\varphi + u - z_o), \tag{38a}$$

$$u^{(1)} = u + \frac{1}{2}v^{(+)}(v^{(+)}\phi^2 - 2z_o\varphi - v^{(-)})/(v^{(+)}\varphi + u - z_o) -$$

$$- \frac{i}{2}(v^{(-)(1)}/b)\mathbb{D}v^{(+)}, \tag{38b}$$

$$v^{(+)(1)} = (u^{(1)^2}-1)/v^{(-)(1)}, \tag{38c}$$

where, however, in this case, the function $\varphi(x)$ is the ratio

$$\varphi(x) = \psi_1(x)/\psi_2(x), \tag{39}$$

and $b=b_-$, $z_o =z_-$. These results provide a technique of construction of explicit solutions of the RH problem (3) from the knowledge of the solution of the same problem for a given matrix $\mathbf{V}(x)$, by repeatedly applying Bäcklund transformations of type $\mathbf{B}^{(+)}$ and $\mathbf{B}^{(-)}$. In fact, one can construct a two-dimensional lattice of solutions by acting $\mathbf{N}_+$ times with $\mathbf{B}^{(+)}$ and $\mathbf{N}_-$ times with $\mathbf{B}^{(-)}$, each elementary Bäcklund transformation entering two arbitrary complex parameters (namely $b_\pm$ and $z_\pm$, see (33)) in the resulting expressions. A simple instance of this construction obtains by starting with $\mathbf{V}=0$, and by applying to it first a Bäcklund transformation $\mathbf{B}^{(+)}$ and then a $\mathbf{B}^{(-)}$ transformation. The final expression we find is

$$v^{(\pm)} = (z_+ - z_-)g^{(\pm)}/\{2i\cosh[2p(x-\xi)]\}, \qquad (40a)$$

$$g^{(\pm)} = \{[2 \mp (z_+ + z_-) +$$

$$(z_+ - z_-)T]/[1 - z_+z_- + (z_+-z_-)T]\}\exp[\pm 2is(x-\lambda)], \qquad (40b)$$

$$T = \mathrm{tgh}[2p(x-\xi)], \qquad (40c)$$

where ξ and λ are arbitrary parameters, and s and p are related to z_+ and z_- by the equations

$$z_\pm = \mathrm{tgh}[\eta(s\pm ip)]. \qquad (40d)$$

We note that the parameters $b_\pm$, see (33), do not play an essential role, and, therefore, while b_+ is absorbed in the parameter p, we have set, for simplicity, $b_- = (z_+ - z_-)/(2i)$. We also notice that the form (40) clearly shows that, in the limit $\eta\to 0$, it reproduces the well-known one-soliton expression for the equation (7) (note that, in this limit, $g^{(\pm)}=2\exp[\pm 2is(x-\lambda)]$). If we use the short-hand notation (N_-,N_+) for the matrix $\mathbf{V}$, which obtains by applying N_- times the transformation $B^{(-)}$, and N_+ times the transformation $B^{(+)}$ on the vanishing matrix, $\mathbf{V}(x)=0$, then the formulae (40) give the solution (1,1), while the N-soliton solution lies at the point (N,N) of the lattice.

4 The evolution equations

Now we derive the class of evolution equations which are associated with RH problem (3). These equations result from the infinitesimal transformations of the class (26) around the unit transformation, which obtains for h=0 and g=1; in fact, by setting

$$\mathbf{V}^{(2)} = \mathbf{V}(x,t), \qquad \mathbf{V}^{(1)} = \mathbf{V}(x,t) + \delta t\mathbf{V}_t(x,t), \qquad (41a)$$

$$g(z) = 1, \qquad h(z) = \tfrac{1}{2}\gamma(z,t)\delta t, \qquad (41b)$$

in the equation (26), and keeping the first order term in δt, there follows that $\mathbf{V}(x,t)$ satisfies the evolution equation

$$\mathbf{V}_t = \sigma\gamma(\mathscr{L},t)\,\mathbf{V} = \sigma[\gamma_o(t) + \gamma_1(t)\,\mathscr{L} + \gamma_2(t)\,\mathscr{L}^2 + \ldots]\mathbf{V}, \quad (42)$$

where $\gamma(z,t)$ is a polynomial of z and $\mathscr{L}$ is the recursion operator, which is the operator Λ, see (23), in the special case $\mathbf{V}^{(1)} = \mathbf{V}^{(2)}$

= **V**, namely

$$\mathcal{L}\ \cdot = i\sigma\{u\mathbb{D}\ \cdot +\frac{1}{2}\ \mathbf{V}\ \mathcal{J}([\mathbf{V},\cdot]/u)\}. \tag{43}$$

If $\mathbf{V}(x,t)$ evolves according to the equation (42), then the results given in the previous sections provide also the evolution of both the Jost solution $\mathbf{F}$ of the RH problem (3), satisfying the boundary condition (12), and the reflection and transmission matrices $\alpha(z,t)$ and $\beta(z,t)$ defined by (30). Indeed, it is easily found that (for the notation, see (22))

$$\mathbf{F}_t = \frac{1}{2}\gamma(z,t)\ [\sigma,\mathbf{F}]+ \frac{1}{2}\{(1-i\mathbb{D})\gamma^{\#}(z,\mathcal{L},t)\mathbf{V} -$$

$$-\frac{1}{2}(1+i\mathcal{J})([\mathbf{V},\gamma^{\#}(z,\mathcal{L},t)\mathbf{V}]/u)\}\mathbf{F}, \tag{44}$$

which, together with the asymptotic behaviour (14), yields the evolution equations

$$\alpha_t = \gamma(z,t)\ \sigma\alpha\ , \qquad \beta_t = 0. \tag{45}$$

It should be pointed out that, in the process of deriving these results, one also shows that the operator $\mathcal{L}$ satisfies the remarkable identity

$$\int_{-\infty}^{+\infty} dx([\mathbf{V},\mathcal{L}^n\mathbf{V}]/u) = 0, \qquad n = 0,1,2,..., \tag{46}$$

which is, in fact, necessary to prove the time-independence of the transmission coefficients.

Three interesting instances of evolution equations which are in the class (42) are the following. If $\gamma(z,t) = icz$, and $v^{(+)} = v$, $v^{(-)} = \varepsilon v$, ε being a real parameter (see (5)), then there obtains an intermediate wave equation, which reads

$$v_t + c(1 + \varepsilon v^2)^{1/2}\mathbb{D}v = 0, \qquad v = v(x,t); \tag{47}$$

this equation is the nonlocal version of the trivial wave equation $\varphi_t + c_o\varphi_x = 0$, to which it reduces in the limit $\eta\to 0$, with $v = \eta\varphi, c = 2c_o/\eta$ and (29c). The equation (47) may be rewritten as

$$\theta_t + c\mathbb{D}\sin\theta = 0, \tag{48}$$

by the change of variable $v = i\ \sin\theta/\sqrt{\varepsilon}$; and in this form (with c=1), it goes, as $\eta\to\infty$ (see (29c)) into the sine-Hilbert equation

$$H\ \theta_t = \sin\theta, \tag{49}$$

whose special classes of solutions have been investigated by spectral techniques [10] and a direct linearization method [11]. If we next set $\gamma(z,t) = iz^2/c$, and impose the reduction $v^{(+)} = \psi$, $v^{(-)} = \varepsilon\psi^*$, ε being a real parameter, we obtain the intermediate nonlinear Schroedinger equation,

$$ic\psi_t + (1+\varepsilon|\psi|^2)^{1/2}\mathbb{D}[(1+\varepsilon|\psi|^2)^{1/2}\mathbb{D}\psi] + \frac{\varepsilon}{2}\psi(|\psi|^2 - |\mathbb{D}\psi|^2) = 0,$$
$$\psi=\psi(x,t), \tag{50}$$

which reduces to the well-known nonlinear Schroedinger equation $i\varphi_t + \varphi_{xx} + 2\varepsilon|\varphi|^2\varphi = 0$ in the limit $\eta\to 0$, with $\psi=\eta\varphi$ and $c = \eta^2/4$. The third equation is the one that reduces to the modified Korteweg-de Vries equation in the limit $\eta\to 0$; it obtains with $\gamma(z,t) = -iz^3/c$, $v^{(+)} = v, v^{(-)} = \varepsilon v$, ε being a real parameter, and reads

$$cv_t + [(1+\varepsilon v^2)^{1/2}\mathbb{D}]^3 v + \frac{\varepsilon}{2}(1+\varepsilon v^2)^{1/2}\mathbb{D}[v^3 - v(\mathbb{D}v)^2] = 0, \quad v=v(x,t). \tag{51}$$

Indeed, setting $c=(\eta/2)^3$, $v=\eta\varphi$, and letting η vanish, yields the mKdV equation $\varphi_t+\varphi_{xxx} + 6\varepsilon\varphi^2\varphi_x = 0$. As suggested by these examples, we note that if $\gamma(z,t)$, which characterizes the evolution equation (42), is polynomial in z, then the r.h.s. of (42) contains only the convolution operator $\mathbb{D}$, see (24), but not the integral operator $\mathcal{J}$, see (25). This property of the operator $\mathcal{L}$, see (43), is the finite-η counter-part of the well-known fact that the evolution equation (42) reduces, when η vanishes, to a partial differential equation.

We conclude with a few remarks on the soliton solution of the evolution equation (42) in the reduced case $v^{(-)} = \varepsilon v^{(+)} = \varepsilon v$, with ε and v real, as for the intermediate wave equation (47) and intermediate mKdV equation (51), and $v^{(-)} = \varepsilon v^{(+)*} = \varepsilon v^*$, as for the intermediate NLS equation (50). In these cases, the one-soliton expression (40) specializes to the forms

$$v = 2\sin(\eta p)\cosh[2p(x-y)]/\{\sinh^2[2p(x-y)]+\cos^2(\eta p)\} \tag{52}$$

if $\varepsilon=1$ (i.e. $v^{(+)} = v^{(-)} = v$), and

$$v = 2\sin(\eta p)\sinh[2p(x-y)]/\{\sinh^2[2p(x-y)] + \sin^2(\eta p)\} \tag{53}$$

if $\varepsilon = -1$ (i.e. $v^{(+)} = -v^{(-)} = v$), where y is a real parameter. Despite the fact that both expressions (52) and (53) are well-behaved bounded functions of x, only the expression (52), corresponding to $\varepsilon=1$, does solve the evolution equation for an appropriate linear dependence of y on time,

$$y=y_o + wt. \tag{54}$$

Thus, for instance, for the evolution equation (47) with $\varepsilon=1$, the velocity of the soliton is found to be

$$w = [c/(2p)]\mathrm{tg}(\eta p) \tag{55}$$

while, for the intermediate mKdV equation (51) with $\varepsilon=1$, the velocity is

$$w = [1/(2pc)]\mathrm{tg}^3(\eta p). \tag{56}$$

On the other hand, the expression (53) for the case $v^{(+)} = -v^{(-)}$, with (54), fails to be a solution of the evolution equation (f.i. of (47) and (51) with $\varepsilon=-1$) because the corresponding function $u(x,t) = [1 - v^2(x,t)]^{1/2}$, see (6), vanishes where v^2 takes its maximum value 1, and the vanishing of $u(x,t)$ makes the recurrence operator $\mathcal{L}$ singular, as is apparent from its expression (43). A similar situation occurs for the reduction $v^{(-)} = \varepsilon v^{(+)*}$, $v^{(+)} = v$: if $\varepsilon=1$ the one-soliton expression (40) yields

$$v = \{2\sin(\eta p)\cos^2(\eta p)\cosh(\eta s)/[\cos^2(\eta p) +\sinh^2(\eta s)]\}\{\cosh\zeta/[\sinh^2\zeta+\cos^2(\eta p)]\} .[1+i\,\mathrm{tg}(\eta p)\mathrm{tgh}(\eta s)\mathrm{tgh}\zeta]\exp(i\theta) \tag{57a}$$

where

$$\zeta = 2p(x-y), \qquad \theta = 2s(x-\lambda), \tag{57b}$$

y and λ being real parameters, while for $\varepsilon=-1$ we obtain

$$v = \{2\sin(\eta p)\cos^2(\eta p)\cosh(\eta s)/[\cos^2(\eta p) + +\sinh^2(\eta s)]\}\{\sinh\zeta/[\sinh^2\zeta+\sin^2(\eta p)]\} .[1 + i\,\mathrm{tg}(\eta p)\mathrm{tgh}(\eta s)\mathrm{cotgh}\zeta]\exp(i\theta) \tag{58}$$

with (57b). Again, only the expression (57), with

$$y = y_o + [t/(pc)]\sin(2\eta p)\sinh(2\eta s)/[\cosh(2\eta s) + \cos(2\eta p)]^2, \tag{59}$$

$$\lambda = \lambda_o + [t/(2sc)][\sinh^2(2\eta s) - \sin^2(2\eta p)]/[\cosh(2\eta s) + + \cos(2\eta p)]^2, \tag{60}$$

solves the intermediate NLS equation (50) with $\varepsilon=1$, while the other version of (50), i.e. for $\varepsilon=-1$, has no soliton solutions because of the existence of zeros of the correspondig function $u(x,t)=[1 - |v(x,t)|^2]^{1/2}$, see (6).

We finally note that these findings are consistent with the fact that, in the case $\varepsilon=-1$ of both reductions considered here, the spectral R-H problem (3) goes, as $\eta \to 0$, into a spectral problem associated with a hermitian differential operator, which, therefore, has no discrete (complex) eigenvalues. Indeed, it is easily seen that both expressions (53) and (58), in the limit $\eta \to 0$, develop a singularity at $x=y$.

References

[1] Degasperis, A. & Santini, P.M. (1983). *Phys. Lett.* **98A**, 240.
[2] Degasperis, A., Santini, P.M. & Ablowitz, M.J. (1985). *J. Math. Phys.* **26**, 2469.
[3] Degasperis, A., Santini, P.M. & Ablowitz, M.J. The inverse spectral problem associated with a Riemann-Hilbert problem, (in preparation).
[4] Ablowitz, M.J. & Segur, H. (1981). *Solitons and the Inverse Scattering Transform.* SIAM: Philadelphia.
[5] Calogero, F. & Degasperis, A. (1982). *Spectral Transform and Solitons I.* North Holland: Amsterdam; and *Spectal Transform and Solitons II,* (in preparation).
[6] Degasperis, A. (1984). Spectral transform approach to Backlund transformations. In *Nonlinear Equations in Classical and Quantum Field Theories,* p. 57, Sanchez, N. (ed.). Lecture Notes in Physics **226**. Springer.
[7] Degasperis, A. Backlund transformations of nonlinear integrodifferential evolution equations, (in preparation).
[8] Konopelchenko, B.G. (1982). *Phys. Lett.* **87A**, 445.
[9] Calogero, F. & Degasperis, A. (1984). *Physica* **14D**, 103.
[10] Santini, P.M., Ablowitz, M.J. & Fokas, A.S. (1987). *J. Math. Phys.* **28**, 2310.
[11] Matsuno, Y. (1987). *J. Phys. A: Math. Gen.* **20**, 3587.

R. Carroll

7. INVERSE SCATTERING AND FORCED NONLINEAR SYSTEMS

1 Introduction

We will deal mainly with forced nonlinear Schrödinger (NLS) equations (*) $iq_t = q_{xx} \pm 2|^2 q$ with $q(x,0) = q_0(x)$ and $q(0,t) = Q(t)$. Our intent is to treat such problems via the inverse scattering transform (IST) in order better to understand soliton dynamics etc. There have been two approaches here; one following D. Kaup and collaborators [11-15] and more recently another technique due to Fokas [8]. We have sketched some of our ideas in [3,6], following the basic formulation of Kaup and report here more fully on these developments. We will comment also on the results of Fokas [8], based on a half line approach, which lead to a nonlinear singular integro-differential equation for the scattering data (cf. also [4,5] for a half line IST); our results also lead to a system of integro-differential equations for the spectral data. The forced problem for the Toda lattice for example seems to be fairly well understood now (cf. [11-15]) but a general theory for NLS and other AKNS systems is still not complete.

2 Basic information

The AKNS generalization of the Zakharov-Shabat systems involves

$$v_{1x} + i\zeta v_1 = qv_2;\quad v_{2x} - i\zeta v_2 = rv_1;\quad v_{1t} = Av_1 + Bv_2;$$

$$v_{2t} = Cv_1 - Av_2;\quad A_x = qC-rB;\quad B_x + 2i\zeta B = q_t - 2Aq; \qquad (1)$$

$$C_x - 2i\zeta C = r_t + 2Ar.$$

For NLS one takes $r = \mp \bar{q}$ and $A = irq+2i\zeta^2$, $B = -iq_x-2\zeta q$, and $C = ir_x-2\zeta r$ leading to (*). We will concentrate on the case $r = \bar{q}$ and $iq_t = q_{xx}-2|q|^2 q$ for various reasons.

One defines generalized eigen-functions of the first two

equations in (1) via $\varphi \sim \binom{1}{0}e^{-i\zeta x}$ and $\hat{\varphi} \sim \binom{0}{-1}e^{i\zeta x}$ $(x \to -\infty)$; $\psi \sim \binom{0}{1}e^{+i\zeta k} = Y_2$ and $\hat{\psi} \sim \binom{1}{0}e^{-i\zeta x} = Y_1$ $(x \to \infty)$. Then $\varphi = a\hat{\psi}+b\psi$, $\hat{\varphi} = -\hat{a}\psi+\hat{b}\hat{\psi}$, $\psi = \hat{b}\varphi-a\hat{\varphi}$, and $\hat{\psi} = b\hat{\varphi} + \hat{a}\varphi$ where a,b, etc. are spectral quantities with $a\hat{a}+b\hat{b} = 1$ (cf. [1,4]). We take $q = r = 0$ for $x < 0$ and a simple argument yields $\psi(0,t) = \binom{\hat{b}(\zeta,t)}{a(\zeta,t)}$. We assume here that $q,r \to 0$ as $x \to \infty$ as rapidly as needed (take even compact support) and set $A_+ = \lim A(x,t,\zeta)$ as $x \to \infty$. It follows that $\psi_t = \begin{pmatrix} A+A_+ & B \\ C & A_+-A \end{pmatrix}\psi$ and taking $P(t) = q_x(0,t)$ (which is not determined by $Q(t) = q(o,t)$) we obtain for $r = \bar{q}$

$$\hat{b}_t = (i|Q|^2+4i\zeta^2)\hat{b} - (iP+2\zeta Q)a;$$
$$a_t = (i\bar{P}-2\zeta\bar{Q})\hat{b} - i|Q|^2a. \qquad (2)$$

Theorem 1. The time evolution of spectral data for the forced NLS with $r = \bar{q}$ is given by the over-determined system (2).

Remark 1. The overdetermination comes from the need to prescribe P and this is connected to the boundary condition $q \to 0$ at ∞. Note for $iq_t = q_{xx} - 2|q|^2q$ with $q \to 0$ suitably as $x \to \infty$ one has, setting $E(t) = \frac{1}{2}\int_0^t ||q(t)||^2 d\tau$ with $||q||^2 = (q, q) = \int_0^\infty |q|^2dx$, $[E(t)-E(0)] = \text{Im}\bar{P}(t)Q(t)$. Hence if P or Q = 0, E(t) = E(0), and this implies unique solutions of the quarter plane problem with $q(x,0) = q_0$ given, along with either $q(0, t)$ or $q_x(0, t)$. In terms of existence consider the linear problem $iq_t = q_{xx}$; a primitive calculation for $q \to 0$ suitably as $x \to \infty$ gives $(F_sq)_t = ik^2(F_sq) - ikQ(t)$ or $(F_cq)_t = ik^2(F_cq) + iP(t)$ ($F_s \sim$ Sine Transform $F_c \sim$ Cosine Transform) while for more general q use the Laplace transform $Lq = \tilde{q}$ to obtain $i\tilde{q}_t = s^2\tilde{q} - sQ - P$. Thus in fact depending on the boundary conditions at $x = \infty$ it may or may not be appropriate to prescribe P along with Q. We note that a similar calculation in [13] using the Fourier transform $F_q = \hat{q}$ leads to $i\hat{q}_t - k^2\hat{q} - P - ikQ = 0$ (here with $q \to 0$ at ∞ so P is overdetermining). Formally and heuristically this leads to a determination of P in forms of Q and q(x,0) but the calculation in [13] should be modified a bit. Thus one obtains (**) $\hat{q}(k,t) = \hat{q}_0(k)\exp(-ik^2t) - i\int_0^t[P + ikQ](s)\exp[ik^2(s-t)]ds$ from which formally $P(t) = \frac{1}{2\pi}\lim_{x\to 0^+}\int_{-\infty}^{\infty} ik\hat{q}e^{ikx}dk$ (to set x = 0 as in [13] could be incorrect). This leads to $P(t) = \frac{1}{2\pi} \lim_{x\to 0^+} [\int_{-\infty}^{\infty} ike^{-ik^2t+ikx}\hat{q}_0(k)dk + \int_{-\infty}^{\infty} ke^{ikx}\int_0^t (P+ikQ)(s)\cdot e^{ik^2(s-t)}dsdk]$ which is an integral equation for P with $\hat{q}_0$ and Q known. Our procedures lead in fact to a similar, but more complicated, determination of P via an integral equation (which we rephrased in a fixed point context in [3,6]). Thus for $iq_t = q_{xx}-2|q|^2q$ we will pick P, Q, solve (2), and recover by inverse scattering an output $\hat{P}$, $\hat{Q}$. For Q given the correct P will correspond to a fixed point of M: (P, Q) → (P, Q) and this can be represented as the solution of an integral equation. In fact we can also express our results in the form of a system of nonlinear integro-differential equations for the spectral data $\hat{b}$ and a; this is in the same spirit as [8], and surely related, but of a different form.

Thus (cf. [1,3,4,6]) one has $\psi = \binom{0}{1}\exp(i\zeta x) + \int_x^{\infty} K(x,s)\exp(i\zeta s)ds$ and $\hat{\psi} = \binom{1}{0}e^{-i\zeta x} + \int_x^{\infty} \hat{K}(x,s)e^{-i\zeta s}ds$. Let C (resp. $\hat{C}$) be suitable contours relative to the zeros of a (resp. $\hat{a}$) and operate on $\hat{\psi}$, ψ resp. by $\frac{1}{2\pi}\int_C e^{i\zeta y}d\zeta$ and $\frac{1}{2\pi}\int_{\hat{C}} e^{-i\zeta y}d\zeta$ for y > x. This gives spectral integrals for the Marcenko kernels which we use to recover q,r via $q = -2K_1(x,x)$ and $r = -2\hat{K}_2(x,x)$ (t is suppressed at times). This approach seems simplest and is consistent with the standard Marcenko equations (cf. also [10]) but we remark that there are also various other equivalent formulas for recovery of q from spectral data via trace formulas or square eigenfunction expansions etc. (cf. [1,4,16,17]). Thus, e.g. for $r = \bar{q}$ we take $C = \hat{C} = (-\infty,\infty)$ and obtain $q(x) = \lim_{y\to x} \frac{-1}{\pi} \int_{-\infty}^{\infty} \psi_1(\zeta,x)e^{-i\zeta y}d\zeta$ and making adjustments for δ functions arising in $q_x(x,t)$ at x = 0 we obtain (cf. [3,6] where the formulas for $r = -\bar{q}$ are also given)

Theorem 2. For $r = \bar{q}$ one has $\hat{Q}(t) = -\frac{1}{\pi} \lim_{x\to 0} \int_{-\infty}^{\infty} \hat{b}(\zeta,t)e^{-2i\zeta x}d\zeta$ and $\hat{P}(t) = +\frac{1}{\pi} \lim_{x\to 0} \int_{-\infty}^{\infty} [Q(1-a)+2i\zeta\hat{b}-\hat{Q}]e^{-2i\zeta x}d\zeta$.

3 Fixed point ideas and integral equations

If one formally integrates in (2) there results, e.g. $\hat{b} = B - \int_0^t \hat{b}md\xi$ (cf. [3,6] for more details) and in order to control the growth of $\hat{b}$ (and a) as $|\zeta| \to \infty$ one requires $P,Q \in C^1$ (or H^1) so that after an integration by parts (***) $B = \hat{b}_0 e^{4i\zeta^2 t + i\int_0^t |Q|^2 d\tau} - (1/4i\zeta^2) a_0 e^{4i\zeta^2 t + i\int_0^t |Q|^2 d\tau} [(iP+2\zeta Q)(0) - (iP+2\zeta Q)(t) e^{-2i\int_0^t |Q|^2 d\tau - 4i\zeta^2 t} + \int_0^t (iP' + 2\zeta Q' - 2i|Q|^2 (iP+2\zeta Q) \cdot e^{-2i\int_0^s |Q|^2 d\tau - 4i\zeta^2 s} ds]$ and $m = [(i\bar{P} - 2\zeta\bar{Q})(\xi)/4i\zeta^2] \cdot e^{4i\zeta^2 t + i\int_0^t |Q|^2 d\tau + i\int_0^\xi |Q|^2 d\tau} [(iP+2\zeta Q)(\xi) e^{-2i\int_0^\xi |Q|^2 d\tau - 4i\zeta^2 \xi} - (iP+2\zeta Q)(t) e^{-4i\zeta^2 t} e^{-2i\int_0^t |Q|^2 d\tau} + \int_\xi^t \{iP' + 2\zeta Q' - 2i|Q|^2 (iP+2\zeta Q)\} \cdot e^{-2i\int_0^s |Q|^2 d\tau - 4i\zeta^2 s} ds$ ($\hat{b}_0$ and $a_0 \sim q(x,0) = q_0(x)$ - the corresponding formulas in [3] have a typographical omission). One shows (cf. [1]) that $\hat{b}$ and a will have now suitable growth as $|\zeta| \to \infty$ and in [3] we sketched heuristically how a sample fixed point theorem $Q \to \hat{Q}$ (with P fixed) could be phrased in H^1 with weak topologies. It is difficult however to get enough mileage from the Riemann-Lebesgue lemma to sustain this argument and we prefer to proceed now somewhat differently (the procedure of [3] having served as a guide). In this direction one assumes P and $Q \in C^\infty$ say and looks at asymptotic expansions as $|\zeta| \to \infty$ of the form $\hat{b} = \sum_1^N b_i/\zeta^i + \cdot e^{4i\zeta^2 t} \sum_1^N \tilde{b}_i/\zeta^i$ and $a = \sum_0^N a_i/\zeta^i + e^{4i\zeta^2 t} \sum_2^N \tilde{a}_i/\zeta^i$ ($D_t\hat{b}$ etc. can also be included). One assumes $a_0 = \sum_0^\infty a_0^i/\zeta^i$ ($a_0^0 = 1$) with $\hat{b}_0 = \sum_1^\infty b_0^i/\zeta^i$ and some calculation shows that in fact $\hat{b} = Q(t)/2i\zeta + O(1/\zeta^2)$ with $a = 1 + O(1/\zeta)$. Thus $2i\zeta\hat{b} - Qa = O(1/\zeta)$ and the δ corrections in $\hat{P}$ in Theorem 2 are not needed (thus simply set $\hat{Q} = Q$ in the $\hat{P}$ formula). Similarly one can compute $D_t\hat{b}$ etc. and e.g. $D_t\hat{Q}$ will have a (calculable) δ function term requiring adjustment, etc. Then one could envision some kind of fixed point theorem in a Frechet space of weighted C^∞ functions and derivatives for example. We do not pursue this here since matters are probably better expressed via integrodifferential equations as in Theorem 3.

First we write, e.g., $\hat{b} = B - M\hat{b}$ (cf. (***)) with $\hat{b} = (1+M)^{-1}B$ and similarly from (2) $a = A - N\hat{b}$ where $A = a_0 e^{-i\int_s^t |Q|^2 d\tau}$ and $Nf = \int_0^t (i\bar{P} - 2\zeta\bar{Q}) e^{-i\int_s^t |Q|^2 d\tau} f(s)ds = \int_0^t nfds$. We prescribe Q and ask for P such that $\hat{Q} = Q$ and $\hat{P} = P$ (cf. Theorem 2 where $\hat{Q} = Q$ in the $\hat{P}$ integral). Then (A) $P = \frac{1}{\pi} \lim_{x\to 0} \int_{-\infty}^{\infty} e^{-2i\zeta x} f(P,Q,\zeta,t)d\zeta$ where $f = (2i\zeta + QN)(1+M)^{-1}B - QA$ with M,N as indicated. Thus a fixed point theorem essentially is equivalent to solving (3). Somewhat more tractable should be the following. Take the equations (2) along with the recovery formulas in Theorem 2 (assuming suitable regularity). Then when everything fits together we must have

$$\hat{b}_t = (i|Q|^2 + 4i\zeta^2)\hat{b} - 2\zeta Qa - iaJ(a,\hat{b});\quad a_t = -i|Q|^2 a - 2\zeta\bar{Q}\hat{b} + i\hat{b}\bar{J}(a,\hat{b});\quad J(a,\hat{b}) = \frac{1}{\pi} \lim_{x\to 0} \int_{-\infty}^{\infty} [2i\zeta\hat{b} - Qa] e^{2i\zeta x} d\zeta. \qquad (3)$$

Theorem 3. The spectral data for the forced problem (*) with $r = \bar{q}$ can be determined by solving (3) with $a(\zeta,0)$, $\hat{b}(\zeta,0)$ given via $q(x,0)$. The fixed point idea for $P,Q \to \hat{P},\hat{Q}$ is embodied in the integral equation (A).

Remark 3. In [13] Kaup remarks (without proof) that via Riemman-Hilbert considerations one can obtain solutions of (2) analytic in appropriate regions. Actually our cut off condition $q(x,t) = 0$ for $x < 0$ implies via Theorem 2 that $B(t,y) = \int_{-\infty}^{\infty} \hat{b}(\zeta,t) e^{-2i\zeta y} d\zeta = 0$ for $y < 0$ (and $A(t,y) = \int_{-\infty}^{\infty} a(\zeta,t) e^{-2i\zeta y} d\zeta = 0$ for $y < 0$ by natural analyticity requirements). Hence one is led to a "new" kind of inverse problem for A,B. Thus one gives $A(0,y) = A_0(y)$ and $B(0,y) = B_0(y)$ and tries to solve (****) $B_t = i|Q|^2 B - 4iB_{yy} - iPA - 2iQA$ and $A_t = i\bar{P}B - 2i\bar{Q}B_y - i|Q|^2 A$ subject to $B(t,0) = -\pi Q(t)$ and $2B_y(t,0) + QA(t,0) = -\pi P(t)$ (with $A = B = 0$ for $y < 0$.

4 Examples and variations

The most elementary consideration of examples leads to some interesting situations. Thus consider $q = f(t)F(x)$ with $f = \exp(iwt)$ and e.g. $F = q_0 \exp(ikx)$ for $x \le R$ say (with $F \to 0$ smoothly for large x) so $w = k^2 + 2|q_0|^2$. Then $Q = q_0 e^{iwt}$ and $p = p_0 e^{iwt}$ where $p_0 = ikq_0$. If one sets $\hat{b} = Q\beta$ in (2) it follows that $\beta_t - i|Q|^2\beta = (4i\zeta^2 - Q'/Q)\beta - (\frac{iP}{Q} + 2\zeta)a$; $a_t + i|Q|^2 a = (i\bar{P}Q - 2\zeta|Q|^2)\beta$ where $|Q|^2 = |q_0|^2$, $Q'/Q = iw$, $P/Q = p_0/q_0 = ik$, and $\bar{P}Q =$

$\bar{p}_0 q_0 = -ik|q_0|^2$. Hence (4) has constant coefficients and e.g. if k = 0 ($p_0 = 0$)

$$\hat{b} = e^{i|q_0|^2 t + 2i\zeta^2 t}[\hat{b}_0 \operatorname{Cos} 2\zeta t Z + \frac{\sin 2\zeta t Z}{Z}(i\zeta\hat{b}_0 - q_0 a_0)] \tag{4}$$

where $Z = (\zeta^2 - |q_0|^2)^{1/2}$. This formula suggests working on a Riemann surface (note however that (4) is even in Z) and dealing with a finite density problem to the left (cf. [2,7,9]). The conceptual picture is not entirely clear but we can suggest a possible mathematical framework (following [2]) to deal with this situation in the context of forced problems. Thus set $Q = q_0 e^{iwt} = \ell$ (with no assumptions on P) and imagine that we are dealing with a situation (*****) $q(x,t) = \ell$ for $x \le 0$ (actually we only use lim $q(x,t) = \ell$ as $x \to -\infty$ for much of the development). If one stipulates (*****) in advance then $q_x = 0$ for $x < 0$ and either $P = 0$ or P is discontinuous at $x = 0$; further clarification is needed here and one imagines (*****) entering as a "forcing" condition in the mathematical analysis (e.g. in the G-L-M theory as in Remark 4a). In any event heuristically now we consider Jost solutions as $x \to -\infty$ (cf. [2]), $\varphi \sim X_1 = \binom{\ell}{i(\zeta-\lambda)}e^{-i\lambda x}$; $\varphi \sim X_2 = \binom{\ell}{i(\zeta+\lambda)}e^{i\lambda x}$ where $\lambda = (\zeta^2 - |\ell|^2)^{1/2}$, $|\ell|^2 = |q_0|^2$, and one works on a 2-sheeted Riemann surface R with the $\lambda > 0$ on the upper sheet R_1 and cuts along $(-\infty, -|\ell|]$ and $[|\ell|, \infty)$ in the ζ plane. One can set $\bar{\varphi} = \varphi$ for $\operatorname{Im}\lambda > 0$ and $\tilde{\phi} = \check{\varphi}$ for $\lambda < 0$ with $\hat{\varphi}(x,\zeta,\lambda) = \tilde{\phi}(x,\xi,-\lambda)$ and write e.g. $\bar{\varphi} = \alpha\hat{\psi} + \beta\psi$ with $\hat{\varphi} = \hat{\beta}\hat{\psi} - \hat{\alpha}\psi$ (so e.g. $\psi = (\hat{\alpha}\varphi - \hat{\beta}\tilde{\phi})/2i\lambda\ell$ on R, etc.) where ψ, $\hat{\psi}$ are defined as in Section 2. We note that the spectral data $\alpha, \beta, \hat{\alpha}, \hat{\beta}$ will be different here from before (and from (4)). Following [2] we write on R_1, $(\tilde{\phi}, \hat{\varphi}) = (\varphi\ \check{\varphi}) = (\hat{\psi}\ \psi)S$ with $S = \begin{pmatrix}\alpha & \hat{\beta} \\ \beta & -\hat{\alpha}\end{pmatrix}$ and let the time evolution be represented via $v_t = Mv$ with $X = (X_1 X_2)$ and $Y + (Y_1 Y_2) = \begin{pmatrix} e^{-i\zeta x} & 0 \\ 0 & e^{i\zeta x}\end{pmatrix}$ (to compare with [2] let $t \to -t$). One obtains $S_t = SX^{-1}(M_- X + X') - Y^{-1}(M_+ Y + Y')S$ where $M_+ = -2i\zeta^2\sigma_3$ and $M_- = \frac{1}{2i\lambda\ell}\begin{pmatrix} -i(2\zeta^2 + |\ell|^2 & 2\zeta\ell \\ 2\zeta\bar{\ell} & i(2\zeta^2 + |\ell|^2)\end{pmatrix}$ ($\sigma_3 = \begin{pmatrix}1 & 0 \\ 0 & -1\end{pmatrix}$). This leads to S =

$e^{-M_+t}S_0e^{Nt}$ where $N = i|\ell|^2I - 2i\zeta\lambda\sigma_3$ and in particular

Theorem 4. The time variation of spectral data based on $\bar{\varphi}$, $\hat{\varphi}$ etc. is given by $\alpha = \alpha_0\exp[i|\ell|^2t + 2i\zeta t(\zeta-\lambda)]$, $\hat{\beta} = \hat{\beta}_0\exp[i|\ell|^2t + 2i\zeta t(\zeta+\lambda)]$, $\hat{\alpha} = \hat{\alpha}_0\exp[i|\ell|^2t - 2i\zeta t(\zeta-\lambda)]$ and $\beta = \beta_0\exp[i|\ell|^2t - 2i\zeta t(\zeta+\lambda)]$.

Remark 4a. The time variation in theorem 4 is not much different from that in (4) but involves Z nontrivially ($\lambda \sim Z$). With some calculation one sees also that α has no zeros on $(-\infty, \infty)$ except possibly at branch points where we assume "good" behaviour. One can produce here also a heuristic variation of the Gelfand-Levitan-Marcenko (G-L-M) theory of [2]. Thus stipulating now $\varphi = X_1$ for $x \le 0$ etc. (i.e. $q(x,t) = \ell$ for $x \le 0$) one has e.g. $\varphi = X_1 + \int_0^x H(x,y)X_1(y)dy$ and for $y > x$ (*) $0 = J(x+y) + \int_0^x H(x,s)J(s+y)ds$ where $J = \begin{pmatrix} -i\ell J_4 \\ J_3+J_4' \end{pmatrix}$ with $J_3(z) = \frac{1}{4\pi i}\int_{-\infty}^{\infty} R_1(\lambda,\zeta)\exp(i\lambda z)d\lambda$ and $J_4(z) = \frac{1}{4\pi i}\int_{-\infty}^{\infty} R_2(\lambda,\zeta)e^{i\lambda z}d\lambda$; here $R_1 = (\beta/\hat{\alpha})(\lambda,\zeta) + (\beta/\hat{\alpha})(\lambda,-\zeta)$ and $R_2 = \frac{1}{\zeta}[(\beta/\hat{\alpha})(\lambda,\zeta) - (\beta/\hat{\alpha})(\lambda,-\zeta)]$. Standard calculation yields also $H(x,x) - \sigma_3H(x,x)\sigma_3 = \tilde{Q}(x) - \tilde{Q}_-$, $H_x + \sigma_3H_y\sigma_3 = \tilde{Q}(x)H(x,y) - \sigma_3H\sigma_3\tilde{Q}_-$, and $H(x,0) = 0$ where $\tilde{Q} = \begin{pmatrix} 0 & q \\ q & 0 \end{pmatrix}$ with e.g. $i\sigma_3(D_x-\tilde{Q}_-)X = \zeta X$. Such equations however need further investigation to determine solvability, meaning, etc.

Remark 4b. The equation of [8] for determination of spectral data is $(\hat{b}_t-4ik^2\hat{b})/4k = -Q(t) + \frac{1}{\pi}\int_{-\infty}^{\infty}\{[(\hat{b}(k')/8k')-(\hat{b}(k)/8k)]/(k'-k)\}D_tH\cdot\log(1+b\hat{b})dk'$ where H denotes the Hilbert transform. Thus our equations (3) do not seem excessively complicated at the present stage of the theory's development (here $\bar{b} \sim -\hat{b}$ and $\hat{b} \ne \hat{b}$ of Theorem 2).

References

[1] Ablowitz, M. & Segur, H. (1981). Solitons and IST. *SIAM*, Philadelphia.

[2] Boiti, M. & Pempinelli, F. (1982). *Nuovo Cimento* **69B**, 212.

[3] Carroll, R.W. (1989). In *Proc. Internatl. Conf. Diff. Eqs.*, p. 134. Ohio Univ. Press.

[4] Carroll, R. (1988). *Mathematical Physics*. North-Holland: Amsterdam.

[5] Carroll, R. (1987). *Applic. Anal.* **26**, 61.

[6] Carroll, R. (1989).In *Lect. Notes Math., 1394*, p. 11. Springer.

[7] Faddeev, L. & Takhtajan, L. (1987). *Hamiltonian methods in the theory of solitons.* Springer: New York.

[8] Fokas, A. (1989). *Physica* **35D**, 167.
[9] Gerdzikov, V. & Kulish, P. (1978). *Bulg. J. Phys.* **5**, 337.
[10] Gerdzikov, V. & Kulish, P. (1979). *Teor. Mat. Fizika* **39**, 69.

[11] Kaup, D. (1986). In *Lect. Appl. Math.*, p. 195; (1987). *Physica* **25D**, 361.
[12] Kaup, D. & Neuberger, H. (1984). *J. Math Phys.* **25**, 282.
[13] Kaup, D. (1984). In *Wave Phenomena*, p. 163. Elsevier; (1984). In *Advances in nonlinear waves*, p. 197. Pitman.
[14] Kaup, D. & Hansen, P. (1986). *Physica* **18D**, 77; (1987). **25D**, 369.
[15] Kaup, D. & Wycoff, D. (1989). *Studies Appl. Math.* **81**, 7.
[16] Newell, A. (1985). Solitons in mathematics and Physics. *SIAM*, Philadelphia.
[17] Novikov, S., Manakov, S., Pitaevskij, L. & Zakharov, V. (1984). *Theory of solitons.* Plenum: New York.

S. De Lillo

8. INITIAL VALUE PROBLEMS FOR THE BURGERS EQUATION

We report the solutions of the Cauchy problems on the line, semi-line and finite interval, for the Burgers equation.

1 Introduction

The Burgers equation

$$u_t = au_{xx} + 2bu_xu = (au_x + bu^2)_x \tag{1}$$

has extensively been studied for its integrability properties and for its applicative relevance as well; indeed it has been used as model equation to describe many realistic situations (see the monograph by Sachdev [5]). In this contribution we summarize the results concerning the initial/boundary value problems (Cauchy problems) on the semi-line and finite intervals with arbitrary boundary conditions [1]. For completeness we also report the solution of the Cauchy problem on the whole line, which can be easily obtained through a standard Hopf-Cole transformation. In our treatment we assume all the quantities to be real and set, without loss of generality,

$$a = b = 1. \tag{2}$$

2 Cauchy problem on the whole line

The initial datum reads as follows:

$$u(x,0) = u_0(x), \quad -\infty < x < \infty, \tag{3a}$$

$$\int_{-\infty}^{+\infty} |u_0(x) < \infty. \tag{3b}$$

The linearizing Hopf and Cole [2,4] transformation reads as follows:

$$v(x,t) = u(x,t) \exp [\int_{\infty}^{x} dx' u(x' t)] \tag{4a}$$

$$u(x,t) = v(x,t)/]1 + \int_{\infty}^{x} dx' v(x,t))] \tag{4b}$$

It is easily seen that this position implies:

$$v_t = v_{xx} \tag{5}$$

$$\int_{-\infty}^{+\infty} dx\ v(x,t) < \infty \tag{6a}$$

$$\int_{-\infty}^{+\infty} dx\ u(x,t) < \infty \tag{6b}$$

The solution of the Cauchy problem for (1) with (3) is given by the following 3 steps:

(i) Evaluate $v(x,0) = v_0(x)$ from $u(x,0) = u_0(x)$ via (4a) with t=0.

(ii) Evaluate $v(x,t)$ from $v(x,0)$ via (5). The explicit formulae read as follows:

$$\hat{v}_0(k) = \int_{-\infty}^{+\infty} dx\ v_0(x) \exp(-ikx) \tag{7a}$$

$$v(x,t) = (2\pi)^{-1} \int_{-\infty}^{+\infty} dk\ \hat{v}_0(k) \exp(ikx-k^2 t) \tag{7b}$$

(iii) Recover $u(x,t)$ from $v(x,t)$ via (4b).

3 Cauchy problem on the semi-line

The initial datum and boundary conditions read as follows:

$$u(x,0) = u_0(x) \quad , \quad 0 \le x < \infty \tag{8a}$$

$$\int_0^{\infty} dx |u_0(x)| < \infty \tag{8b}$$

$$u(0,t) = F(t) \quad , \quad 0 \le t < \infty \tag{9}$$

with

$$u_0(0) = F(0). \tag{10}$$

The generalized Hopf-Cole linearizing transformation [1] appropriate to solve this problem reads:

$$v(x,t) = C(t)u(x,t)\ \exp\Big[\int_0^x dx'u(x't)\Big], \qquad (11a)$$

$$u(x,t) = v(x,t)/\Big[C(t) + \int_0^x dx'v(x't)\Big], \qquad (11b)$$

$$C(0) = 1 \qquad (11c)$$

It is easily seen that this transformation implies the relations:

$$v_t = v_{xx} \qquad , \quad 0 \le x < \infty \qquad (12a)$$

$$v(0,t) = C(t)\ F(t)\ , \quad 0 \le t < \infty \qquad (12b)$$

$$\dot{C}(t) = v_x(0,t), \qquad (13)$$

$$\int_0^x dx'\,|v(x',t)| < \infty, \qquad (14a)$$

$$\int_0^x dx'\,|u(x',t)| < \infty. \qquad (14b)$$

To describe the solution of the Cauchy problem for (1) with (8)-(10) let us assume temporarily that C(t) is known. Then u(x,t) can be obtained by the following 3 steps:

(i) Evaluate v(x,0) from u(x,0) via (11a) (with t=0) and (11c).

(ii) Evaluate v(x,t) from v(x,0) via (12). The explicit formulae read as follows:

$$\hat{v}_0(k) = \int_0^x dx\ v_0(x)\sin(kx), \qquad (15a)$$

$$\hat{v}(k,t) = \hat{v}_0(k)\ \exp(-k^2t) + k\int_0^t dt'C(t')F(t')\exp[k^2(t'-t)], \qquad (15b)$$

$$v(x,t) = (2/\pi)\int_0^\infty dk\ \hat{v}(k,t)\sin(kx). \qquad (15c)$$

(iii) Recover u(x,t) from v(x,t) via (11b).

The remaining task is to compute C(t). The relevant equations are (13) with (11c); hence the main question is to find a suitable expression for $v_x(0,t)$ in terms of known quantities.

These can be done [1], obtaining the following integral equation for C(t):

$$C(t) = 1 + H(t) - \int_0^t dt' h(t-t')F(t')C(t') \tag{16}$$

where

$$H(t) = \int_0^\infty dx\ j(x,t)u_0(x)\exp[\int_0^x dy\ u_0(y)], \tag{17a}$$

$$j(x,t) = \pi^{-1/2}(x/2)\int_0^t dt'(t')^{-3/2}\exp[-x^2/4t'], \tag{17b}$$

$$h(t) = (\pi t)^{-1/2} \tag{17c}$$

The Volterra integral equation (16) determines C(t). In the special case when the boundary datum is time independent, F(t)=F(0)=F, the integral equation is solvable by quadratures:

$$C(t) = 1 + H(t) - F\int_0^t dt' R(t-t')[1 + H(t')], \tag{18a}$$

$$R(t) = (\pi t)^{-1/2} + \exp(F^2t)[-1 + \mathrm{erf}(Ft^{1/2})]. \tag{18b}$$

4 Cauchy problem on the finite interval

The initial/boundary value problem for the Burgers equation (1) on the finite interval [0,1] (that could be translated and rescaled to any other finite interval by trivial transformations) is characterized by the following initial and boundary data:

$$u(x,0) = u_0(x)\ ,\quad 0 \le x \le 1, \tag{19a}$$

$$u(0,t) = F_0(t)\ ,\quad 0 \le t < \infty, \tag{19b}$$

$$u(i,t) = F_1(t)\ ,\quad 0 \le t < \infty, \tag{19c}$$

with

$$u_0(0) = F_0(0); \tag{19d}$$

to solve this problem it is expedient to introduce, in addition to the quantity C(t)=v(0,t)/u(0,t), the quantity

$$B(t) = v(1,t)/u(1,t), \tag{20a}$$

$$B(t) = C(t)\exp[\int_0^1 dx\ u(x,t)], \tag{20b}$$

$$B(t) = C(t) + \int_0^1 dx\ v(x,t). \tag{20c}$$

The last two expressions follow of course from (11a) and (11b); the first of them implies the initial condition

$$B(0) = \exp\,[\int_0^1 dx\ u_0(x)]; \tag{21}$$

The second of them yields, via (13) and (12a), the equation

$$\dot{B}(t) = v_x(1,t). \tag{22}$$

Proceeding in close analogy to the treatment of the preceding Section 3, we now assume temporarily that the two quantities B(t) and C(t) are known. Then u(x,t) is evaluated by the following 3 steps:

(i) Determine the initial and boundary data in the interval [0,1] for v(x,t):

$$v(x,0) \equiv v_0(x) = u_o(x)\ \exp[\int_0^x dx' u.(x')],\ 0 \le x \le 1, \tag{23a}$$

$$v(0,t) \equiv f_0(t) = C(t)\ F_0(t)\ ,\quad 0 \le t < \infty, \tag{23b}$$

$$v(1,t) \equiv f_1(t) = C(t)\ F_1(t)\ ,\quad 0 \le t < \infty; \tag{23c}$$

(ii) Obtain v(x,t) by solving the initial/boundary value problem in the interval [0,1] for the linear ("heat") equation (12a). The explicit formulae read as follows:

$$v(x,t) = 2 \sum_{m=1}^{\infty} \sin(m\pi x)\ \hat{v}_m(t) \tag{24a}$$

$$\hat{v}_m(t) = \hat{v}_m(0)\ \operatorname{ex}(-m^2\pi^2 t)+m\pi\int_0^t dt'[(C(t')F_0(t')-(-)^m B(t')F_1(t')$$
$$\exp[m^2\pi^2(t-t')] \tag{24b}$$

$$\hat{v}_m(0) = \int_0^1 dx\ \sin|m\pi x)u_0(x)\ \exp[\int_0^x dx\ u_0(x')] \tag{24c}$$

(iii) Recover u(x,t) from v(x,t) via (11b).

The remaining task is to determine B(t) and C(t). This can be done by a procedure analogous to that described in the preceding section 3. The final equations read:

$$C(t) = 1 + H_0(t) +$$
$$\int_0^t dt'\{[1+h_1(t-t')]F_1(t')B(t')-[1+h_0(t-t')]F_0(t')C(t')\}, \tag{25a}$$

$$B(t) = B(0) + H_1(t) +$$

$$\int_0^t dt'\{[1+h_0(t-t')]F_1(t')B(t')-[1+h_1(t-t')]F_0(t')C(t')\}, \tag{25b}$$

where

$$H_j(t) = \int_0^1 dx\ \gamma_j(x,t)u_0(x)\exp[\int_y^x dy\ u_0(y)], \quad j = 0,1, \tag{25c}$$

$$\gamma_j(x,t) =$$

$$-\int_0^t dt' dg_j(x,t')/dx = 2 \sum_{m=1}^{\infty} (-j^{jm}(m\pi)^{-1}\sin(m\pi x)[1-\exp(-m^2\pi^2 t)], \tag{25d}$$

$$g_j(x,t) = 2 \sum_{m=1}^{\infty} 1-j^{jm}\cos(m\pi x)\ \mathrm{ex}(-m^2\pi^2 t), \quad j = 0,1, \tag{25e}$$

$$h_j(t) = g_j(0,t) = 2 \sum_{m=1}^{\infty} 1-j^{jm}\exp(-m^2\pi^2 t), \quad j = 0,1 \tag{25f}$$

and of course B(0) is defined by (21).

Note that the functions $H_j(t)$ are explicitly defined in terms of the initial datum $u_0(x)$, while $g_j(x,t)$ (as well, of course, as $h_j(t)$ are known special functions:

$$g_0(x,t) = \theta_3(\pi x/2|i\pi t) - 1\ , \tag{26a}$$

$$g_1(x,t) = \theta_4(\pi x/2|i\pi t) - 1\ , \tag{26b}$$

where $\theta_3|z|\tau)$ and $\theta_4(z|\tau|)$ are elliptic theta functions [3].

The coupled Volterra integral equations (25) determine the two functions B(t) C(t) for $t \geq 0$. The initial/boundary problem for the Burgers equation (1) in the finite interval [0,1] is thereby solved. Note that, in the special case with time-independent boundary data, $F_j(t) = F_j(0)$, the integrals in (25a,b) are convolutions, so that these integral equations can be solved by quadratures via Laplace transform. An additional simplification occurs if one, or both, of the boundary data vanish identically.

References

[1] Calogero, F. & De Lillo, S. (1989). The Burgers equation on the semi-finite and finite intervals. *Nonlinearity* **2**, 37.

[2] Cole, J.D. (1950). On the quasilinear parabolic equation occurring in aerodynamics. *Quart. appl. Math.* **9**, 225.
[3] Gradshteyn, I.S. & Ryzhik, I.M. (1965). *Tables of integrals, series, and products.* Academic Press: New York.
[4] Hopf, E. (1950). The partial differential equation $u_t+u_{xx}=uu_{xx}$. *Comm. Pure Appl. Math.* **3**, 201.
[5] Sachdev, P.L. (1987). *Nonlinear diffusive waves.* Cambridge University Press: Cambridge.

Marek Jaworski

9. CAUCHY PROBLEM FOR THE SINE-GORDON EQUATION WITH PERIODIC INITIAL CONDITIONS

1. Introduction

It is well known that the inverse scattering transform (IST) can be used to solve the Cauchy problem for a variety of nonlinear partial differential equations (see, e.g. [5,1,2,10]).

The IST method can be applied in a natural way if the initial conditions decay sufficiently fast at infinity. On the other hand, for periodic initial data the problem becomes much more complicated and direct generalization of the IST to this latter case is rather difficult [9].

For periodic cases the question is usually formulated as follows [6,4]: How to construct a multiperiodic solution (or a class of such solutions) for given spectral structure of the associated linear eigenvalue problem? Unfortunately, little attention has been paid in this context to the problem of time evolution for arbitrary periodic initial data.

In this paper we present an effective method for solving the Cauchy problem in the case of periodic initial conditions. The sine-Gordon (sG) equation is studied in detail, however the method is applicable also to other completely integrable nonlinear evolution equations.

2. Spectral structure of the associated eigenvalue problem

Let us consider the sG equation in laboratory coordinates

$$u_{xx} - u_{tt} = \sin u. \tag{1}$$

The associated eigenvalue problem can be written as [7,4]:

$$\left[\begin{pmatrix}0 & -1\\ 1 & 0\end{pmatrix}\frac{d}{dx} + \frac{i}{4}w\begin{pmatrix}0 & 1\\ 1 & 0\end{pmatrix} + \frac{1}{16\lambda}\begin{pmatrix}e^{iu} & 0\\ 0 & e^{-iu}\end{pmatrix} - \lambda\right]\begin{bmatrix}\psi_1\\ \psi_2\end{bmatrix} = 0 \qquad (2)$$

where $w = u_x + u_t$, $\lambda = E^{1/2}$ denotes a spectral parameter.

If u and w are periodic functions of x then the spectrum of (2) is piecewise continuous, i.e. consists of bands, endpoints of which either occur on the negative real E axis or are complex conjugate. As shown by Forest & McLaughlin [4] and Dubrovin [3], for arbitrary periodic u and w the spectrum consists of:

(i) a finite number of bands in the left half-plane E (highly nonlinear excitation such as soliton and/or breather trains),

(ii) infinitely many bands growing off the positive E axis (quasi-linear small-amplitude oscillations).

The number of bands is equal to the number of phases (periodic components) in a multiperiodic solution. Thus, it is clear that for arbitrary periodic initial data the number of phases is infinite and the solution goes beyond the class of so-called finite-gap solutions.

Fortunately, for higher quasi-linear components the corresponding bands tend to shrink [4,3]. Owing to this, one can approximate initial conditions by a finite-gap solution retaining the whole spectrum in the left half-plane and the most dominant

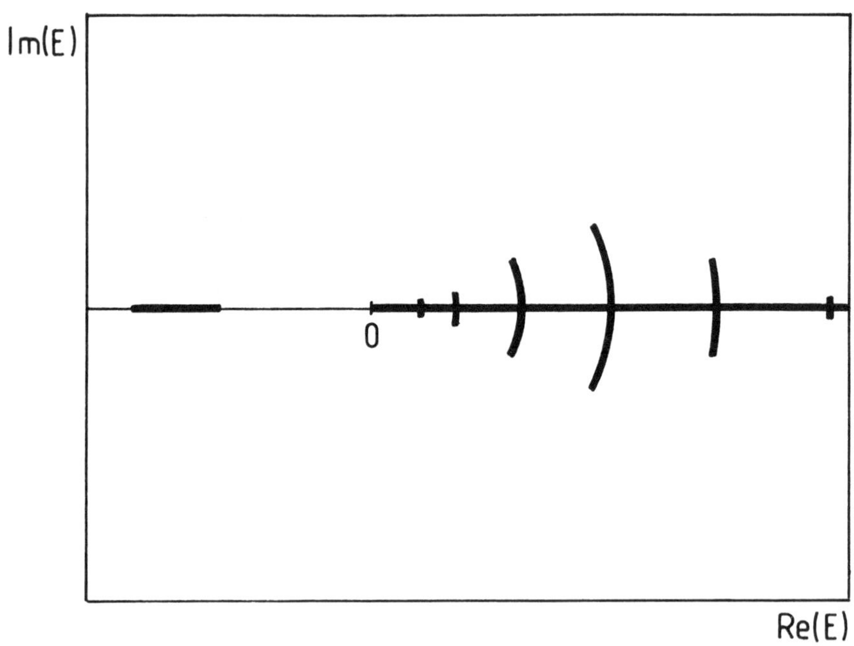

Figure 1. Example of the spectral structure for periodic potentials u and w.

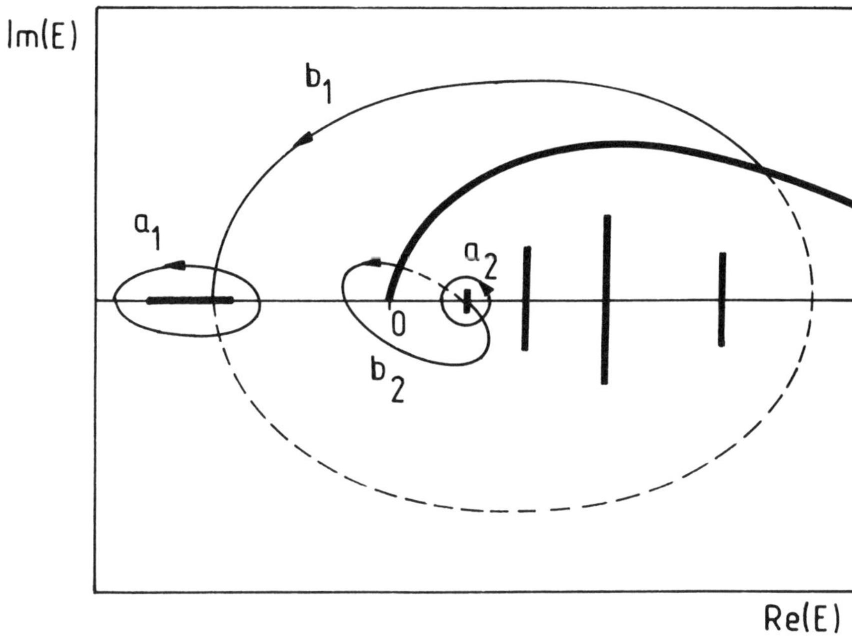

Figure 2. Two-sheet hyperelliptic Riemann surface of genus g = 5; heavy line - branch cuts, a_i and b_i - canonical contours, solid line - upper sheet, dashed line - lower sheet. For clarity, only the contours for i = 1,2 are shown.

bands in the right half-plane. In particular, a hyperelliptic Riemann surface of finite genus g can be constructed, simply by identifying the branch points with the ends of the spectrum bands. It should be noted, however, that the spectrum (which is determined uniquely) may differ from the branch cuts (which are arbitrary provided the branch points and band ends coincide).

A typical spectral structure in the E-plane and corresponding branch cuts on the Riemann surface are shown in Figs. 1 and 2, respectively. Figure 2 shows also the canonical contours of integration.

3 An approximate analytical finite-gap solution

For a given Riemann surface with properly chosen branch cuts and integration contours, one can calculate appropriate Abelian integrals in order to approximate the initial condition by an expression involving g-dimensional Riemann theta functions [6,4,8]:

$$u(x) = 2i \ln \frac{\theta(\mathbf{z}+\mathbf{e}/2|B)}{\theta(\mathbf{z}|B)}\bigg|_{t=0}, \tag{3}$$

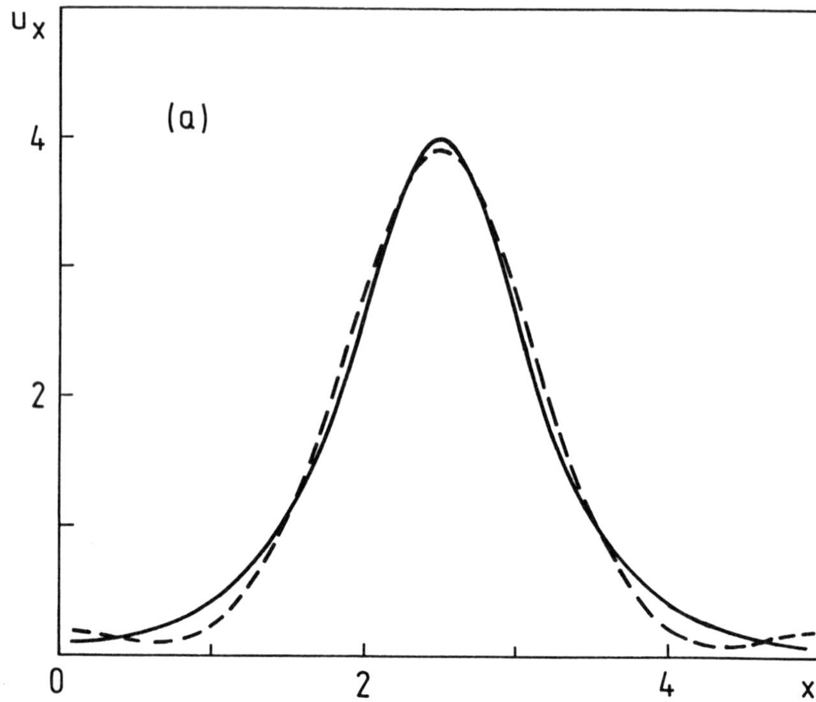

Figure 3. Time dependence of the approximate analytical solution (dashed line) and finite-difference numerical solution (solid line).
Figure 3a. t = 0.

where B is a $g \times g$ Riemann matrix, e denotes the g-dimensional unit vector, and $z \in \mathbb{C}^g$ is a linear function of space and time coordinates

$$z_i = \kappa_i x + \omega_i t + z_i^o, \quad x,t \in \mathbb{R}\ , \ \mathbf{z}^o \in \mathbb{C}^g\ , \ i = 1,2....g.$$

The Riemann theta function is given by

$$\theta(\mathbf{z}|B) = \sum_{\mathbf{n}\in\mathbb{Z}^g} \exp(i2\pi\mathbf{n}\circ\mathbf{z} + i\pi\mathbf{n}\circ B\mathbf{n}), \tag{4}$$

while B and **z** are related to Abelian integrals in the following way:

$$B = F\ H^{-1}\ , \quad \mathbf{z} = H^{1t}\mathbf{r} + \mathbf{z}^o, \tag{5}$$

where $\mathbf{r} = 2i\ [(x-t)/16P^{1/2},0,...0,(x+t)]^t\ , \quad g > 1,$

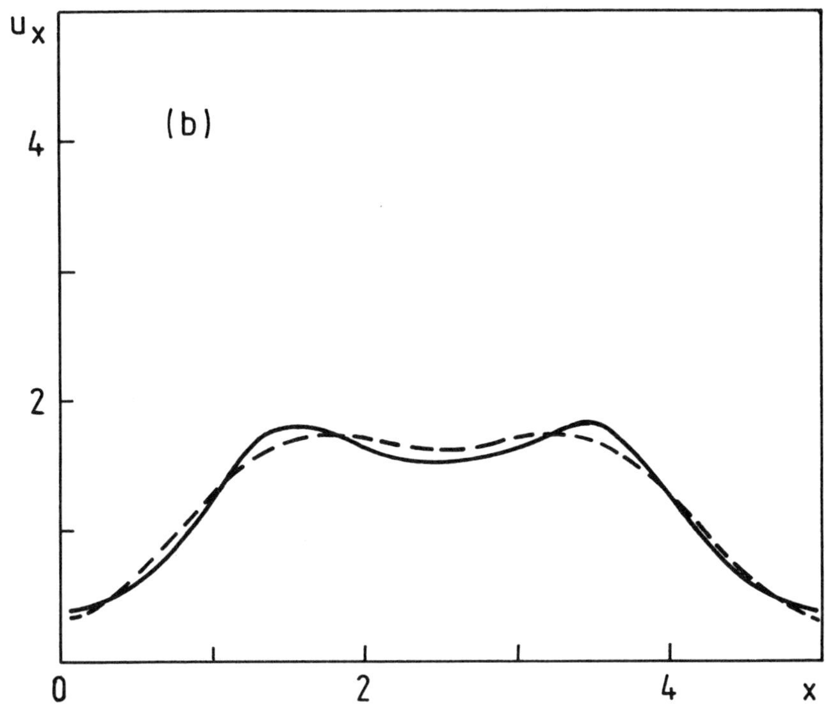

Figure 3b. t = 1.14.

$$P = E_1E_2 \ldots\ldots E_{2g},$$

and $E_k (k=1,2\ldots.2g)$ are the branch points on the Riemann surface.

The matrix elements F_{ij}, H_{ij} are given by

$$F_{ij} = \int_{b_i} d\Omega_j, \quad H_{ij} = \int_{a_i} d\Omega_j, \tag{6}$$

where a_i, b_j denote canonical contours (see Fig. 2), $d\Omega_j = E^{j-1}dE/R(E)$, and $R(E)$ generates the two-sheet hyperelliptic Riemann surface

$$[R(E)]^2 = E(E-E_1)(E-E_2)\ldots\ldots(E-E_{2g}). \tag{7}$$

The constants z_i^o denote phase-shifts of the corresponding periodic components of the solution and can be determined by fitting eq. (3) to the initial conditions. One can consider, for example, minimization of the following functional with respect to

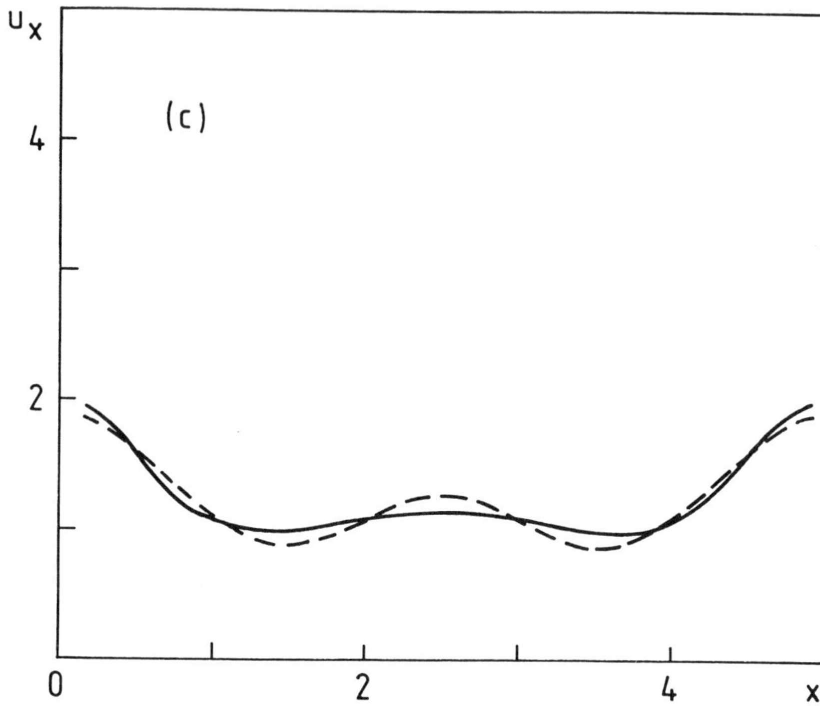

Figure 3c. t = 2.28.

z_i^o:

$$\mathscr{L} = \int_0^L |u^o(x) - u(x)| \, dx \tag{8}$$

where L denotes the spatial period, $u^o(x)$ - the initial condition, and $u(x)$ is the g-periodic solution (3) taken at $t = 0$ and implicitly dependent on z_i^o.

Note that calculating parameters of the solution for $t = 0$, we have obtained also the frequencies ω_i, hence the time dependence of eq. (3) has already been fixed. In other words, the full solution $u(x,t)$ satisfying the sG equation (1) with initial conditions $u^o(x)$, $w^o(x)$ can be found simply by allowing eq. (3) to evolve in time.

4 Numerical example

As an example, let us consider the following initial conditions:

(i) $u^o(x)$ in the form of one-periodic solution of (1) taken at

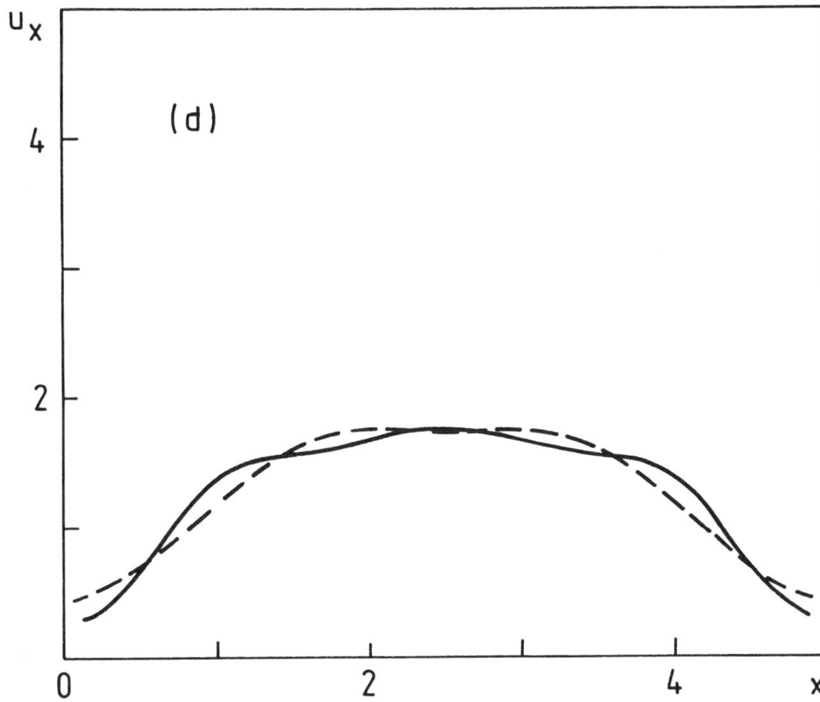

Figure 3d. t = 3.42.

t=0 and having the wavenumber κ=2,

(ii) $u_t^o(x)=0$.

In other words, we consider a one-periodic soliton train which does not satisfy eq. (1) since its amplitude and velocity are incompatible.

The spectral structure of the associated eigenvalue problem corresponds qualitatively to that shown in Fig. 1. The spectrum consists of one 'soliton' band along the negative real axis and infinitely many bands in the right half-plane. (Note that the positive real semiaxis also belongs to the spectrum.)

Constructing a hyperelliptic Riemann surface we confine our attention to genus g=5, i.e. we consider only the soliton band and four most dominant bands corresponding to quasi-linear oscillations. The approximate analytical solution (3) has been plotted in Fig. 3a-e at five consecutive time steps. The time evolution obtained this way has been compared with direct integration of eq. (1) by the finite-difference scheme. One can see that the agreement is satisfactory for merely four quasi-linear oscillations taken into account.

The sequence 3a-e covers one time period of the most dominant component (T_1=4.56). One can observe a recurrence phenomenon since

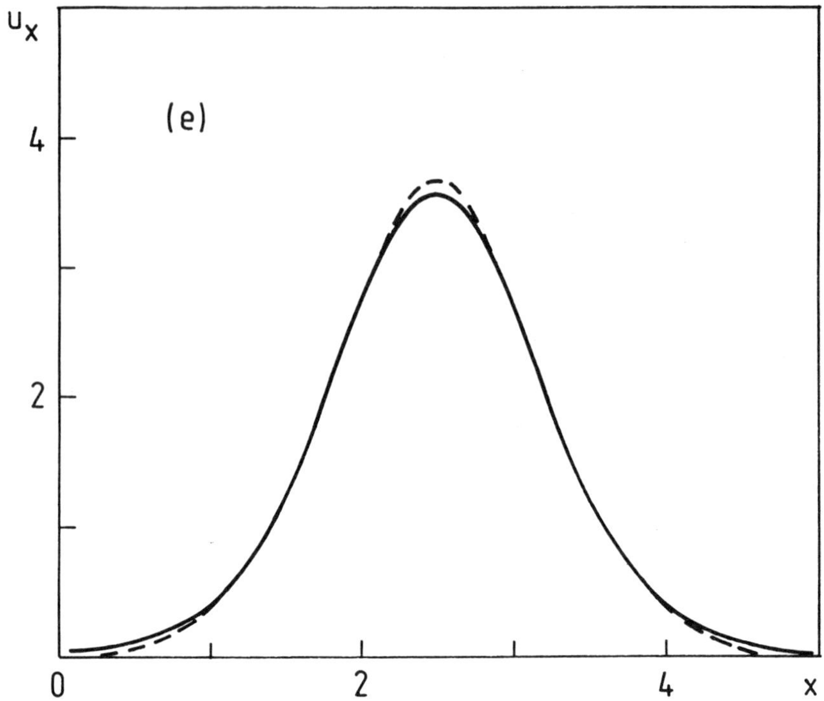

Figure 3e. t = 4.56.

the plot for $t=T_1$ is very close to the initial condition. A more detailed analysis shows, however, that the time periods of higher quasi-linear components are incommensurate with T_1. In other words, the time depenendence of eq. (3) is quasi-periodic (rather than strictly periodic).

Acknowledgements

The author wishes to thank Dr. J. Zagrodziński for many valuable discussions.

References

[1] Ablowitz, M.J., Kaup, D.J., Newell, A.C. & Segur, H. (1973a). Method for solving the sine-Gordon equation. *Phys. Rev. Lett.* **30**, 1262.

[2] Ablowitz, M.J., Kaup, D.J., Newell, A.C. & Segur, H. (1973b). Nonlinear evolution equations of physical significance. *Phys. Rev. Lett.* **31**, 125.

[3] Dubrovin, B.A. (1983). Matrix finite-gap operators. In *Contemporary Problems of Mathematics*, vol. 23, p. 33 (in Russian).

[4] Forest, M.G. & McLaughlin, D.W. (1982). Spectral theory for the periodic sine-Gordon equation: a concrete viewpoint. *J. Math. Phys.* **23**, 1248.
[5] Gardner, C.S., Greene, J.M., Kruskal, M.D. & Miura, R.M. (1967). Method for solving the Korteweg-deVries equation. *Phys. Rev. Lett.* **19**, 1095.
[6] Kozel, V.O. & Kotlarov, V.P. (1976). On quasi-periodic solutions of the sine-Gordon equation. *Doklady Akad. Nauk UkrSSR* **A10**, 878 (in Ukrainian).
[7] Takhtajian, L.A. & Faddeev, L.D. (1974). Essentially nonlinear one-dimensional model of classical field theory. *Theor. & Math. Phys.* **21**, 1046.
[8] Zagrodziński, J. (1984). Identification of the periodic processes in Josephson junctions. *Phys. Rev. B* **29**, 1500.
[9] Zakharov, V.E., Manakov, S.V., Novikov, S.P., Pitaevski, L.P. (1980). *Theory of Solitons. The Method of the Inverse Scattering Problem.* Nauka: Moscow (in Russian).
[10] Zakharov, V.E. & Shabat, A.B. (1974). A scheme for integrating the nonlinear equations of mathematical physics by the method of the inverse scattering problem, I. *Funct. Anal. & Applic.* **8**, 226.

J. Léon and F. Pempinelli

10. SINGULAR GENERAL EVOLUTIONS IN 1+1 AND 2+1 DIMENSIONS

General integrable evolutions are constructed in 1+1 and 2+1 dimensions by using the $\bar{\partial}$-approach of the inverse spectral transform scheme with singular dispersion relations.

1 Introduction

Among the physically relevant nonlinear evolution equations integrable through the "inverse spectral transform" [10] there is one particular class that has the characteristic to be related to singular dispersion relations. By "singular dispersion relation" we understand a function $\Omega(k)$ of the complex variable k (momentum or spectral parameter) which is not analytic in k.

When $\Omega(k)$ presents a discontinuity on the real k-axis and is analytic elsewhere, the following interesting integrable problems have been studied: the self-induced transparency equations (SIT) of McCall & Hahn [8] shown to be integrable by Lamb Jr. [17,18] and studied by Ablowitz et al. [2], the reduced Maxwell Bloch (RMB) equation in Eilbeck et al. [11] and the caviton equation in a plasma solved by Kaup [15]. The standard approach to these equations having singular dispersion relations has been developed by Kaup & Newell [16] where the main tool is the generalized wronskian relation of Calogero & Degasperis [9].

A new method for considering such equations has been introduced [19] and used to obtain and study a new solvable evolution [22], the generalized coupled Bloch system relevant for nonlinear optics (propagation of polarization domains). The method starts with the $\bar{\partial}$-problem formulation of the inverse spectral transform as introduced by Beals & Coifman [3] and further considered as the starting tool by Jaulent & Manna [14].

The use of the $\bar{\partial}$-approach for studying the nonlinear evolutions possessing non-analytic dispersion relations has been proved successful with the following positive points:

(i) The evolution equation is deduced from the given linear evolution of the spectral transform (which we prefer to call here the $\bar{\partial}$-**matrix**). This is unlike the standard approach which proceeds with first the construction of the Lax pair. In the case of singular dispersion relations, this construction is somewhat artificial, and moreover the time-evolution of the spectral transform (which can be referred to as the transfer matrix) is obtained through a sequence of complicated steps which make explicit use of the boundary conditions of the fields. These points are discussed in [20].

(ii) The method provides a very general integrable evolution which produces readily the particular evolutions related to a particular spectral problem.

(iii) The $\bar{\partial}$-approach allows us to construct a new class of solutions named step-solitons which are discontinuous in $x = 0$ and have a nonzero time-dependent asymptotic behaviour as described in [21]. These solutions are related to a rational reflection coefficient and it is shown in [20] that the standard transfer matrix fails in this case because, due to the asymptotics, some divergences occur when computing the time evolution of the transmission coefficient. These solutions provide then one example (the only one to our knowledge) where it is necessary to consider the $\bar{\partial}$-approach of the spectral transform method rather than the usual transfer-matrix approach.

(iv) As proved by Boiti et al. [4] the $\bar{\partial}$-approach gives the possibility of extending the theory of singular dispersion relations to evolutions in 2+1 dimensions. As it will be seen hereafter, it appears that in 2+1 dimensions the same phenomena as self induced transparency or radiation amplification [23] may still occur.

(v) Finally, probably the main advantage of the method is that it produces a unified view of the question of singular dispersion relations without prescribing the spectral operator (or the Lax pair) hence giving the possibility of extending the spectral transform to non-analytic spectral operators. This question will not be examined here and we report to future work.

In Section 2 we establish the general statement for the case of a non-analytic dispersion relation in a domain $\mathcal{D} \subset \mathbb{C}$ which vanishes (or at least has a prescribed behaviour) on the boundary $\partial\mathcal{D}$.

In Section 3 we consider the case of 2+1 dimensions for the $\bar{\partial}$-problem of Fokas [12] related to the Zakharov-Shabat [25,1] spectral problem.

2 1+1 dimensions, general statement

Let $U(k,x)$ be a given matrix valued function of $\mathbb{C} \times \mathbb{R} \longrightarrow \mathbb{C}$ with an x-independent bound $U_o(k)$ as $x \longrightarrow -\infty$ defining a **solvable spectral problem** that is:

Definition: $U(k,x)$ defines a solvable spectral problem if there exists a bijection between $U(k,x)$ and $R(k,x)$ through the couple of matrices $\phi(k,x)$, $B(k,x)$, solutions of the two following problems

$$\partial_x \phi(k,x) = U(k,x)\, \phi(k,x), \quad x \in \mathbb{R}, \tag{1a}$$

$$B^{-1}\phi \longrightarrow 1, \quad x \longrightarrow -\infty \tag{1b}$$

and

$$\partial_{\bar{k}} \phi(k,x) = \phi(k,x)\, R(k,x), \quad k \in \mathbb{C}, \tag{2a}$$

$$\phi\, B^{-1} \longrightarrow 1, \quad |k| \longrightarrow \infty. \tag{2b}$$

when B(k,x) is defined by (1b), (2b) and

$$B_x(k,x) = U_o(k)\, B(k,x). \tag{3}$$

The problem (1) is called the spectral problem and (2) the $\bar{\partial}$-problem ($\partial_{\bar{k}}$ in (2a) or $\bar{k}$ in indices hereafter, stand for $\frac{\partial}{\partial k_R}$ + $i\frac{\partial}{\partial k_I}$ with $k = k_R + ik_I$). We consider here neither the question of the existence of the bijection between the "potential" U(k,x) and its "$\bar{\partial}$-matrix" R(k,x) nor that of the uniqueness of the couple (ϕ,B). However, for the particular choices that we shall make for U in the next section, these questions are now well understood.

Now the general statement is:

The following set of equations

$$\partial_t\, U(k,x,t) = \int_{\mathbb{C}} d\lambda \wedge d\bar{\lambda} \left[\Phi(\lambda,x,t), \frac{U(\lambda,x,t) - U(k,x,t)}{\lambda - k}\right], \tag{4}$$

$$\partial_x\, \Phi(k,x,t) = [U(k,x,t),\, \Phi(k,x,t)] \tag{5}$$

with the initial boundary value problem (data: U(k,x),F(k,t))

$$U(k,x,0) = U(k,x) \text{ and } \lim_{x \to -\infty} \Phi(k,x,t) = F(k,t) \tag{6}$$

verifying

$$[U_o(k),\, F(k,t)] = 0 \tag{7}$$

$$\left[U_o(k),\, \int_{\mathbb{C}} \frac{d\lambda \wedge d\bar{\lambda}}{\lambda - k}\, F(\lambda,t)\right] = 0 \tag{8}$$

is **integrable** through the evolution

$$\partial_t\, R(k,x,t) = [R(k,x,t),\, \Omega(k,t)] \tag{9}$$

where the **dispersion relation** Ω is given by

$$\Omega(k,t) = 2i\pi \int_{\mathbb{C}} \frac{d\lambda \wedge d\bar{\lambda}}{\lambda - k}\, B^{-1}(\lambda,x)\, F(\lambda,t)\, B(\lambda,x). \tag{10}$$

The word "integrable" means here that the solution (U,Φ) of the nonlinear system (4) (5) for the initial datum and boundary value (6) can be reduced to the solution of the linear integral equation

$$\phi(k,x,t)\, B^{-1}(k,x) - 1 =$$

$$= \frac{1}{2i\pi} \int_{\mathbb{C}} \frac{d\lambda \wedge d\bar{\lambda}}{\lambda - k}\, \phi(\lambda,x,t) \left[R(\lambda,x,t) + B^{-1}(\lambda,x)\, \partial_{\bar{\lambda}} B(\lambda,x) \right] B^{-1}(\lambda,x) \tag{11}$$

for R and B known distributions.

We give hereafter altogether the method of solution and some algebraic identities which provide the proof of our statement. The procedure consists of the following seven numbered steps:

(i) Given U(k,x,0) we must first solve the spectral problem (1) (and we have assumed that it is feasible). To do this, we compute the solution B(k,x) of (3) up to an arbitrary x-independent normalization; we solve then the ordinary differential equation (1a) with the boundary (1b) and finally we fix the normalization of B(k,x) by requiring that the solution ϕ does obey the behaviour (2b). Then the $\bar{\partial}$ matrix R(k,x,0) is obtained out of (2a) just by computing $\partial_{\bar{k}} \phi$.

(ii) Given the boundary value F(k,t) of the unknown $\Phi(k,x,t)$, the dispersion relation $\Omega(k,t)$ is obtained by computing the integral in (10). One fundamental property is that

$$\frac{\partial \Omega}{\partial x} = 0 \tag{12}$$

which is readily obtained from (7). We may remark that (7) smiply results from (5) and the fact that the behaviour of Φ as $x \longrightarrow -\infty$ has been arbitrarily chosen to be x-independent. Even if it were not the case, the property (12) would still hold thanks to (5) taken as $x \longrightarrow -\infty$.

We remark also that (7) and (8) both imply

$$\partial_t B(k,x) = 0 \tag{13}$$

as obtained by taking the limit of (4) as $x \longrightarrow -\infty$ and using (3).

(iii) We arrive now at the central point of our method which results in noticing first that the evolution law (9) implies the relation

$$\partial_{\bar{k}} V(k,x,t) = -\phi(k,x,t)\, \partial_{\bar{k}} \Omega(k,t)\, \phi^{-1}(k,x,t), \tag{14}$$

where the matrix V is defined as

$$V = \phi_t\, \phi^{-1} - \phi\, \Omega\, \phi^{-1}. \tag{15}$$

Second we note also that the property (12) implies

$$U_t - V_x + [U,V] = 0 \tag{16}$$

which is just the usual "zero curvature" equation for the differential equations (1a) and (15).

(iv) We may now compute V out of (14) if we use in (15) that Ω (given in (10)) is $O(\frac{1}{k})$ for large k, that $\{\phi\ B^{-1} - 1\}$ is also $O(\frac{1}{k})$ and that B is t-independent as mentioned in (13), to obtain that V is $O(\frac{1}{k})$ too. Then the solution of (14) reads

$$V(k,x,t) = -\frac{1}{2i\pi}\int_{\mathbb{C}} \frac{d\lambda \wedge d\bar{\lambda}}{\lambda - k}\ \phi(\lambda,x,t)\ \partial_{\bar{\lambda}}\Omega(\lambda,t)\ \phi^{-1}(\lambda,x,t) \tag{17}$$

(v) We check also that the property (12) implies the following relation

$$\partial_x\left(\phi\ \Omega_{\bar{k}}\ \phi^{-1}\right) = \left[U,\ \phi\ \Omega_{\bar{k}}\ \phi^{-1}\right] \tag{18}$$

readily obtained by remembering that ϕ is precisely a solution of (1a).

(vi) Now with the help of (18) we simply insert V given in (17) into equation (16) to obtain (forgetting the (x,t)-dependence for a while):

$$\partial_t\ U(k) = \int_{\mathbb{C}} d\lambda \wedge d\bar{\lambda}\ \left[\frac{1}{2i\pi}\ \phi(\lambda)\ \Omega_{\bar{\lambda}}\phi^{-1}(\lambda),\ \frac{U(\lambda) - U(k)}{\lambda-k}\right] \tag{19}$$

(vii) Comparing finally (18) with (5), and (19) with (4) we will have reached our purpose if we prove that

$$\Phi(k,x,t)= \frac{1}{2i\pi}\ \phi(k,x,t)\ \Omega_{\bar{k}}(k,t)\ \phi^{-1}(k,x,t). \tag{20}$$

The two quantities appearing in the above equation are solutions of the same linear first order differential equation ((5) or (18)), hence (20) holds if they possess the same asymptotic behaviours. This last statement is proved by differentiating (10) with respect to $\bar{k}$ to obtain

$$F(k,t) = \frac{1}{2i\pi}\ B(k,x)\ \Omega_{\bar{k}}(k,t)\ B^{-1}(k,x), \tag{21}$$

which then implies (20).

The result of the seven properties listed above is that the solution $\phi(k,x,t)$ of the $\bar{\partial}$-problem (2), or equivalently the solution $\phi(k,x,t)$ of the integral equation (9) from R(k,x,o) computed in step 1, allows us to build the solution of the system (4) (5) as follows: $\Phi(k,x,t)$ is immediately obtained from (20) which can be rewritten with the help of (21) as

$$\Phi(k,x,t) = (\phi\ B^{-1})F\ (\phi\ B^{-1})^{-1} \tag{22}$$

and U(k,x,t) is simply obtained through (1a).

In all that precedes, we have seen that it is essential that the datum U(k,x,o) allows the definition of a **solvable** spectral problem. Up to now this constraint can be stated explicitly only for some U(k,x,o) having a **prescribed k-dependence** which actually makes eq. (4) k-independent and therefore reduces the dimension of our system (the k-dependence in (5) becomes parametric).

However we can make a weaker statement which does not presume the existence of a solvable spectral problem. This is the following: **the data of R(k,x,o) and B(k,x) allow the building of solutions of the system (4) (5) with the initial boundary value problem (6) through the solution of first (9) (10) and then (11).** We defer the study of possible extensions in this direction of the inverse spectral transform to future work.

Examples of application will be found in [20,21,22]. We list below two interesting solvable evolutions:

$$i\, q_t = -q_{xx} + 2|q|^2 q - 2\int_{-\infty}^{+\infty} d\lambda\ \nu(\lambda,t)\ u(\lambda,x,t)\ v(\lambda,x,t) \tag{23}$$

$$u_x(k,x,t) = -ik\ u(k,x,t) - q(x,t)\ v^*(k^*,x,t) \tag{24}$$

$$v_x(k,x,t) = -ik\ v(k,x,t) + q(x,t)\ u^*(k^*,x,t) \tag{25}$$

$$u\ e^{ikx} \longrightarrow 1,\quad v \longrightarrow 0,\quad q \longrightarrow 0,\quad x \longrightarrow -\infty. \tag{26}$$

and

$$S_t = \frac{1}{2i}\ [S,\ S_{xx}] + [S,\ \langle W\rangle] \tag{27}$$

$$\langle W\rangle \doteq \int_{-\infty}^{+\infty} d\lambda\ \frac{g(\lambda,t)}{\lambda}\ W(\lambda,x,t) \tag{28}$$

$$\text{with } W_x = \frac{i}{\lambda}\ [S,\ W],\quad W \longrightarrow \sigma_3,\quad S \longrightarrow \sigma_3 \text{ as } x \longrightarrow -\infty. \tag{29}$$

3 The general method and one example in 2+1 dimensions

Let T_1 be a differential linear operator (with respect to (x,y) space variables), $\phi(k,x,y)$ its spectral eigenfunction and E(k,x,y) some given function which solves

$$\left(\lim_{x\to\infty} T_1\right)\ E = 0 \tag{30}$$

The spectral problem analogue of (1) is written

$$T_1\ \phi = 0,\quad \phi(k,x,y)\ E^{-1}(k,x,y) \longrightarrow 1 \text{ as } x \to -\infty. \tag{31}$$

We assume that this spectral problem can be solved as a $\bar\partial$-

problem on a complex domain $\mathcal{D}$ with boundary $\partial\mathcal{D}$

$$\frac{\partial}{\partial \bar{k}} \phi(k,x,y) = \int\int_{\mathcal{D}} d\ell \wedge d\bar{\ell}\ \phi(\ell,x,y)\ R(k,\ell), \quad k \in \mathcal{D} \tag{32}$$

$$\phi(k,x,y)\ E^{-1}(k,x,y) \sim 1 \quad \text{as} \quad k \longrightarrow \partial\mathcal{D}.$$

With respect to 1+1 dimension (see (2), the $\bar{\partial}$-problem becomes now non-local in the spectral parameter.

The matrix $R(k,\ell)$ is a distribution in $\mathcal{D} \times \mathcal{D}$ and E is analytic in $\mathcal{D}$ We note that $E(k,x,y)$ allows us to introduce the k-dependence of the eigenfunction ϕ in (31) or its (x,y) dependence in (32).

The solution ϕ of the $\bar{\partial}$-problem is obtained as the solution of an integral equation by the Cauchy-Green formula

$$\phi(k,x,y) = E(k,x,y) + \frac{1}{2\pi i} \int\int_{\mathcal{D}} \frac{d\zeta \wedge d\bar{\zeta}}{\zeta - k} \frac{\partial\phi}{\partial\bar{\zeta}} E^{-1}(\zeta,x,y)\ E(k,x,y). \tag{33}$$

The parametric time dependence is introduced through R by prescribing the following linear evolution (compare with (9))

$$R_t(k,\ell;t) = R(k,\ell;t)\ \Omega(k;t) - \Omega(\ell;t)\ R(k,\ell;t) \tag{34}$$

where $\Omega(k;t)$, the "dispersion relation", is a given matrix function on $\mathcal{D} \times \mathcal{R}$.

To build a Lax pair (T_1,T_2), which satisfies the "weak" compatibility condition [6]

$$T_1\ \phi \doteq 0, \quad [T_1,T_2]\ \phi = 0, \tag{35}$$

we must construct the auxiliary spectral operator T_2, which we seek in the form

$$T_2 = \partial_t - W. \tag{36}$$

$W(k;x,y,t)$ is a matrix to be determined, according to the requirement

$$T_2\ \phi = \phi\ \Omega \tag{37}$$

which ensures that the weak compatibility condition (6) is satisfied.

By differentiating (37) with respect to $\bar{k}$ and using (34) we get for $k \in \mathcal{D}$

$$\frac{\partial W}{\partial \bar{k}} = -\phi(k) \frac{\partial\Omega}{\partial\bar{k}} \phi^{-1}(k) + \int\int_{\mathcal{D}} d\ell \wedge d\bar{\ell}[W(\ell)-W(k)]\ \phi(\ell)\ R(k,\ell)\ \phi^{-1}(k). \tag{38}$$

So the operator T_2 is obtained by solving a linear singular integral equation (38), unlike in 1+1 dimensions, where the auxiliary operator is explicitly given (17).

In summary, for a non-analytic dispersion relation Ω, we are able to build from the $\bar{\partial}$-problem a couple of operators T_1 and T_2, which satisfies the weak compatibility condition in $\mathcal{D}$. This condition reads

$$\{T_{1,t} + [T_1,W]\}\phi = 0 \tag{39}$$

and it furnishes the k-independent non-linear evolution equation

$$\lim_{k\to\partial\mathcal{D}} \{(T_{1,t}\phi + [T_1,W]\phi)\phi^{-1}\} = 0. \tag{40}$$

Let us now work out an explicit example.

We choose $\mathcal{D} = \mathbb{C}$, $\partial\mathcal{D} = \{|k| = \infty\}$, $E(k,x,y) = e^{ik(\sigma_3 x-\epsilon y)}$, $\epsilon^2 = \pm 1$ and using the $\bar{\partial}$-approach we can reconstruct the spectral operator T_1

$$T_1\phi = 0 \ , \quad T_1 = \partial_x + \epsilon^3 \sigma_3 \partial_y + Q \tag{41}$$

where $Q = i\left[\varphi^{(1)}(x,y),\ \sigma_3\right] \cdot \varphi^{(1)}$ is the first order of the asymptotic expansion of ϕ as $|k| \longrightarrow \infty$

$$\phi(k,x,y) \sim \left\{1 + \sum_{n=1}^{\infty} k^{-n}\ \varphi^{(n)}(x,y)\right\} e^{ik(\sigma_3 x-\epsilon y)} \ . \tag{42}$$

(41) is the Zakharov & Shabat [25,1] spectral problem in 2+1 dimensions (see also [12], [13]). We restrict here our study to the hyperbolic case ($\epsilon = +1$). For the elliptic case ($\epsilon = -i$), see Boiti et al. [4].

Let us choose Ω diagonal and such that W vanishes on the boundary $\partial\mathcal{D}$

$$\Omega = \omega(k,t)\ \sigma_3 \ , \quad \omega \sim 0\left(\frac{1}{k}\right) \text{ as } |k| \longrightarrow \infty \tag{43}$$

We have in particular from (36), (37) $W \sim -\Omega$ as $|k| \longrightarrow \infty$ and polynomial terms are excluded in Ω (for more general behaviour see [4]).

The weak compatibility condition [39] becomes

$$Q_t + (\partial_x + \sigma_3\ \partial_y)W + [Q,W] + [\sigma_3,W]\ G = 0 \tag{44}$$

where $G \equiv \phi_y\ \phi^{-1}$ satisfies the equation

$$Q_y + (\partial_x + \sigma_3\ \partial_y)G + [Q,G] + [\sigma_3,G]\ G = 0 \tag{45}$$

with the asymptotic behaviour

$$G \sim -ik + O\left(\frac{1}{k}\right) \text{ as } |k| \longrightarrow \infty. \tag{46}$$

Let us now introduce the distributions

$$\Phi = \frac{\partial W}{\partial \bar{k}}, \quad \Gamma = \frac{\partial G}{\partial \bar{k}} \tag{47}$$

the operator I_k inverse of $\partial_{\bar{k}}$

$$I_k\, f(k) = \frac{1}{2\pi i} \int\int \frac{d\ell \wedge d\bar{\ell}}{\ell - k}\, f(\ell) \tag{48}$$

and the average

$$\langle f \rangle = \frac{1}{2\pi i} \int\int d\ell \wedge d\bar{\ell}\, f(\ell). \tag{49}$$

By looking at (44) and (45) for large $|k|$, we easily obtain

$$Q_t + i[\sigma_3, \langle\Phi\rangle] = 0 \tag{50}$$

$$Q_y + i[\sigma_3, \langle\Gamma\rangle] = 0. \tag{51}$$

By differentiating (44) and (45) with respect to $\bar{k}$, we get

$$(\partial_x + \sigma_3\partial_y)\Phi + [Q - ik\,\sigma_3, \Phi] = -[\sigma_3, \Phi]\, I_k\Gamma - [\sigma_3, I_k\Phi]\Gamma \tag{52}$$

$$(\partial_x + \sigma_3\partial_y)\Gamma + [Q - ik\,\sigma_3, \Gamma] = -[\sigma_3, \Gamma]\, I_k\Gamma - [\sigma_3 I_k\Gamma]\Gamma. \tag{53}$$

The set (50 – 53) represents the general form of the non-linear evolution equations associated with the 2x2 hyperbolic Zakharov & Shabat spectral problem in 2+1 dimensions with singular dispersion relation. Note that the three matrices Q, Φ and Γ are obtained as a solution of the system (50), (52), (53). Eq. (51) is then automatically satisfied.

Finally we must associate boundary conditions in the (x,y) plane. For $Q \longrightarrow 0$ as $x^2+y^2 \longrightarrow \infty$, we can show that $\Phi \longrightarrow -\dfrac{\partial\Omega}{\partial\bar{k}}$ as $x \longrightarrow -\infty$, at least in the sense of the distributions.

Note that in 2+1 dimensions we have, besides the "t-evolution" of Q related to Φ as in 1+1 dimensions, a "y-evolution" of Q related to Γ. Φ and Γ are the analogue of the squared eigenfunctions in 1+1 dimension and are furnished by the set (52-53).

The inverse spectral transform allows us to solve the set of eqs. (50) $\div$ (53) for the initial datum $Q(x,y,o)$ and the asymptotic behaviour $-\dfrac{\partial\Omega}{\partial\bar{k}}$ of Φ as $x \longrightarrow -\infty$ for any $t > 0$. This is achieved as follows:

(i) Solve the direct problem to obtain $\phi(k,x,y,o)$ and $R(k,\ell,o)$ from $Q(x,y,o)$.

(ii) Obtain $R(k,\ell,t)$ from $R(k,\ell,o)$ by eq. (34) with

$$\Omega(k,t) = \frac{1}{2\pi i}\int\!\!\int \frac{d\ell \wedge d\bar{\ell}}{\ell - k}\,\frac{\partial\Omega}{\partial\bar{\ell}}\,(\ell,t) \tag{54}$$

(iii) Solve the inverse problem by an appropriate integral equation to obtain $Q(x,y,t)$ and $\phi(k,x,y,t)$.

(iv) ϕ and Γ are obtained from

$$\Phi = \frac{\partial}{\partial\bar{k}}\left(\phi_t\phi^{-1} - \phi\,\Omega\,\phi^{-1}\right) \tag{55}$$

$$\Gamma = \frac{\partial}{\partial\bar{k}}\left(\phi_y\,\phi^{-1}\right). \tag{56}$$

We summarize here the principal results (see [4,5]) for the choice

$$\frac{\partial\Omega}{\partial\bar{k}} = g(k,t)\,\sigma_3\,\delta(k_I). \tag{57}$$

We get

$$R(k,\ell) = \frac{i}{2}\begin{pmatrix} 0 & r_1(k,\ell)\,\delta(\ell_I+o)\,\delta(k_I+o) \\ r_2(k,\ell)\,\delta(\ell_I-o)\,\delta(k_I-o) & 0 \end{pmatrix}, \tag{58}$$

$$\Gamma(k) = \delta(k_I)\,\Gamma_o(k),$$

$$\Gamma_o(k) = \int_{-\infty}^{+\infty} d\ell\,\phi_y(\ell)\,R_1(k,\ell)\,\phi^{-1}(k)$$

$$-\,\phi_y\,\phi^{-1}(k)\int_{-\infty}^{+\infty} d\ell\,\phi(\ell)\,R_1(k,\ell)\phi^{-1}(k). \tag{59}$$

$$\Phi(k) = \delta(k_I)\,\Phi_o(k),$$

$$\Phi_o(k) = -g(k)\,\phi(k)\,\sigma_3\phi^{-1}(k) + \int_{-\infty}^{+\infty} d\ell[W(\ell)-W(k)]\,\phi(\ell)R_1(k,\ell)\phi^{-1}, \tag{60}$$

where $R_1 = \begin{pmatrix} 0 & r_1 \\ r_2 & 0 \end{pmatrix}$. Note that Φ_o and Γ_o are functions and the set (50 – 53) of solvable equations can be written

$$Q_t + i[\sigma_3, \langle\Phi_o\rangle] = 0 \tag{61}$$

$$Q_y + i[\sigma_3, \langle\Gamma_o\rangle] = 0 \tag{62}$$

$$(\partial_x+\sigma_3\partial_y)\Phi_o + [Q-ik\sigma_3,\Phi_o] = -[\sigma_3,\Phi_o]I\Gamma_o - [\sigma_3,I\Phi_o]\,\Gamma_o \tag{63}$$

$$(\partial_x+\sigma_3\partial_y)\Gamma_o + [Q-ik\sigma_3,\Gamma_o] = -[\sigma_3,\Gamma_o]I\Gamma_o - [\sigma_3,I\Gamma_o]\Gamma_o \tag{64}$$

where $\langle\Phi_o\rangle = -\frac{1}{\pi}\int_{-\infty}^{+\infty} d\ell\ \Phi_o(\ell)$, $I\ \Phi_o = -\frac{1}{\pi}\int_{-\infty}^{+\infty}\frac{d\ell}{\ell-k}\ \Phi_o(\ell)$.

The time evolution of R_1 is given by

$$R_1(k,\ell,t) = R_1(k,\ell,o)\ \exp\left[\int_o^t dt\begin{pmatrix}\omega^-(k)+\omega^-(\ell) & 0\\ 0 & -\omega^+(k)-\omega^+(\ell)\end{pmatrix}\right] \tag{65}$$

with

$$\omega^{\pm}(k,t) = \mp\ ig(k,t) - \frac{1}{\pi}\int_{-\infty}^{+\infty}\frac{d\ell}{\ell-k_R}\,g(\ell,t),\ k = k_R \pm io. \tag{66}$$

Depending on g(k), the time evolution of R_1 can present exponential growth or damping and therefore such phenomena as self-induced transparency or radiation amplification in two level media can also occur in 2+1 dimensions.

Finally let us note that we can also write Bäcklund transformations and superposition formulae, following Boiti et al. [7], and get in this way explicit soliton solutions (see [4]). In particular, a very interesting solution, a soliton localized in the plane with exponential decay in all directions, has been found [5] (see also the paper "Localized solitons in the plane" by Boiti et al. in this volume).

References

[1] Ablowitz, M.J, & Haberman, R. (1975). *Phys. Rev. Lett.* **35**, 1185.

[2] Ablowitz, M.J., Kaup, D.J. & Newell, A.C. (1974). *J. Math. Phys.* **15**, 1852.

[3] Beals, R. & Coifman, R.R. (1986). *Physica* **D18**, 242.

[4] Boiti, M., Leon, J.J.-P., Martina, L. & Pempinelli, F. (1988). *J. Physics A* **21**.

[5] Boiti, M., Leon, J.J.-P., Martina, L. & Pempinelli, F. (1988). *Phys. Lett. A* **132**, 432.

[6] Boiti, M., Leon, J.J.-P., Manna, M. & Pempinelli, F. (1986). *Inverse Problems* **2**, 271.

[7] Boiti, M., Konopelchenko, B.G. & Pempinelli, F. (1985). *Inverse Problems* **1**, 33.

[8] McCall, S.L. & Hahn, E.L. (1969). *Phys. Rev.* **183**, 457.

[9] Calogero, F. & Degasperis, A. (1976). *Nuovo Cimento B* **32**, 201.

[10] Calogero, F. & Degasperis, A. (1982). *Spectral Transform and Solitons.* North Holland.

[11] Eilbeck, J.C., Gibbon, J.D., Caudrey, P.J. & Bullough, R.K. (1973). *J. Physics A* **6**, 1337.

[12] Fokas, A.S. (1983). *Phys. Rev. Lett.* **51**, 3.

[13] Fokas, A.S. & Ablowitz, M.J. (1984). *J. Math. Phys.* **25**, 2494.

[14] Jaulent, M. & Manna, M. (1986). *Phys. Lett. A* **117**, 62.

[15] Kaup, D.J. (1987). *Phys. Rev. Lett.* **59**, 2063.

[16] Kaup, D.J. & Newell, A.C. (1979). *Adv. Math.* **31**, 67.

[17] Lamb, Jr. G.L. (1973). *Phys. Rev. Lett.* **31**, 196.

[18] Lamb, Jr. G.L. (1974). *Phys. Rev. A* **9**, 422.

[19] Leon, J.J.-P. (1987). *Phys. Lett. A* **123**, 65.

[20] Leon, J.J.-P. (1988). In *Some Topics on Inverse Problems,* Sabatier, P.C. (ed.). World Scientific: Singapore.

[21] Leon, J.J.-P. (1988). *Phys. Lett. A* **131**, 79.

[22] Leon, J.J.-P. (1988). *J. Math. Phys.* **29**, 2012.

[23] Manakov, S.V. (1982). *Sov. Phys. JETP***56**, 37.

[24] Tratnik, M.V. & Sipe, J.E. (1987). *Phys. Rev. Lett.* **58**, 1104.

[25] Zakharov, V.E. & Shabat, A.B. (1979).

D. Maison

11. GEROCH GROUP AND INVERSE SCATTERING METHOD

1 Introduction

The Geroch group emerged in Geroch's study [1] of the 2-dimensional equations describing stationary, axially symmetric solutions of Einstein's vacuum field equations. Starting from any given solution Geroch constructed an infinite set of potentials and used them to produce a family of new solutions of these equations from the given one. He conjectured this process could be expressed as the action of some huge transformation group parametrized by 1-dimensional curves, i.e. some kind of loop group. What was missing was a constructive procedure actually to perform these transformations for a given "seed" metric. Later Kinnersley & Chitre [2] constructed a related set of potentials and an infinite dimensional Lie algebra of infinitesimal transformations of these potentials easily recognizable as a loop (Kac-Moody) algebra. Using a generating function for the potentials they were able actually to perform finite transformations in simple cases.

On the other hand Belinski & Zakharov [3] and Maison [4] constructed linear spectral problems yielding the non-linear 2-dimensional field equations of Geroch as integrability condition indicating the complete integrability of these equations. Both these approaches were inspired by the non-linear σ-model structure underlying these equations. This approach initiated the construction of Bäcklund transformations for the Ernst equation [5], which is just another (complex) form of the Geroch equations. Reformulating the Kinnersley-Chitre transformations as a Riemann-Hilbert problem Ernst & Hauser [6] could materialize the original idea of Geroch of an infinite dimensional group acting on the solutions. At this point it became clear that there was a tight connection between the implementation of the Geroch group and the "Inverse Scattering Method" (ISM). This aspect was elucidated in a joint work with P. Breitenlohner [7] and will be the main subject of this talk. In the last part I will indicate applications of this connection between the ISM and the action of a Kac-Moody group to the "classical" completely integrable equation like the KdV equation.

2 Stationary solutions of Einstein's equations

For the description of stationary solutions of Einstein's equations it is convenient to introduce an adapted time coordinate t, such that the metric is independent of t. In addition the metric is decomposed according to the corresponding 3 + 1-split of the tangent space

$$g_{\alpha\beta} = \begin{pmatrix} \frac{1}{\Delta}h_{ab} + \Delta k_a k_b & \Delta k_b \\ \Delta k_a & \Delta \end{pmatrix}. \qquad (1)$$

where the indices a,b run over the three space dimensions. Einstein's vacuum field equation $T_{\alpha\beta} = 0$ implies the "Maxwell" equation for k_a

$$\nabla_a\left(\frac{\Delta^2}{2}k^{ab}\right) = 0 \qquad (2)$$

with $k_{ab} = \partial_a k_b - \partial_b k_a$. This equation can be considered as a Bianchi identity for the dual potential $\tilde{\psi}$ (the twist potential) defined by

$$k^{ab} = \frac{1}{\sqrt{h}}\epsilon^{abc}\frac{1}{\Delta^2}\partial_c\tilde{\psi} \qquad (3)$$

Using the (pseudo-)scalar potential $\tilde{\psi}$ instead of k_a in addition to Δ and the 3-metric h_{ab} the "dimensionally reduced" three dimensional theory describing stationary solutions of Einstein's equations may be expressed by the Langrangean

$$\mathcal{L}^{(3)} = \sqrt{h}\left(\frac{1}{2}R^{(3)} - \frac{1}{2}h^{ab}\left(\frac{1}{2\Delta^2}(\partial_a\Delta\partial_b\Delta + \partial_a\tilde{\psi}\partial_b\tilde{\psi})\right)\right) \qquad (4)$$

The form of this Lagrangean looks like 3-dimensional gravity coupled to "matter", where the matter part has the structure of a so-called non-linear σ-model [8] for the coset space SL(2)/SO(2). The complex potential $\varepsilon = \Delta + i\tilde{\psi}$ is the well-known Ernst potential [6].

A similar structure is obtained for "dimensionally reduced" forms of other theories extending Einstein's theory. A typical example is the original 5-dimensional Kaluza-Klein theory yielding a σ-model for the coset space SL(3)/SO(2,1). Other important cases are the Einstein-Maxwell theory leading to SU(2,1)/S(U(1,1) x

U(1)) or N = 8 supergravity with $E_{8(+8)}/SO^*(16)$. All these coset spaces are non-compact (pseudo-)riemannian symmetric spaces.

3 **Structure of the symmetric space SL(2)/SO(2)**

The Lie algebra sl(2) can be decompose into a direct sum sl(2) = $\mathcal{H} \oplus \mathcal{K}$ with $\mathcal{H}$ = so(2) and $\mathcal{K}$ = sl(2) $\ominus$ so(2). The subalgebra $\mathcal{H}$ is distinguished by its invariance under the involutive automorphism

$$\tau:\quad l \longrightarrow -l^T \qquad l \in sl(2) \tag{5}$$

On SL(2) the involution τ is given by $\tau:\ g \longrightarrow (g^T)^{-1}$. One may imbed the coset space SL(2)/SO(2) as exp($\mathcal{K}$) into SL(2) and can therefore parametrize it by symmetric, unimodular 2 x 2 matrices:

$$M = \begin{pmatrix} \Delta + \dfrac{\tilde{\psi}^2}{\Delta} & \dfrac{\tilde{\psi}}{\Delta} \\ \dfrac{\tilde{\psi}}{\Delta} & \dfrac{1}{\Delta} \end{pmatrix}. \tag{6}$$

The invariant metric of the symmetric riemannian space SL(2)/SO(2) is given by

$$ds^2 = \frac{1}{4}\mathrm{Tr}(M^{-1}dM)^2 = \frac{d\Delta^2 + d\tilde{\psi}^2}{2\Delta^2}\ . \tag{7}$$

Comparing the Lagrangean (4) with this expression justifies its interpretation as a non-linear σ-model for the coset space SL(2)/SO(2).

The matrix M may be factorized as $M = P^TP$ with

$$P = \begin{pmatrix} \sqrt{\Delta} & 0 \\ \dfrac{\tilde{\psi}}{\sqrt{\Delta}} & \dfrac{1}{\sqrt{\Delta}} \end{pmatrix} \tag{8}$$

Whereas the matrix M transforms linearly under SL(2)

$$M \longrightarrow M_g = g^TMg \tag{9}$$

the group acts in a non-linear fashion on the triangular representatives P

$$P \longrightarrow P_g = h(P,g)Pg^{-1} \qquad g \in SL(2) \tag{10}$$

where $h(P,g) \in SO(2)$ restores the triangular form of P_g.

The Lie algebra vauled 1-form dPP^{-1} can be decomposed according to

$$dPP^{-1} = \mathcal{A} + \mathcal{T} = (\mathcal{A}_k + \mathcal{T}_k)dx^k \qquad \begin{array}{l}\tau(\mathcal{A}) = \mathcal{A} \\ \tau(\mathcal{T}) = -\mathcal{T}\end{array} \tag{11}$$

with the transformation laws under SO(2) (induced by (10))

$$\begin{aligned}\mathcal{A} &= \tfrac{1}{2}(dP\,P^{-1} + \tau(dP\,P^{-1})) \longrightarrow h\mathcal{A}h^{-1} + dh\,h^{-1} \\ \mathcal{T} &= \tfrac{1}{2}(dP\,P^{-1} - \tau(dP\,P^{-1})) \longrightarrow h\mathcal{T}h^{-1}\end{aligned} \tag{12}$$

i.e. $\mathcal{A}$ can be interpreted as connection for H whereas $\mathcal{T}$ transforms H-covariantly and both are G-invariant.

4 Stationary, axially symmetric solutions

Specializing from stationary to stationary, axially symmetric solutions simplifies things insofar as it is then possible to parametrize the metric h_{ab} as

$$ds^2 = h_{ab}dx^a dx^b = \lambda^2((dx^1)^2 + (dx^2)^2) + \rho^2 d\varphi^2 \tag{13}$$

by a suitable choice of coordinates. λ and ρ are functions of (x^1, x^2) only. In this parametrization one of the field equations for the Lagrangean eq. (4) is

$$\partial^i \partial_i \rho = 0 \tag{14}$$

i.e. $\rho(x)$ is a harmonic function. Introducing a conjugate harmonic function $z(x)$ through $\partial z = -{*}\partial\rho$ we may choose the gauge (Weyl's canonical coordinates) $\rho = x^1$ and $z = x^2$. In these coordinates the equation for the conformal factor λ of the 2-dimensional metric reads

$$\lambda^{-1}\partial_\pm \lambda = \mp\frac{i\rho}{4}\mathrm{Tr}(\mathcal{T}_\pm \mathcal{T}_\pm) \tag{15}$$

with $\partial_\pm = \partial_\rho \mp i\partial_z$. From this equation one may obtain λ through integration from $\mathcal{T}$. In addition we have the field equation for M

$$\partial^i(\rho M^{-1}\partial_i M) = 0 \tag{16}$$

which may also be written in terms of the triangular parametrization P

$$\partial^i(\rho\partial_i P\,P^{-1}) - \partial^i(\rho\tau(\partial_i P\,P^{-1})) + \rho[\partial^i P\,P^{-1}, \tau(\partial_i P\,P^{-1})] = 0 \tag{17}$$

or in terms of the currents

$$\mathcal{D}^i(\rho\mathcal{T}_i) \equiv \partial^i(\rho\mathcal{T}_i) - \rho[\mathcal{A}^i, \mathcal{T}_i] = 0 \tag{18}$$

with the SO(2)-covariant derivative $\mathcal{D}_i = \partial_i - [\mathcal{A}_i,]$.

5 The linear spectral problem

One of the typical properties of completely integrable non-linear field equations is the existence of a linear system of equations with a complex parameter (spectral parameter), whose integrabiity conditions reproduce the non-linear field equations. Classical cases are the KdV equation or the Sine-Gordon equation [11]. Amazingly also the equations (16) resp. (17) have this property. The linear spectral problem (LSP) in this case reads

$$\partial_{\pm}\mathcal{P}(x,t) = (\mathcal{A}_{\pm} + \frac{1 \mp it}{1 \pm it}\mathcal{T}_{\pm})\mathcal{P}(x,t) \tag{19}$$

where $\mathcal{A}(x)$ and $\mathcal{T}(x)$ are the currents derived from P(x). $\mathcal{P}(x,t)$ is a SL(2) matrix obeying the normalization condition $\mathcal{P}(x,0) = P(x)$. As a consequence of the explicit ρ-dependence of the field equations (16) resp. (17) the "spectral parameter" t is a function of $x = (\rho,z)$ satisfying

$$\partial_{\pm}t = t\frac{1 \mp it}{1 \pm it} \tag{20}$$

Upon integration one obtains the quadratic equation

$$\frac{1}{t} - t = \frac{2}{\rho}(w - z) \tag{21}$$

with a constant of integration w. The quadratic equation for t has the solutions

$$t_{\pm}(w,x) = \frac{1}{\rho}(z - w) \pm \sqrt{(z - w)^2 + \rho^2}) = -\frac{1}{t_{\mp}}. \tag{22}$$

The pairs (w,t) solving eq. (21) define for each given x with $\rho \neq 0$ a two-sheeted Riemann surface $\mathcal{R}_x$ with the branch points $w = z \pm i\rho$. The replacement $t \longrightarrow -\frac{1}{t}$ exchanges the two sheets. w plays the role of an x-independent spectral parameter. At points x where P(x) is regular the SL(2)-valued function $\mathcal{P}(x,t)$ is analytic in a certain domain $\mathcal{T}_+^x$ on the Riemann surface $\mathcal{R}_x$ containing the point $(w = \infty, t = 0)$. This implies in particular that $\mathcal{P}(x,t)$ has an expansion in positive powers of t at t = 0. Together with the triangularity of $\mathcal{P}(x,0) = P(x)$ this can be interpreted as a "triangularity" property of $\mathcal{P}$ considered as an element of a formal "Kac-Moody group" of SL(2)-valued formal power series in t and 1/t [12].

One can easily verify that the LSP (19) is invariant under the transformation

$$\tau^{(\infty)}:\mathcal{P}(t,x) \longrightarrow \tau(\mathcal{P}(-\frac{1}{t},x)). \tag{23}$$

The involution $\tau^{(\infty)}$ is a natural extension of the involution τ acting on sl(2) to the infinite dimensional Kac-Moody algebra. Because of the invariance of the LSP $\tau^{(\infty)}\mathcal{P}(t,x)$ is a solution of the same LSP as $\mathcal{P}(t,x)$. Its domain of analyticity $\mathcal{T}_{-}^{x}$ contains a neighbourhood of $(w = \infty, t = \infty)$ on $\mathcal{R}_x$. The intersection $\mathcal{T}^{x}$ of the domains $\mathcal{T}_{\pm}^{x}$ contains a contour C^x invariant under $t \longrightarrow -\frac{1}{t}$. Guided by eq. (8) we can construct a new function

$$\mathcal{M}(t,x) \equiv \tau^{(\infty)}(\mathcal{P}^{-1}(t,x))\mathcal{P}(t,x) = \mathcal{P}^{T}(-\frac{1}{t},x)\mathcal{P}(t,x) \tag{24}$$

analytic in $\mathcal{T}^{x}$ for fixed x. Since $\mathcal{P}(t,x)$ and $\tau(\mathcal{P}(-\frac{1}{t},x))$ are solutions of the same linear differential equation we find, as a consequence, that $\mathcal{M}(t,x)$ is independent of x, i.e. only a function of w, which we denote by $\mathcal{M}(w)$ for simplicity. $\mathcal{M}(w)$ plays the role of the scattering or monodromy matrix of the LSP used in the "Inverse Scattering Method" (ISM) [11]. The analyticity properties of $\mathcal{M}(w)$ in the complex w-plane reflect the properties of $\mathcal{P}(t(x,w),x)$ as a function of x and lastly those of P(x). For solutions P(x), which are smooth for large enough $|x|$ and asymptotically flat it turns out that $\mathcal{M}(w)$ is analytic at $w = \infty$ and $\mathcal{M}(\infty) = \mathbf{1}$.

The ISM is based on the possibility of reconstructing $\mathcal{P}(t(x,w)),x)$ from $\mathcal{M}(w)$ solving the factorization problem (Riemann-Hilbert problem)

$$\mathcal{M}(w(x,t) = \tau^{(\infty)}(\mathcal{P}^{-1}(t,x))\mathcal{P}(t,x) \tag{25}$$

where w(x,t) is obtained solving eq. (21) for w. The factorization has to be performed such that $\mathcal{P}(t,x)$ is analytic in a domain $\mathcal{T}_{+}^{x}$ containing $t = 0$ and triangular at $t = 0$. Correspondingly $\tau^{(\infty)}\mathcal{P}(t,x)$ is analytic in a domain (obtained through $t \longrightarrow -\frac{1}{t}$) containing $t = \infty$. A standard method to solve a Riemann-Hilbert problem of the type considered here is to reduce it to a matrix Fredholm integral equation using Cauchy's integral formula [13]. However, in cases where $\mathcal{M}(w)$ is rational the factorization can be obtained in a purely algebraic way, solving a system of linear equations. Once the factorization is achieved the factor $\mathcal{P}(t,x)$ analytic at $t = 0$ automatically solves the LSP, i.e. eq. (19) with $\mathcal{A}$ and $\mathcal{T}$ derived from $P(x) \equiv \mathcal{P}(0,x)$ and as a consequence P(x) solves the field equation (17).

6 The Geroch group

Bäcklund transformations play an important role in the theory of solitons of the classical integrable systems [14]. A first attempt to understand them as elements of a "Kac-Moody group" of transformations acting on special solutions of completely integrable systems was made in [15] (compare also [12]). We shall construct a group of transformations, the Geroch group, containing the Backlund transformations as a subset. This group acts in a canonical way on the solutions of the LSP making essential use of the analyticity properties of $\mathcal{P}(t,x)$.

For any solution $\mathcal{P}(t,x)$ of the LSP and any SL(2)-valued function $g(w)$ the function $\mathcal{P}'(t,x) = \mathcal{P}(t,x)g(w)^{-1}$ is also a solution. Here $g(w(t,x))$ may be interpreted as a constant, i.e. x-independent element of the set of SL(2)-valued functions $g(t,x)$. Two properties of the so obtained new solution of the LSP are essential for it to qualify as a bona fide solution corresponding to some acceptable $P'(x) = \mathcal{P}'(0,x)$:

(i) $\mathcal{P}'$ has to have an appropriate domain of analyticity in t;

(ii) $\mathcal{P}'$ has to be "triangular";

(iii) $\mathcal{P}'(t(w,x),x) \longrightarrow \mathbf{1}$ for $|x| \longrightarrow \infty$ (asymptotic flatness).

Condition (i) requires analyticity of $g(w)$ in a domain containing $w = \infty$, whereas condition (iii) requires the normalization $g(\infty) = \mathbf{1}$. Condition (ii) is in general violated. From eq. (21) we see that $w(t,x)$ introduces negative powers of t and even a w-independent g would in general destroy the triangularity of $P(x)$. In order to restore the "triangularity" of $\mathcal{P}'(t,x)$ we may multiply it from the left by some function $h(\mathcal{P},g)$ invariant under $\tau^{(\infty)}$. Since this does not change the corresponding monodromy matrix $\mathcal{M}_g = (g^T)^{-1}\mathcal{M}g^{-1}$ the transformation

$$\mathcal{P}(t,x) \longrightarrow h(\mathcal{P},g)\mathcal{P}(t,x)g(w)^{-1} \equiv \mathcal{P}_g(t,x) \tag{26}$$

yields another solution of the LSP (19). Observe the remarkable similarity to the transformation properties of coset representatives P and M discussed in section 3!

In view of the properties required for $g(w)$ we arrive at a preliminary definition of the Geroch group ("Kac-Moody group") as the set of SL(2)-valued functions which are holomorphic in a domain containing $w = \infty$ normalized to $g(\infty) = \mathbf{1}$. In addition we have to require that $g(w)$ is real for real w. In order to obtain a group (by pointwise multiplication) we have to take the maximal connected domain of $g(w)$ containing $w = \infty$. As will be discussed in the next section the actual Geroch group has to be taken as a central extension of the group just defined.

The main problem in the implementation of the transformation eq. (26) lies in the construction of $h(\mathcal{P},g)$. One possibility for solving this problem is to apply the ISM. The essential observation is the simplicity of the transformation behaviour of

$\mathcal{M}(w)$. From $\mathcal{M}_g$ we can obtain $\mathcal{P}_g$ solving the Riemann-Hilbert factorization problem discussed above. Hence we end up with the following sequence of steps to implement the Geroch group on the solutions of eq. (17)

$$P \longrightarrow \mathcal{P} \longrightarrow \mathcal{M} \longrightarrow \mathcal{M}_g \longrightarrow \mathcal{P}_g \longrightarrow P_g. \tag{27}$$

Let us consider a simple example where the factorization can be obtained by inspection. As the initial solution we take Minkowski space given by $\mathcal{P}(t,x) = \mathbf{1}$ and for $g(w)$ we take

$$g(w) = \begin{pmatrix} \sqrt{\frac{w-m}{w+m}} & 0 \\ 0 & \sqrt{\frac{w+m}{w-m}} \end{pmatrix}. \tag{28}$$

leading to

$$\mathcal{M}_g(w) = \begin{pmatrix} \frac{w-m}{w+m} & 0 \\ 0 & \frac{w+m}{w-m} \end{pmatrix}. \tag{29}$$

Factorizing

$$\begin{aligned} w - m &= -\frac{\rho}{2t_1}(t-t_1)(\frac{1}{t}+t_1) \text{ with } t_1 = \frac{1}{\rho}\left((z-m) + \sqrt{(z-m)^2 + \rho^2}\right) \\ w + m &= -\frac{\rho}{2t_2}(t-t_2)(\frac{1}{t}+t_2) \text{ with } t_2 = \frac{1}{\rho}\left((z+m) + \sqrt{(z+m)^2 + \rho^2}\right) \end{aligned} \tag{30}$$

allows us to separate the poles of $\mathcal{M}_g(t,x)$ in the complex t-plane and to obtain

$$\mathcal{P}_g(t,x) = \begin{pmatrix} \sqrt{\frac{t_2}{t_1}\frac{t-t_1}{t-t_2}} & 0 \\ 0 & \sqrt{\frac{t_1}{t_2}\frac{t-t_2}{t-t_1}} \end{pmatrix} \tag{31}$$

The corresponding solution of the field equation (16) is $\tilde{\psi} = 0$ and $\Delta = \frac{t_1}{t_2}$, which is nothing but the Schwarzschild solution in Weyl's canonical coordinates. Hence the Schwarzschild solution can be regarded as some kind of soliton obtained through a Bäcklund transformation from the trivial solution (Minkowski space).

In addition to the transformations of the Geroch group there acts also a non-linear realization of the conformal group of the complex plane (resp. the Riemann sphere). For any holomorphic function f(w) leaving the point at infinity fixed we may put $\mathcal{M}_f(w) \equiv \mathcal{M}(f^{-1}(w))$. Again there corresponds a transformation $\mathcal{P} \longrightarrow \mathcal{P}_f$ obtained through factorization of $\mathcal{M}_f$. The infinitesimal form of these transformations on the solutions of the Ernst equation, a realization of the so-called Virasoro algebra, was introduced in [16]. Its combination with the Kac-Moody algebra has the structure of a semi-direct product, well-known from "String theory".

7 **The central extension**

The full 4-dimensional metric requires the knowledge of the conformal factor λ(x) (compare (13)) in addition to M(x). Although λ can be computed through eq. (15) once M is known it would be more satisfactory to know its change under the transformation (26) directly. An analysis of this change under infinitesimal transformations reveals that λ gives rise to a central extension of the Lie algebra of sl(2)-valued formal Laurent series in the spectral parameter. This central extension is given in terms of a 2-cocycle ω on the Lie algebra. ω is characterized by the identities

$$\omega(A,B) = -\omega(B,A) \tag{32}$$

$$\omega([A,B],C) + \omega([B,C],A) + \omega([C,A],B) = 0 \tag{33}$$

In the present case the relevant 2-cocycle is given by

$$\omega(\delta g_1(w),\delta g_2(w)) = \frac{1}{2\pi i} \oint \mathrm{Tr}(\delta g_1(w) d\phi g_2(w)). \tag{34}$$

Lie algebra 2-cocycles are related to inhomogeneous group 2-cocycles obeying

$$\Omega(b,c) - \Omega(ab,c) + \Omega(a,bc) - \Omega(a,b) = 0 \tag{35}$$

The relation between ω and Ω is given by

$$\omega(A,B) = \frac{d^2}{dv^2}\Omega(e^{vA},e^{vB})\Big|_{v=0} \tag{36}$$

leading to the expression for Ω

$$\Omega(a,b) = \int_\Delta \omega(g^{-1}dg,g^{-1}dg) \tag{37}$$

where Δ is a geodesic triangle in the group with vertices {1,a,ab}.

Such a group 2-cocycle defines a group extension through

$$(a,e^{\alpha}) \circ (b,e^{\beta}) = (ab,e^{\alpha+\beta+\Omega(a,b)}) \tag{38}$$

Combining λ and $\mathcal{P}$ it is possible to express its behaviour under the transformation (26) with the group extension corresponding to (34):

$$(\mathcal{P},\lambda^{-1}) \longrightarrow (h(\mathcal{P},g),1) \circ (\mathcal{P},\lambda^{-1}) \circ (g,e^{\gamma})^{-1}. \tag{39}$$

To $\mathcal{M}$ corresponds the x-independent pair (using $\tau^{(\infty)}\lambda = \lambda^{-1}$)

$$(\mathcal{M},\mu) \equiv \tau^{(\infty)}((\mathcal{P},\lambda^{-1})^{-1}) \circ (\mathcal{P},\lambda^{-1}) = (\mathcal{M},\lambda^{-2}e^{\Omega(\tau^{(\infty)}\mathcal{P}^{-1},\mathcal{P})}) \tag{40}$$

from which one obtains a nice explicit formula for λ:

$$\lambda^2 = e^{\Omega(\tau^{(\infty)}\mathcal{P}^{-1},\mathcal{P})}. \tag{41}$$

8 Application to KdV

The LSP of the KdV equation is the Schrödinger equation

$$-\psi_{xx} + u(x)\psi = k^2\psi \tag{42}$$

which we conveniently write as a first order system

$$\begin{pmatrix}\psi\\ \psi_x\end{pmatrix}_x = \begin{pmatrix}0 & 1\\ u-k^2 & 0\end{pmatrix}\begin{pmatrix}\psi\\ \psi_x\end{pmatrix} = L\begin{pmatrix}\psi\\ \psi_x\end{pmatrix}. \tag{43}$$

Assuming $u(x) \longrightarrow 0$ for $|x| \longrightarrow \infty$ there exists a basis (φ,ψ) of solutions behaving as e^{ikx} for $x \longrightarrow -\infty$ resp. $x \longrightarrow +\infty$. $\bar{\varphi}$ and ψ have an analytic extension to Imk < 0.

Next we introduce as an analogue of $\mathcal{P}$ the matrix

$$\Phi = \begin{pmatrix}\psi & \varphi\\ \psi_x & \bar{\varphi}_x\end{pmatrix} \quad \text{obeying} \quad \Phi_x = L\Phi \tag{44}$$

Similarly there is an involution leaving the LSP invariant

$$\tau^{(\infty)}\Phi(k,x) = \bar{\Phi}(\bar{k},x)\sigma_1 \quad \text{with} \quad \sigma_1 = \begin{pmatrix}0 & 1\\ 1 & 0\end{pmatrix}. \tag{45}$$

The analogue of $\mathcal{M}$ is given by $S(k) = \tau^{(\infty)}\Phi(k,x)^{-1}\Phi(k,x)$. The x-independence follows again from the fact that with Φ also $\tau^{(\infty)}\Phi$ solves eq. (44) with the same (real) value of k. The matrix

$$S(k) = \begin{pmatrix} \frac{1}{a} & \frac{\bar{b}}{a} \\ -\frac{b}{a} & \frac{1}{a} \end{pmatrix} \tag{46}$$

is the scattering matrix of the Schrödinger equation. It has the following properties

(i) $S^{+} = S^{-1}$,

(ii) $\bar{S}(\bar{k}) = S(-k)$,

(iii) $S(\infty) = \mathbf{1}$ and

(iv) $(\sigma_1 S)^{T} = \sigma_1 S$.

There is again a Kac-Moody group of transformations induced by matrices g(k) depending on the spectral parameter. The S-matrix is transformed as

$$S_g = \sigma_1^{-1} g^{T} \sigma_1 S g. \tag{47}$$

In order to preserve the properties (i)-(iv) of the S-matrix g(k) has to have the properties $g^{+} = g^{-1}, \bar{g}(\bar{k}) = g(-k)$ and $g(\infty) = \mathbf{1}$.

Let us consider some simple examples:

(i) Starting from $S = \mathbf{1}$ corresponding to $u = 0$ we may take

$$g(k) = \begin{pmatrix} \sqrt{\frac{k+i\kappa}{k-i\kappa}} & 0 \\ 0 & \sqrt{\frac{k+i\kappa}{k-i\kappa}} \end{pmatrix} \tag{48}$$

leading to $a(k) = \frac{k+i\kappa}{k-i\kappa}$ and $b = 0$. Solving the factorization problem yields the 1-soliton solution of the KdV equation, i.e. we have performed a Bäcklund transformation.

(ii) Taking

$$g = \begin{pmatrix} e^{4ik^3 t} & 0 \\ 0 & e^{-4ik^3 t} \end{pmatrix} \tag{49}$$

results in the transformation of the scattering data

$$a(k) \longrightarrow a(k) \quad \text{and} \quad b(k) \longrightarrow b(k)e^{8ik^3 t} \tag{50}$$

describing the time evolution $u(x,0) \longrightarrow u(x,t)$ of the KdV equation.

(iii) Taking

$$g(k) = \begin{pmatrix} e^{2ikt} & 0 \\ 0 & e^{-2ikt} \end{pmatrix} \tag{51}$$

results in a space translation $u(x) \longrightarrow u(x + t)$ of the solution. More generally, taking

$$g_l = \begin{pmatrix} e^{i(2k)^{2l-1}t} & 0 \\ 0 & e^{-i(2k)^{2l-1}t} \end{pmatrix} \tag{52}$$

induces the lth flow of the KdV hierarchy.
(iv) Finally we may combine the translation described in (iii) with the conformal transformation $k \longrightarrow f(k)$ with some holomorphic function $f(k)$ to produce the whole KdV hierarchy.

References

[1] Geroch, R. (1972). *J. Math. Phys.* **13**, 394.
[2] Kinnersley, W. & Chitre, D. (1977). *J. Math. Phys.* **18**, 1538; (1978). **19**, 1926; (1978). **19**, 2037.
[3] Belinskii, V.A. & Zakharov, V.E. (1978). *Zh. Eksp. Teor. Fiz.* **75**, 1955; (1979). **77**, 3.
[4] Maison, D. (1978). *Phys. Rev. Lett.* **41**, 521.
[5] Harrison, B.K. (1978). *Phys. Rev. Lett.* **41**, 1197; G. Neugebauer (1979). *J. Phys.* **A 12**, L67.
[6] Ernst, F. & Hauser, I. (1981). *J. Math Phys.* **22**, 1051; Hauser, I., in [9]; Ernst, F., in [9]; Ernst, F. & Hauser, I., in [10].
[7] Breitenlohner, P. & Maison, D. (1987). *Ann. Inst. Henri Poincaré* **46**, 215.
[8] Weinberg, S. (1968). *Phys. Rev.* **166**, 1568; Meetz, K. (1969). *J. Math. Phys.* **10**, 65; Coleman, S., Wess, J. & Zumino, B. (1969). *Phys. Rev.* **177**, 2239.
[9] Hoenselaers, C. & Dietz, W. (eds.) (1984). *Solutions of Einstein's Equations: Techniques and Results.* Springer: Berlin.
[10] Morris, H. & Dodd, R. (eds.). *Solitons in General Relativity*, to appear. Plenum: New York.
[11] Novikov, S., Manakov, S.V., Pitaevskii, L.P. & Zakharov, V.E. (1984). *Theory of Solitons.* Consultants Bureau: New York.
[12] Segal, G.B. & Wilson, G. (1985). *Pub. Math. I.H.E.S.* **61**, 5.
[13] Muskhelishvili, N.I. (1953). *Singular Integral Equations.* P. Nordhoff N.V.: Groningen.
[14] Bullough, R.K. & Caudrey, P.J. (eds.) (1980). *Solitons.* Springer: Berlin; Eilenberger, G. (1981). *Solitons.* Springer: Berlin.
[15] Date, E., Jimbo, M., Kashiwara, M. & Miwa, T. (1983). In *Non-linear Integrable Systems - Classical Theory and Quantum Theory.* Jimbo, M. & Miwa, T. (eds.). World Scientific: Singapore.
[16] Hou, B.Y. & Li, W. (1987). *Lett. Math. Phys.* **13**, 1.

Gu Chaohao and Zhou Zixiang

12. EXPLICIT FORM OF BÄCKLUND TRANSFORMATIONS FOR GL(N), U(N) AND O(2N) PRINCIPAL CHIRAL FIELDS

The explicit form of Darboux matrices of Bäcklund transformations for GL(N), U(N) and O(2N) principal chiral fields is obtained. The algorithm is purely algebraic and can be continued successively, provided that a seed solution and a fundamental solution of the Lax pair is given. It is noted that a Bäcklund transformation does not create or kill solitons in the case of principal chiral fields.

1 **Darboux matrix method for GL(N) principal chiral field**

Let g(t,x) be a GL(N)-valued function and

$$S[g] = \int \mathrm{tr}(g_\xi g^{-1} g_\eta g^{-1}) d\xi d\eta \tag{1}$$

an action functional. Here ξ, η are light-cone coordinates

$$\xi = x+t, \qquad \eta = x-t \tag{2}$$

A principal chiral field is a solution of the variational problem

$$\delta S[g] = 0 \tag{3}$$

and is called a harmonic map mathematically.
The Euler equation is

$$A_\xi + B_\eta = 0 \tag{4}$$

Here

$$A = g_\eta g^{-1}, \qquad B = g_\xi g^{-1} \tag{5}$$

Hence we have

$$A_\xi - B_\eta + [A,B] = 0 \tag{6}$$

(4) and (6) are equations of GL(N) principal chiral field.

Equations (4) and (6) admit a Lax pair

$$\phi_\eta = (1+\mu)^{-1}A\phi, \qquad \phi_\xi = (1-\mu)^{-1}B\phi \tag{7}$$

or

$$\phi_\eta = \lambda A\phi, \qquad \phi_\xi = \lambda(2\lambda-1)^{-1}B\phi \tag{8}$$

The zero-curvature condition

$$\lambda A_\xi - \lambda(2\lambda-1)^{-1}B_\eta + \lambda^2(2\lambda-1)^{-1}[A,B] = 0 \tag{9}$$

is exactly (4) and (6) [8].

The Bäcklund transformations for principal chiral fields were considered in [9]. We are going to construct the Darboux matrix explicitly. It should be noted that the Darboux matrix method is very powerful for AKNS system (e.g. [1,2,4,5,6,7,10]) and the Cauchy problem for the principal chiral fields of compact group in $R^{1,1}$ was first solved in [3].

Suppose that a solution (A,B) and a fundamental solution $\phi=\phi(\lambda,\eta,\xi)$ of (8) are given. The Darboux matrix of first degree has the form

$$S = I + \alpha\lambda \tag{10}$$

where α is an N x N matrix. Let $\phi_1 = S\phi$. From

$$\phi_{1\eta} = \lambda A_1\phi_1, \qquad \phi_{1\xi} = \lambda(2\lambda-1)^{-1}B_1\phi_1 \tag{11}$$

we obtain

$$\begin{aligned} \alpha_\eta\alpha &= \alpha A - A\alpha \\ \alpha_\xi\alpha + 2\alpha_\xi &= B\alpha - \alpha B \end{aligned} \tag{12}$$

and

$$\begin{aligned} A_1 &= \alpha A\alpha^{-1} = A + \alpha_\eta \\ B_1 &= (\alpha+2)B(\alpha+2)^{-1} = B - \alpha_\xi \end{aligned} \tag{13}$$

It is seen that (12) is a completely integrable system for α and (13) is an explicit form of the Bäcklund transformation provided α is known.

The general solution α can be constructed in the following way.

Choose N numbers λ_i $(i=1,2,\ldots,N)(\lambda_i \neq 0, 1/2)$, and let $h_1(\lambda_1),\ldots\ldots,h_N(\lambda_N)$ be N column solutions of the Lax pair which can be obtained from the fundamental solution $\phi(\lambda)$. Moreover, construct a matrix

$$H = [h_1(\lambda_1),\ldots\ldots,h_N(\lambda_N)]$$

We choose $h_i(\lambda_i)$ such that $\det H \neq 0$.

It is seen that

$$\alpha = -H\Lambda^{-1}H^{-1} \tag{14}$$

with

$$\Lambda = \begin{pmatrix} \lambda_1 & & 0 \\ & & \\ 0 & & \lambda_N \end{pmatrix} \tag{15}$$

is a solution of the system (12). In fact, we have

$$\begin{aligned} H_\eta &= AH\Lambda \\ H_\xi &= BH\Lambda(2\Lambda-1)^{-1} \end{aligned} \tag{16}$$

Hence

$$\begin{aligned} \alpha_\eta\alpha &= H_\eta\Lambda^{-2}H^{-1} - H\Lambda^{-1}H^{-1}H_\eta\Lambda^{-1}H^{-1} \\ &= \alpha A - A\alpha \\ \alpha_\xi\alpha + 2\alpha_\xi &= H_\xi\Lambda^{-2}H^{-1} - H\Lambda^{-1}H^{-1}{}_\xi\Lambda^{-1}H^{-1} - 2H_\xi\Lambda^{-1}H^{-1} + 2H\Lambda^{-1}H^{-1}H_\xi H^{-1} \\ &= B\alpha - \alpha B \end{aligned}$$

Thus (12) is satisfied.

It is also noted that all solutions of (12) which are similar to diagonal matrices at a point (hence at all points) are obtained if the diagonal elements are neither zero nor (-2).

Moreover, from (11), we know that the corresponding principal chiral field is

$$g_I = (I + \alpha)g \tag{17}$$

if $\lambda_i \neq 1$ $(i=1,2,\ldots\ldots,N)$.

2 Example

For simplicity, we choose N = 2. Take a seed solution (A,B) such that A,B are diagonal,

$$A = \begin{bmatrix} \varphi_1 & 0 \\ 0 & \varphi_2 \end{bmatrix}, \quad B = \begin{bmatrix} \psi_1 & 0 \\ 0 & \psi_2 \end{bmatrix} \tag{18}$$

Since A and B commute each other, the seed solution consists of linear waves

$$A = \begin{bmatrix} \varphi_1(\eta) & 0 \\ 0 & \varphi_2(\eta) \end{bmatrix}, \quad B = \begin{bmatrix} \psi_1(\xi) & 0 \\ 0 & \psi_2(\xi) \end{bmatrix} \tag{19}$$

where φ_1, φ_2, ψ_1, ψ_2 are arbitrary functions. Moreover, $\varphi_1(\eta)$, $\varphi_2(\eta)$ are waves going to the right and $\psi_1(\xi)$, $\psi_2(\xi)$ are waves going to the left. If these functions contain several impulses, we have "linear multi-solitons".

We assume that $\varphi_a(\eta)$, $\psi_a(\xi)$ (a=1,2) approach zero rapidly as $\eta,\xi\to-\infty$, then a fundamental solution to the Lax pair is

$$\phi(\lambda) = \begin{bmatrix} P_1Q_1 & 0 \\ 0 & P_2Q_2 \end{bmatrix} \tag{20}$$

where

$$\begin{aligned} P_a &= \exp(\lambda\textstyle\int_{-\infty}^{\eta}\varphi_a(\eta)d\eta) \\ Q_a &= \exp(\lambda(2\lambda-1)^{-1}\textstyle\int_{-\infty}^{\xi}\psi_a(\xi)d\xi) \end{aligned} \qquad (a=1,2) \tag{21}$$

Let

$$h_1(\lambda_1) = \begin{bmatrix} aP_1(\lambda_1)Q_1(\lambda_1) \\ bP_2(\lambda 1)Q_2(\lambda_1) \end{bmatrix} \quad h_2(\lambda_2) = \begin{bmatrix} cP_1(\lambda_2)Q_1(\lambda_2) \\ dP_2(\lambda_2)Q_2(\lambda_2) \end{bmatrix} \tag{22}$$

where a,b,c,d are constants such that

$$\det H = \det\,[h_1(\lambda_1),\ h_2(\lambda_2)] \neq 0$$

Then

$$\alpha = -(adP_1(\lambda_1)Q_1(\lambda_1)P_2(\lambda_2)Q_2(\lambda_2) - bcP_1(\lambda_2)Q_1(\lambda_2)P_2(\lambda_1)Q_2(\lambda_1))^{-1}\cdot$$

$$\cdot\begin{bmatrix} aP_1(\lambda_1)Q_1(\lambda_1)\lambda_1^{-1} & cP_1(\lambda_2)Q_1(\lambda_2)\lambda_2^{-1} \\ bP_2(\lambda_1)Q_2(\lambda_1)\lambda_1^{-1} & dP_2(\lambda_2)Q_2(\lambda_2)\lambda_2^{-1} \end{bmatrix}\cdot$$

$$\cdot\begin{bmatrix} dP_2(\lambda_2)Q_2(\lambda_2) & -cp_1(\lambda_2)Q_1(\lambda_2) \\ -bP_2(\lambda_1)Q_2(\lambda_1) & aP_1(\lambda_1)Q_1(\lambda_1) \end{bmatrix}$$

Thus, from (13), we obtain a new solution for which A_1, B_1 do not commute each other. Moreover, if $\varphi_a(\eta)$, $\varphi_a(\xi)$ approach zero rapidly as $\eta \to \infty$ and $\xi \to \infty$, then derivatives of

$$\int_{-\infty}^{\eta}\varphi_a(\eta)d\eta, \qquad \int_{-\infty}^{\xi}\psi_a(\xi)d\xi$$

tend to zero as η and $\xi\to\infty$. The nonlinear solution approaches the original one as $t\to\infty$. In this sense, we see that the Bäcklund transformation may not create or kill solutions for the principal chiral fields, while, as is well-known, the Bäcklund transformations do create or kill solitons for KdV, MKdV, NLS equations, etc.

3 U(N) principal chiral field

In this section, we consider U(N) principal chiral field, i.e. g takes value in U(N), while A,B satisfy $A^+ = -A, B^+ = -B$. Here + denotes complex conjugate transpose.

The Darboux matrix S is constructed as in Section 2, with two additional requirements.

(i) λ_i can be either λ_0 or $\bar{\lambda}_0$, where λ_0 is a complex number which is not real.

(ii) At some point (η_0, ξ_0), $h_1, \ldots, h_N$ are linearly independent, and $(h_i^+h_j)|_{(\eta_0\xi_0)} = 0$ if $\lambda_i \neq \lambda_j$ (i.e. $\lambda_i = \bar{\lambda}_j$).

We know from (8) that $(h_i^+h_j)_\eta = (h_i^+h_j)_\xi = 0$ if $\lambda_i = \bar{\lambda}_j$, and hence $h_i^+h_j = 0$ everywhere. Thus, the linear independence of $\{h_i|\lambda_i = \lambda_0\}$ and $\{h_i|\lambda_i = \bar{\lambda}_0\}$ implies the linear independence of all $\{h_i\}$. Therefore, H is nondegenerate everywhere, and $\alpha = -H\Lambda^{-1}H^{-1}$ is globally defined.

From

$$\alpha[h_1, \ldots, h_N] = -[\lambda_1^{-1}h_1, \ldots, \lambda_N^{-1}h_N] \tag{23}$$

we have

$$h_i^+(\alpha^+ + \alpha)h_j = -(\bar{\lambda}_i^{-1} + \lambda_j^{-1})h_i^+h_j = -(\lambda_0^{-1} + \bar{\lambda}_0^{-1})h_i^+h_j \tag{24}$$

which implies

$$\alpha^+ + \alpha = -(\lambda_0^{-1} + \bar{\lambda}_0^{-1})I \tag{25}$$

since $\{h_i\}$ is a basis of R^N. Hence

$$\begin{aligned}\alpha^+\alpha &= (-(\lambda_0^{-1} + \bar{\lambda}_0^{-1})I-\alpha)\alpha \\ &= H((\lambda_0^{-1} + \bar{\lambda}_0^{-1})I-\Lambda^{-1})\Lambda^{-1}H^{-1} \\ &= |\lambda_0|^{-2}I\end{aligned} \tag{26}$$

and

$$(I + \alpha)^+(I+\alpha) = |1-\lambda_0^{-1}|^2 I \tag{27}$$

Considering the transformation (13), we have

$$A_1^+ = -A_1, \qquad B_1^+ = -B_1 \tag{28}$$

since

$$\alpha_\eta^+ = -\alpha_\eta, \qquad \alpha_\xi^+ = -\alpha_\xi \tag{29}$$

by (25). Therefore, the Darboux matrix given above keeps the reduction.

The transformation on g is

$$g_1 = (I + \alpha)g\Lambda(\Lambda-1)^{-1} \tag{30}$$

We take a right multiplier $\Lambda(\Lambda-1)^{-1}$ such that g_1 takes value in U(N) by (27).

Moreover, if we consider SU(N) principal chiral field, the construction of the Darboux matrix is the same, because the transformation (13) keeps trace zero.

4 O(2N) principal chiral field

For O(2N) model, A,B are real matrices with $A^T = -A, B^T = -B$. The Darboux matrix is dealt with in a similar way as in Section 3. However, the matrix α should be real.

Given $\lambda_0 = (\mu + \nu i)^{-1}$ $(\nu \neq 0)$, let

$$J = \begin{bmatrix} \mu I & \nu I \\ -\nu I & \mu I \end{bmatrix} \tag{31}$$

where I is N x N identity matrix. Then, choose a nondegenerate real matrix R such that $R^T R$ has the form

$$\begin{bmatrix} W & 0 \\ 0 & W \end{bmatrix}$$

for some N x N matrix W.

Let $\alpha_0 = -RJR^{-1}$, then

$$\alpha_0^T \alpha_0 = R^{T-1} J^T R^T R J R^{-1}$$

$$= (\mu^2 + \nu^2)\ R^{T-1} \begin{bmatrix} W & \\ & W \end{bmatrix} R^{-1} = (\mu^2 + \nu^2) I_{2N} \tag{32}$$

Hence α_0 is a multiple of an orthogonal matrix, which is certainly diagonizable.

We can write $\alpha_0 = -K\Lambda^{-1}K^{-1}$ where

$$\Lambda = \begin{bmatrix} \lambda_1 & & 0 \\ & & \\ 0 & & \lambda_{2N} \end{bmatrix}$$

with $\lambda_i = \lambda_0$ or $\bar{\lambda}_0$, and $K = [k_1, \ldots, k_{2N}]$ nondegenerate. Inserting this expression of α_0 into (32), we have

$$(\mu^2 + \nu^2)\Lambda^+ K^+ K \Lambda = K^+ K \tag{33}$$

i.e.

$$(\bar{\lambda}_i \lambda_j - |\lambda_0|^2) k_i^+ k_j = 0 \quad (i, j = 1, \ldots 2N) \tag{34}$$

which implies $k_i^+ k_j = 0$ if $\lambda_i \neq \lambda_j$.

Let $h_i(\lambda_i, \eta, \xi)$ be the solution of

$$\begin{cases} h_{i\eta} = \lambda_i A h_i \\ h_{i\xi} = \lambda_i (2\lambda_i - 1)^{-1} B h_i \\ h_i(\eta_0, \xi_0) = k_i \end{cases} \tag{35}$$

and $H = [h_1, \ldots, h_{2N}]$, the $h_i^+ h_j = 0$ if $\lambda_i \neq \lambda_j$, and $\det H \neq 0$. From

Section 3, we only need to prove that $\alpha = -H\Lambda^{-1}H^{-1}$ takes real values.

Write $\alpha = F + Gi$, where F,G are real-matrix-valued functions. Let $Z = \{(\eta,\xi)\in R^{1,1} | G(\eta,\xi)=0\}$, then $(\eta_0,\xi_0)\in Z$. Near a point $(\eta_1,\xi_1)\in Z$, F, F+2 are nondegenerate since α has no real eigenvalues. Here 2 is the abbreviation of the matrix $2I_{2N}$.

Considering (12), we have

$$\begin{aligned} F_\eta F - G_\eta G &= [F,A] \\ F_\eta G + G_\eta F &= [G,A] \\ F_\xi (F+2) - G_\xi G &= [B,F] \\ F_\xi G + G_\xi (F+2) &= [B,G] \end{aligned} \tag{36}$$

Eliminating F_η, F_ξ, we have

$$\begin{aligned} G_\eta (F+GF^{-1}G) &= GA - FAF^{-1}G \\ G_\xi (F+2+G(F+2)^{-1}G) &= (F+2)B(F+2)^{-1}G - GB \end{aligned} \tag{37}$$

G=0 is clearly a solution of this group of equations. Moreover, since the coefficient matrices $F + GF^{-1}G$, $F + 2 + G(F + 2)^{-1}G$ are nondegenerate near (η_1,ξ_1), the uniqueness of (37) implies G=0 near (η_1,ξ_1), which means Z is open. Hence G = 0 everywhere and α has real value.

The Darboux transformation on A,B is given by (13), and

$$g_1 = (I + \alpha)g|1-1/\lambda_0|^{-1}$$

is an O(2N) principal chiral field.

Acknowledgement

This work was supported by the Chinese Fund of Natural Science and the French University Council when the first author was visiting the University of Paris VI.

A part of the results of the present paper was obtained when the first author was visiting the University of Paris VI and Dijon University. He is grateful to Prof. Y. Choquet-Bruhat and Prof. M. Flato for their hospitality, attention to the results and valuable discussions. He is indebted to Dr. E. Taflin for interesting discussions.

References

[1] Gu, C.H. (1986). On the Bäcklund Transformations for the Generalized Hierarchies of Compound MKdV-SG Equations. *Lett. Math. Phys.* **12**, 31.

[2] Gu, C.H. (1987). On the Darboux Form of Bäcklund Transformations. *Proc. 1987 Symp. on Integrable Systems, Nankai Univ.* World Press.

[3] Gu, C.H. (1980). On the Cauchy Problem for Harmonic Maps Defined on Two-dimensional Minkowski Space. *Comm. Pure Appl. Math.* **33**, 727.

[4] Gu, C.H. & Hu, H.S. (1986). A Unified Explicit Form of Bäcklund Transformations for Generalized Hierarchies of KdV equations. *Lett. Math. Phys.* **11**, 325.

[5] Li, Y.S., Gu, X.S. & Zou, M.R. (1987). Three kinds of Darboux Transformations for the Evolution Equations which connect with the AKNS Eigenvalue Problem. *Acta Math. Sinica* **3**, 143.

[6] Gu, C.H. & Zhou, Z.X. (1987). On the Darboux Matrices of Bäcklund Transforations for the AKNS Systems. *Lett. Math. Phys.* **13**, 179.

[7] Neugerbauer, C. & Meinel, R. (1984). General N-soliton Solution of the AKNS Class on Arbitrary Background. *Phys. Lett.* **100A**, 467.

[8] Novikov, S., Manakov, S.V. Pitaevskii, L.P. & Zakharov, V.E. (1984). *Theory of Solitons* (English translation). Consultants Bureau: New York.

[9] Ogielski, A.T., Prasad, M.K., Sinha, A. & Chau Wang, L.L. (1980). Backlund Transformations and Local Conservation Laws for Principal Chiral Fields. *Phys. Lett.* **91B**, 387.

[10] Sattinger, D.H. & Zurkowski, V.D. (1987). Gauge Theory of Bäcklund Transformations II. *Physica* **26D**, 225.

N.C. Freeman and C.R. Gilson

13. A STUDY OF TWO EVOLUTION EQUATIONS USING THE WRONSKIAN TECHNIQUE

1 Introduction

Hirota's direct method of solution of nonlinear evolution equations is well-known and has proved very successful in deriving n-soliton solutions for many of the common integrable equations. The work of Jimbo & Miwa [5], subsequently JM, has shown that this technique has wide application and their catalogue of equations which have such solutions proves invaluable in the search for new integrable equations of practicable value. However, the matching of such equations to those listed in the catalogue is not always straightforward and much ingenuity is often needed to achieve success.

In this paper, examples of two equations will be given which belong to the class of equations studied by JM. The first, the Classical Boussinesq equation has a long history and its derivation from the theory of water waves has been given by Whitham [9]. Its relation to the JM hierarchies was shown by Hirota [3,4] using a sophisticated limiting procedure. Here, we shall describe this approach using a technique developed by one of the authors [1] which makes use of the representation of the solution in the form of Wronskian determinants.

A more direct route to the solution of this equation can be made with an entirely different approach using Wronskians which is anticipated in JM's work on two-component τ functions. This topic is also examined in this volume by Nimmo [8].

Secondly, we investigate, using similar techniques, the soliton solutions of Mel'nikov's equations [6,7]. Solutions to these equations have been obtained using inverse scattering theory which can be shown to have the double Wronskian form and hence are related to two component τ-functions. It is also possible to derive an 'm-component' Wronskian to satisfy Mel'nikov's system [6] of m + 1 equations.

2 The Classical Boussinesq equation

The Classical Boussinesq equation [3] may be written in the form

$$u_t + \{(1+u)v\}_x - v_{xxx} = 0$$
$$v_t + (u + \tfrac{1}{2}v^2)_x = 0. \qquad (1)$$

By introducing new variables τ and τ'

$$u = -1 - 2(\log \tau\tau')_{xx}, \qquad v = 2i \log(\tau'/\tau)_x,$$

the equations (1) can be written in terms of Hirota derivatives as

$$(iD_t + D_x^2)\tau\cdot\tau' = 0, \qquad (iD_xD_t + D_x^3)\tau\cdot\tau' = 0 \qquad (2)$$

where the Hirota derivative is defined by

$$D_x^m D_t^n \,\tau\cdot\tau' = \left(\frac{\partial}{\partial x} - \frac{\partial}{\partial x'}\right)^m \left(\frac{\partial}{\partial t} - \frac{\partial}{\partial t'}\right)^n \tau(x,t)\tau'(x',t')\Bigg|_{\substack{x=x'\\ t=t'}}.$$

Equations (2) are the Hirota bilinear forms for the Classical Boussinesq equations, they can be related to the class of equations introduced by JM and referred to as the Modified K-P Hierarchy. The first two equations of this hierarchy are

$$(D_1^2 + D_2)\tau\cdot\tau' = 0 \qquad (3)$$

$$(D_1^3 - 4D_3 - 3D_1D_2)\tau\cdot\tau' = 0 \qquad (4)$$

where D_i denotes the Hirota derivative in the variable x_i and τ,τ' depend on the phase variables $(p_j-q_j)x_1 + (p_j^2 - q_j^2)x_2 + (p_j^3-q_j^3)x_3$, $j = 1, \ldots, n$.

Hirota [3,4] has shown that, by using the 'pq = c' reduction (i.e. assuming p_jq_j is independent of j or $q_j = c/p_j$), equations (3,4) become

$$(D_1^2 + D_2)\tau\cdot\tau' = 0 \qquad (5)$$

$$(D_1^3 + D_1D_2 + 4cD_1)\tau\cdot\tau' = 0 \qquad (6)$$

$$(D_3 + D_1D_2 + cD_1)\tau\cdot\tau' = 0. \qquad (7)$$

In the case pq = c = 0 the equations reduce to the Classical Boussinesq equations (2).

The n-soliton solutions to equations (3,4) can be expressed in terms of Wronskian determinants [2,1]

$$\tau = (\widehat{n-1}), \qquad \tau' = (\tilde{n}) \tag{8}$$

where the notation indicates that τ and τ' are composed of determinants with the jth columns of the form $[\phi_i^{(j)}]$ and the subscript $_{(j)}$ denotes the jth x_1-derivative. For τ, ^ denotes inclusion of the derivatives 0 to n − 1 and, for τ', ~ denotes derivatives 1 to n. The functions in the determinants take the form

$$\phi_i = P_i \exp\left(\sum_{r=1}^{3} (-p_i)^r x_r\right) + Q_i \exp\left(\sum_{r=1}^{3} q_i^r x_r\right) \quad i = 1, \ldots, n$$

these represent the single soliton solutions.[1]

It may be shown, in the notation of Freeman [1], that

$$(D_1^3 + D_1 D_2 + 4cD_1)\tau\cdot\tau' = 4(D_3 + D_1 D_2 + D_1)\tau\cdot\tau'$$

$$= 4(-1)^n\left(\begin{vmatrix} \widehat{n-2} & \cdot & n-1 & n & n+1 & n+2 \\ \cdot & \widehat{n-3} & n-2 & n-1 & n & n+1 \end{vmatrix} + c\begin{vmatrix} \widehat{n-2} & \cdot & n-1 & n \\ \cdot & \widetilde{n-1} & n & n+1 \end{vmatrix}\right). \tag{9}$$

Using the reduction $(-p_i)q_i = c$ in the form $-(p_i + \frac{c}{p_i})\phi_i^{(j)} = \phi_i^{(j+1)} + c\phi_i^{(j-1)}$ the determinantal expression on the right hand side of equation (9) can be shown to be zero using elementary methods of row and column manipulation of the determinants.

3 The 'pq = 0' reduction

If we choose $P_i = (-p_i)^{-r}$ and $Q_i = (q_i)^{-r}$ then this has the effect of shifting all the derivatives r places down so that τ and τ' now take the form

$$\tau = ((\tilde{r})^-, (\widehat{n-r-1})), \qquad \tau' = ((\widetilde{r-1})^-, (\widehat{n-r})) \tag{10}$$

where $(\tilde{k})^-$ denotes all the 'negative' derivatives (powers) from k to 1 with

$$\phi_i = e^{\xi_i} + e^{\eta_i}$$

---

[1]Note that to relate with earlier work on the Wronskian method the sign of p_i has been changed.

and

$$\xi_i = \sum_{r=1}^{3} (-p_i)^r x_r , \quad \eta_i = \sum_{r=1}^{3} q_i^r x_r .$$

Now if we let $q_i \to 0$ $(c \to 0)$ then the dominant behaviour comes from the inverse powers of q_i and the exponential factor e^{η_i} in (8) is unity. So, we obtain:

$$\tau = ([(1/\tilde{q})^r],(\hat{n-r-1})), \qquad \tau' = ([(1/\tilde{q})^{r-1}],(\hat{n-r})), \qquad (11)$$

with the functions in the 'positive' derivative part becoming $\phi_i = 1 + e^{\xi_i}$, where $[(1/q)^r]$ is the column $\{(1/q_i)^r\}^T$.

Using the fact that $pq = c$ this becomes:

$$\tau = ([(-\tilde{p})^r],\hat{n-r-1}), \qquad \tau' = ([(-\tilde{p})^{r-1}],\hat{n-r}). \qquad (12)$$

A certain symmetry will be necessary if the physical variables u and v (equation (1)) are to be real. It will be sufficient to say that we require τ and τ' to be complex conjugates as far as possible. This may be achieved by putting $n = 2N$ and $r = N$. Equations (12) are then:

$$\tau = ([(-\tilde{p})^N],\hat{N-1}), \qquad \tau' = ([(-\tilde{p})^{N-1}],\hat{N}), \qquad (13)$$

now, choosing $p_{i+N} = -p_i^*$ we get $\phi_{i+N} = 1 + e^{\xi_{i+N}} = 1 + e^{-\xi_i^*}$.

A little manipulation shows that

$$\tau' = \tau^* \prod_i e^{(\xi_i - \xi_i^*)} = \tau^* \exp[\sum_i(\xi_i - \xi_i^*)] \qquad (14)$$

thus our aim has been achieved apart from an exponential factor.

The equation (6) (with $c = 0$) now becomes

$$(D_1^3 + D_1D_2)\tau\cdot\tau' = 4\begin{vmatrix} [(-\tilde{p})^N] & \hat{N-2} & \cdot & \cdot & N-1 & N & N+1 & N+2 \\ \cdot & \cdot & [(-\tilde{p})^{N-1}] & \hat{N-3} & N-2 & N-1 & N & N+1 \end{vmatrix}$$

which can be shown to be zero by elementary row and column manipulation of the determinant. Thus, we have shown that the functions τ and τ' given by equations (13) satisfy the equations

$$(D_1^2+D_2)\tau\cdot\tau' = 0, \qquad (D_1^3 + D_1D_2)\tau\cdot\tau' = 0. \qquad (15)$$

If we choose $x_1 = x$, $x_2 = -it$ and $x_3 = 0$, then we obtain $\xi = -px - ip^2t$, $\eta = qx - iq^2t$. Equations (6-7) become the Hirota form of the Classical Boussinesq equations (2).

Since τ and τ' are complex conjugates except for an exponential factor with an argument purely imaginary and linear in x, we find that the physical variables u and v satisfying equation (1) are real but v is shifted by an added constant. This factor can readily be removed by a Galilean transformation.

4 Two component τ-function or double Wronskian formulation

The procedures adopted in the previous sections have left the τ-functions (13) with a two block structure. It is not difficult to transform this into a double Wronskian formulation. A factor $e^{1/2\xi_i}$ can be extracted from each of the rows of the two functions τ and τ'. After a redefinition of variables $P = 1/2p$, $X_1 = x_1$, $X_2 = 2x_2$, we get the representation

$$\tau = (\tilde{N}, \hat{N-1}), \qquad \tau' = (\tilde{N-1}, \hat{N}) \tag{16}$$

where the first Wronskian block is formed by differentiation of the function $\exp(PX_1 - P^2X_2)$ [2], and the second is formed by differentiation of $\exp((-PX_1 + (-P)^2X_2)$. (It should be noted that the first term in the second block of both τ and τ', i.e. the undifferentiated terms, are in fact made up of the sum of both these two functions so that strictly τ should be written as the sum of two Wronskians, viz, $\tau = \{(-1)^N(\hat{N}, \tilde{N-1}) + (\tilde{N}, \hat{N-1})\}$ but since derivatives of the undifferentiated term <u>always</u> give terms which can be removed by column subtraction in the subsequent determinants this complication can be ignored). Equations (15) now become

$$[(D_1 + \hat{D}_1)^2 + 2D_2 + 2\hat{D}_2]\tau\cdot\tau' = 0 \tag{17}$$

$$[(D_1 + \hat{D}_1)^3 + 2(D_1 + \hat{D}_1)(D_2 + \hat{D}_2)]\tau\cdot\tau' = 0 \tag{18}$$

using notation introduced by JM where 'D' denotes differentiation of the first block only and '$\hat{D}$' denotes differentiation of the second block. Unlike them, however, we will use the **same** independent variables in the two blocks.

We can write (17) as

$$[2(D_1^2 + D_2) + 2(\hat{D}_1^2 + \hat{D}_2) - (D_1 - \hat{D}_1)^2]\tau\cdot\tau' = 0. \tag{19}$$

---

[2]Note that the signs in the first block do not coincide with those used by JM.

It is immediately obvious from the form of τ and τ' that the first two groups of terms are zero because of the first equation in the modified K-P Hierarchy (noting the change of sign in the first block).

Using the elementary identity in the theory of determinants

$$|a_1 \ldots a_n| \sum_{i=1}^{n} x_i = \sum_{i=1}^{n} |a_1 \ldots xa_i \ldots a_n| \tag{20}$$

where a_i denotes $(a_{i1} \ldots a_{in})^T$ and xa_i is $(x_1a_{i1}, \ldots, x_na_{in})^T$, we see that

$$(D_1 - \hat{D}_1)W\cdot W' = ((\partial_1-\hat{\partial}_1)W)W' - W(\partial_1-\hat{\partial}_1)W' \quad (\text{with } \partial_1 = \partial/\partial x_1)$$

$$= (\Sigma P_i)WW' - W(\Sigma P_i)W'$$

$$= 0 \tag{21}$$

for <u>any</u> double Wronskians W and W' with the functions $\exp(PX_1 - P^2X_2)$ and $\exp(PX_1 + (-P)^2X_2)$ in the first and second blocks respectively.

Since $(\hat{D}_1 - D_1)^2\tau\cdot\tau'$ is made up of $(\hat{D}_1 - D_1)$ acting on a sum of products of such Wronskians the third term in equation (19) is also zero.

Equation (18) can be written as

$$[(D_1^3+D_1D_2) + (\hat{D}_1^3+\hat{D}_1\hat{D}_2) + 3(\hat{D}_1D_1^2+\hat{D}_1D_2) + 3(D_1\hat{D}_1^2+D_1\hat{D}_2)$$
$$- (\hat{D}_1-D_1)(D_2-\hat{D}_2)]\tau\cdot\tau' = 0. \tag{22}$$

The last term is obviously zero using (21). Now $\hat{D}_1(D_1^2+D_2)\tau\cdot\tau'$ may be written as

$$\hat{D}_1[(\widetilde{N-2},N,N+1;\widehat{N-1})(\widetilde{N-1};N) - (\widetilde{N-1},N+1;\widehat{N-1})(\widetilde{N-2},N;\hat{N})$$
$$+ (\tilde{N},\widehat{N-1})(\widetilde{N-2},N+1,\hat{N})]$$

$$= \begin{vmatrix} (\widetilde{N-2})_1 & (\widehat{N-2,N})_2 & \cdot & \cdot & (N-1)_1 & (N)_1 & (N+1)_1 & (N-1)_2 \\ \cdot & \cdot & (\widetilde{N-2})_1 & (\widehat{N-2,N})_2 & (N-1)_1 & (N)_1 & (N+1)_1 & (N-1)_2 \end{vmatrix}$$

$$- \begin{vmatrix} (\widetilde{N-2})_1 & (\widehat{N-1})_2 & \cdot & \cdot & (N-1)_1 & (N)_1 & (N+1)_1 & (N+1)_2 \\ \cdot & \cdot & (\widetilde{N-2})_1 & (\widehat{N-1})_2 & (N-1)_1 & (N)_1 & (N+1)_1 & (N+1)_2 \end{vmatrix} = 0. \tag{23}$$

by elementary operations on the determinant. A similar result follows for $D_1(\hat{D}_1^2+\hat{D}_2)\tau\cdot\tau' = 0$. Now

$$(D_1^3+D_1D_2)\tau\cdot\tau' = (D_1-\hat{D}_1)(D_1^2+D_2)\tau\cdot\tau' + \hat{D}_1(D_1^2+D_2)\tau\cdot\tau' = 0$$

and similarly $(\hat{D}_1^3+\hat{D}_1\hat{D}_2)\tau\cdot\tau' = 0$.

We thus see that the equations

$$\hat{D}_1(D_1^2+D_2)\tau\cdot\tau' = 0, \qquad D_1(\hat{D}_1^2+\hat{D}_2)\tau\cdot\tau' = 0$$

are members of the two component hierarchy of JM. These, together with the identities obtained from (21) are sufficient to verify that the equations (15) are satisfied.

5 Mel'nikov's equations

Mel'nikov's equations are a set of coupled equations

$$\begin{aligned} &3u_{yy} - [u_t + (3u^2+u_{xx}+8k\phi\psi)_x]_x = 0 \\ &i\phi_y = u\phi + \phi_{xx} \\ &-i\psi_y = u\psi + {}_{xx} \qquad\qquad (k^2 = 1). \end{aligned} \tag{24}$$

The first of these equations is the K-P equation with an interaction term added on. The second two equations are the Schrödinger equation with potential u. In the theory of waves ϕ is usually taken to be the complex conjugate of ψ, i.e. $\phi = k\psi^*$ (* = complex conjugate).

Solutions to these equations have been found using inverse scattering techniques [6,10]. Mel'nikov's solution takes the following form

$$f = |F| = \begin{vmatrix} \delta_{ij} + \dfrac{e^{\Omega_i}e^{\Sigma_j}}{(\omega_i-\sigma_j)} & \dfrac{e^{\Omega_i}e^{\Sigma_k}}{(\omega_i-\sigma_k)} & 0 \\ 0 & \delta_{lk} & -\dfrac{e^{-4(\omega_1^3-\sigma_m^3)t}}{(\omega_1^3-\sigma_m^3)} \\ \dfrac{e^{\Omega_n}e^{\Sigma_j}}{(\omega_n-\sigma_j)} & \dfrac{e^{\Omega_n}e^{\Sigma_k}}{(\omega_n-\sigma_k)} & \delta_{nm} \end{vmatrix}$$

$$\Phi = \left|\begin{array}{c:c:c:c} 0 & 0 & 0 & e^{4\sigma_m^3 t} \\ \hline e^{\Omega_i} & & & \\ \hline 0 & & F & \\ \hline e^{\Omega_n} & & & \end{array}\right| \qquad \Psi = \left|\begin{array}{c:c:c:c} 0 & e^{\Sigma_j} & e^{\Sigma_k} & 0 \\ \hline 0 & & & \\ \hline e^{-4\omega_1^3 t} & & F & \\ \hline 0 & & & \end{array}\right|$$

where $\Omega_i = \omega_i x - i\omega_i^2 y - 4\omega_i^3 t$, $\Sigma_j = -\sigma_j x + i\sigma_j^2 y + 4\sigma_j^3 t$ with $1 \le i,j \le r_1$, $r_1 < k,l \le r_1+r_2$, $r_1+r_2 < n,m \le r_1+r_2$, where the u,ϕ,ψ in Mel'nikov's equations are given by

$$u = 2\frac{\partial^2}{\partial x^2}(\ln f), \qquad \phi = \Phi/f, \qquad \psi = \Psi/f.$$

(F is an $(r_1+r_2+r_2) \times (r_1+r_2+r_2)$ matrix, Φ and Ψ each have one more row and column than F).

In terms of Hirota derivatives the equations (24) can be rewritten as

$$iD_y(\Phi\cdot f) - D_x^2(\Phi\cdot f) = 0$$

$$iD_y(\Psi\cdot f) + D_x^2(\Psi\cdot f) = 0 \tag{25}$$

$$(3D_y^2 - D_x D_t - 4D_x^4) f\cdot f - 8\Phi\cdot\Psi = 0.$$

6 Wronskian formulation

Interchanging the last two blocks of f we get (suitably relabelling the parameters)

$$f = (-)^{r_2} \left|\begin{array}{ccc} \delta_{ij} + \dfrac{e^{\Omega_i} e^{\Sigma_j}}{(\omega_i - \sigma_j)} & \dfrac{e^{\Omega_i} e^{\Sigma_k}}{(\omega_i - \sigma_k)} & 0 \\ \dfrac{e^{\Omega_l} e^{\Sigma_j}}{(\omega_l - \sigma_j)} & \dfrac{e^{\Omega_l} e^{\Sigma_k}}{(\omega_l - \sigma_k)} & \delta_{lm} \\ 0 & \delta_{nk} & -\dfrac{e^{-4(\omega_n^3 - \sigma_m^3)t}}{(\omega_n^3 - \sigma_m^3)} \end{array}\right|$$

From the form of Mel'nikov's solutions it is possible to recast the solution as a Wronskian. Let

$$M = \left\{\frac{1}{\omega_i - \sigma_j}\right\} \qquad 1 \le i,j \le r_1+r_2 \tag{26}$$

using symmetric function theory M may be related to Vandermonde-type matrices

$$V = \{(-\omega_j)^{i-1}\},\ W = \{(-\sigma_j)^{i-1}\}$$

we have

$$M = P^{-1}V^{-1}WQ$$

where

$$P^{-1} = \{\delta_{ij}\prod_{\substack{p=1\\p\neq i}}^{r_1+r_2}(\omega_p-\omega_i)\}, \qquad Q^{-1} = \{\delta_{ij}\prod_{p=1}^{r_1+r_2}(\omega_p-\sigma_i)\}.$$

Similarly for the cubic term we will let

$$N = \left\{\frac{1}{(\omega_i^3-\sigma_j^3)}\right\}.$$

Then

$$N = RU^{-1}YS^{-1}$$

where U and Y are Vandermondes and R and S are diagonal matrices. If we let

$$E_1 = \{\delta_{ij}e^{\Omega_j}\} \qquad F_1 = \{\delta_{nm}e^{-4\omega_m^3t}\}$$

$$E_2 = \{\delta_{ij}e^{\Sigma_j}\} \qquad F_2 = \{\delta_{nm}e^{4\sigma_m^3t}\}$$

then by multiplying f by

$$\begin{vmatrix} VPE_1^{-1} & 0 \\ 0 & UR^{-1}F_1^{-1} \end{vmatrix} \tag{27}$$

where the first block is (r_1+r_2) rows and the second is (r_2) rows we get

$$\left|\begin{array}{ccc} VPE_1^{-1}+WQE_2 & WQE_2 & VPE_1^{-1}I_3 \\ \hline 0 & UR^{-1}F_1^{-1} & -YS^{-1}F_2 \end{array}\right|$$

where $I_3 = \{z_{ij}\}$ $\quad \begin{cases} z_{ij} = 1 & \text{for } i = r_1 + j \\ z_{ij} = 0 & \text{otherwise} \end{cases}$

The top (r_1+r_2) rows form a Wronskian in x and the bottom r_2 rows form Wronskian in $t' = -4t$, i.e.:

$$f = (r_1+\hat{r}_2-1,(\hat{r}_2-1)_{t'}).$$

It is also worth noting that in the top part the functions depend on x,y and t while in the bottom they depend only on t.

Φ and Ψ can be dealt with similarly, by pulling out the factor (27) we are left with

$$\Phi = \left|\begin{array}{c|c|c} 0 & 0 & e^{4\sigma_m^3 t} \\ \hline e_i & & \\ \hline 0 & F & \end{array}\right| \qquad r_1+r_2 < m \le r_1+2r_2$$

$$\begin{cases} e_i = 0 \\ e_{r_1+r_2} = 1. \end{cases} \qquad 1 < i < r_1+r_2-1$$

$$\Psi = \left|\begin{array}{c|c|c} 0 & e^{\Sigma_j} & \\ \hline 0 & & \\ \hline f_l & & \end{array}\right| \qquad \begin{cases} f_l = 0 \\ f_{r_1+2r_2} = 1. \end{cases} \qquad r_1+r_2 < l \le r_1+2r_2-1$$

Both these can be expanded about the first column. Using symmetric functions in Φ the top row can be turned into an extra row on the bottom block and in Ψ the top row can be turned into an extra row in the first block. This gives us the Wronskian representation

$$\Phi=((r_1+\hat{r}_2-2),(\hat{r}_2)_{t'}), \qquad \Psi=((r_1+\hat{r}_2),(\hat{r}_2-2)_{t'}).$$

Thus the solution can be expressed as a set of double Wronskians.

7 The two component hierarchy

By rescaling x,y and t we can relate our equations to JM's hierarchies. Let $x_1 = x$, $x_2 = iy$, $x_3 = -4t$. Then the equations (25) become

$$(D_2+D_1^2)\Phi\cdot f = 0 \tag{28a}$$

$$(D_2+D_1^2)f\cdot\Psi = 0 \tag{28b}$$

$$(D_1^4-4D_1D_3+3D_2^2)f\cdot f + 8\Phi\cdot\Psi = 0. \tag{28c}$$

Now if we take a general set of double Wronskians

$$f = (\widehat{n-1},(\widehat{m-1})_{x_3})$$

$$\Phi = (\widehat{n-2},(\hat{m})_{x_3}), \qquad \Psi = (\hat{n},(\widehat{m-2})_{x_3}) \tag{29}$$

where $(\ )_{x_3}$ indicates that the second component depends on x_3 only we can see that equations (28a) and (28b) come straight from the modified K-P hierarchy and hence are automatically satisfied. Equation (28c) can be written in terms of Hirota derivatives that act on the first components D_1, D_2 and D_3 and the Hirota derivative that acts only on the second component $\hat{D}_3$, so we have

$$(D_1^4-4D_1D_3+3D_2^2)f\cdot f - 4(D_1\hat{D}_3f\cdot f-2\Phi\cdot\Psi) = 0. \tag{30}$$

The first part of (30) is the KP equation thus it must be zero for f given in (29). The second part may be written

$$-8[(\widehat{n-2},n;\widehat{m-2},m)(\widehat{n-1};\widehat{m-1})-(\widehat{n-2},n;\widehat{m-1})(\widehat{n-1};\widehat{m-2},m)-(\widehat{n-2};\hat{m})(\hat{n};\widehat{m-2})]$$

$$= 4\begin{vmatrix} \widehat{n-2} & \widehat{m-2} & \cdot & \cdot & n & n-1 & m & m-1 \\ \cdot & \cdot & \widehat{n-2} & \widehat{m-2} & n & n-1 & m & m-1 \end{vmatrix}$$

which is obviously zero and so this belongs to the 2-component K-P hierarchy equation as given by JM

$$D_1\hat{D}_3f\cdot f = 2\Phi\cdot\Psi.$$

Hence, the general set of Wronskians (29) satisfies Mel'nikov's equations.

Although our solution appears quite general to make it relate back to Mel'nikov's original system we want

$$f = f^*, \quad \Phi = \Psi^*.$$

This condition restricts the number of free parameters considerably. Mel'nikov takes

$$\omega_i = -\sigma_i^*,$$

in our case the choice of conditions will depend on the precise form of the Wronskians used.

8 Multicomponent solutions

We can define a multicomponent Wronskian as follows

$$f = (\hat{n-1}, (m_1\hat{-1})_t, (m_2\hat{-1})_t \ldots\ldots (m_r\hat{-1})_t)$$

then if we introduce

$$\Phi_i = (\hat{n-2}, (m_1\hat{-1})_t \ldots\ldots (\hat{m_i})_t, (m_{i+1}\hat{-1})_t \ldots)$$

$$\Psi_i = (\hat{n}, (m_1\hat{-1})_t \ldots\ldots (m_1\hat{-2})_t, (m_{i+1}\hat{-1})_t \ldots)$$

these will obviously satisfy the following equations

$$(D_2+D_1^2)\Phi_i\cdot f = 0., \qquad (D_2+D_1^2)f\cdot\Psi_i = 0$$

they also satisfy

$$(D_1^4-4D_1D_3+3D_2^2)f\cdot f + 8\Sigma\Phi_i\cdot\Psi_i = 0$$

because as before the equation can be written in two parts

$$(D_1^4-4D_1D_3+3D_2^2)f\cdot f - 4[D_1\hat{D}_3f\cdot f-2\Sigma\Phi_i\cdot\Psi_i] = 0 \qquad (31)$$

where the ^ acts on all except the first component. Again, the first term in (31) is from the K-P hierarchy while the second term is a slightly more generalized case of the 2-component hierarchy and is also zero.

Acknowledgements

The authors are indebted to Dr. J.J.C. Nimmo for valuable assistance in the course of this work. C.R.G. would like to thank S.E.R.C. for financial support during the period of this research.

References

[1] Freeman, N.C. (1984). Soliton solutions of nonlinear evolution equations. *I.M.A. J. Appl. Math.* **32**, 125.

[2] Nimmo, J.J.C. & Freeman, N.C. (1983). A method of obtaining the N-soliton solution of the Boussinesq equation in terms of a Wronskian. *Phys. Lett.* **95A**, 4.

[3] Hirota, R. (1985). Classical Boussinesq equation is a reduction of the modified KP equation. *J. Phys. Soc. Jap.* **54**, 2409.

[4] Hirota, R. (1986). Solutions of the classical Boussinesq equation and the spherical Boussinesq equation: the Wronskian technique. *J. Phys. Soc. Jap.* **55**, 2137.

[5] Jimbo, M. & Miwa, T. (1983). Solitons and infinite dimensional Lie algebras. *Publ. Res. Inst. Math. Sci., Kyoto University* **19**, 943.

[6] Mel'nikov, V.K. (1983). On equations for wave interactions. *Lett. Math. Phys.* **7**, 129.

[7] Mel'nikov, V.K. (1987). A direct method for deriving a multisoliton solution for the problem of interaction of waves on the x,y plane. *Comm. Math. Phys.* **112**, 639.

[8] Nimmo, J.J.C. (1988). In this volume.

[9] Whitham, G.B. (1974). *Linear and Nonlinear Waves*, Ch. 13. Wiley-Interscience: New York.

[10] Zakharov, V.E. & Kuznetsov, E.A. (1986). Multi-scale expansions in the theory of systems integrable by the inverse scattering transform. *Physica* **18D**, 455.

J.J.C. Nimmo

14. WRONSKIANS AND MULTI-COMPONENT KP HIERARCHIES

1. Introduction

Solutions to a large number of completely integrable nonlinear evolution equations may be expressed in terms of wronskian determinants. Such solutions may either be obtained through an inverse method [1,17] which are based on ideas of [3] (see also [18]) or by direct means using Hirota's method [8].

The latter approach has been developed by Freeman and the author (see for example [4,6,7]) and has more recently been used by other authors (e.g. [9]).

Whilst earlier work dealt with specific examples, more recently we have been concerned with developing the technique and investigating its scope [16]. Here we would like to extend this work and discuss wronskians within the algebraic framework developed by the group working at Kyoto. The aspect of their studies relevant to this work is reviewed in a paper by Jimbo and Miwa [11], hereafter called JM.

First, definitions will be given for the bilinear KP and multicomponent KP hierarchies. In JM a one parameter family of functions $\tau_s(x)$, $s \in \mathbb{Z}$ where $\mathbf{x} = \{x_k\}$, $k \in \mathbb{Z}\backslash\{0\}$ stands for an infinite sequence of independent variables, is defined.

Those Hirota equations involving these τ-functions constitute the KP hierarchy (in JM's terminology this also includes the modified KP hierarchies plus equations involving time evolutions singular at zero).

The simplest nontrivial member of this bilinear KP hierarchy is

$$(D_{x_1}^4 + 3D_{x_2}^2 - 4D_{x_3}D_{x_1})\,\tau_s \cdot \tau_s = 0,$$

which is the Hirota form of the KP equation

$$(u_t + 6uu_x + u_{xxx})_x + 3\alpha^2 u_{yy} = 0,$$

when $x_1 = x$, $x_2 = \alpha^{-1}y$, $x_3 = -4t$ and $2\frac{\partial^2}{\partial x_1^2} \log \tau_s = u$.

The n-component KP hierarchy is defined in a similar way. One defines a (2n-1)-parameter family of functions

$$\tau^{N_1 \cdots\cdots N_n}_{s_1 \cdots\cdots s_n} (\mathbf{x}^{(1)}, \mathbf{x}^{(2)}, \ldots, \mathbf{x}^{(n)}),$$

depending on n sequences of independent variables $\mathbf{x}^{(k)} = \{x_j^{(k)}\}, j \in \mathbb{Z}\backslash\{0\}$ where $s_1, \ldots, s_n, N_1, \ldots, N_n \in \mathbb{Z}$ and $N_1 + \ldots + N_n$ is a fixed value for each member of this family. The Hirota equations involving these τ-functions belong to the n-component KP hierarchy.

This definition (for n = 2) together with generating functions for the members of the KP and 2-component KP hierarchies are given in JM.

The fundamental example from the 2-component KP hierarchy is a two-dimensional nonlinear Schrödinger equation: Consider the three 2-component KP hierarchy τ-functions $F \equiv \tau^{N,M}_{0,0}$, $G \equiv \tau^{N+1,M-1}_{0,0}$ and $G^* \equiv \tau^{N-1,M+1}_{0,0}$. The 2-component KP hierarchy contains the five Hirota equations

$$\left(D_{x_2^{(1)}} - D^2_{x_1^{(1)}}\right) F\cdot G = 0,$$

$$\left(D_{x_2^{(1)}} - D^2_{x_1^{(1)}}\right) G^*\cdot F = 0,$$

$$\left(D_{x_2^{(2)}} - D^2_{x_1^{(2)}}\right) G\cdot F = 0,$$

$$\left(D_{x_2^{(2)}} - D^2_{x_1^{(2)}}\right) F\cdot G^* = 0,$$

$$\left(D_{x_2^{(1)}} - D_{x_1^{(2)}}\right) F\cdot F = 2GG^*.$$

If we let $x_1^{(1)} = x+iy$, $x_2^{(1)} = -2it$, $x_j^{(2)}$ be complex conjugate to $x_j^{(1)}$ (j = 1, 2) and choose *F* to be real and G^* complex conjugate to *G* then we get

$$(iD_t+D_x^2-D_y^2)\ G\cdot F = 0,$$

$$(D_x^2+D_y^2)\ F\cdot F = 8GG^*,$$

the Hirota form of the Davey-Stewartson equation

$$iu_t+u_{xx}-u_{yy}+4u\Sigma-8u|u|^2 = 0$$

$$\Sigma_{xx}+\Sigma_{yy} = 4|u|^2_{xx}$$

where $u = \frac{G}{F}$ and $\Sigma = (\log F)_{xx}$.

2. Wronskians as τ-functions

In this section we shall describe how some rather general wronskian-type determinants may be used as τ-functions for the n-component KP hierarchy. First we recall the situation for the KP hierarchy itself.

The wronskian determinant

$$W_s(\mathbf{x}) = \det(\partial^s\varphi,\ \partial^{s+1}\varphi,\ldots,\partial^{N+s-1}\varphi)$$

where $\mathbf{x}$ is as defined in the introduction, $\partial \equiv \dfrac{\partial}{\partial x_1}$ and $\varphi = \Big(\varphi_1(\mathbf{x}),\ldots,\varphi_N(\mathbf{x})\Big)^T$ is such that $\dfrac{\partial}{\partial x_k}\varphi = \dfrac{\partial^k}{\partial x_1^k}\varphi\ \forall\ k \in \mathbb{Z}\backslash 0$, is a τ-function for the KP hierarchy. More explicitly $\tau_s = W_s(\mathbf{x})$ satisfies the KP hierarchy [7,8,16].

For the multicomponent hierarchies the situation is similar: we define the MxM determinant.

$$W^{N_1,\ldots N_n}_{s_1,\ldots s_n}(\mathbf{x}^{(1)},\ldots,\mathbf{x}^{(n)})$$

$$=\det\left(\partial^{(1)^{s_1}}\varphi^{(1)},\ldots,\partial^{(1)^{s_1+N_1-1}}\varphi^{(1)};\ldots;\partial^{(n)^{s_n}}\varphi^{(n)},\partial^{(n)^{s_n+N_n-1}}\varphi^{(n)}\right)$$

where $\mathbf{x}^{(i)}$ are as in the introduction, $\partial^{(i)} = \dfrac{\partial}{\partial x_1}$ and $\varphi^{(i)} = (\varphi_1^{(i)}(\mathbf{x}^{(i)}))^T$ is such that $\dfrac{\partial}{\partial x_k^{(i)}}\varphi^{(i)} = \dfrac{\partial}{\partial x_1^{(i)k}}\varphi^{(i)}$.

Here is it evident that $N_1+\ldots+N_n = M$. These determinants are τ-functions for the n-component KP hierarchy i.e.

$$\tau^{N_1,\ldots N_n}_{s_1,\ldots s_n} = W^{N_1,\ldots N_n}_{s_1,\ldots s_n}(\mathbf{x}^{(1)},\ldots,\mathbf{x}^{(n)}),$$

and we call them n-component wronskians.

With reference to these n-component wronskians the subscripts and superscripts on the τ-functions have a natural interpretation:- s_i is the number a derivatives at the Start of the ith component, $N_1+\ldots+N_n$ = constant requires that the determinants all be of the same size.

One obtains solutions to any of the equations belonging to these hierarchies by specifying functions $\varphi^{(i)}_j(\mathbf{x}^{(i)})$ satisfying the appropriate conditions. Here we shall only consider soliton solutions.

For the KP hierarchy one takes

$$\varphi_j = \alpha_j e^{\xi(\mathbf{x},p_j)} + \beta_j e^{\xi(\mathbf{x},q_j)}$$

where $\xi(\mathbf{x},k) = \sum_{i\in\mathbb{Z}\setminus\{0\}} k^i x_i$ and α_j, β_j, p_j and q_j are arbitrary constants. In general, to obtain solutions to the nonlinear evolution equation that are physical (e.g. real and bounded) we may need to impose restrictions on the values of the arbitrary constants, but these considerations will not be pursued here (see [7]).

In the n-component hierarchies we take

$$\varphi^{(i)}_j = \alpha^{(i)}_j \, e^{\xi(\mathbf{x}^{(i)},p^{(i)}_j)}$$

where $\alpha^{(i)}_j$ and $p^{(i)}_j$ are arbitrary constants.

Clearly, in both cases we could choose much more general sums of exponentials for the α's (because we only require that they satisfy linear equations), but this restriction we take gives us solitons and further is necessary to describe the reduction procedure that follows.

In this context, the term reduction refers to the "admissable" elimination of independent variables from τ-functions. For wronskians this reduction may be achieved by choosing the φ's in an appropriate manner. This is most easily demonstrated by example.

Consider the choice of φ_j given above for the KP hierarchy (and in particular for the KP equation). If we wish to obtain solutions to the KdV hierarchy we must eliminate dependence on variables with even index (in particular remove the x_2 or y dependence in the KP equation). We do this not by taking $x_{2k} = 0$ which would not be admissible since it does not preserve the requirement that $\frac{\partial\varphi_j}{\partial x_{2k}} = \frac{\partial^{2k}\varphi_j}{\partial x_1^{2k}}$, rather we choose $q_j = p_j$. Now φ_j still satisfies the appropriate linear p.d.e.'s and we have

$$\varphi_j = e^{\xi_e(\mathbf{x},p_j)}\left(\alpha_j e^{\xi_o(\mathbf{x},p_j)} + \beta_j e^{-\xi_o(\mathbf{x},p_j)}\right)$$

where $\xi_e(\mathbf{x},k) = \sum_{i \text{ even}} k^i x_i$ and $\xi_o(\mathbf{x},k) = \sum_{i \text{ odd}} k^i x_i$ and hence

$$W_s(\mathbf{x}) = \left(\prod_j e^{\xi_e(\mathbf{x},p_j)}\right) . \; W_s(\mathbf{x})\big|_{x_{2k}} = 0.$$

Using the fact that any Hirota equation is invariant under a transformation $\tau \longrightarrow a\tau$ where a is a linear exponential, we deduce that

$$W_s(\mathbf{x})\Big|_{\substack{x_{2k}=0 \\ q_j=-p_j}}$$

is a τ-function for the reduced (KdV-) hierarchy. For later reference we note also that in this reduced case

$$W_{s+2}(\mathbf{x}) = \left(\prod_j p_j^2\right) W_s(\mathbf{x}).$$

For the 2-component KP hierarchy a similar reduction may be achieved; we first take

$$\varphi_j^{(1)} = \alpha_j e^{\xi(\mathbf{x}^{(1)},p_j^{(1)})} \quad \text{and} \quad \varphi_j^{(2)} = \beta_j e^{\xi(\mathbf{x}^{(2)},\, -p_j^{(1)})}.$$

Then let $x_i^{(1)} = x_i + y_i$, $x_i^{(2)} = (-1)^{i+1}\, x_i - y_i)$, so that

$$W_{s_1,s_2}^{N_1,N_2}(\mathbf{x}^{(1)},\mathbf{x}^{(2)}) = \left(\prod_j e^{\xi(\mathbf{y},p_j^{(1)})}\right) W_{s_1,s_2}^{N_1,N_2}(\mathbf{x},\tilde{\mathbf{x}}) .$$

where $\tilde{\mathbf{x}} = \{(-1)^{i+1}x_i\}$, $i \in \mathbb{Z}\backslash\{0\}$.

Again the invariance $\tau \longrightarrow a\tau$ implies that

$$W_{s_1,s_2}^{N_1,N_2}(\mathbf{x},\tilde{\mathbf{x}}\;\Big|_{p_j^{(2)}=-p_j^{(1)}}$$

is a τ-function for the reduced 2-component KP hierarchy. We note that under this reduction we get solutions to the nonlinear Schrödinger equation from the two dimension equation (1). In the n-component case (n>2) one can carry out a similar reduction process although in those cases one does not get such simple results.

3. Examples

Finally we shall discuss three examples of nonlinear evolution equations whose Hirota forms belong to KP or 2-component KP hierarchies. In each case one puts the equations into Hirota form and then identifies the hierarchy that the resulting Hirota equations belong to. Once this identification has been made, one automatically has solutions in terms of wronskian determinants. A partial list of the members of the KP and reduced 2-component KP hierarchies is found in JM but for our purposes it has proved more convenient to obtain the members of the hierarchies using the fact that their solutions are wronskians determinants. Then straightforward computer algebraic techniques may be used to generate the Hirota equations involving particular τ-functions. This work will be discussed elsewhere.

First, consider the equation of dispersive water waves,

$$q_t = (-q_x+2qr)_x, \tag{2a}$$

$$r_t = (r_x+r^2-2q)_x, \tag{2b}$$

[2,12,15]. If we let

$$q = -\frac{\partial^2}{\partial x^2}\log A \text{ and } r = \frac{\partial}{\partial x}\log\frac{B}{A}$$

then (2) may be written as

$$(D_2+D_x^2)\ B\cdot A = 0$$

$$D_x(D_t+D_x^2)\ B\cdot A = 0$$

We identify this pair as belonging to the reduced 2-component KP hierarchy provided $x = x_1$, $t = \frac{1}{2}x_2$,

$$A = \alpha\tau^{N-1,M+1}_{1,0} + \beta\tau^{N,M}_{0,1}$$

and

$$B = \beta\tau^{N+1,M-1}_{0,1} - \alpha\tau^{N,M}_{1,0}$$

where α and β are arbitrary constants (see also [5]).

A second, related system

$$u_t = (u_x+2uv-u^2)_x \tag{3a}$$

$$v_t = -(v_x+2vu-v^2)_x \tag{3b}$$

[2,12,13,14] may be dealt with in the same way. Let

$$q = -2\frac{BG^*}{FA}$$

$$r = 2\frac{AG}{FB}$$

so that (3) has Hirota form,

$$D_x F\cdot A = -2BG^*, \tag{4a}$$

$$D_x F\cdot B = 2AG, \tag{4b}$$

$$D_x^2 F\cdot F = 8GG^*, \tag{4c}$$

$$(D_t + D_x^2)\, F\cdot A = 0, \tag{4d}$$

$$(D_t - D_x^2)\, F\cdot B = 0. \tag{4e}$$

The five Hirota equations are once again in the reduced 2-component KP hierarchy with x, t, A and B as above and

$$F = \tau_{0,0}^{N,M}\ ,\ G = \tau_{0,0}^{N+1,M-1} \text{ and } G^* = \tau_{0,0}^{N-1,M+1}\ .$$

In this example especially, the bilinearization process is rather tricky, but guessing at the outset that the Hirota equations should be in the 2-component KP hierarchy, and using this knowledge to guess parts of the Hirota form, is a great help in the calculation. The system (2) and (3) are related via a Miura transformation [15]

$$q = u-v\ ,\ r = u_x + uv\ .$$

In terms of the changes of variable to reduce the systems to Hirota form this transformation is almost trivial:

$$u-v = -2\left(\frac{BG^*}{FA} + \frac{AG}{FB}\right) = \frac{\partial}{\partial x}\log\frac{F}{A} - \frac{\partial}{\partial x}\log\frac{F}{B}$$

$$= \frac{\partial}{\partial x}\log\frac{B}{A} = q \qquad \text{(from (4a,b))}$$

$$u_x + uv = \frac{\partial^2}{\partial x^2}\log\frac{F}{A} - 4\,\frac{GG^*}{F^2}$$

$$= -\frac{\partial^2}{\partial x^2}\log A + \frac{1}{2F^2}\,(D_x^2\, F\cdot F - 8GG^*)$$

$$= r \qquad \text{(using (4c))}$$

The final example is a system presented in [10]:

$$u_t = \tfrac{1}{4}\,(u_{xxx} + 6u_x^2) + (1+u_y)v_x$$

$$v_t = \frac{1}{4}(v_{yyy}+6v_y^2) + (1+v_x)u_y$$

with the constraint

$$(1+u_y)(1+v_x) = 1 .$$

The use of independent variables x,y and t suggests that this is a system in (2+1) dimensions, however from what follows it will become evident that it is really in (1+1) dimensions whilst the third variable is auxilliary.

If we let $u = 2\frac{\partial}{\partial x}\log F$ and $v=2\frac{\partial}{\partial y}\log G$ we find the Hirota form

$$(D_xD_t-\frac{1}{4}D_X^4)\; F\cdot F = D_y^2\; G\cdot\hat{G} ,$$

$$(D_yD_t-\frac{1}{4}D_y^4)\; G\cdot G = D_x^2\; F\cdot\hat{F} ,$$

$$D_xD_y\; F\cdot F = 2(F^2-G\hat{G}) ,$$

$$D_xD_y\; G\cdot G = 2(G^2-F\hat{F}) ,$$

where $\hat{F}$ and $\hat{G}$ are functions introduced to effect the bilinearization and the constraint reads $GF = \hat{G}\hat{F}$.

These Hirota equations belong to the KdV hierarchy (and hence can only really be in (1+1) dimensions) when $x_1 = x$, $x_{-1} = y$ $x_3 = x_{-3} = t$ and

$$F = \tau_0, \hat{F} = \tau_2 = \alpha\tau_0$$

$$G = \tau_1,\; \hat{G} = \tau_{-1} = \frac{1}{\alpha}\tau_1$$

where for soliton solutions α is $\left(\prod_j p_j^{\,2}\right)$ or unity for rational solutions. Note how the constraint is satisfied in a trivial manner.

If one uses the same identification of variables to look for a generalization of the modified KdV analogous to the above, one obtains

$$D_X^2\; F\cdot G = 0,$$

$$D_y^2\; F\cdot G = 0,$$

$$(D_t-D_x^3+D_y^3)\; F\cdot G = 0,$$

$$D_x\, D_y\; F\cdot F = 2(F^2-G\hat{G}),$$

$$D_x D_y\, G\cdot G = 2(G^2-F\hat{F}),$$

which is the Hirota form of

$$u_t = u_{xxx}+2u_x^3-u_{yyy}-2u_y^3,$$

$$u_{xy} = 2\sin u,$$

where $u = i\log\frac{G}{\sqrt{\alpha F}}$.

Here again x or y is an auxilliary variable, and this is a (highly non local) evolution equation in (1+1) dimensions. It is interesting to note that the presence of auxilliary variables in Hirota equations is a commonplace occurrence - one only has to look at the fifth order KdV equation for the simplest example - but in evolution equations themselves it is less usual.

4. Conclusions

We have described how the wronskian representation of solutions to nonlinear evolution equation fits into the algebraic framework of the Kyoto school; an n-component wronskian is a τ-function for the n-component KP hierarchy. Some unusual examples of nonlinear evolution equations whose Hirota forms belong to these hierarchies have been presented. The intention of this is to illustrate the richness of variety of interesting systems - other than "just" the KP and KdV equations - that lie within these hierarchies.

References

[1] Adler,M. and Moser,J. (1978) *Commun. Math. Phys.*, **61**, 1-30.

[2] Antonowicz,M. and Fordy, A.P, (1989) *Commun. Math. Phys.* **124**, 465-86.

[3] Crum,M.M. (1955) *Quart. J. Math.*, ser. 2, **6**, 121-127.

[4] Freeman,N.C (1984) *IMA J. Appl. Math.* **32**, 125-145.

[5] Freeman,N.C and Gilson,C.R. (1988) in this volume.

[6] Freeman,N.C., Horrocks,G. & Wilkinson,P. (1981) *Phys. Lett.A*, **81**, 305-307.

[7] Freeman,N.C. and Nimmo,J.J.C. (1983) *Proc. R. Soc. Lond. A*, **389**, 319-329.

[8] Hirota,R. (1980) in *Solitons* eds. R.K.Bullough and P.J.Caudrey, Springer-Verlag. Berlin, Heidelberg, New York.

[9] Hirota,R. (1986) *J. Phys. Soc. Jpn.* **55**, 2137-2150.

[10] Jaulent,M., Manna,M.A and Martinez Alonso,L. (1988) preprint.

[11] Jimbo,M. and Miwa,T. (1983) Publ. RIMS, Kyoto Univ., **19**, 943-1001.

[12] Kuperschmidt,B.A. (1985) *Commun. Math. Phys.* **99**, 51-73.

[13] Levi,D. (1981) *J. Phys. A* **14**, 1083-1098.

[14] Levi,D., Neugebauer,G. and Meinel,R. (1984) *Phys. Lett.* **102A**, 1-6.

[15] Levi,D., Sym,A. and Wojciechowski,S. (1983) *J. Phys. A* **16**, 2423-2432.

[16] Nimmo,J.J.C. (1988) in "Nonlinear Evolutions" ed Jérome J.P.

Leon, World Scientific. Singapore.
[17] Satsuma,J. (1979) *J. Phys. Soc. Jpn.*, **46**, 359-360.
[18] Wadati,M., Sanuki,H. and Konno,K. (1975) *Prog. Theor. Phys.* **53**, 419 436.

J. Gibbons and Y. Kodama

15. EXACT SOLUTIONS OF THE ZABOLOTSKAYA-KHOKHLOV EQUATION

Exact local solutions of the Benney equations may be obtained by considering the conservation laws and Riemann invariants of reductions of the system which correspond to dispersionless Lax equations. These are given implicitly; the space and time variables are given in terms of known functions of the dynamical variables. This method is related to that of Tsarev, but is more explicit, since much more is known about this hierarchy, in particular its symmetries and conservation laws, than the general class of systems discussed by him.

1 Introduction

The equation

$$u_{xt} = \frac{1}{2}(u^2)_{xx} + u_{yy} \tag{1}$$

arises in the theories of both unsteady transsonic flow, where it was introduced by Timman [1], and nonlinear acoustics, where Zabolotskaya & Khokhlov derived it [2]. In either case, the nonlinear term models the advective nonlinearity, while the u_{yy} term is a reasonable approximation to the effects of diffraction (in the flow problem one must assume the lateral disturbance is small, for instance as in flow past a slender body, while in acoustics one considers an almost unidirectional beam of sound).

The system is intimately related to the Benney equations [3,4]

$$\frac{\partial A_i}{\partial t_2} = \frac{\partial A_{i+i}}{\partial x} + i\, A_{i-1} \frac{\partial A_0}{\partial x} \tag{2}$$

and the next equation in this hierarchy, described below, which gives the dependence of the A_i on another, independent, "time" variable t_3. As this hierarchy of Hamiltonian systems possesses infinitely many symmetries, generated by its conservation laws, it is best to regard the moments A_i as depending on infinitely many such "times" t_n. If u is identified with A_0, y with t_2, and t with t_3, then the equations of the Benney hierarchy imply (1), as was shown in [6]. The object of this paper is to construct, explicitly, those solutions which are invariant under any one of the symmetries of the hierarchy. Evidently, it does not make sense to look for solutions in which the A_i are given explicitly as functions of the variables x and t_n, as may be seen by considering the y-independent case of (1). The correct procedure is rather to exchange the roles of the dependent and independent variables; this is the hodograph transformation. Although the resulting equations are not in general linear, it is possible, after some work, to give a large class of solutions in a particularly simple and elegant form, in which the variables t_n are given by the conserved densities of (2) and their fluxes. This work was started in 1987, and an early version was reported in [7].

The construction given here is intimately related to the method of Tsarev [8]; however, it seems to be easier to construct solutions using the techniques given below, as well as in related work. The recent work of Krichever [9], which contains a construction of some solutions to (1), should also be noted.

The paper is arranged as follows. In Section 2 the relevant facts fromm the theory of Benney's equations will be stated, in particular the conservation laws, Riemann invariants and characteristic velocities. Then, in Section 3, the hodograph equations will be derived, while their solutions will be given in Section 4. Finally, some possible further developments will be discussed in Section 5.

2 **The Benney hierarchy**

Benney's equations were originally obtained as a description of long waves on a shallow perfect fluid [3]. In that paper, Benney showed that the system possessed an infinite family of conservation laws. In [4], Kupershmidt & Manin obtained the Hamiltonian structure of the system, and hence its symmetries, while in [5], Lebedev & Manin obtained a formal Lax representation of the system. This constitutes the simplest language in which to discuss it.

Let us consider, following Kupershmidt & Manin, the formal Laurent series in powers of p:

$$\lambda = p + \sum_{0}^{\infty} A_i \, p^{-(i+1)} \tag{3}$$

and its unique inverse of the same form

$$p = \lambda - \sum_{0}^{\infty} H_i \, \lambda^{-(1+i)} \tag{4}$$

Here the H_i are polynomials in the variables A_i; they are the conserved densities of the hierarchy. We allow λ to evolve according to the equations:

$$\frac{\partial\lambda}{\partial t_n} = \frac{1}{n}\{(\lambda^n)_+, \lambda\} \tag{5}$$

Here braces denote the canonical Poisson bracket with respect to x and p, while $(\lambda^n)_+$ denotes the part of the formal series λ^n which is polynomial in p. It is easy to check that these equations make sense as evolution equations for the A_i, and, with little more difficulty, that they commute. One may also compute directly that A_0 satisfies the Zabolotskaya-Khokhlov equation:

$$\frac{\partial^2 A_0}{\partial t_3 \partial x} = \frac{1}{2}\frac{\partial^2}{\partial x^2}(A_0{}^2) + \frac{\partial^2 A_0}{\partial t_2{}^2} \tag{6}$$

To find special solutions of the equation, as with any dynamical system with symmetry, it is natural to seek solutions invariant under one of the symmetries in the hierarchy. For instance, let us suppose that the A_i are all independent of t_2. Then we must have:

$$\{(\lambda^2)_+ , \lambda\} = 0 \tag{7}$$

so that λ is a function of $(\lambda^2)_+$; that is, of $p^2 + 2A_0$. In general this function is fairly arbitrary, but the choice

$$\lambda = \sqrt{(p^2 + 2A_0)} \tag{8}$$

is general enough for the purposes of this paper. We then obtain

$$\frac{\partial}{\partial t_3}\sqrt{(p^2 + 2A_0)} = \{\frac{p^3}{3} + A_0 p, \sqrt{(p^2 + 2A_0)}\} \tag{9}$$

which reduces to

$$\frac{\partial A_0}{\partial t_3} = A_0\frac{\partial A_0}{\partial x} \tag{10}$$

This equation is easily solved if we consider x as a function of A_0, depending parameterically on t_3, so that (10) becomes:

$$\frac{\partial x}{\partial t_3} + A_0 = 0 \tag{11}$$

The solution of this is $x = F(A_0) - A_0\, t_3$, where F is an arbitrary smooth function; such a function is a conserved density of (10).

The next simplest case, with the A_i independent of t_3, is solved by a similar technique, treating x and t_2 as functions of A_0 and A_1, which would depend parametrically on t_4, t_5, t_7, t_8, and so on. In this case it is t_2 which turns out to be an arbitrary conserved density of the reduced system, just as x was in the t_2-independent case, while x is now its flux. The object of this paper is to generalize this construction to the case with the A_i independent of t_N, for arbitrary N. To do this, we take λ as in (8), namely:

$$\lambda^N = (\lambda^N)_+ = p^N + N\,A_0\,p^{N-2} + \ldots = \Lambda(p) \tag{12}$$

It will be convenient below to deal with the polynomial Λ, rather than the formal series λ; the equations of motion are:

$$\frac{\partial\Lambda}{\partial t_n} = \left[\lambda^{n-1}\frac{\partial\lambda}{\partial p}\right]_+\frac{\partial\Lambda}{\partial x} - \left[\lambda^{n-1}\frac{\partial\lambda}{\partial x}\right]_+\frac{\partial\Lambda}{\partial p} \tag{13}$$

To proceed further, we need the Riemann invariant form of the equations; to get this, we want the second term on the right-hand side of (13) to vanish. If we write

$$\frac{\partial\lambda}{\partial p} = N\prod_{i=1}^{N-1}(p - p_i) \tag{14}$$

then we may easily see that the Riemann invariants are the values $\Lambda(p_i)$, which are advected with speeds which are polynomials of degree (n-1) in p_i,

$$v^{(n-1)}(p_i) = \left.\left[\lambda^{n-i}\frac{\partial\lambda}{\partial p}\right]_+\right|_{p=p_i} \tag{15}$$

We may obtain a useful recurrence relation for these polynomials, and hence a generating function for them, starting with the identity:

$$\lambda^n\frac{\partial\lambda}{\partial p} = v^{(n)}(p) + O(p^{-2}) \tag{16}$$

and a rearrangement of (4),

$$\lambda = p + \sum_{i=0}^{\infty} H_i\,\lambda^{-(i+1)} \tag{17}$$

Now if we multiply both sides of (17) by $\lambda^n\,\partial\lambda/\partial p$, we get:

$$\lambda^{(n+1)} \frac{\partial\lambda}{\partial p} = p\,\lambda^{n} \frac{\partial\lambda}{\partial p} + \sum_{i=0}^{\infty} H_i\, \lambda^{n-i-1} \frac{\partial\lambda}{\partial p} \tag{18}$$

which leads, on taking the polynomial part, to the recurrence:

$$v^{(n+1)} = p\, v^{(n)} + \sum_{i=0}^{n-1} H_i\, v^{(n-i-1)} \tag{19}$$

This is equivalent to:

$$\sum_{i=0}^{\infty} v^{(i)} = 1/(1 - p - \sum_{i=0}^{\infty} H_i) \tag{20}$$

if terms on either side are assigned weights as follows; $|p| = 1$, $|v^{(i)}| = i$, and $|H_i| = i+2$. Further, one may easily verify that

$$\sum_{m=0}^{\infty} v^{(m)}(p_i) - v^{(m)}(p_j)$$

$$= (p_i - p_j)/((1 - p_i - \sum_{k=0}^{\infty} H_k)(1 - p_j - \sum_{k=0}^{\infty} H_k))$$

$$= (p_i - p_j) \sum_{\ell=0}^{\infty} v^{(\ell)}(p_i) \sum_{m=0}^{\infty} v^{(m)}(p_j) \tag{21}$$

Taking p_i and p_j to be distinct roots of $v^{(N-1)}(p)$, we then find, on comparing terms of weight N-1 on either side, and dividing by $(p_i - p_j)$, that:

$$\sum_{\ell=0}^{N-2} v^{(\ell)}(p_i)\, v^{(N-\ell-2)}(p_j) = 0 \tag{22}$$

Similarly we may obtain the result:

$$\sum_{\ell=0}^{N-2} v^{(\ell)}(p_i)\, v^{(N-\ell-2)}(p_i) = v^{(N-1)}(p_i) \tag{23}$$

These results, (20-23), were derived by George Wilson, and are crucial in what follows.

Now the Riemann invariant form of the equations is:

$$\frac{\partial\Lambda(p_i)}{\partial t_n} = v^{(n-1)}(p_i) \frac{\partial\Lambda(p_i)}{\partial x} \tag{24}$$

We will denote the ith Riemann invariant $\Lambda(p_i)$ as Λ_i, and the corresponding speeds $v^{(n)}(p_i)$ as $v^n{}_i$, for brevity. We also have infinitely many conservation laws, and it is convenient to write them in terms of the Riemann invariants:

$$\frac{\partial H}{\partial t_n} = \sum_{i=1}^{N-1} \frac{\partial H}{\partial \Lambda_i} v^{n-1}{}_i \frac{\partial \Lambda_i}{\partial x}$$

$$= \frac{\partial F_n}{\partial x} = \sum_{i=1}^{N-1} \frac{\partial F_n}{\partial \Lambda_i} \frac{\partial \Lambda_i}{\partial x} \qquad (25)$$

We will see below that the conserved densities H ($=F_1$) and their fluxes F_n, with the ansatz $t_n = F_{N-n}$, give the implicit solutions required. We note, on comparing coefficients of $\partial\Lambda_i/\partial x$, the result $\partial F_n/\partial\Lambda_i = v^{n-1}{}_i \ \partial H/\partial\Lambda_i$, needed below.

3 The Hodograph transformation

If we consider the equations of motion in Riemann invariant form (24), and exchange the roles of the dependent variables Λ_i and independent variables t_n ($n = 1,\ldots N-1$), then they become, as was reported in [7]:

$$(-1)^{n+1} \det_{\substack{m\neq n\\ j\neq i}} \left(\frac{\partial t_m}{\partial \Lambda_j}\right) = (-1)^{1+i} v^{n-1} \det_{\substack{m\neq 1\\ j\neq i}} \left(\frac{\partial t_m}{\partial \Lambda_j}\right) \qquad (26)$$

Here and below, in these (N-2)x(N-2) determinants, m and j both run from 1 to (N-1), and one row and one column are deleted. Now, if we substitute the ansatz $t_n = F_{N-n}$ into this equation, on the left-hand side we get:

$$(-1)^{n+i} \det_{\substack{m\neq n\\ j\neq i}} \left(\frac{\partial F_{N-m}}{\partial \Lambda_j}\right)$$

which is the same as:

$$(-1)^{n+i} \det_{\substack{m\neq n\\ j\neq i}} \left(v^{N-m-i}{}_j \frac{\partial H}{\partial \Lambda_j}\right) \qquad (27)$$

while the right-hand side of (26) is similarly:

$$(-1)^{i+1} v^{n-1}{}_i \det_{\substack{m\neq 1\\ j\neq i}} \left(v^{N-m-1}\right) \frac{\partial H}{\partial \Lambda_j}) \qquad (28)$$

We will now prove that these two expressions are identically equal; the factor of $(-1)^i$ and the (N-1) factors $\partial H/\partial\Lambda_j$ are common to both, and may be divided out. It remains to be proved that:

$$(-1)^{n+i} \det_{\substack{m\neq n\\ j\neq i}} (v^{N-m-i}{}_j) = v^{n-1}{}_i \det_{\substack{m\neq 1\\ j\neq i}} (v^{N-m-1}{}_j) \tag{29}$$

This is a consequence of (22), for that result implies that the inverse of the (N-1)x(N-1) matrix $(v^n{}_i)$ has rows which are proportional to the columns of $(v^n{}_i)$ reversed. On multiplying each element of a row of $(v^n{}_i)^{-1}$ by det $(v^n{}_i)$, which vanishes only if some of the characteristic speeds p_i are equal, we obtain (29).

The consequence of this calculation is that any dispersionless Lax hierarchy possesses infinitely many implicit solutions which may be expressed in terms of known polynomials, namely the conserved densities of the Benney hierarchy and their fluxes. A surprising feature of the calculation is that although the proof is easiest in the Riemann invariant picture, using the Λ_i and $v^n{}_i$, the result is most easily expressed in terms of the conserved densities and fluxes, which are polynomials in the moments A_i.

4 **Other methods**

These results may also be obtained by Tsarev's method [8]; he observes, essentially, that the condition that the two systems

$$\frac{\partial \Lambda_i}{\partial t} = v_i(\Lambda) \frac{\partial \Lambda_i}{\partial x} \tag{30}$$

and

$$\frac{\partial \Lambda_i}{\partial s} = w_i(\Lambda) \frac{\partial \Lambda_i}{\partial x} \tag{31}$$

should commute is:

$$\frac{\partial v_i}{\partial \Lambda_k} (w_i - w_k) = \frac{\partial w_i}{\partial \Lambda_k} (v_i - v_j) \tag{32}$$

which, given the v_i, is a linear system for the w_i. This equation is related to the classical problem of the construction of orthogonal curvilinear coordinate systems in R^n. Given any solution of (32), an implicit solution of (30) is given by:

$$w_i = x + v_i t \tag{33}$$

as may be checked by a direct computation. In the Benney problem, infinitely many sets of characteristic speeds $v^n{}_i$ are known a priori, as are many of their properties, in particular (22). If we solve an extended version of (33), namely:

$$v^{N-1}{}_i = x + \sum_{n=1}^{N-2} v^n{}_i \, t_{n+1} \qquad (34)$$

we obtain, after some work, the solutions found in Section 3.

One question to be asked is whether these solutions are in any sense special, apart from the symmetry we have imposed on them, or whether there are solutions of these dispersionless Lax equations which are unrelated to the conservation laws. The most general conserved density of a dispersionless Lax equation is expressible in the form

$$H = \oint \Phi(\Lambda) \, \Lambda^{n/N} \, dp/p \qquad (n=1,\ldots,N-1) \qquad (35)$$

where the integral is taken around a large circle in the complex p-plane, and Φ is an arbitrary entire function of Λ; clearly these densities form (N-1) distinct series, depending on the value of n. Thus the method described generates solutions depending on N-1 arbitrary functions, while the order of the p.d.e.'s is also N-1.

5 Further problems

The theory given in this paper is very closely tied to a particular class of reduction of the hierarchy, corresponding to dispersionless Lax equations. Other reductions are possible; one important open problem concerns the classification of those ansätze for $\lambda(p)$ which are invariant under the equations of motion, and hence provide valid reductions. The best known of these, due to Zakharov, is:

$$\lambda = p + \sum_{1}^{N} h_i/(p - u_i) \qquad (36)$$

We have been investigating the relationship of hodograph solutions of this and other reductions to the conservation laws of the hierarchy [11]. It is now known that for any integrable system of hydrodynamic type, one may associate the hodograph solutions with the conservation laws, though the relationship is usually less simple than it is here. An important outstanding problem of this type is to apply the same techniques to the problems of modulated g-gap KdV wavetrains [10]; this problem is mentioned by Tsarev [8], and is included in the large class of problems discussed by Krichever in [9]. It does seem possible that a more effective method could be derived, however. The work of Levermore [12], and Serre [13], should be very helpful here.

References

[1] Timman, R. (1964). *Symposium Transsonicum,* p. 394, Oswatitsch, K. (ed.). Sringer-Verlag: Berlin.

[2] Zabolotskaya, E.A. & Khokhlov, R.V. (1969). *Sov. Phys. Acoustics* **15**, 35.
[3] Benney, D.J. (1973). *Stud. Appl. Math.* **52**, 45.
[4] Kupershmidt, B.A. & Manin, Yu.I. (1978). *Funct. Anal. Appl.* **11**, 188; (1978). *Funct. Anal. Appl.* **12**, 20.
[5] Lebedev, D.R. & Manin, Yu.I. (1980). *Phys. Lett.* **74A**, 154.
[6] Gibbons, J. (1985). *Dynammical Problems in Soliton Systems,* p. 36, Takeno, S. (ed.). Springer-Verlag: Berlin.
[7] Gibbons, J. & Kodama, Y. (1988). *Proc. NEEDS 1987,* p. 97, Leon, J. (ed.). World Scientific: Singapore.
[8] Tsarev, S.P. (1985). *Sov. Math. Dokl.* **31**, 488.
[9] Krichever, I.M. (1988). *Funkts. Analiz Pril.* **22**, 37.
[10] Flaschka, H., Forest, M.G. & McLaughlin, D.W. (1980). *Comm. Pure and Appl. Math.* **33**, 739.
[11] Gibbons, J. & Kodama, Y. *Phys. Lett. A.*, to appear; Kodama, Y. *A Method for Solving the Disperionless KP Hierarchy,* preprint.
[12] Levermore, C.D. (1988). *Comm. in P.D.E.'s* **13**, 495.
[13] Serre, D. *Systemes Hyperboliques Riches de Lois de Conservation*, preprint.

Part 2

PAINLEVÉ AND SINGULARITY MANIFOLDS

B. Fuchssteiner and S. Carillo

16. THE SOLITON-SINGULARITY TRANSFORM

A common approach to singularity analysis and soliton structure is introduced. The Painlevé analysis is modified in such a sense that it carries over to cases where usually truncation at the lower end causes difficulties. A method to recover the Painlevé series from its constant level term is exhibited. The soliton-singularity transform is introduced and recognized to be connected to the Möbius group. This gives rise to a Darboux-like result for the spectral properties of the recursion operator. These connections explain why poles of soliton equations move like trajectories of interacting solitons. Furthermore it is explicitly computed how solutions of singularity equations behave under the effect of this soliton-singularity transform. This then leads to the result that only for scaling degrees $\alpha=-1$ and $\alpha=-2$ the usual Painlevé analysis can be carried out. A new invariance principle, connected to kernels of differential operators, is discovered. This new invariance, for example, connects the explicit solutions of the Liouville equation with the Miura transform.

1 **Introduction**

In order to motivate the considerations which are to follow we consider, for example, the Painlevé expansion for the Korteweg-deVries equation [14]

$$u = -2\frac{\phi_x^2}{\phi^2} + 2\frac{\phi_{xx}}{\phi} + \left(-\frac{1}{2}\frac{\phi_{xxx}}{\phi_x} + \frac{1}{4}\frac{\phi_{xx}^2}{\phi_x^2}\right) \tag{1}$$

and we pay special attention to its constant level term

$$\tilde{u}(\phi) = \left(-\frac{1}{2}\frac{\phi_{xxx}}{\phi_x} + \frac{1}{4}\frac{\phi_{xx}^2}{\phi_x^2}\right). \tag{2}$$

Using the substitution

$$s=\phi_x \tag{3}$$

we rewrite this term as

$$\hat{u} = -\frac{1}{2}\frac{s_{xx}}{s} + \frac{s_x^2}{4s^2} \tag{4}$$

There are a number of remarkable links between these representations. Let us list some of them:

(i) We find

$$\hat{u}\left(\frac{\phi_x}{\phi^2}\right) = u.$$

Thus a replacement of ϕ in (2) leads to a recovery of the Painlevé series (1) from its constant level term.

(ii) It is well known that the constant level term $\tilde{u}(\phi)=\hat{u}(s)$ is again a solution of the KdV if u is such a solution. Thus there must be some well defined dynamical behaviour of the quantity s given by (3). Remarkably, this dynamical behaviour does not involve any mixed derivatives in x and t since one easily obtains

$$s_t = s_{xx} - 2s_x s_x/s.$$

Now, since (4) constitutes a descending sequence in negative powers of a quantity s with well defined dynamical behaviour, one may ask why this series is not considered to be the singularity expansion.
There are other surprising properties for the quantities s and ϕ to which we would like to draw the attention of the reader. For example:

(iii) For equation (4) there are far more invariances than one might expect by considering only the known invariances for the time evolution of the singularity manifold (Möbius group).
(iv) Some of the invariances mentioned under (iii) are indeed connected to solutions of the Liouville equation

$$h_{xt} = e^{-2h}. \tag{5}$$

(v) A simple analysis shows that the quantity s related to u by (4) is annihilated by the recursion operator of the KdV, i.e.

$$\Phi(u)s = (D^2+2u+2DuD^{-1})s=0. \tag{6}$$

What we claim in this paper is that all these remarkable properties are intimately connected and can be easily understood, and generalized, by extending the Painlevé test. In addition it turns out that the structural results leading to these properties indeed show that the usual Painlevé test can only be carried out for scaling degrees $\alpha=-1$ and $\alpha=-2$. An extended version of this paper will appear elsewhere [7], also there can be found an abundance of examples for the interrelations demonstrated here.

2 Painlevé test versus Expansion test

We consider solutions of the evolution equation

$$u_t = K(u) \tag{7}$$

where K(u) is a polynomial in u and its derivatives. Eventually we will also allow infinite polynomials (Taylor series). We make the ansatz

$$u = \sum_{n=0}^{\infty} \psi^{-n} F_{(n)}(\psi_x, \psi_{xx}, \ldots) \tag{8}$$

in negative descending powers of a quantity ψ. We say that equation (7) passes the expansion test if this ansatz is always compatible with a time evolution of ψ given by

$$\psi_t = \sigma(\psi) \tag{9}$$

where σ is only depending on ψ and its spatial derivatives and where it is assumed that in the Laurent expansion of σ, with respect to its dependence on ψ, only negative powers occur. These requirements of course are severe restrictions since in general an ansatz like (8) when introduced into (7) yields the equation

$$K\Big(\sum_{n=0}^{\infty} \psi^{-n} F_{(n)}\Big) = \sum_{n=0}^{\infty} \psi^{-n}\{-n\psi^{-1} F_{(n)} + F_{(n)}'\}\sigma(\psi) \tag{10}$$

where very many mixed derivatives occur. In this formula the prime in F′ stands for the variational derivative of F with respect to ψ. Formula (10) gives, by comparison of powers in ψ, rise to an algorithmic determination of $F_{(n)}$ once the starting point $F_{(0)}$ has been chosen. There are two important cases for this algorithmic procedure.

Case 1 *(Painlevé test):*
We assume that $\sigma(\psi)$ is only depending on the spatial derivatives of ψ, i.e. no ψ without x-derivative occurs. This corresponds to saying that the t-derivatives affect the singularities given by the zeros of ψ in the same way as x-derivatives. This, of course, is the requirement one usually tacitly assumes in the Painlevé test. Hence this assumption singles out the Painlevé expansion. Looking at lowest order in ψ^{-1} in the expression (10) one discovers in this case as a consequence

$$F_{(0)t}=K(F_{(0)}), \tag{11}$$

which proves that

$$\tilde{u}=F_{(0)} \tag{12}$$

again has to be a solution of (7). Looking at the next order one finds the well known fact that

$$K'(\tilde{u})[F_{(1)}] = F_{(1)}(\psi)'[\sigma] = F_{(1)t} \tag{13}$$

which implies that $F_{(1)}$ is a symmetry generator for (7) at the manifold point $\tilde{u}$.

Case 2 *(constant highest order case, soliton test):*
In order to distinguish the field variable in this case from the previous one we use s instead of ψ. If $u=\lambda$ (λ a constant) does satisfy equation (7) then we put $\sigma(s)=D^{-1}K(u)D$ and $F_{(0)}=\lambda$ where we let vary the value of the constant λ. This is possible since any constant obviously satisfies the condition on the highest order term because the variational derivatives of $F_{(0)}$ disappears. In this case it does not matter at all whether or not negative powers of ψ appear explicitly in (9).

Remark: The series which results for $\lambda=0$ out of the case 2 expansion test, we call the **special expansion series**. However, it has to be observed that the coefficients $F_{(n)}$ of this series may have a certain analytical structure with respect to the variable λ such that the choice of $\lambda=0$ results in a value where the $F_{(n)}$ are singular with respect to λ. In general this means that the special expansion series has to be the analytical continuation of (10) with respect to λ at $\lambda=0$. An example where this happens is provided by the members of the mKdV hierarchy (see [7]).

For reasons which become obvious in the next section, we say that, if the Case 2 expansion-test is successful, then equation (7) passes the **soliton-test**. In principle, both tests could be run through by starting with $F_{(0)}$ (or $F_{(1)}$ in the special expansion test) as initial data. However, in case 1 this is not a realistic alternative since usually $\tilde{u}=F_{(0)}$ is related to u by a complicated Bäcklund transformation. Therefore in this case, for computational reasons, one has to start with the highest order term which is only possible if the series truncates also on the lower end. This, of course, is not necessary for case 2 since there the constant-level term is known in advance.

For this reason we do not need in that situation a truncation at the lower end. So in principle these expansion series can be infinite series. From the computational viewpoint it seems remarkable that case 2 reverses the order of case 1. Therefore it is a surprise that nevertheless both cases almost give the same information.

3 Interacting solitons

We assume that $\Phi(u)$ is a hereditary recursion operator [4] for (7). Then the potentials s of its eigenvectors

$$\Phi(u)s_x = \lambda s_x \tag{14}$$

are called interacting solitons [5],[6]. Their dynamics is given by

$$s_{xt} = K'(u)[s_x] = \frac{\partial}{\partial\varepsilon|\varepsilon=0} K(u+\varepsilon s_x). \tag{15}$$

Usually it is not difficult to express u in terms of the quantity s (see [5])

$$u=\Gamma(s). \tag{16}$$

We assume that our equations are of such a form that Γ only depends on s and its spatial derivatives. For example, for the KdV this representation of u in terms of s is

$$u = \frac{1}{4s^2}(\lambda s^2 - 2ss_{xx} + s_x^2). \tag{17}$$

it should be remarked that (4) is just the special case of (17) for $\lambda=0$. Insertion of the relation (16) into (15) leads to the dynamics for s. In case of the KdV this yields

$$s^2 s_t = s^2 s_{xxx} - 3ss_x s_{xx} + \frac{3}{2}s_x^3 + \frac{3}{2}\lambda\, s^2 s_x \tag{18}$$

of which the dynamics for (4) again is a special case.

In the following we only consider equations with negative scaling $\alpha<0$. Scaling α means that the replacement

$$x\longrightarrow mx, \quad u\longrightarrow m^{\alpha}u, \quad t\longrightarrow m^{\beta}t \text{ (some } \beta) \tag{19}$$

leaves the equation invariant. As a consequence of negative scaling we obtain that an expansion of (16) in terms of powers of s must be of the form

$$u = c+ \frac{1}{s}\Gamma_{(1)}+\frac{1}{s^2}\Gamma_{(2)}\dots, \quad c = \text{constant} \tag{20}$$

with $\Gamma_{(i)}$ polynomials in s_x, s_{xx} etc. This is the form which was required in case 2 of the expansion test. Since the members of the expansion test are uniquely and algorithmically determined we know that this expansion is the unique representation for case 2. This justifies that we gave this expansion the name "soliton expansion".

It should be observed that now all equations for which the dynamics of the interacting soliton is known can serve as examples for case 2 of the expansion test (see [5],[6]). Among these examples one finds several where the series does not truncate at the lower end (mKdV and sine-Gordon for $\lambda\neq0$).

Now, we compare the Painlevé series for (7) with the soliton expansion. We look at the constant level term $\tilde{u}=F_{(0)}(\phi)$ which is a homogeneous rational function in the derivatives of ϕ. We consider an expansion in negative powers of the lowest derivative $\phi_{(N)}$ in the denominator of that expression

$$\tilde{u} = \sum_{n=1}^{\infty} \phi_{(N)}{}^{-n} G^{(n)}(\phi_{(N+1)}, \phi_{N+2)}, \ldots). \tag{21}$$

The zero order term of that has to disappear because it could only be a constant (independent of ϕ) which would be in contradiction to negative scaling degree. Obviously, that expansion again is a case 2-expansion for $\tilde{u}$ (instead of u), since only ϕ_N has to be renamed by s. Since the case 2 expansions were unique we know that (21) must coincide with the special expansion for $\tilde{u}$. It remains to compute the N (order of derivative). For this we look at the next order term in the Painlevé series. This term is of the form

$$F_{(1)}(\phi_x, \phi_{xx}, \ldots)/\phi. \tag{22}$$

Since any rescaling $\phi\longrightarrow a\phi$ leaves the whole Painlevé series invariant we know that $F_{(1)}$ is of first order, and because of the scaling degree it has to bear $-\alpha$ derivatives. Hence we obtain

$$F_{(1)} = C\phi_{(-\alpha)} \tag{23}$$

where, without loss of generality, C=1 can be chosen. Since we know that $F_{(1)}$ is a symmetry around $\tilde{u}$ and that the eigenvector $\tilde{s}_x$ of $\Phi(\tilde{u})$ has the same dynamic as a symmetry generator around $\tilde{u}$ we find the crucial identity

$$\phi_{(-\alpha)} = \tilde{s}_x. \tag{24}$$

Hence we found the N to be

$$N=-\alpha-1. \tag{25}$$

4 Soliton-singularity transform

Let us resume the results of the preceding sections in a more systematic way. Again we consider

$$u_t = K(u) \tag{7}$$

with negative scaling $\alpha<0$. We write the Painlevé series and the special expansion series for u respectively as

$$u = PE(\phi) \text{ (Painlevé expansion)} \tag{26}$$
$$u = SE(s) \text{ (special expansion)} \tag{27}$$

If $\tilde{u}$ denotes the constant level term in the Painlevé expansion for u then we also consider the expansion for this solution $\tilde{u}$ of (7).

$$\tilde{u} = PE(\tilde{\phi}) \tag{28}$$

$$\tilde{u} = SE(\tilde{s}) \tag{29}$$

The map going from u,s,ϕ to $\tilde{u},\tilde{s},\tilde{\phi}$ we denote by SoSi$(\cdot)$, i.e.

$$\tilde{u}=\mathrm{SoSi}(u)$$
$$\tilde{s}=\mathrm{SoSi}(s) \tag{30}$$
$$\tilde{\phi}=\mathrm{SoSi}(\phi)$$

The fundamental result of the last section was the

DUALITY:

$$\tilde{s}_x = \phi_{(-\alpha)}, \quad s_x = \tilde{\phi}_{(-\alpha)} \tag{31}$$

So SoSi$(\cdot)$ changes, up to derivatives, solitons in singularities and vice versa. This explains the name, since SoSi is meant to be an abbreviation for Soliton-Singularity transform.

Interchanging the role between u and $\tilde{u}$ in the respective series shows that SoSi is an involution, i.e.

$$\mathrm{SoSi}^2 = I \quad \text{(identity)}. \tag{32}$$

Now we want to compute SoSi explicitly, at least the effect it has on s and ϕ. Then its effect on u can be computed by considering the series obtained from one of the tests. In order to compute SoSi() we make the ansatz

$$s = \frac{\nu}{\phi^k} \tag{33}$$

where ν is of the form

Soliton singularity transform

$$\nu = \nu_0(\phi_x,\phi_{xx},\ldots)+\phi\nu_1(\phi_x,\phi_{xx},\ldots)+\phi^2\ldots. \tag{34}$$

Insertion in u=SE(s) yields

$$u = SE(\nu/\phi^k). \tag{35}$$

Since the special expansion series was homogeneous in s we see that only those parts of the derivatives of s contribute to powers of zero order in ϕ where the ϕ's are left unaffected by the derivatives. Since for those terms the powers of ϕ cancel, we easily obtain the constant level term $\tilde{u}$ as

$$\tilde{u} = SE(\nu). \tag{36}$$

Comparison with the special expansion for $\tilde{u}$ leads to

$$\nu = \beta\tilde{s},\ \beta \text{ constant} \tag{37}$$

or

$$s = SoSi(\tilde{s}) = \beta\frac{\tilde{s}}{\phi^k} = \beta\frac{\tilde{s}}{(D^{\alpha+1}(\tilde{s}))^k}. \tag{38}$$

Since SoSi(·) has to be reciprocal we find that only the values α=-2 and α=-1 are possible, because for other values this transformation never can be an involution. In both cases we find for ϕ that

$$\tilde{\phi} = SoSi(\phi) = 1/\phi \tag{39}$$

Hence the substitution $\phi\longrightarrow1/\phi$ is a canonical invariance for the equation of the singularity manifold. Since only derivatives occur in this equation another invariance must be

$$\phi\longrightarrow\phi+c,\quad c \text{ constant}.$$

These two invariances combined yield the Möbius group

$$\phi\longrightarrow\frac{a\phi+b}{c\phi+d},\quad ad-bc \neq 0.$$

A consequence of this is that the full Painlevé series is always obtained from its constant level term. This is seen from

$$u = SE(s) = SE(SoSi(\tilde{s})) = SE(SoSi(D^{-1-\alpha}\phi)) = PE(\phi). \tag{40}$$

So substitution of

$$s = SoSi(D^{-1-\alpha}\phi) \tag{41}$$

into the special expansion u=SE(s) yields the Painlevé expansion u=PE(ϕ) for u.

The effect of the SoSi-transform on the equations under consideration is exhibited in Fig. 1. The notation is self-explanatory. For example KdV(u) means the KdV equation for the field variable u, and so on. And in order to avoid confusion when arrows have to cross each other then instead of drawing bridges we denote that by circles. Since, by a reciprocal transformation, the singularity equation for KdV is related to the Harry Dym equation [7,13] SoSi must affect this equation as it can be transferred via the Bäcklund transformations shown in this figure.

Strictly speaking, the Painlevé analysis amounts to a representation of u in terms of its zero soliton. Therefore we believe that the information of the Painlevé test is somewhat restricted since it corresponds to only one point of the possible spectrum of the recursion operator. If one desires to transfer SoSi to other points of the spectrum of the recursion operator one has to proceed differently: replace (7) by its translated equation

$$u_t = K(u) - \lambda u_x \tag{42}$$

and assign to the variable parameter λ a scaling such that the scaling of the equation does not change. Then the special expansion corresponds to the eigenvectors of $\Phi(u)$ with eigenvalues λ. Application of SoSi then leads to the following Darboux-like result:

Theorem:

Let s be an eigenvector of the recursion operator $\Phi(u)$ such that

$$u = SE(s) \tag{43}$$

can be expressed by some homogeneous expression in s. Then

$$\tilde{s} = SoSi(s) \tag{44}$$

is an eigenvector of

$$\tilde{u} = SE(\tilde{s}) \tag{45}$$

with the same eigenvalue.

One should observe that the procedure for going over to other points of the spectrum of the recursion operator sometimes leads to a dramatic change in the behaviour of the expansions under consideration. For example, for the mKdV, due to the analytical structure with respect to λ, it turns out that the series for $\lambda=0$ are finite whereas those for $\lambda\neq0$ have to be infinite (for details see [7]).

5 Kernel transformations

Looking at the last result in section 4 one is reminded of the classical Darboux method [1,2,3] for generating eigenvectors of the Schrödinger equation. This especially since there is a connection between the recursion operator of the KdV and the Schrödinger operator. However, any attempt to use SoSi(·) in order to prove this classical result fails. So we must have overlooked some invariances and indeed that is the case. Let us demonstrate that for the KdV.

Recall from the interacting soliton theory [5] that if u evolves according to the KdV and if

$$u = SE(s) \tag{46}$$

then s evolves in the following way

$$s_{xt} = K(u)'[s_x] = K'(SE(s)[s_x]. \tag{47}$$

But obviously (46) is a differential equation for s. of second order, having a second solution. Changing from s to this second solution does not change u, therefore the evolution (47) is not changed. Hence going from s to the other solutions of (46) must yield an invariance of (47). These transformations within the solution space of (46) we call **kernel-transformations**, since they are transformations within the kernel of $(\Phi(u)-\lambda I)$, the recursion operator of the translated equation. A general transformation group may now be obtained by combining SoSi and the kernel transformations. This group then also comprises the classical Darboux theorem. Therefore we call it the **Darboux-Möbius group**. Let us demonstrate that again in the KdV case:

Given s then the the second solution usually can be found by variation of constants. In the KdV-case this procedure is easily carried out. Equation (46) may be rewritten in the form

$$u = SE(s) = \frac{\lambda}{4} - \frac{(\sqrt{s})_{xx}}{\sqrt{s}} . \tag{49}$$

Hence, as it is well known, $\sqrt{s}$ fulfils a linear differential equation of second order. For this the second solution is easily obtained from the given one by variation of constants. This second solution has the general form

$$\sqrt{s_2} = a\sqrt{s}+\sqrt{s}\ (D^{-1}(1/s)). \tag{50}$$

Thus for the solutions of (47) we have found the invariance

$$s \longrightarrow a^2 s+2asD^{-1}(1/s)+s(D^{-1}1/s)^2. \tag{51}$$

A special case of this is

$$s \longrightarrow s(D^{-1}(1/s))^2 = ke(s) \tag{52}$$

which must leave (47) invariant. Hence for the singularity equation we have found the new invariance

$$\phi \longrightarrow a^2\phi+2aD^{-1}(\phi_x D^{-1}(1/\phi_x))+D^{-1}(\phi_x(D^{-1}(1/\phi_x))^2). \tag{53}$$

We like to remark that for the special expansion of the mKdV the kernel transformation and SoSi coincide. So the question arises what has happened to the kernel transformation by going from KdV to mKdV. The answer is simple. We recall that the Miura transformation relates the KdV to the mKdV. Obviously that is a differential equation for the field quantity v of the mKdV in terms of u having not only one solution but a one-parameter family of solutions. So there must be a kernel transformation for the mKdV which does not affect the KdV. The determination of the kernel transformations of the Miura transformation is not that simple, but actually this is not necessary any more, because we already know the result. Going from the KdV to its zero soliton via the Miura transformation we find the following transformation from mKdV to the zero soliton of KdV which can be seen from the transformation table:

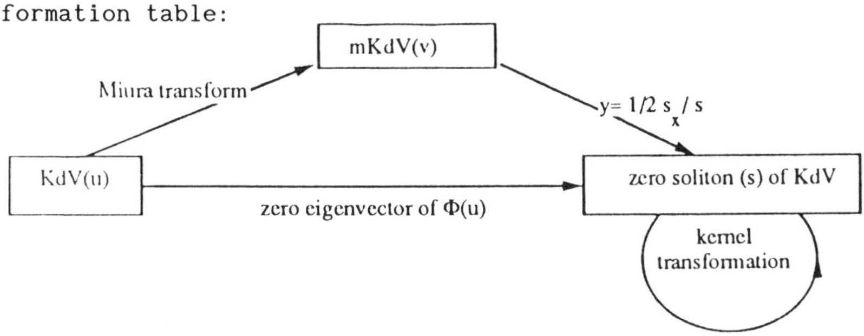

Because of the commutativity of the diagram the kernel transformation of the zero-soliton induces then a transformation for mKdV which is annihilated by the Miura transform. Hence, this yields the kernel transformation of the Miura transformation. Then

$$v = \frac{1}{2}\frac{s_x}{s} \Leftrightarrow s = e^{-2D^{-1}(v)} \tag{54}$$

implies (by using (52)) that

$$v \longrightarrow \frac{1}{2}\frac{d}{dx}\ln\big(a^2 e^{-2D^{-1}(v)} + 2ae^{-2D^{-1}(v)} D^{-1}(e^{2D^{-1}(v)}) + e^{-2D^{-1}(v)}(D^{-1}(e^{2D^{-1}(v)}))^2\big) \tag{55}$$

leaves the mKdV invariant, and that this transformation substituted into the Miura transform leads to the same u as before.

This last statement seems interesting from another viewpoint. If one tries to find the kernel of the Miura transform directly, one would proceed in the following way: at first one looks for infinitesimal transformations which leave this transform invariant, i.e. one looks for $v \longrightarrow v + \varepsilon\tilde{v}$ such that this leads, up ε^2-terms, to the same u. This gives for $\tilde{v}$ the condition

$$-\tilde{\nu}_x + 2\tilde{\nu}\nu = 0 \tag{56}$$

or

Soliton singularity transform

$$\tilde{v} = e^{-D^{-1}2v}. \tag{57}$$

Hence the evolution equation

$$v_t = e^{-2D^{-1}v} \tag{58}$$

is the equation which describes the kernel of the Miura transform. By $h_x = v$ we rewrite that as

$$h_{xt} = e^{-2h} \tag{59}$$

which is the well-known Liouville equation. Although the solution of that looks difficult we know it already; it is given by formula (55) which describes a one-parameter manifold invariant under that equation. The solutions obtained this way are the ones usually found in the literature ([10, p. 254], [8, p. 71]).

An interesting feature of the transformation SoSi, and also the kernel transformation, is that they often make out of zeros of interacting solitons poles of solutions for the interacting soliton equation. This explains why usually poles behave like trajectories of solitons (a detailed analysis of this will appear elsewhere).

6 Comparisons with other work

On the structural level the new results of this paper are all connected to the newly introduced expansion test, a generalization of the fundamental Painlevé test for integrability proposed by Weiss, Tabor & Carnevale [14] which gave a new insight in the study of nonlinear partial differential equations. Since its introduction the Painlevé test has been applied by many authors for a large number of nonlinear systems recognizing that they possess the Painlevé property. Now there are excellent surveys on this subject (for example [12] or [11]) so we do not have to go into a detailed analysis of these very many results.

The expansion test shares with the Painlevé test its two most important aspects: namely its wide applicability and its algorithmic structure. The generalized test we propose, besides giving meaningful results whenever the Painlevé test does, also is applicable in the study of those nonlinear systems for which the usual test can say nothing. In fact, by definition, a partial differential equation possesses the Painlevé property whenever it admits solutions expressed in the form of a finite truncated expansion in terms of its singularity manifold function ϕ. The requirement that the expansion truncates after a finite number of terms is both restrictive and not needed for integrability. Furthermore it is the most difficult task to achieve in performing the test. For instance to prove that the sine-Gordon equation does possess the Painlevé property *ad hoc* considerations are needed. Such is also the case for other equations. On the contrary, as far as our test is concerned, it does not make any difference whether the expansion has a finite or infinite number of terms [7].

However, since the possibility of infinite series prevents us from starting at the bottom, we have to start at the top of the series thus needing a simple solution of the equation to start with. This means that the two tests cannot coincide and that the algorithmic procedure has been reversed.

This new algorithmic strategy allows the revelation of the intimate connection between the structure of the singularity equation and the corresponding interacting soliton [5,6]. In fact, a particular case of the Expansion test, we referred to as the special expansion, is represented by the study of the interacting soliton for $\lambda=0$, so that the singularity equation and the corresponding zero-soliton equation turn out to be linked by an involutive Bäcklund transformation. For the KdV and the AKNS hierarchy such an interpretation of the singularity function in terms of squared eigenfunctions of course is well known (see for example [11]). However, a new finding is that this gives rise systematically to an involutory SoSi-transformation for all equations under consideration and, in addition, that the effect of SoSi on the singularity function indeed can be explicitly computed independently of the equation under consideration. This observation then leads to a wide variety of results (impossibility of the Painlevé test for certain scaling degrees, recovery of the Painlevé series from its constant level term, Darboux-Möbius transform, etc.). Another point also is clarified which was missing so far in the literature, namely that contrary to the KdV case, in general a relation between the singularity function and the square eigenfunctions with nonzero eigenvalue is not possible. Insofar as the expansion test really goes beyond the Painlevé test, it has therefore to be expected that there is a wide variety of nonintegrable equations passing the Painlevé test where the general expansion test fails.

Anyway the results show a close relation between the recursion operator and the expansion test. An immediate consequence is that a successful expansion test delivers the recursion operator whenever it does exist and that given the recursion operator both the singularity and the zero-soliton equation can be easily obtained without going through the whole expansion test.

The involutivity of the SoSi transformation between the zero-soliton and the singularity equation implies that the special expansion and the Painlevé expansion can be obtained one from the other in a straightforward way. This provides a structural reason why no nonlinear system with α different from -1 or -2 has been proved to possess the Painlevé property. This is in accordance with the examples which can be found in the literature. To the same conclusion by different arguments, recently, came R.B. King [9] who was interested in finding necessary conditions which a nonlinear partial differential equation must satisfy in order to possess the Painlevé property.

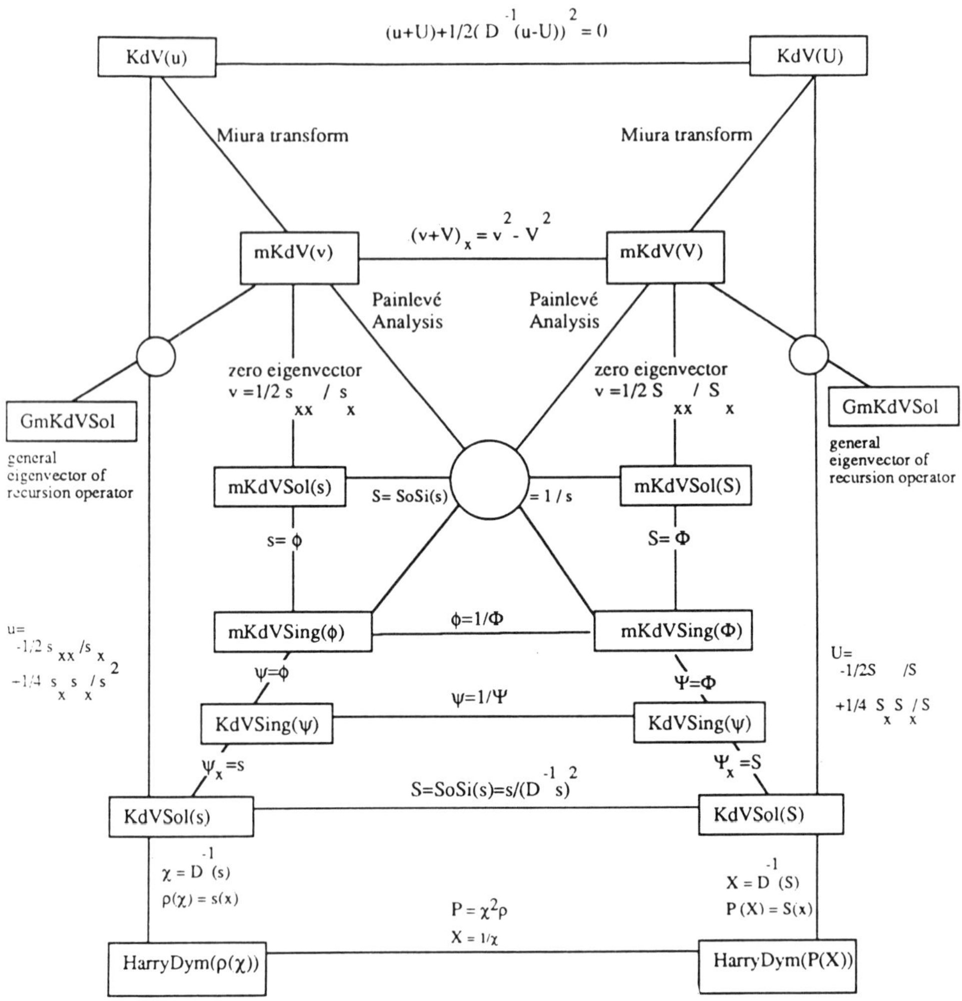

Figure 1.

References

[1] Crum, M.M. (1955). Associated Sturm-Liouville equations. *Q.J. Math.* **6**, 121.

[2] Darboux, G. (1882). Sur une proposition relative aux équations linéaires. *Comptes Rendus* **94**, 1449.

[3] Deift, P. & Trubowitz, E. (1980). Some remarks on the Korteweg-de Vries and Hill's equations. In *Nonlinear Dynamics*, vol. 357, p. 55. Hellemann, R.H.G. (ed.). New York Academy of Science.

[4] Fuchssteiner, B. (1979). Application of Hereditary Symmetries to Nonlinear Evolution equations. *Nonlinear*

Analysis TMA **3**, 849.

[5] Fuchssteiner, B. (1987). Solitons in Interaction. *Prog. Theoret. Phys.* **78**, 1022.

[6] Fuchssteiner, B. (1988). The dynamical behavior of intereacting solitons. In *Proceedings of the workshop of nonlinear Evolution equations and Dynamical Systems.* World Scientific Publications: Singapore.

[7] Fuchssteiner, B. & Carillo, S. (1988). Soliton Structure versus Singularity Analysis: Third-order Completely Integrable Nonlinear Differential Equations in 1+1-dimensions. *Physica A* **154**, 467.

[8] Goursat, E. (1964). *A course in mathematical Analysis, Vol. III, part 1.* Dover Publications, Inc.: New York.

[9] King, R.B. (1986). Systematics of strongly self-dominant higher-order differential equations based on the Painlevé analysis of their singularities. *J. Math. Phys.* **27**, 966.

[10] Lamb, G.L. (1980). *Elements of Soliton Theory, Pure and Applied Mathematics.* Wiley Interscience Publ.: New York-Chichester-Brisbane-Toronto.

[11] Newell, A.C., Ratiu, T., Tabor, M. & Yunbo, Z. (1986). Soliton Mathematics. *Séminaire de Mathematiques Supérieures, Université de Montreal* **103**.

[12] Ramani, A., Grammaticos, B. & Bountis, T. (1988).The Painlevé property and singularity analysis of integrable and non-integrable systems. Preprint.

[13] Rogers, C. & Nucci, M.C. (1986). On Reciprocal Bäcklund transformations and the Korteweg-de Vries Hierarchy. *Physica Scripta* **33**, 289.

[14] Weiss, J., Tabor, M. & Carnevale, G. (1983). The Painlevé property for partial differential equations. *J. Math. Phys.* **24**, 522.

M. Musette

17. PAINLEVÉ-BÄCKLUND TRANSFORMATION FOR THE KURAMOTO-SIVASHINSKY EQUATION

The Painlevé analysis developed by Weiss et al. for nonlinear partial differential equations is ordinarily performed to obtain the Lax pairs associated with completely integrable equations. Here, it is applied to an equation which is known to be partially integrable and passes the Painlevé test in a weak sense, in order to find new particular analytical solutions. Obtaining the solutions of the Painlevé-Bäcklund (P.-B.) equations is made easy by the introduction of two elementary invariants under the Möbius group, which are closely connected to linear partial differential equations. Contrarily to the Korteweg-de Vries (KdV) case where it is possible to define an iteration from the P.-B. set, due to the presence of a spectral parameter, and to obtain the N-soliton solutions, we show that here no iteration can be defined. The solutions of the P.-B. set reduce to only one type, namely the already known kink-shaped solitary wave.

1 **Introduction**

The Kuramoto-Sivashinsky (KS) equation is a model partial differential equation (PDE) frequently encountered in the study of continuous media which exhibits a chaotic behaviour. In its conservative form, it reads:

$$u_t + uu_x + u_{xx} + u_{xxxx} = 0 \tag{1}$$

It describes for instance the fluctuation of position of a flame front, or the motion of a fluid going down a vertical wall, or a spatially uniform oscillating chemical reaction in a homogeneous medium. A recent review can be found in a Les Houches lecture [1]. This equation also arises from the minimal ingredients necessary to observe interesting bifurcations in a simplified equation for a complex amplitude in fluid dynamics [2].

The purpose of this work is to look for new analytical solutions using the Painlevé-Bäcklund transformation, which will be described below. Particularly, some special solitary waves which have been numerically observed [3-4] cannot be represented by the known kink-shaped analytical solution of Kuramoto & Tsuzuki [5].

We refer the reader who is not familiar with the Painlevé analysis for nonlinear PDE to the paper of Weiss et al. [6]. In Section 2, we recall some generalities on this technique and the results of its application to the KdV equation. The use of two elementary invariants under homographic transformation in the Painlevé-Bäcklund equations and their close connection with the linear scattering equations shorten greatly the attainment by iteration of the N-soliton solutions.

In Section 3, we perform the Painlevé analysis on KS and obtain a transformation between two solutions of this equation by truncating the Painlevé-Laurent expansion. Here also, the replacement of the four Painlevé-Bäcklund equations by four invariant expressions under the Möbius group and the connection of the two elementary invariants with linear partial differential equations make easy the attainment of the solutions of the Painlevé-Bäcklund set. We show that unfortunately due to the absence of spectral parameter, no iteration can be defined from this set and the solutions reduce to only one class. The solutions of KS correspond to the kink-shaped stationary waves mentioned above, obtained from the Kuramoto & Tsuzuki solution by making a Galilean transformation. The Painlevé analysis does not allow us the finding of new analytical solutions.

2 Painlevé-Bäcklund transformation for KdV

The application of the Painlevé analysis to the KdV equation [6,7]:

$$u_t + uu_x + u_{xxx} = 0 \qquad (2)$$

consists of looking for the general solution of the PDE in the form:

$$u(x,t) = \varphi(x,t)^p \sum_{j=0}^{\infty} u_j(x,t)\, \varphi(x,t)^j \qquad (3)$$

where $p = -2$, $\varphi(x,t) = 0$ is the equation of a non-characteristic ($\varphi_x\varphi_t \neq 0$) singular manifold, and the function u_j's have to be determined by substitution of expansion (3) in (2) which becomes:

$$\sum_{j=0}^{\infty} E_j(u_o,\ldots,u_j,\varphi_x,\varphi_t \text{ and their derivatives})\, \varphi^{j-5} = 0 \quad (4)$$

The first equation E_o determines u_o

$$u_o = -12\varphi_x^2 \tag{5}$$

and for j>0, one has:

$$\forall\ j{:}E_j \equiv (j+1)P(j)\varphi_x^3 u_j + Q_j(u_o,\ldots u_{j-1},\varphi_x,\varphi_t \text{ and derivatives}) = 0 \tag{6}$$

where $P(j) = (j-4)(j-6)$.

The two resonances (zeros of P), $j_r = 4$ and $j_r = 6$, are "compatible": after substitution in (6) of the previoùsly computed u_l's, $l \leq j_r-1$, the function Q_{j_r} is identically zero.

The Painlevé property is characterized by the fact that p is a negative integer and all resonances occur at positive integer values of j and are compatible.

We now consider the truncated expansion for u at the "constant" level in φ:

$$u(x,t) = \varphi^{-2} \sum_{j=0}^{2} u_j\ \varphi^j \tag{7}$$

The expression (5) for u_o and that one obtained for u_1 from the equation $E_1 = 0$:

$$u_1 = 12\varphi_{xx} \tag{8}$$

implies that:

$$u = 12(\ln\varphi)_{xx} + u_2 \tag{9}$$

It relates two solutions $u(x,t)$ and $u_2(x,t)$ of (2) if the 3 equations are satisfied:

$$j = 2{:}\ \varphi_x\varphi_t + \varphi_x^2 u_2 + 4\varphi_x\varphi_{xxx} - 3\varphi_{xx}^2 = 0 \tag{10}$$

$$j = 3{:}\ \varphi_{xt} + \varphi_{xx}u_2 + \varphi_{xxxx} = 0 \tag{11}$$

$$j = 5{:}\ \mathrm{KdV}(u_2) = 0 \tag{12}$$

As these equations depend only on the derivatives of the function $\varphi(x,t)$, it is possible to express them in terms of 3 functions, namely two invariants under an arbitrary homographic transformation ($H{:}\ \varphi \rightarrow \frac{a\varphi+b}{c\varphi+d}$)

$$S \equiv \{\varphi, x\} = \frac{\varphi_{xxx}}{\varphi_x} - \frac{3}{2}\left(\frac{\varphi_{xx}}{\varphi_x}\right)^2 \tag{13}$$

called the Schwarzian derivative, and

$$C = -\frac{\varphi_t}{\varphi_x} \tag{14}$$

which has the dimension of a velocity, and a non invariant ratio:

$$D = -\frac{\varphi_{xx}}{2\varphi_x} \tag{15}$$

Moreover, one must take into account that the 2-elementary invariants S and C are linked by the cross-derivative condition $(\varphi_t)_{xxx} = (\varphi_{xxx})_t$ which reads:

$$S_t + C_{xxx} + 2\,C_x\,S + C\,S_x = 0 \tag{16}$$

An important simplification occurs when one makes the transformation

$$\varphi_x = \Psi^{-2} \tag{17}$$

Taking into account that

$$D = (\ln\Psi)_x \tag{18}$$

$$D^2 = -\frac{S}{2} - (\ln\Psi)_{xx} \tag{19}$$

and defining new H-invariant expressions $\tilde{E}_j$, $3 \le j \le 5$, by

$$\tilde{E}_j = \sum_{k=3}^{j} (-D)^{j-k} \frac{(5-k)!}{(5-j)!\,(j-k)!} E_k \tag{20}$$

(E_4 representing the compatibility condition for the resonance $j_r = 4$) the expressions (10), (11) and (12) can be replaced by

$$u_2 = C + 2S + 12(\ln\Psi)_{xx} \tag{21}$$

$$\tilde{E}_3 = 0: \quad -12(S-C)_x = 0 \tag{22}$$

$$\tilde{E}_5 = 0: \quad (S + C/2)\,\tilde{E}_3 + 2\tilde{E}_{3,xx} + (C-S)_t = 0 \tag{23}$$

$$(\tilde{E}_4 \equiv 0: \quad \tilde{E}_4 = \tilde{E}_{3,x})$$

From (22) and (23), we deduce the relation:

$$S-C = \lambda \tag{24}$$

and, from the compatibility condition (16), we obtain

$$S = \frac{1}{3}(u'+\lambda) \tag{25}$$

$$C = \frac{1}{3}(u'-2\lambda) \tag{26}$$

where u' is a solution of (2).

From this set and the relation (21), one can build by iteration the N-solitons, choosing negative values for λ. Let us first remark that due to the transformation (17):

$$S = -2\Psi^{-1}\Psi_{xx} \tag{27}$$

and the relation (25) becomes

$$\Psi_{xx} + \frac{1}{6}(u'+\lambda)\Psi = 0 \tag{28}$$

On the other side, making use of the definition (14) for C, the transformation (17) and the relation (26), we obtain:

$$\Psi_t = \frac{1}{3}(2\lambda-u')\Psi_x + \frac{1}{6}u'_x\Psi \tag{29}$$

(28) and (29) correspond to the classical scattering equations [8] associated with the solution u' of KdV. The N-soliton solutions are obtained by the following procedure (see also ref. [9]):

first iteration: $u' = 0\ ,\quad \lambda = -\frac{3}{2}k_1^2\ ,\quad C + 2S = 0$

Noting by $\Psi^{(1)}$ the solution of the system

$$\begin{cases} \Psi_{xx} - \dfrac{k_1^2}{4}\Psi = 0 \\ \Psi_t = -k_1^2\Psi_x \end{cases} \tag{30}$$

we obtain:

$$u_2^{(1)} = 12(\ln\Psi^{(1)})_{xx} = 12[\ln(1+e^{\theta_1})]_{xx},\ \theta_1 = k_1x - k_1^3t + \eta_1^{(o)}$$

($\eta_1^{(o)}$ arbitrary constant) (31)

second iteration: $u' = u_2^{(1)}\ ,\quad \lambda = -\frac{3}{2}k_2^2\ (k_2>k_1>0)$,

$$C + 2S = 3k_1^2 \mathrm{csh}^{-2}\left(\frac{\theta_1}{2}\right) = u_2^{(1)}.$$

Noting by $\Psi^{(2)}$ the solution of the system

$$\left\{\begin{array}{l} \Psi_{xx} + [-\frac{k_2^2}{4} + \frac{k_1^2}{2}\left(\mathrm{csh}^{-2}\left(\frac{\theta_1}{2}\right)\right)]\Psi = 0 \\ \Psi_t = [-k_2^2 - k_1^2\left(\mathrm{csh}^{-2}\left(\frac{\theta_1}{2}\right)\right)\Psi_x - \frac{k_1^3}{2}\left(\mathrm{csh}^{-3}\frac{\theta_1}{2}\right)\mathrm{sh}\left(\frac{\theta_1}{2}\right)\Psi \end{array}\right. \tag{32}$$

we obtain:

$$u_2^{(2)} = 12(\ln\frac{\Psi^{(2)}}{\Psi^{(1)}})_{xx} = 12[\ln(1 + e^{\theta_1} + e^{\theta_2} + K_{12}e^{\theta_1+\theta_2})]_{xx}$$

$$\theta_i = k_i x - k_i^3 t + \bar{\eta}_i \qquad (\bar{\eta}_i\text{: arbitrary constant}) \tag{33}$$

which corresponds to the expression of the 2-soliton solution.

3 Painlevé-Bäcklund transformation for KS equation

We find that for eq. (1) the dominant terms are uu_x and u_{xxxx}, the leading power is $p = -3$ and the first coefficient is:

$$u_o = 120\varphi_x^3 \tag{34}$$

In this case,

$$P(j) = (j - 6)(j^2 - 13j + 60) \tag{35}$$

and the resonances are $j = 6$ and $j = \frac{13}{2} \pm i\frac{\sqrt{71}}{2}$. Since the three resonances are not all located at positive integers, there is no way to obtain by this procedure the general solution of eq. (1). However, we may still obtain a solution depending on two arbitrary functions. In order to examine the nature of the resonance located at $j = 6$, we must compute the coefficients u_j up to $j = 6$ and we list here the next 2 coefficients:

$$u_1 = -180\varphi_x\varphi_{xx} \tag{36}$$

$$u_2 = 60\varphi_{xxx} + \frac{60}{19}\varphi_x \tag{37}$$

As already observed by Fournier & Spiegel [10], the resonance $j = 6$ is compatible. We can even show that the compatibility condition, which is identically satisfied, has the following nice structure

$$E_6 \equiv -(\varphi_x^{-1}E_5)_x - \frac{1}{2}(\varphi_x^{-1}(\varphi_x^{-1}E_4)_x)_x - \frac{1}{6}(\varphi_x^{-1}(\varphi_x^{-1}E_3)_x)_x)_x \tag{38}$$

From the above analysis we can say that the KS equation passes the Painlevé test not in the strict sense (all resonances occur at positive integers and are compatible), but in a weaker sense (all resonances which appear at positive integers are compatible).

The Painlevé expansion (2), truncated at j = 3 (constant level in φ) represents a solution of eq. (1) of the following form:

$$u = \frac{60}{19}(\ln\varphi)_x + 60(\ln\varphi)_{xxx} + u_3 \tag{39}$$

provided u_3 and φ satisfy the five equations

$$E_j \text{ (derivatives of } \varphi,\ u_3, \text{ derivatives of } u_3) = 0 \qquad j = 3 \text{ to } 7 \tag{40}$$

Let us note that $E_7 = 0$ is nothing other than the KS equation for u_3. E_6, which represents the compatibility condition of the resonance j = 6, is identically zero. Then equation (39) defines a transformation between 2 solutions, u_3 and u, of the same equation if the set of solutions (φ, u_3) of the 4 following equations:

$$E_j = 0 \qquad \text{for } j = 3,\ 4,\ 5,\ 7 \tag{41}$$

is not empty.

By the transformation

$$\tilde{E}_j = \sum_{k=4}^{j} (-D)^{j-k} \frac{(7-k)!}{(7-j)!(j-k)!} E_k \qquad j = 4 \text{ to } 7 \tag{42}$$

and the relations (18), (19) and

$$D^3 = \frac{S_x}{4} - \frac{S}{2}(\ln \Psi)_x + \frac{1}{2}(\ln \Psi)_{xxx} \tag{43}$$

the system (43) becomes:

$$u_3 = C + 15\, S_x + \frac{60}{19}(\ln \Psi)_x + 60\, (\ln \Psi)_{xxx} \tag{44}$$

$$\tilde{E}_4 \equiv 120\, (6\, S_{xx} + 4\, S^2 + \frac{20}{19} S - 2\, C_x - \frac{11}{19^2}) = 0 \tag{45}$$

$$\tilde{E}_5 \equiv -60\, (7\, S_{xx} + 3\, S^2 + \frac{25}{19} S - 3\, C_x)_x = 0 \tag{46}$$

$$\tilde{E}_7 \equiv C_t + CC_x + \frac{177}{38}C_{xx} + 45\, SC_{xx} + 112\, C_{xxxx} + 30\, C_x S_x + f(\tilde{E}_4, \tilde{E}_5) = 0 \tag{47}$$

where $f(\tilde{E}_4,\tilde{E}_5)$ is an expression vanishing with $\tilde{E}_4$ and $\tilde{E}_5$.

In the appendix of [11], it is shown that the solutions of the system $\{\tilde{E}_j = 0,\ j = 4,\ 5,\ 7\}$ compatible with the condition (16) come to:

$$C = c \tag{48}$$

$$S^2 + \frac{5}{19} S - \frac{11}{4.(19)^2} = 0 \tag{49}$$

where c is an arbitrary constant and S has only two values $S_1 = \frac{1}{38}$ and $S_2 = -\frac{11}{38}$.

Knowing the solution of the linear system

$$\begin{cases} \Psi_{xx} + \dfrac{S_i}{2} \Psi = 0 \quad (i = 1,2) \\ \Psi_t = -c\Psi_x \end{cases} \tag{50}$$

$$\Psi = A\, e^{\sqrt{-\frac{S_i}{2}}\xi} + B\, e^{-\sqrt{-\frac{S_i}{2}}\xi}, \quad \text{with } \xi = x - ct \tag{51}$$

(A and B arbitrary constants), we have from (44)

$$u_3 = c + \frac{60}{19}(\ln \Psi)_x + 60\ (\ln \Psi)_{xxx} \tag{52}$$

and, as S can only take constant fixed values, it is impossible to define an iteration procedure from (48), (49) and (52).

Aside from the constant trivial solution when A.B = 0, the solution obtained from the Painlevé-Bäcklund equations is simply the steady wave solution of Kuramoto & Tsuzuki after having performed a Galilean transformation of velocity c:

$$u = c + (\frac{30}{19}k_i - 15\ k_i^3)\ \text{th}\ (\frac{\bar{\xi}_i}{2}) + 15\ k_1^3\ \text{th}^3\ (\frac{\bar{\xi}_i}{2}) \tag{53}$$

$$\text{with } k_i^2 = -2\ S_i,\quad \bar{\xi}_i = k_i(x - ct) + \ln \frac{B}{A}$$

We notice that equation (53) represents three distinct types of real solutions, namely:

k_i real, $\frac{B}{A} > 0$: kink-shaped regular

k_i real, $\frac{B}{A} < 0$: kink-shaped singular

k_i purely imaginary: periodic singular.

k_i real, $\frac{B}{A} > 0$: kink-shaped regular

k_i real, $\frac{B}{A} < 0$: kink-shaped singular

k_i purely imaginary: periodic singular.

The solution which comes out of the Painlevé analysis is in agreement with the result [12] that KS equation has no solution which is a rational fraction of more than one exponential whose argument is linear in x and t.

References

[1] Manneville, P. (1988). The Kuramoto-Sivashinsky equation: a progress report. Les Houches, March 10-18, 1987. *Propagation in systems far from equilibrium,* Wesfreid, J., Brand, H.R., Manneville, P., Albinet, G. & Boccara, N. (eds.). Springer Proceedings in Physics.

[2] Coulet, G., Gil, L. & Lega, J. (1989). Defect-mediated turbulence. *PRL* **62**, 1619.

[3] Chang, H.-C. (1986). Travelling waves on fluid interfaces: normal form analysis of the Kuramoto-Sivashinsky equation. *Physics of Fluids* **29**, 3142.

[4] Toh, S. (1987). Statistical model with localized structures describing the spatio-temporal chaos of Kuramoto-Sivashinsky equation. *J. Phys. Soc. Jpn.* **56**, 949.

[5] Kuramoto, Y. & Tsuzuki, T. (1976). Persistent propagation of concentration waves in dissipative media far from thermal equilibrium. *Prog. Theor. Phys.* **55**, 356.

[6] Weiss, J., Tabor, M. & Carnevale, G. (1983). The Painlevé property for partial differential equations. *J. Math. Phys.* **24**, 522.

[7] Weiss, J. (1983). The Painlevé property for partial differential equations II: Bäcklund transformation, Lax pair, and the Schwarzian derivative. *J. Math. Phys.* **24**, 1405.

[8] Gardner, C.S., Greene, J.M., Kruskal, M.D. & Miura, R.M. (1967). Method for solving the Korteweg-de Vries equation. *Phys. Rev. Lett.* **19**, 1095.

[9] Wadati, M., Sanuki, H. & Konno, K. (1975). Relationships among Inverse Method, Bäcklund Transformation and an Infinite Number of Conservation Laws. *Prog. Theor. Phys.* **53**, 419.

[10] Fournier, J.-D. & Spiegel, E.A. (1986). Meromorphic integrals of the equations of phase chaos. Preprint.

[11] Conte, R. & Musette, M. (1989). Painlevé analysis and Bäcklund transformation in the Kuramoto-Shivashinsky equation. *J. Physics A: Math. Gen.* **22**, 169.

[12] Cornille, H. & Gervois, A. (1983). Unpublished.

R. Conte

18. PAINLEVÉ ANALYSIS AND BÄCKLUND TRANSFORMATION OF THE NONINTEGRABLE KPP PARTIAL DIFFERENTIAL EQUATION

We give a detailed example of all the information which can be obtained from the Painlevé technique when applied to a nonintegrable PDE, namely the Kolmogorov-Petrovsky-Piskunov equation. This information is expressed in the form invariant under an arbitrary homographic transformation of the singular manifold function. The simplicity of invariant equations allows us to find their general solution, thus leading to a 2-parameter family of analytic solutions.

1 Introduction

The technique [15] of the Painlevé analysis of PDE's is to look for the general solution under the form of an expansion:

$$u(x,t) = \varphi(x,t)^{p} \sum_{j=0}^{+\infty} u_j(x,t)\ \varphi(x,t)^{j} \tag{1}$$

where the leading power p is a negative integer, $\varphi(x,t)=0$ the equation of a singular manifold, and u_j's unknown functions to be determined by the condition that expansion (1) satisfies the PDE, which becomes:

$$\sum_{j=0}^{+\infty} E_j(u_0,\dots,u_j,\varphi)\ \varphi^{j+q} = 0 \tag{2}$$

where q is some negative integer constant. E_j depends on φ only by its derivatives.

A second kind of information can be obtained: the auto-Bäcklund transformation [15], i.e. a transformation between two solutions u and u_{-p} of the same PDE by truncating expansion (1) to order $j = -p$. The equations defining this transformation are:

$$\begin{cases} E_j(u_0,\dots,u_j,\varphi) = 0, & j = 0,\dots,-p \quad (3a) \\ E_j(u_0,\dots,u_{-p},0,\dots,0,\varphi) = 0, & j=-p+1,\dots,-q \quad (3b) \end{cases}$$

After substitution of $u_0,\dots,u_{-p}$ by their value derived from (3a), equations (3b), which we will from now on call Painlevé-Bäcklund (PB) equations, depend only on the derivatives of φ and the arbitrary functions $u_{a,r}$ introduced at the integer resonances $r \in [0,-p]$. In particular, last equation j-q is the condition that u_{-p} satisfies the PDE and it should not be omitted from the set (3b).

As detailed elsewhere [2], all the results of the Painlevé analysis of a PDE consist of three informations for each couple (leading power, branch):

(i) Yes or no to the question "Does it pass the such and such form of the Painlevé test for arbitrary values of the PDE parameters?",

(ii) the set of compatibility conditions at "non compatible" resonances,

(iii) the set of Painlevé-Bäcklund equations.

In a series of papers [10-15], Weiss noticed that some Painlevé-Bäcklund equations could be put in a form invariant under the group $\mathcal{H}$ of homographic transformations of φ:

$$\varphi \to \frac{a\ \varphi + b}{c\ \varphi + d} \qquad a\ d - b\ c \neq 0 \tag{4}$$

This has been generalized recently [3] and one can prove that all the Painlevé analysis is $\mathcal{H}$-invariant.

It is accordingly natural to introduce the two following elementary homographic invariants:

$$S = \frac{\varphi_{xxx}}{\varphi_x} - \frac{3}{2}\left(\frac{\varphi_{xx}}{\varphi_x}\right)^2 \tag{5}$$

$$C = -\frac{\varphi_t}{\varphi_x} \tag{6}$$

The invariant S is called the schwarzian (see e.g. [6]) and C has the dimension of a velocity. Their cross-derivative condition $(\varphi_t)_{xxx} = (\varphi_{xxx})_t$ is $\mathcal{H}$-invariant and reads:

$$S_t + C_{xxx} + 2\ C_x\ S + C\ S_x = 0 \tag{7}$$

Consequently, an arbitrary polynomial in the derivatives of φ, with positive or negative powers of φ_x and φ_t, invariant under the subgroup $\varphi \to a\ \varphi + b$, can be uniquely expressed as a polynomial in $-\frac{\varphi_{xx}}{2\ \varphi_x}$ with polynomial coefficients in S, C and their derivatives, with possible negative powers of C.

2 Application to a nonintegrable equation

We now detail the method in the case of a particular nonintegrable equation, the KPP equation [8]. It reads:

$$E \equiv u_t - u_{xx} - u + 2\,u^3 = 0 \tag{8}$$

There is one leading power p = -1, and one finds $u_0^2 = \varphi_x^2$, i.e. taking into account the symmetry $x \to -x$, only one branch $u_0 = \varphi_x$. The only resonance is 4. The results of the analysis consist of:

- the singular part $u_0 \varphi^{-1}$, expressed as a linear differential operator of order 1 acting on Log φ,
- the value of the constant term u_1, expressed as a polynomial of degree 1 in the non invariant variable $-\dfrac{\varphi_{xx}}{2\,\varphi_x}$ with invariant coefficients,
- the condition $Q_4 = 0$ for the resonance 4 to be compatible, in its invariant form,
- the set of 2 Painlevé-Bäcklund equations $E_2 = 0$, $E_3 = 0$ in their invariant form.

We find:

$$u - u_1 = (\mathrm{Log}\ \varphi)_x\ ,\quad u_1 = -\frac{C}{6} + \left(-\frac{\varphi_{xx}}{2\,\varphi_x}\right) \tag{9}$$

$$\begin{cases} Q_4 \equiv \dfrac{2\,C}{9}\,(\dfrac{1}{3}\,C^3 - \dfrac{3}{2}\,C - C\,C_x - C_t) = 0 & (10) \\ E_2 \equiv -1 - S + \dfrac{1}{6}\,C^2 - C_x = 0 & (11) \\ E_3 \equiv \dfrac{1}{6}\,(C + 3\,S\,C - \dfrac{1}{18}\,C^3 + 3\,S_x - C_t + 4\,C_{xx}) = 0 & (12) \end{cases}$$

and we must add the constraint (7) between S and C.

Any solution of the PB equations also satisfies the compatibility condition, due to the identity:

$$Q_4 \equiv C\,E_3 + E_{3,x} + (\frac{1}{2}\,C^2 + \frac{1}{6}\,C_x)\,E_2 + \frac{5}{6}\,C\,E_{2,x} - \frac{1}{6}\,E_{2,t} + \frac{1}{2}\,E_{2,xx} \tag{13}$$

but the inverse is not true.

Let us now solve the PB equations (11), (12); they imply an ODE for C:

$$C_{xx} - C\,C_x + \frac{1}{9}\,C^3 - \frac{C}{2} = 0 \tag{14}$$

and this ODE, which is easily seen to have the Painlevé property, is linearizable by the change $C = -3\,\frac{y'}{y}$. Hence the general solution (S,C) of PB equations:

$$\left\{\begin{array}{ll} C = -\dfrac{3}{\sqrt{2}}\,\dfrac{-a_1 e^{-\frac{x}{\sqrt{2}}} + a_2 e^{\frac{x}{\sqrt{2}}}}{a_1 e^{-\frac{x}{\sqrt{2}}} + a_2 e^{\frac{x}{\sqrt{2}}} + a_3 e^{\frac{3t}{2}}} & (15) \\ S = -1 + \frac{1}{6} C^2 - C_x & (16) \end{array}\right.$$

depending on two arbitrary constants.

The problem is now to go back from the invariants (S,C) to φ. Two methods are available:

(i) Integrate the third order nonlinear ODE (5) (strictly speaking, it is a PDE, but it does not depend on t), which leads to a function φ depending on 3 arbitrary functions of t; one then has to express that this function φ also satisfies the first order linear PDE (6), which determines a set of first order ODE's for these 3 functions of t, leading to a final function φ depending on 3 more arbitrary constants.

(ii) Integrate the first order linear PDE (6), which leads to a function φ equal to an arbitrary function Φ of F, where F(x,t) is a particular solution of (6); one then has to express that this function Φ also satisfies the third order nonlinear ODE (5), which determines Φ as a precise function of F depending on three arbitrary constants.

These two methods of course give the same final result, but, depending on the particular couple (S,C) under consideration, one method may be "better" than the other. For the Kuramoto-Sivashinsky equation [4], the first one was used, although both were easy. In the present case, we apply the second method: it appears that the 1-form dx - C dt, with C given by (15), admits an integrating factor [1], leading to the following general solution φ of the first order linear PDE (6):

$$\varphi = \Phi(F(x,t)),\quad F(x,t) = \frac{-2\sqrt{2}\,a_2\,e^{-\frac{3t}{2}} - \sqrt{2}\,a_3\,e^{-\frac{x}{\sqrt{2}}}}{a_2^2\,e^{\frac{x}{\sqrt{2}}} - a_1 a_2\,e^{-\frac{x}{\sqrt{2}}}} \qquad (17)$$

where Φ is an arbitrary function of one variable.

Then the ODE for the function Φ reads:

$$\frac{\Phi'''}{\Phi'} - \frac{3}{2}\left(\frac{\Phi''}{\Phi'}\right)^2 = F_x^{-2}\left(-1+\frac{1}{6}F_x^{-2}F_t^2+F_x^{-1}F_{xt}-F_x^{-2}F_tF_{xx}-F_x^{-1}F_{xxx}+\frac{3}{2}F_x^{-2}F_{xx}^2\right) \qquad (18)$$

and, when we use the value of F as given by (17), the right-hand side vanishes and the general solution φ of PB equations is:

$$\varphi = \frac{\alpha + \beta F}{\gamma + \delta F}, \quad \alpha\delta - \beta\gamma \neq 0 \qquad (19)$$

which depends on 5 arbitrary constants $\frac{a_1}{a_3}$, $\frac{a_2}{a_3}$, $\frac{\gamma}{\delta}$ a_3, $\frac{\beta}{\delta}$, $\frac{\alpha\delta - \beta\gamma}{\delta^2}\, a_3$.

The general solution for u and u_1 (eq. (9)) immediately follows:

$$\begin{cases} u = \dfrac{F_t}{6\,F_x} - \dfrac{F_{xx}}{2\,F_x} + \dfrac{\beta\,F_x}{\alpha + \beta\,F} & (20) \\[2ex] u_1 = \dfrac{F_t}{6\,F_x} - \dfrac{F_{xx}}{2\,F_x} + \dfrac{\delta\,F_x}{\gamma + \delta\,F} & (21) \end{cases}$$

i.e. for both u and u_1, the expression:

$$u = \left[\mathrm{Log}\left(1 + e^{-\frac{x}{\sqrt{2}} + \frac{3\,t}{2} + c_1} + e^{\frac{x}{\sqrt{2}} + \frac{3\,t}{2} + c_2}\right)\right]_x \qquad (22)$$

where c_1 and c_2 are two arbitrary complex constants.

This solution is a bisoliton in the Cornille & Gervois sense [5], i.e. without coupling between the two exponentials, but another of its features is that the two values of (ω, k), which satisfy the dispersion relation, are quantized, just like in the Kuramoto-Sivashinsky equation [4].

The particular solution $\beta\delta = 0$ is the known stationary solution

$$u = \frac{F_t}{6\,F_x} - \frac{F_{xx}}{2\,F_x} = \frac{1}{\sqrt{2}} \tanh\left(\frac{x}{\sqrt{2}} + \frac{1}{2}\,\mathrm{Log}\,\frac{a_2}{a_1}\right) \qquad (23)$$

Another particular case of this solution is the solitary wave

$$u = \pm\,\frac{1}{2\sqrt{2}}\left(1 - \tanh\left(\frac{x}{2\sqrt{2}} - \frac{3\,t}{4}\right)\right) \qquad (24)$$

As already noticed [1,9], it is unfortunately impossible to use iteratively the auto-Bäcklund transformation (9) in order to generate new solutions.

This solution (22) is already known [7] but the clear advantage of the present method is that it is algorithmic until a very advanced stage.

Conclusion

This completely soluble example of a set of Painlevé-Bäcklund equations outlines the general method for finding particular analytic solutions of a nonintegrable PDE.

References

[1] Cariello, F. & Tabor, M. (1989). Painlevé expansions for nonintegrable evolution equations. *Physica-D* **39**, 77.

[2] Conte, R. (1988). Universal invariance properties of Painlevé analysis and Bäcklund transformation in nonlinear partial differential equations. *Phys. Lett. A* **134**, 100.

[3] Conte, R. (1989). Invariant Painlevé analysis of partial differential equations. *Phys. Lett. A* **140**, 383.

[4] Conte, R. Musette, M. (1989). Painlevé analysis and Bäcklund transformation in the Kuramoto-Sivashinsky equation. *J. Physics A* **22**, 169.

[5] Cornille, H. & Gervois, A. (1983). Connection between the existence of bisolitons for quadratic nonlinearities and a factorization of the associated linear operator. *J. Math. Phys.* **24**, 2042.

[6] Hille, E. (1976). *Ordinary differential equations in the complex domain.* John Wiley and Sons.

[7] Kawahara, T. & Tanaka, M. (1983). Interactions of travelling fronts: an exact solution of a nonlinear diffusion equation. *Phys. Lett.* **A97**, 311.

[8] Kolmogorov, A.N., Petrovsky, I.G. & Piskunov, N.S. (1937). *Bulletin of Moscow University, Section A, Vol. I,* 26 pp.

[9] Musette, M. (1989). Painlevé-Bäcklund transformation for the Kuramoto-Sivashinsky equation. In this volume.

[10] Weiss, J. (1983). The Painlevé property for partial differential equations. II: Bäcklund transformation, Lax pairs, and the Schwarzian derivative. *J. math. Phys.* **24**, 1405.

[11] Weiss, J. (1984). On classes of integrable systems and the Painlevé property. *J. math. Phys.* **25**, 13.

[12] Weiss, J. (1984). The sine-Gordon equations: complete and partial integrability. *J. math. Phys.* **25**, 2226.

[13] Weiss, J. (1985). The Painlevé property and Bäcklund transformations for the sequence of Boussinesq equations. *J. math. Phys.* **26**, 258.

[14] Weiss, J. (1985). Modified equations, rational solutions, and the Painlevé property for the Kadomtsev-Petviashvili and Hirota-Satsuma equations. *J. math. Phys.* **26**, 2174.

[15] Weiss, J. (1986). Bäcklund transformation and the Painlevé property. *J. math. Phys.* **27**, 1293.

[16] Weiss, J., Tabor, M. & Carnevale, G. (1983). The Painlevé property for partial differential equations. *J. math. Phys.* **24**, 522.

A.Ramani, B.Grammaticos and H.Yoshida

19. RIGOROUS NON-INTEGRABILITY RESULTS RELATED TO SINGULARITY ANALYSIS

ABSTRACT

Rigorous results on the non-integrability of Hamiltonian systems based on a theorem of Ziglin are presented, leading to a new justification of the singular point ('Painlevé') analysis.

1 Introduction

Integrability has been under the spotlight of interest, in the domain of dynamical systems, for over a decade. Powerful constructive methods have provided scores of partial differential equations integrable through the Inverse Scattering Transform. Curiously, the results have been much more scant for finite dimensional systems. A systematic and effective approach for the prognostication of integrability was initiated by Ablowitz, Ramani and Segur [1]. Resurrecting a century old idea of Kowalevskaya, this method has related the single-valuedness of the singularities of the equations of motion (the 'Painlevé property') to integrability. The integrable character of the 'Painlevé' candidates was established by the explicit construction of the constants of motion, thus reinforcing the ties between integrability and single-valuedness. The cases for which integrability was not proven remained in the limbo of unclassified ones. For some of them, the appearance of chaos could be interpreted as an undeniable indication of non-integrability. What was lacking till recently was rigorous proofs. This has been remedied recently thanks to the pioneering work of Ziglin [2] (Before proceeding any further, we should mention, in passing, another non-integrability theorem based on the self-intersection of separatrices and which is known as the Melnikov method [3]). Ziglin's theorem has established the necessary condition for an analytic Hamiltonian system to possess single-valued first integrals. The main idea of Ziglin was to relate the existence of analytic constants of motion for the complete Hamiltonian to that of a reduced system (obtained by variation around a given trajectory) and ultimately to the monodromy group of the latter system.

Ziglin's approach has produced recently an interesting spin-off: a new proof of a theorem relating local singularity analysis with integrability [4]. In an attempt to provide a rigorous justification to the singular ('Painlevé') analysis, weighted homogeneous differential systems have been considered. The Kowalevski exponents (KE's) which characterize the singular behaviours of the solutions (hence the relation to the Painlevé approach) were introduced. This theorem, roughly stated, says that the presence of irrational or complex KE's is incompatible with 'algebraic' integrability [4]. The original formulation of the theorem, however, gave rise to certain criticisms [5,6], as the notion of 'algebraic' integrability was too restrictive for the theorem to be of practical usefulness. The recent proof concerns only a certain type of (homogeneous) Hamiltonian systems [7] but now uses the same definition of integrability as Ziglin's theorem on which it is based.

In what follows, we are going to present a sketch of the proofs of Ziglin's theorem and of the theorem of Ref[7], together with a review of the results on non-integrability studies is the extension of the Toda system to one where the kinetic energy is exponential in the momenta. In the conclusion of our review, we stress the need for a nonlinear extension to Ziglin's method.

2 Ziglin's theorem

In the original formulation, Ziglin's theorem furnishes necessary conditions for the existence of additional analytic first integrals for a given N-dimensional Hamiltonian system H and in particular for the existence of N integrals (including H) i.e. the Liouville integrability of the system.

However, the presentation can be greatly simplified if we restrict ourselves to the case of N=2 dimensions, where the existence of only one additional integral is investigated. In what follows, we will illustrate the various stages of the algorithm using the Hénon-Heiles cubic Hamiltonian

$$H = (p_x^2+p_y^2)/2+Cx^2y+y^3/3 \qquad (1)$$

We start from the equations of motion:

$$\ddot{x} = -2Cxy$$

$$\ddot{y} = -y^2-Cx^2 \qquad (2)$$

and we look for particular solutions, and more specifically, straight-line ones. Solutions of this kind always exist for homogeneous potentials. Their advantage will become clear later. In the present case, we obtain two kind of straight-line solutions:

(i) $$x = 0 \text{ (and } p_x = 0)$$

(ii) $$y+\mu x = 0 \text{ with } \mu^2 = C/(2C-1) \qquad (3)$$

(remark that for C = 1/2 only one straight-line solution exists).

On the type i) straight-line solution, the equation of motion for y reduces to:

$$\ddot{y}_0 = -y_0^2$$

which can be integrated to

$$\frac{1}{2}\dot{y}_0^2+\frac{1}{3}y_0^3+K = 0 \qquad (4)$$

The latter equation defines a Jacobi elliptic function $y_0(t;K,t_0)$, with two generically complex periods and one double pole per unit-period parallelogram. This situaton is typical for the reduced equations of motion in homogeneous potentials: one obtains multiply periodic functions (generally defined on some complicated Riemann surface).

In the case ii) we get

$$\ddot{y}_0 = -2C/\mu\ y_0^2$$

which leads to an equation similar to (4).

Next, we linearize around the solution x=0, $y=y_0$ for case i) and analogously for case ii). We put $y=y_0+\eta$, $x=0+\xi$ and obtain:

$$\ddot{\eta} = -2y_0\eta \qquad (5a)$$

$$\ddot{\xi} = -2Cy_0\xi \qquad (5b)$$

Here, the equations for η and ξ separate. This is the advantage of the straight-line solutions. The variational equations of motion always separate into the tangential (here (5a)) and the normal (here (5b)) parts through a fixed (i.e. time dependent) rotation. The solution of the tangential variational equation are $\eta_1=\partial y_0/\partial K$ and $\eta_2=\partial y_0/\partial t_0=-\partial y_0/\partial t$.

The main ingredient in the proof of Ziglin's theorem is that an integral of the initial Hamiltonian (other than H itself) leads, when expanded around the straight-line solution, to an integral for the normal variational equations of motion. Thus a necessary condition for the integrability of the initial Hamiltonian is the integrability of the NVE.

In the case at hand, a second integral exists for instance for C=1, given by

$$I = p_xp_y+\frac{1}{3}x^3+xy^2 \qquad (6)$$

Linearizing around type i) solutions, we find that this constant leads to

$$J = \dot{y}_0\dot{\xi}+y_0^2\xi$$

while for type ii) (here μ=1)

$$J = \dot{y}_0\dot{\xi}+2y_0^2\xi$$

where y_0 is given by the appropriate equation in each case. In the present example the integral J is linear in ξ, $\dot{\xi}$, but this need not always be the case. If, however I is analytic and independent of the Hamiltonian, there exists some finite higher order term in its expansion which does not vanish is an integral of the NVE.

Looking at equation (5b) we remark that it is a Hill's equation, with periodic coefficient y_0. Its evolution along a path Γ in t-plane connecting point t_0 to any other point (maybe t_0 itself) where both y_0 and $\dot{y}_0$ recover the same value, defines an area preserving mapping:

$$\begin{pmatrix} \xi \\ \dot{\xi} \end{pmatrix}_{final} = M_\Gamma \begin{pmatrix} \xi \\ \dot{\xi} \end{pmatrix}_{initial} \qquad (7)$$

The mappings defined above form a group, induced by the fundamental group of the Riemann surface defined by y_0, called the monodromy group of the NVE, and they are represented by matrices M of determinant unity, which are called monodromy matrices. Note that because the Riemann surface defined by y_0 is generally not simply connected a closed path in t plane need not be represented by the unit matrix.

One definition is needed at this point. Let us assume that M is diagonalizable (the non-diagonalizable case will be treated apart) with eigenvalues ρ and ρ^{-1}. We call M non-resonant if ρ is not a root of unity. (The extension of the definition of non-resonance for N dimensions is straightforward: the eigenvalues $\rho_1, \rho_1^{-1}, \rho_2, \rho_2^{-1}, \ldots, \rho_{N-1}, \rho_{N-1}^{-1}$, must not satisfy any relation $\rho_1^{k_1} \rho_2^{k_2} \ldots \rho_{N-1}^{k_{N-1}} = 1$ with integer k's unless all the k's vanish).

We can now formulate Ziglin's theorem in its original version.

Theorem 1:

Suppose that a two-dimensional Hamiltonian system has an additional integral which is analytic (at least in a neighbourhood of a given straight line solution of the equations of motion). Furthermore suppose that there exists a non-resonant monodromy matrix M_1. Then any other monodromy matrix M_2 must

- either commute with M_1 i.e. $M_1 M_2 M_1^{-1} M_2^{-1} = \mathbf{1}$
- or be such that $M_1 M_2 M_1^{-1} M_2^{-1} = M_1^2$

(which means that M_2 permutes the eigenvectors of M_1 in which case Tr M_2=0)

A generalisation of the theorem, given in [8], concerns the extension to non-diagonizable matrices. The extended theorem can be formulated as follows:

Theorem 2:

With the same assumptions as Theorem 1 above, suppose that there exists a non-diagonalizable monodromy matrix M_1. Any other monodromy matrix M_2 must be resonant and moreover, in a basis where M_1 is upper (resp. lower) triangular, it must also be upper (resp. lower) triangular.

Below, we give a sketch of the proof of Ziglin's theorem. We have seen earlier that an analytic integral I of the original Hamiltonian induces an integral for the NVE. Thus:

Lemma 1:

If the Hamiltonian system under consideration has an additional analytic integral then there exists a nontrivial homogeneous polynomial

$$J(\xi,\dot{\xi},t) = \sum_{j=0}^{n} \psi_j(t)\xi^j\dot{\xi}^{n-j} \tag{8}$$

which is an integral of the NVE, where the $\psi_j(t)$'s depend on t <u>only</u> through y_0 and $\dot{y}_0$ and therefore are single valued on Γ.

This polynomial $J(\xi,\dot{\xi},t)$ is invariant under the action of the monodromy group. Indeed, suppose we now fix a point $(y_0,\dot{y}_0)$ on Γ and integrate the equations of motion over a loop in Γ. The quantities ξ and $\dot{\xi}$ are thus modified by the action of the monodromy group while the ψ_k's recover their initial values, and J has kept its value constant throughout. Thus, we have

Lemma 2:

With the same assumptions as above, there exists a homogeneous polynomial

$$\Phi(\xi,\dot{\xi}) = \sum_{j=0}^{n} \varphi_j\xi^j\dot{\xi}^{n-j} \tag{9}$$

which is invariant under the action of the monodromy group of the NVE. Using the quantity Φ we can prove Ziglin's theorem.

Suppose that there exists some matrix M_1 which is diagonalizable and non-resonant. Changing to the basis of the eigenvectors of M_1, the new coordinates being ζ and η, we get:

$$\Phi = \sum_{j=0}^{n} \omega_j\zeta^j\eta^{n-j} \tag{10}$$

The action of M_1 changes Φ to

$$\Phi = \sum_{j=0}^{n} \omega_j \rho^{2j-n} \zeta^j \eta^{n-j} \tag{11}$$

But ρ^{2j-n} cannot be unity unless $n=2j$, as M_1 is non-resonant. Thus for Φ to be invariant under M_1 we must have:

$$\Phi = (\zeta\eta)^j \tag{12}$$

Now, we act on Φ with some other member M_2 of the monodromy group, which, in the basis which diagonalizes M_1, can be written:

$$M_2 = \begin{pmatrix} a & b \\ c & d \end{pmatrix} \tag{13}$$

(recall that $\det M_2 = ad-bc=1$). Since M_2 must also leave Φ invariant we find:

$$[(a\zeta+b\eta)(c\zeta+d\eta)]^j = (\zeta\eta)^j \tag{14}$$

which entails:

$$ac=bd = 0 \quad \text{and} \quad (ad+bc)^j = 1$$

The only solutions are:

(i) $b=c=0$ and $ad = 1$ i.e. M_2 commutes with M_1 thus $M_1M_2M_1^{-1}M_2^{-1}=\mathbf{1}$

(ii) $a=d=0$ and $bc=-1$ i.e. M_2 permutes the eigenvectors of M_1 thus the 'commutator' $M_1M_2M_1^{-1}M_2^{-1}$ is equal to M_1^2.

In the latter case, $\mathrm{Tr}\, M_2=0$, which implies that the eigenvalues of M_2 are i and $-i$ and M_2 is thus resonant. Therefore, if two non-resonant matrices exist, they must commute for an additional integral to exist.

We now turn to non-diagonalizable matrices. In an appropriate basis, M_1 can be written:

$$M_1 = \pm \begin{pmatrix} 1 & 0 \\ \alpha & 1 \end{pmatrix} \tag{15}$$

with $\alpha \neq 0$. As a consequence, in this frame, Φ can only assume the form $\Phi=\zeta^n$. For the matrix M_2 to leave ζ^n invariant, we must have $a^n=1$, $b=0$ and $ad=1$. If a is neither 1 nor -1, the two eigenvalues a and d are distinct and M_2 is therefore diagonalizable, and moreover resonant. If however $a=d=\pm 1$, then M_2 is not diagonalizable (unless $c=0$, in which case it is just $\pm\mathbf{1}$) and assumes a lower triangular form <u>in the same basis</u> as M_1. The 'commutator' $M_1M_2M_1^{-1}M_2^{-1}$ is unity only if $a=d=\pm 1$, but its trace is always 2.

Ziglin's original theorem was also formulated in the N-dimensional case [2].

Theorem 3

Suppose that a given Hamiltonian H has N-1 analytic integrals, which are functionally independent together with H, and moreover that there exists a non-resonant matrix M. Then it is necessary that any other monodromy matrix M' either: a) commute with M or b) permute the eigenspaces of M. In particular whenever M' is also non-resonant it must also necessarily commute with M.

The problem which arises at this point is that of the practical applicability of Ziglin's theorem. The knowledge of the monodromy matrices is necessary. Of course, one can always compute them numerically with some precision. For N>2, however, the numerically obtained eigenvalues are of little help if one wants to investigate the resonance condition. Still several possibilities exist. One of the simplest ones can be implemented whenever the reduced equation of motion for the straight-line solution leads to a doubly periodic trajectory with only one singularity per unit period parallelogram. In this case, the path corresponding to the 'commutator' $M_s=M_1M_2M_1^{-1}M_2^{-1}$ is homotopic to a closed loop around the singularity. Now, as was explained in [5] and [9], the eigenvalues of M_s can be obtained <u>explicitely</u> from singularity analysis considerations. (This statement will become clearer in the next section dealing with the theorem of Ref [4]). The explicit knowledge of the 'commutator' can be of great help. Of course, $M_s=\mathbf{1}$ the M_1,M_2 matrices are not constrained in any way. On the other hand, if $M_s\neq\mathbf{1}$ (even if $M_s=-\mathbf{1}$, a case that often arises with polynomial potentials), then we can prove non-integrability by numerically computing the traces of M_1 and M_2. If any of them has a trace depending on the energy, then by continuity it cannot be always resonant, nor can the (energy independent) commutator always be its square [9]. Similiarly, if at least one of the two matrics has a trace larger than 2 in absolute value it is certainly not resonant, and one can check numerically whether the commutator can be its square.

3 Explicit computation of the monodromy matrics

As we have seen in the previous section, one must compute the monodromy matrices, so as to be able to conclude on the resonance condition. The ideal situation is when we can get the monodromy matrices explicitely. It turns out that this can be done for a large class of Hamiltonians, in particular for Hamiltonians involving homogeneous potentials [10]. This result is obtained by transforming the NVE into a Gauss hypergeometric equation, the monodromy properties of which are exactly known. Let us illustrate this calculation through a specific example [11].

We start with the Hamiltonian:

$$H = (p_x^2 + p_y^2)/2 + \rho + \kappa y \tag{16}$$

where $\rho = \sqrt{x^2+y^2}$. The equations of motion are:

$$\ddot{x} = x/\rho$$

$$\ddot{y} = -y/\rho - \kappa \tag{17}$$

The straight-line solutions are $x=0$, $\ddot{y}=-(\kappa\pm1)$, the $\pm$ sign being due to the square root in the definition of ρ). We rescale the time so as to obtain $\ddot{y}=h-t^2/2$ (with the appropriate initial conditions and $h\neq0$). The NVE is then written as:

$$\ddot{\xi}+\lambda\xi/y = 0 \tag{18}$$

where $\lambda=(1\pm\kappa)^{-1}$. We introduce a new independent variable:

$$z = (1+t\sqrt{2h})/2 \tag{19}$$

and obtain for the NVE:

$$z(1-z)d^2\xi/dz^2+2\lambda\xi = 0 \tag{20}$$

This is a special case of Gauss hypergeometric equation:

$$z(1-z)d^2\xi/dz^2+[c-(a+b+1)z]d\xi/dz-ab\xi = 0 \tag{21}$$

From the monodromy properties of the Gauss hypergeometric equation we can conclude that, for integrability, λ must be of the form $n(n-1)/2$. Moreover, we can use the fact that λ is given by $1/(1+\kappa)$ in one case and $1/(1-\kappa)$ in the other. Thus the only possible cases are $1/(1+\kappa)=n(n-1)/2$ and $1/(1-\kappa)=m(m-1)/2$, with n and m integer. This is clearly impossible unless $\kappa=0$ (or infinite). Thus, the system is never integrable apart from these trivial cases.

In fact, the transformation to an equation of the type (21) can be performed for any homogeneous potential $V(x_1,x_2)$ of degree k. Indeed, we start by looking for straight-line solutions of the form $x_1=c_1\phi(t)$, $x_2=c_2\phi(t)$. We find a solution where c_1,c_2 and $\phi(t)$ satisfy:

$$(c_1,c_2) = \text{grad } V(c_1,c_2) \tag{22a}$$

and

$$\ddot{\phi}+\phi^{k-1} = 0 \tag{22b}$$

which can be integrated to (by rescaling the integration constant, namely the energy, to $1/k$):

$$t = \sqrt{\frac{k}{2}} \int \frac{d\phi}{\sqrt{1-\phi^k}} \tag{23}$$

The denominator of the integral defines a Riemann surface, the branch points of which are $w_n = e^{2i\pi n/k}$ $(n=0,\ldots,k-1)$. The NVE takes the form:

$$\ddot{\xi} + \lambda\phi^{k-2}\xi = 0 \tag{24}$$

with λ given by:

$$\lambda = \nabla^2 V(c_1, c_2) - (k-1) \tag{25}$$

Introducing the change of independent variable $z=\phi^k$ we transform (24) into a Gauss hypergeometric equation (21) with

$$a+b = 1/2-1/k, \quad ab = -\lambda/2k, \quad c = 1-1/k \tag{26}$$

The basic monodromy matrices of the Gauss hypergeometric equation correspond to anticlockwise loops around the points z=0 and z=1. Their expressions are [12]:

$$M_0 = \begin{pmatrix} 1 & e^{-2i\pi b} - e^{-2i\pi c} \\ 0 & e^{-2i\pi c} \end{pmatrix} \tag{27}$$

and

$$M_1 = \begin{pmatrix} e^{2i\pi(c-a-b)} & 0 \\ 1-e^{2i\pi(c-a)} & 1 \end{pmatrix}$$

We now consider two closed circuits on the Riemann surface, C_1 and C_2 enclosing respectively (counterclockwise) 1 and $e^{2i\pi/k}$, and $e^{-2i\pi/k}$ and 1, with a common base point w on the real axis, 0<w<1. The circuits are mapped in the complex z plane through $z=\phi^k$ and we get the corresponding monodromy matrices:

$$M(C_1) = M_1 M_0 M_1 M_0^{-1}$$

$$M(C_2) = M_0^{-1} M_1 M_0 M_1$$

From the above expressions it is fairly easy to implement Ziglin's theorem. We have

$$E(\lambda)=\mathrm{Tr}M(C_1)=\mathrm{Tr}M(C_2)=2\cos\left(\frac{2\pi}{k}\right)+4\cos^2\left(\frac{\pi}{2k}\sqrt{(k-2)^2+8k\lambda}\right) \tag{28}$$

and moreover one can check that $M(C_1)$ commutes with $M(C_2)$ only when $k=\pm2$ or $E(\lambda)=2$ or $2\cos(2\pi/k)$. So if λ is such that both $M(C_1)$ and $M(C_2)$ are non-resonant then we can conclude on the non-integrability of H. This occurs in particular whenever $|E(\lambda)|>2$. Using this inequality, we can define regions for λ in which no additional analytic integral exists [10]. We obtain thus non-integrability in the regions:

- for $k\geq3$ $\quad\lambda<0, 1<\lambda<k-1, k+2<\lambda<3k-2, \ldots$
- for $k=1$ $\quad\mathbb{R}-\{0,1,3,6,10,\ldots\}$
- for $k=-1$ $\quad\mathbb{R}-\{1,0,-2,-5,-9,\ldots\}$
- for $k\leq-3$ $\quad\lambda>1, 0>\lambda>k+2, k-1>\lambda>3k+3, \ldots$

(The case $k=-2$ is integrable for all values of λ).
We remark that for $|k|>3$ we obtain whole regions in which non-integrability <u>cannot</u> be proved. However, as we have seen for Hamiltonian (16), by using information from several straight-line solutions we can further limit the possible regions for integrability and in some cases constrain the problem to the point where an exact answer is reached. For the potential

$$V = x^4+y^4+2\varepsilon x^2y^2$$

we find that the non-integrability region is:

$$\lambda<0, 1<\lambda<3, 6<\lambda<10, \ldots$$

where λ is defined for the two different straight-line solutions as

$$\lambda=\varepsilon \quad\text{and}\quad \lambda=\frac{3-\varepsilon}{\varepsilon+1}.$$

It is easy to check that only the well known integrable cases $\varepsilon=0,1,3$ survive. For all other values of ε the above quartic potential is not integrable, (in particular for ε infinite which corresponds to a well-known ergodic potential).

For the Hénon-Heiles case we introduced at the beginning of this paper, however, we cannot reach a clear-cut conclusion. Implementing the non-integrability condition on both straight-line solutions, we find:

$$C^{-1}<1, 1<C^{-1}<2, 2<C^{-1}<3, 6<C^{-1}<8, 13<C^{-1}<16, \ldots$$

Thus, apart from the discrete values $C=1$, $C=1/2$ we have whole regions on which non-integrabilitiy cannot be deduced although possible integrable cases have zero measure even in these regions. This is indeed a typical situation in Ziglin's approach.

In Ref. [13] it was shown that the conclusions on the non-integrability regions reached for homogeneous potentials can be extended to the case of <u>rational</u> k.

4 Non-integrability in terms of Kowalevski exponents

The main difficulty with Ziglin's theorem lies in the fact that the resonance condition cannot be practically exploited: one must check whether some expressions are the cosines of angles rationally related. Even in the best case where expressions are explicitely known, this is a transcendental problem.

This difficulty has been overcome by a new version, given in Ref.[7], of a theorem of Ref.[4]. The first step is to obtain a new condition for the non-integrability regions through a different choice of paths C_1 and C_2 on the Riemann surface which will lead to simpler expressions for the traces of the monodromy matrices.

In what follows, we will give a detailed description of the paths C_1 and C_2 in the case of a homogeneous potential of degree k even and larger than two. We start from equation (22b) and integrate it once with appropriate initial conditions to

$$\dot{\phi}^2 = 2(1-\phi^k)/k$$

We choose a (real) point t_0 where $\phi=0$ and $\dot{\phi}=\sqrt{2/k}$. We will now consider two singularities of ϕ close to t_0. For $t-t_0=-e^{\pm i\pi/k}\tau$, with τ real positive, ϕ is also of the form $-e^{\pm i\pi/k}\chi$, with real positive χ, as $\dot{\phi}$ is a real positive number larger than $\sqrt{2/k}$. As τ increases, $\dot{\phi}$ and χ (and thus ϕ), related through

$$\dot{\phi}^2 = 2(1+\chi^k)/k$$

increase faster and faster and reach for finite value τ_* of τ a singularity of the form $\phi \propto (\tau-\tau_*)^{-g}$ with $g=2/(k-2)$. The two singularities we will consider are situated at $t-t_0=e^{\pm i\pi/k}\tau_*$.

The path C_1 is built in the following way (see Fig.1). Start from a point on the real axis $t_i<t_0$ close enough to t_0 that $-1<\phi(t_i)<0$. (Remember $\dot{\phi}$ is real positive).

(a) make an arc of a small circle around t_0, counterclockwise, of an angle π/k.

(b) follow the straight line $t-t_0=-e^{i\pi/k}\tau$, τ real positive, up to a value close to τ_*.

(c) make 2(k-2) clockwise turns around the singularity at the point $t-t_0=-e^{i\pi/k}\tau_*$. Note that even through the function ϕ is not single-valued at the singularity, after (k-2) turns one goes back on the initial Riemann sheet.

(d) retrace b) backwards

(e) retrace a) backwards to the point of origin.

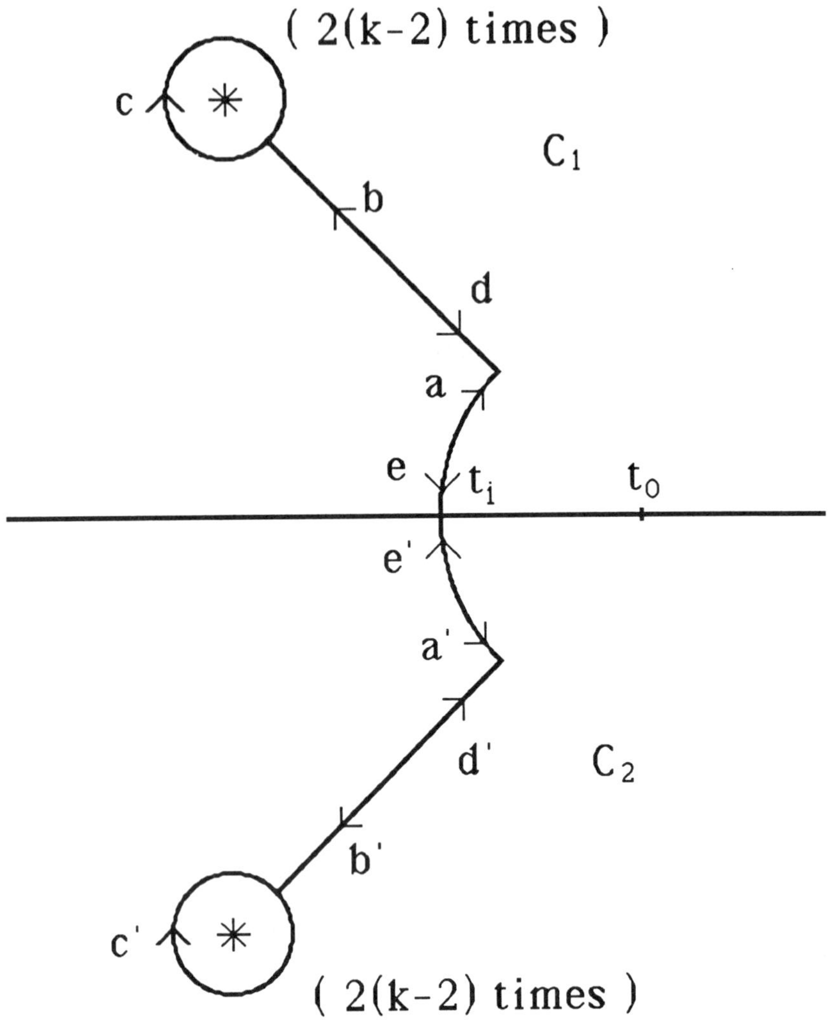

Figure 1 : Paths C_1 and C_2 in t-plane. The asterisks correspond to the locations of the singularities.

We will now follow this path in z-plane, where $z=\phi^k$ is the independent variable of the hypergeometric equation (21) (see Fig.2). Since k is even, the initial point $z_i=\phi(t_i)^k$ lies between 0 and 1.

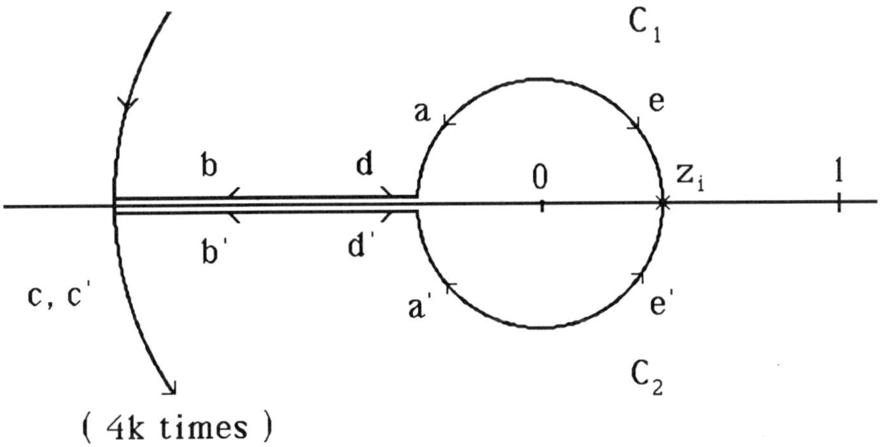

Figure 2 : Paths C_1 and C_2 in z-plane.

(a) z describes a (small) counterclockwise half-circle above 0, to a small negative real value.

(b) z follows the negative real axis, up to a large (negative) value.

(c) z describes 4k <u>large</u> circles counterclockwise. Indeed, the singularity is of the form $\phi \propto (\tau-\tau_*)^{-2/(k-2)}$ thus 2(k-2) clockwise turns in t-plane become 4 counterclockwise turns in ϕ-plane and 4k in z-plane.

(d) back along the real axis to a small negative value

(e) back to z_i by a half circle above 0.

Summarizing, z goes 4k times counterclockwise around 0 and 1, encircling 0 first. Thus the corresponding monodromy matrix is:

$$M(C_1) = (M_0M_1)^{4k}$$

The path C_2 starts from the same initial point t_1, and is similar to C_1 with the following modifications:

(a') make an arc of a small circle around t_0, <u>clockwise</u>, of an angle π/k.

(b') follow the straight line $t-t_0=-e^{-i\pi/k}\tau$, τ real positive, up to a value close to τ_*.

(c') make 2(k-2) (still) clockwise turns around the singularity at the point $t-t_0=-e^{-i\pi/k}\tau_*$.

(d') retrace b') backwards

(e') retrace a') backwards to the point of origin.

In z-plane, the only changes with respect to C_1 is that the half-circles a') and e') now go <u>below</u> 0. Thus z goes 4k counterclockwise around 0 and 1, but encircles 1 first. The monodromy matrix is thus

$$M(C_2) = (M_1M_0)^{4k} .$$

With adequate modifications, corresponding paths C_1 and C_2 can be found for all other values of k except k=2, resulting to the same expressions for all monodromy matrices.

The advantage of this choice lies in the fact that the common value of the traces of these matrices is:

$$2\cos(2\pi\sqrt{(k-2)^2+8k\lambda}) \tag{29}$$

and moreover $M(C_1)$ and $M(C_2)$ commute only when their trace is ±2, or when k=±2. The non-resonance condition is now easier to check. What is more interesting is that this theorem can be expressed in a particular simple way in terms of the KE's, and thus furnishes a further proof of the theorem originally proposed in [4, part II], (although here in a slightly less general context).

In particular we consider a Hamiltonian of the form

$$H = p_i^2/2+V(x_1x_2,\dots,x_N) \tag{30}$$

with V(x) a homogeneous, algebraic potential of degree k. The equations of motion admit singular solutions of the form

$$x_j = c_j\phi(t) \tag{31}$$

where the c_j can be chosen as a solution of

$$(c_1,c_2,\dots,c_N) = \text{grad } V(c_1,c_2,\dots,c_N)$$

and $\phi(t)$ is a singular solution αt^{-g} of equation (22a) for ϕ where $g=2/(k-2)$ and $\alpha=(kg^2/2)^{g/2}$. The KE's are introduced when one looks for a variational solution around (31) through

$$x_j = (c_j+\chi_j)\alpha t^{-g} \tag{32}$$

Linearizing, one obtains for χ solutions of the form t^{ρ_i} where the ρ_i's are the KE's. Due to the Hamiltonian character of the equations of motion the KE's come by pairs ρ_i and ρ_{i+N}. To obtain them we compute the eigenvalues λ_i of the Hessian matrix $\partial^2V(x_1,x_2,\dots x_N)/\partial x_j\partial x_m$ evaluated at $x_j=c_j$. The set of the KE's ρ_i and ρ_{i+N} are given by the roots of the equation

$$\rho^2-(2g+1)\rho+g(g+1)(1-\lambda_i) = 0$$

We denote by $\Delta\rho_i$ the difference of the two roots, namely $\Delta\rho_i=\sqrt{1+8k\lambda_i/(k-2)^2}$.

Note that in equation (29) the expression that appears is just

$$2\cos(2\pi(k-2)\Delta\rho). \tag{33}$$

This is by no means a coincidence. Indeed, the paths C_1 and C_2 correspond to going 2(k-2) times around the singularity in t-plane. Along this path, the eigensolution (for a given KE ρ_i) of the linearized equation acquires a phase factor $e^{4i\pi(k-2)\rho_i}$ which is equal to $e^{2i\pi(k-2)\Delta\rho_i}$. Equation (33) expresses just this fact.

We define the rational independence of the $\Delta\rho_i$'s if no relation of the form $\sum m_i\Delta\rho_i=0$ exists for integer m_i's unless all $m_i=0$. We thus have the following theorem:

Theorem 4

If the $\Delta\rho_i$ are rationally independent the N-dimensional Hamiltonian system cannot possess any additional analytic integral besides the Hamiltonian itself.

In the two-dimensional case, there are only two $\Delta\rho_i$ one of which is always rational. Thus rational independence just means that the other $\Delta\rho_i$ is not a rational number. Moreover the sum $\rho_i+\rho_{i+N}=2g+1$ is always rational. This allows the following corollary to the above theorem:

If there exists an irrational or complex KE the two-dimensional Hamiltonian cannot have any additional analytic integral.

The main advantage of this theorem is that the singularity analysis leads directly to the result. No transcendental non-resonance condition need be checked. On the other hand Hamiltonians which are not of the form (30) are not covered by this theorem. In particular, Rod et al [6] have presented the example of:

$$H = p_1(p_1^2+\mu_1x_1^2)+\varepsilon x_1(p_2^2+\mu_2x_2^2) \tag{34}$$

which has irrational KE's $1\pm2i\omega$ $(\omega^2=\varepsilon^2\mu_2/\mu_1^2)$ while still being integrable. Still, no contradiction arises as the monodromy group is a commutative one [7]. It might well turn out that this is always the case whenever irrational or imaginary KE's appear for integrable Hamiltonians. On the other hand the Hamiltonian

$$H = (p_x^2+p_y^2)/2+p_xxy+p_y(ax^2+by^2) \tag{35}$$

studied by Roekaerts and Yoshida [14] possess a non-commutative group. However no explicit analytic expression for the monodromy matrices of the associated monodromy group exists yet. Still, the known integrable cases indeed correspond to values of the parameters a and b for which the KE's are rational.

One last remark is in order at this point. In the original formulation of the theorem of Ref.[4], the issue was the algebraic integrability of the system. Now it has become clear that this notion is more restrictive than the one of Ziglin. The latter requires only the analyticity of the additional integrals. For a weighted homogeneous Hamiltonian the above mentioned theorem demands that the additional integral I be a weighted homogenous rational function of p and x (thus compatible with Ziglin's requirement). Moreover, algebraic integrability in the sense of Ref.[4] requires that the 'angle' Ω associated to the 'action' I also be weighted homogeneous. It is in that sense only that the original theorem of Ref.[4] is true, the existing 'counterexamples' [5,6] leading to angles Ω involving logarithms. Algebraic integrability in the sense of Adler-van Moerbeke [15] is still more constrained as it poses requirements on the behaviour of the solutions themselves.

5 Extensions to non-homogeneous Hamiltonians

In the case of homogeneous Hamiltonians analytic expressions for the monodromy matrices can often be obtained, thus allowing the application of Ziglin's theorem. In the non-homogeneous case, the situation becomes more difficult as no closed-form results exist. Still, the results on the homogeneous potentials can be of great help [16,17]. By taking the infinite-(or zero-)energy limit of the equations of motion (through appropriate scalings) one can reduce the system to one where only the highest degree homogeneous part contributes. (Similarly, at the zero-energy limit we recover the lowest-degree part). Thus, the two limits can be treated exactly and if either one turns out to be non-integrable, then the same holds for the whole potential. Moreover, numerical computations can be of some help as an energy dependence of the trace of a monodromy matrix indicates that it is non-resonnant for a dense set of values of the energy.

Apart from numerics, there exist cases where this energy dependence is established through a perturbation expansion of the solutions of the NVE around an equilibrium point [9] or by very general considerations [18]. Thus Ito [9] and Rod [18] examined the Hénon-Heiles potential with isotropic quadratic part of the form:

$$V = (x^2+y^2)/2+Cx^2y+y^3/3 \qquad (36)$$

They have shown that this potential is not integrable for <u>all</u> values of C except 0, 1/6,1/2 and 1. Indeed, the integrability of V is established for C=0,1/6, and 1 as an additional invariant is known in these cases. The case C=1/2 is presumably <u>not</u> integrable but Ziglin's analysis cannot deal with it. In fact, the NVE is the same as the one resulting from the (integrable) potential

$$V' = (x^2+y^2)/2+(x^2+y^2)^{3/2}/3 \qquad (37)$$

and no linear analysis can distinguish between the two.

As a further example, we can consider the potential

$$V = 1/(x^2+\lambda y^2+\mu)^{k/2} \qquad (38)$$

Taking the low-energy limit we can put μ to zero and thus we are left with an homogeneous potential of degree -k. Two straight line solutions exist which give constraints on the possible integrability domain. In particular we can prove that the potential is <u>not</u> integrable when $\lambda<1$ and also when $\lambda>1$. This leaves $\lambda=1$ as the only possible integrable value, in which case integrability is trivial.

One situation, typical for polynomial potentials, is when the lowest order term is quadratic, of the $V_2=ax^2+by^2$. In this limit, the trace of one monodromy matrix is just

$$\mathrm{Tr}M = 2\cos 2\pi\sqrt{a/b}.$$

If one can compute also the high energy limit and if the trace of the monodromy matrix for the 'same' path [17] (which is not always that easy to ascertain) is different in this limit from the one computed at zero energy, then this shows that the trace depends on the energy and that the matrix is non-resonant over a dense set of energy values h. Various results have been obtained using a combination of the above methods. A typical example is the truncated Toda lattice (see next Section).

More results can be obtained using still different techniques. Among them we can quote the potentials:

$$V = W(x,y)+1/\rho \tag{39}$$

$$V = W(x,y)+1/\rho^2 \tag{40}$$

where W is a homogeneous potential and $\rho^2=x^2+y^2$. For example, in the case of the perturbed Kepler potential (39) [19] we can consider a W of the form $W=Ax^n+By^n$. It turns out that the only possible integrable cases are

(i) n=1, A,B free

(ii) n=2, A=B

and

(iii) n=2, A=4B (or B/4)

which are indeed integrable.

For the perturbed centrifugal potential (40) [20] we write W as $\rho^n\nu(\theta)$. We have proven that it cannot be integrable unless

$$\nu''(\theta_0)/\nu(\theta_0) = (n+2)^2 m(m-1)/2$$

with integer m for all angles θ_0 for which $\nu'(\theta_0)=0$. In particular, the potential

$$Ax^n+By^n+1/r^2$$

is integrable only for n=2 and A=B.

6 Toda system revisited : the exponential Toda

Ziglin's analysis has been applied also to various Toda systems. Among them we have examined a generalised Toda lattice of the form [8]:

$$H = (p_x^2+p_y^2)/2+e^{\alpha x-y}+e^{-\alpha x-y} \tag{41}$$

Straight-line solutions exist for this type of Hamiltonians. In particular taking $x=p_x=0$, we get the NVE:

$$\ddot{\xi}+2\alpha^2 e^{-y}\xi = 0 \tag{42}$$

with y given by the equation

$$\ddot{y}+2e^{-y}y = 0 \qquad (43)$$

By introducing a new independent variable $z=2e^{-y}/h$ (where h is the energy) we can transform the NVE into a Gauss hypergeometric equation with

$$a+b = 1/2, \quad ab = -\alpha^2/2 \quad \text{and } c = 1.$$

The monodromy matrices obtained (for a particular path) can be simultaneously cast into upper triangular form provided

$$\alpha^2 = n(n-1)/2$$

with integer n. The integrable cases can only be found among these values of α, which include the known ones $\alpha^2=0,1,3$.

Another interesting problem related to the Toda Hamiltonian deals with truncations of the three particle system. By reducing to the center of mass frame, we get:

$$H = (p_x^2+p_y^2)/2+e^{\sqrt{3}x-y}+e^{-\sqrt{3}x-y}+e^{2y} \qquad (44)$$

Suppose we expand the exponential potential in Taylor series and truncate it at some order, leading to a polynomial potential [21]. The integrability of this truncated system has been investigated from Ziglin's point of view in [17], where it was proven that by examining the higher order term we can prove the non-integrability at any order of truncation except quadratic. Only the fourth-order truncation necessitates a special study (performed in [22]) as the fourth order term alone is integrable and only its combination with the lower order terms leads to a non-integrable potential. (The non-integrability of the fourth-order truncation has also been confirmed recently by detailed numerical calculations [23] exhibiting large scale chaos for the proper choice of the value of energy). Thus, the Toda Hamiltonian does not possess any integrable truncation apart from quadratic one.

The last remark had spurred a new direction of investigation: the quadratic kinetic energy part could also be the only integrable truncation of some term exponential in the momenta. We have thus tried to find an integrable Toda potential with exponential kinetic energy. This resulted to the following Hamiltonian which we give here in N-dimensions $(x_{N+1}=x_1)$:

$$H = \sum\exp(p_i)+\sum\exp(x_i-x_{i-1}) \qquad (45)$$

Note that this Hamiltonian does not involve p and x in a symmmetrical way. However, by the transformation

$$p_i = \pi_i-\pi_{i+1}, \quad \xi_i = x_i-x_{i-1},$$

which is canonical, we can rewrite it as

$$H = \sum\exp(\pi_i-\pi_{i+1})+\sum\exp(\xi_i) \qquad (45')$$

which shows that coordinates and momenta fundamentally play the same role. Hamiltonian (45) is in fact integrable. Its integrability can be deduced from its transformation to the Kac-van Moerbeke system [24]. Indeed by the change of variables $\{u_{2i}=\exp(p_i),\ u_{2i-1}=\exp(x_i-x_{i-1})\}$, the equation of motion for p and x become respectively:

$$\dot{u}_{2i} = u_{2i}(u_{2i+1}-u_{2i-1})$$
$$\dot{u}_{2i-1} = u_{2i-1}(u_{2i}-u_{2i-2}) \qquad (46)$$

which is just the Kac-van Moerbeke system.

Thus, the Toda system is the quadratic truncation in p of a system exponential in both p and x (the quadratic truncation in x would be the same token be integrable) which would constitute an extension of the classical one reminiscent of the recently proposed 'relativistic' generalisation [25].

7 Conclusion and outlook

In the previous sections, we have reviewed Ziglin's method for obtaining rigorous proofs of non-integrability, and recent developments which, at least for homogeneous systems, greatly simplify this approach. In particular, for the systems where the theorem of Ref.[4] can be applied, the study may be reduced to the computation of the KE's and the investigation of their rationality. The main limitation (which is also an advantage) of Ziglin's method is that it deals with first order expansions (i.e. the NVE is a linear equation) and thus cannot investigate departures from integrability which appear at higher orders. A typical example of such a case is the Hénon-Heiles Hamiltonian (36) for C=1/2. No Ziglin-based method would ever distinguish it from a genuinely integrable one. Thus the need for a non-linear extension of Ziglin arises.

As a matter of fact, an attempt towards such a nonlinear approach already exists. Indeed Kruskal has introduced the poly-Painlevé method [26] which is based on the singularity analysis (Painlevé method) but has non-local features built in it as it takes care of the 'interaction' between singularities. As in the case of Ziglin, multivaluedness is incompatible with integrability unless some 'commutation' properties hold. No precise theorem has been formulated to date, related to this method (although this can undoubtedly be taken care of), moreover neither do we have a general algorithm for its implementation. Still, some results already exist [1] and in any case it constitutes a possible way of extending the present non-integrability approaches to the fully non-linear domain.

Acknowledgements

The authors want to thank J. Gibbons, J. Hietarinta and M.D. Kruskal for interesting discussions. They are also grateful to L. Martinet for communicating them his numerical results on the fourth-order truncated Toda potential.

References

[1] For an extensive review of the singularity analysis approach and its applications, see:
A.Ramani, B.Grammaticos and T.Bountis, *Physics Reports*

180,159 (1989).
[2] S.L.Ziglin, *Func. Anal. Appl.* **16**,181 and **17**,6 (1983).
[3] M.Liebermann and A.Lichtenberg 'Regular and stochastic motion' (Springer, Berlin 1983).
[4] H.Yoshida, *Celestial Mech.* **31**,363 and 381 (1983).
[5] H.Yoshida, B.Grammaticos and A.Ramani, *Acta Appl. Math.* **8**,75 (1987).
[6] M.Kummer, R.C.Churchill and D.L.Rod, 'On Kowalevski exponents', preprint.
[7] H.Yoshida, 'Non-existence of an additional analytic integral in Hamiltonian systems with an N-dimensional homogeneous potential' (1988), preprint.
[8] H.Yoshida, A.Ramani, B.Grammaticos and J.Hietarinta, *Physica* **144A**,310 (1987).
[9] H.Ito, *Kodai Math. J.* **8**,120 (1985).
[10] H.Yoshida, *Physica* **21D**,163 (1986) and **29D**,128 (1987).
[11] B.Grammaticos, A.Ramani and H.Yoshida, *Phys. Lett.* **124A**,65 (1987).
[12] J.Plemelj, 'Problems in the sense of Riemann and Klein' *Interscience*, New York (1962), chap.6.
[13] R.C.Churchill and D.L.Rod, *Journal of Diff. Equations* **76**,91 (1988).
[14] D.Roekaerts and H.Yoshida, *J.Phys. A.* **21**,3547 (1988).
[15] M.Adler and P.van Moerbeke, *Comm. Math. Phys.* **83**,83 (1982).
[16] H.Yoshida, 'Ziglin's analysis for proving non-integrability of Hamiltonian systems', in 'Finite Dimensional Integrable Nonlinear Dynamical Systems', p. 74, ed. P.G.L.Leach and W.H.Steeb, *World Scientific,* Singapore (1988).
[17] H.Yoshida, *Comm. Math. Phys.* **116**,529 (1988).
[18] D.L.Rod, 'On a theorem of Ziglin in Hamiltonian dynamics', to appear in "Proceedings of a Conference on Hamiltonian Systems", K.Meyer and D.Saari, (Eds.) *American Mathematical Society*, Providence, R.I., 1988.
[19] H.Yoshida, *Phys. Lett.* **120A**,388 (1987).
[20] J. Hietarinta, H.Yoshida, A.Ramani and B.Grammaticos, *Europhys. Lett.* **5**,195 (1988).
[21] G.Contopoulos and C.Polymilis, *Physica* **24D**,328 (1987).
[22] H.Yoshida, A.Ramani and B.Grammaticos, *Physica* **30D**,151 (1988).
[23] L.Martinet, private communication.
[24] M.Kac and P.van Moerbeke, *Adv. Math.* **16**,160 (1975).
[25] M.Bruschi and O.Ragnisco, *Phys. Lett.* **129A**,21 (1988). S.N.M.Ruijsenaars and H.Schneider, *Ann. Phys.* (N.Y.) **170**,370 (1986).
[26] M.D.Kruskal, private communication. See also Ref.[1] for the only (so far) published account of the method.

M.C. Nucci

20. PAINLEVE PROPERTY AND PSEUDOPOTENTIALS FOR NONLINEAR EVOLUTION EQUATIONS (TABLES)

The link between the singularity manifold equation and the pseudopotential equations is found for nonlinear evolution equations both in 1+1 and 2+1 dimensions. Many examples are given: Burgers, Korteweg-de Vries (KdV), modifed KdV, Harry-Dym, higher-order KdV, Sawada-Kotera, Kadomtsev-Petviashvili (KP), Kaup-Kupershmidt, Boussinesq, Modified KP, and (2+1dim.)-Sawada-Kotera equations. Here the results are shown via tables only. The original paper will be published somewhere else.

Figure 1 shows well-known properties of the Riccati equation on the dependent variable v *and independent variable* x*: the cross ratio of four solutions* x_i (i=1,2,3,4) *is a constant; under a Möbius transformation* (a,b,c,d *are arbitrary constants such that* ad-bc≠0), *one gets a formally invariant Riccati equation on the dependent variable* v^**; there is a transformation to a linear second order differential equation on the dependent variable* r, *which can be transformed into a third order differential equations, on the dependent variable* w, *involving the Schwarzian derivative. Note that* k *is a constant, and the linear term of the Riccati equation is missing to apply the transformation which leads to the same third order differential equations in* w.

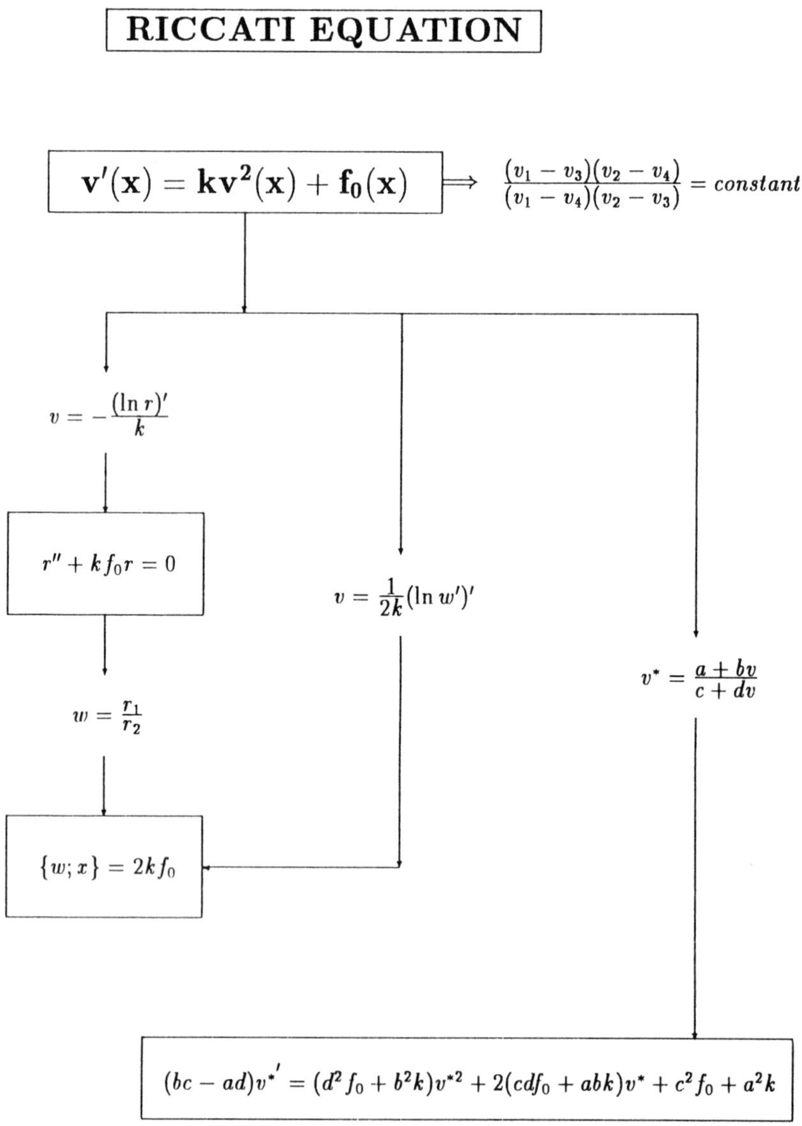

Figure 2 shows the way to get the Lax equations, the singularity manifold equation, and the auto-Bäcklund transform for a generic nonlinear evolution equation in 1+1 dimensions by means of u, *a pseudopotential which is given by two coupled equation of Riccati-type. Note that* G *is a polynomial of second order in* u, u^* *satisfies the same equation as* u, *and* q^* *is another solution of the nonlinear evolution equation.*

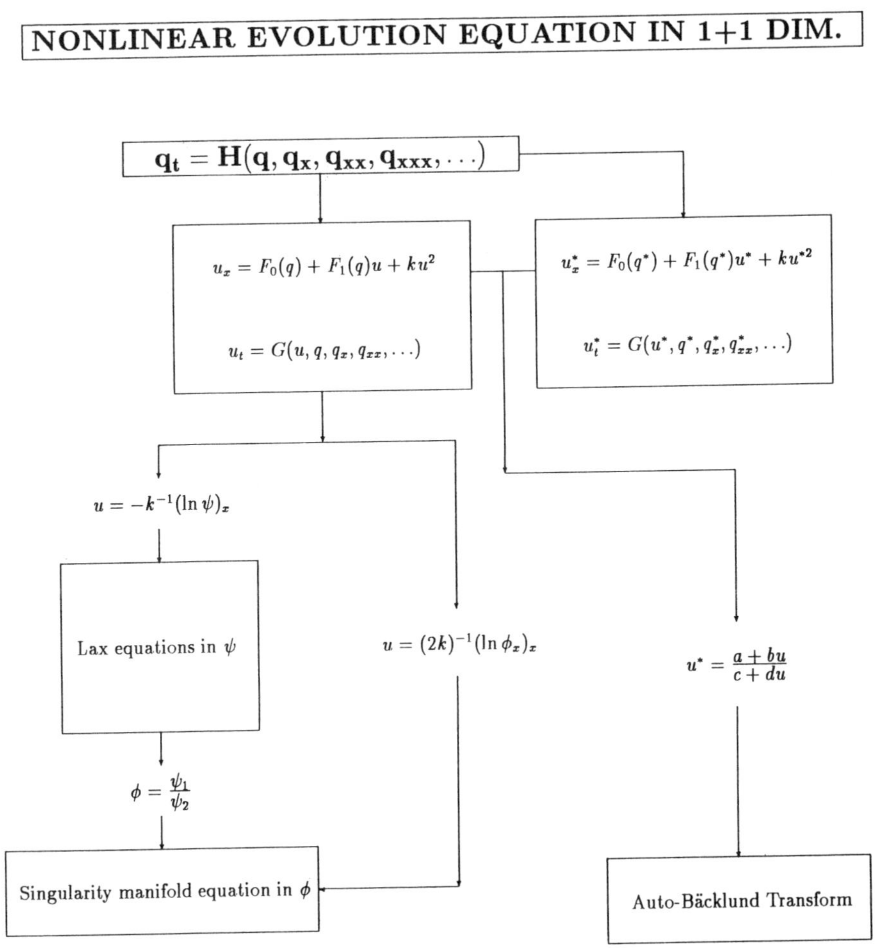

Figure 3 shows the results obtained for the Burgers equation. Note that u *satisfies the Burgers equation, and* $\bar{u}$ *is another pseudopotential which leads to the Lax equations with spectral functions* v.

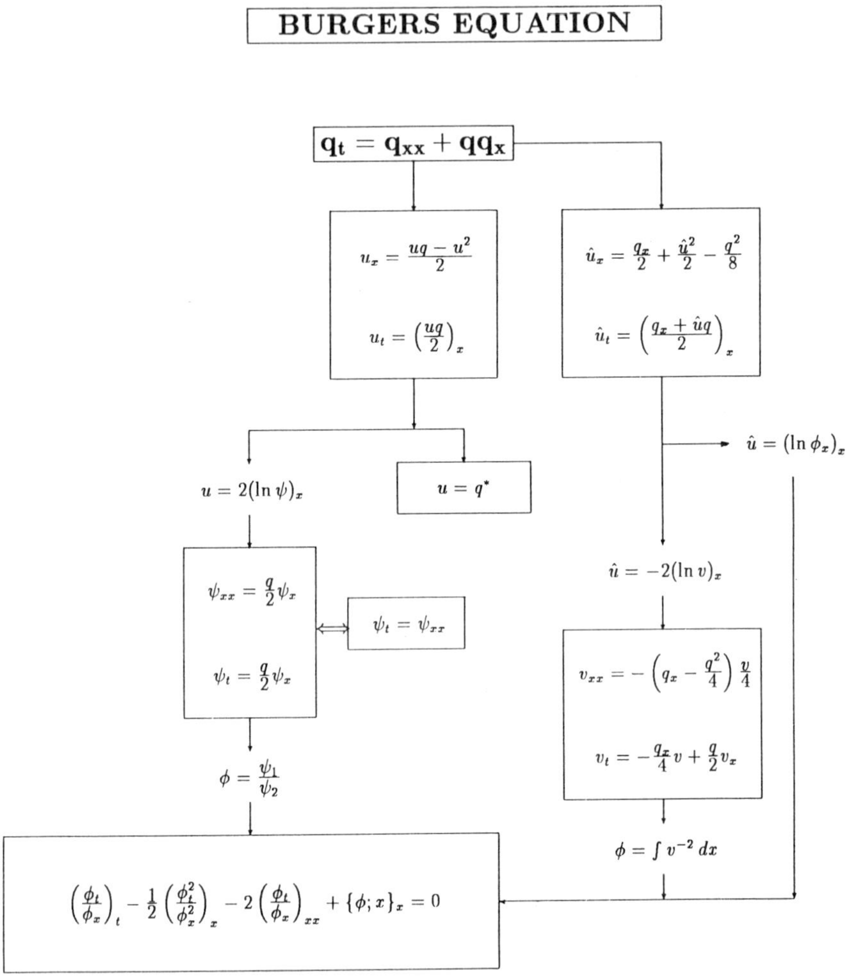

Figure 4 shows the results obtained for the KdV equation. Note that the equation satisfied by u is the modified KdV equation.

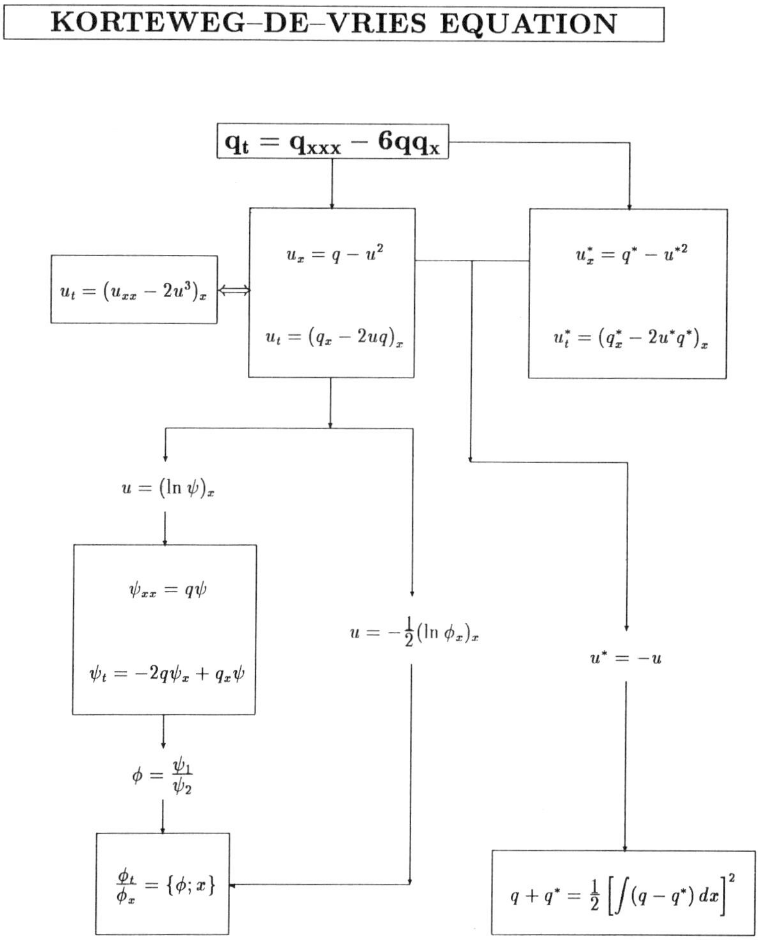

Figure 5 shows the results obtained for the modified KdV equation. Here the singularity manifold equation is the equation satisfied by ψ.

MODIFIED KORTEWEG–DE–VRIES EQUATION

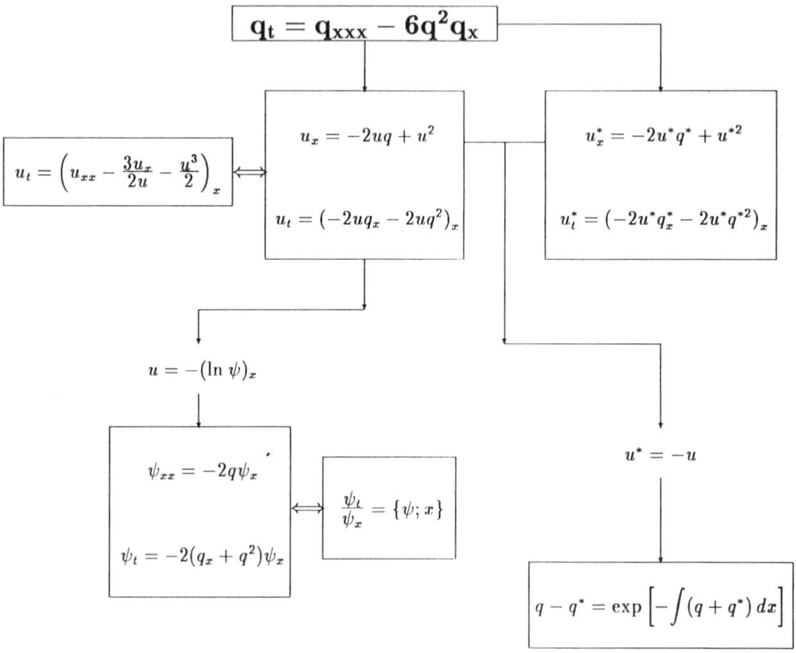

Figure 6 shows the results obtained for the Harry Dym equation. Only trivial auto-Bäcklund transforms are found.

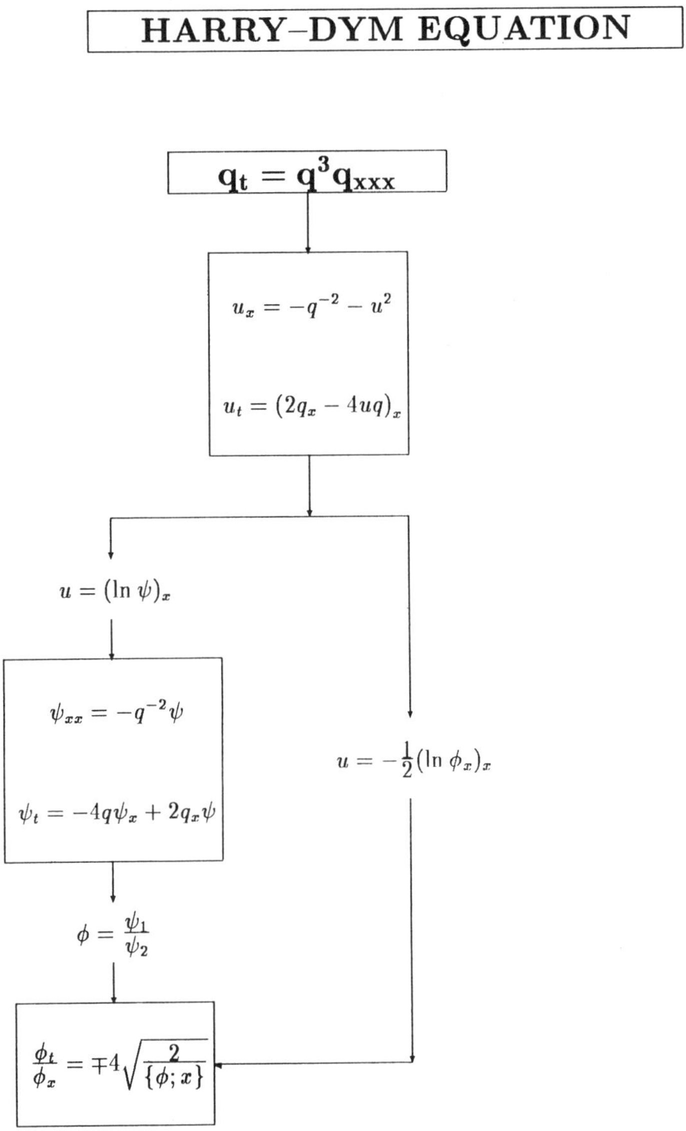

Figure 7 shows the results obtained for a higher-order KdV equation.

A HIGHER–ORDER KORTEWEG–DE–VRIES EQUATION

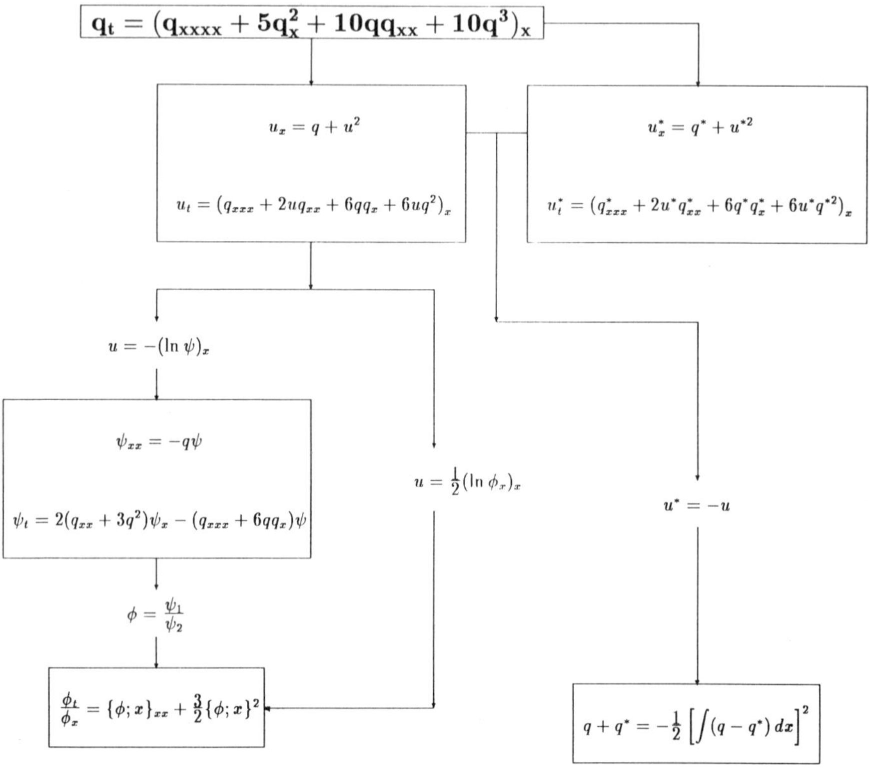

Figure 8 shows the results obtained for the SK equation. The singularity manifold equation is given by θ. The equation satisfied by ϕ is the singularity manifold equation of the KK equation: there is a cross-result due to the fact that u *is a common pseudopotential between SK and KK equations. The auto-Bäcklund transform of SK is given through its singularity manifold equation which is invariant under the Möbius group. Note that* $Q_x=q$ *and* $Q^*_x=q^*$.

SAWADA–KOTERA EQUATION

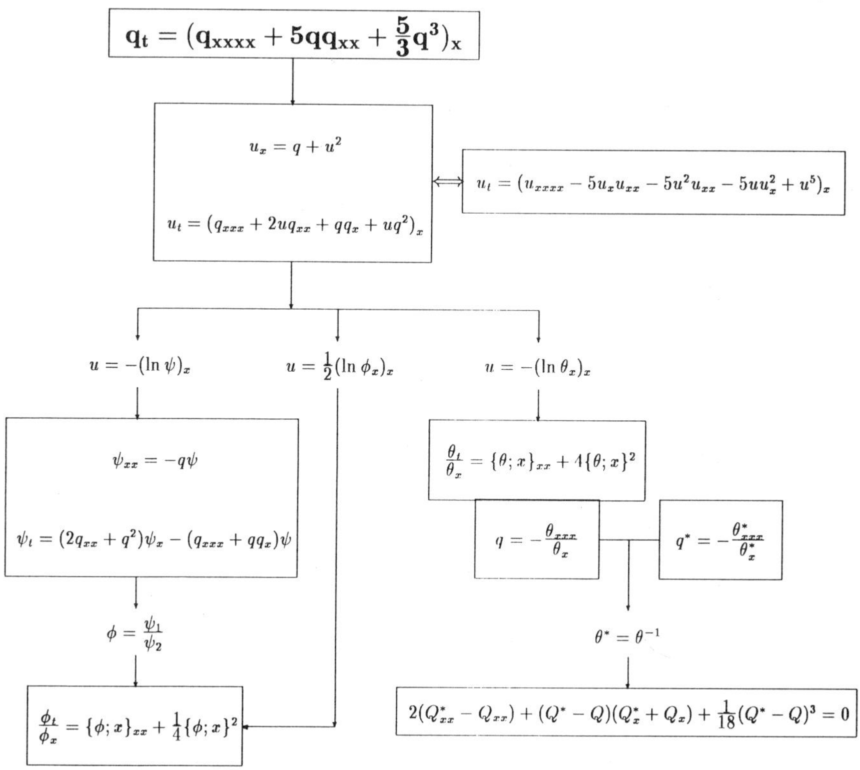

Figure 9 shows the results obtained for the KK equation. Here the spectral function is $\bar{\psi}$.

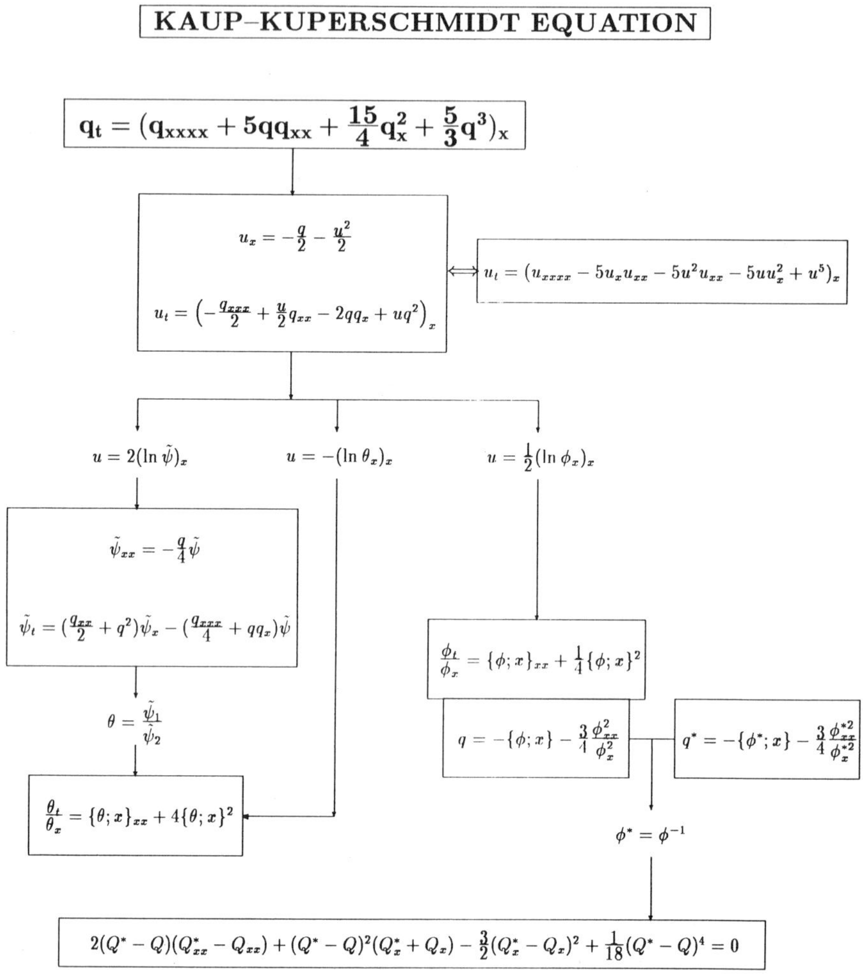

Figure 10 shows the way to get the Lax equations, the singularity manifold equation, and the auto-Bäcklund transform for a generic nonlinear evolution equation in 2+1 dimensions. Here F *is a polynomial of second order in* u *with* k *the leading coefficient;* G *is linear in* u, *of course. For other symbols, see Figure 2.*

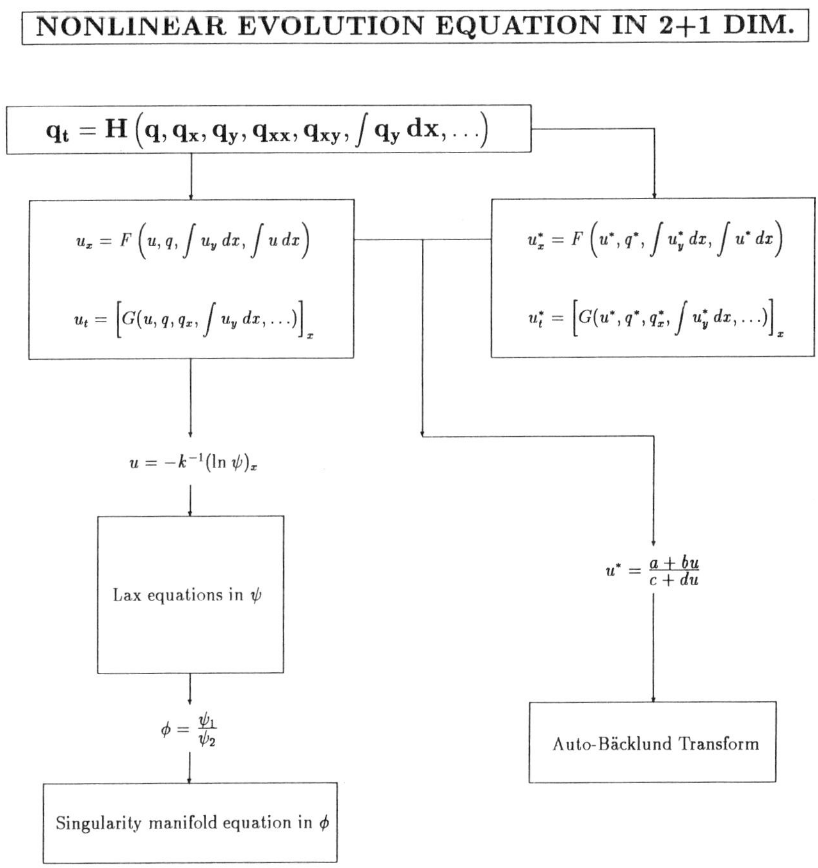

Figure 11 shows the results of the KP equation.

KADOMTSEV–PETVIASHVILI EQUATION

$$\mathbf{q_t = q_{xxx} - 6qq_x + 3\int q_{yy}\,dx}$$

$$u_x = q - u^2 - \int u_y\,dx$$

$$u_t = \left(q_x - 2uq - 4u_y - 4u\int u_y\,dx + 3\int q_y\,dx\right)_x$$

$$u^*_x = q^* - u^{*2} + \int u^*_y\,dx$$

$$u^*_t = \left(q^*_x - 2u^*q^* + 4u^*_y + 4u^*\int u^*_y\,dx - 3\int q^*_y\,dx\right)_x$$

$$u = (\ln\psi)_x$$

$$\psi_{xx} = q\psi - \psi_y$$

$$\psi_t = -2q\psi_x + \left(q_x + 3\int q_y\,dx\right)\psi - 4\psi_{xy}$$

$$u^* = -u$$

$$(Q^* - Q)^2 - 2(Q^*_x + Q_x) - 2\int(Q^*_y - Q_y)\,dx = 0$$

$$\phi = \frac{\psi_1}{\psi_2}$$

$$\left[\frac{\phi_t}{\phi_x} - \{\phi; x\} - \frac{3}{2}\left(\frac{\phi_y}{\phi_x}\right)^2\right]_x - 3\left(\frac{\phi_y}{\phi_x}\right)_y = 0$$

Figure 12 shows the results for the Boussinesq equation. They were obtained through the KP equation, by imposing t=0 *and* y=τ.

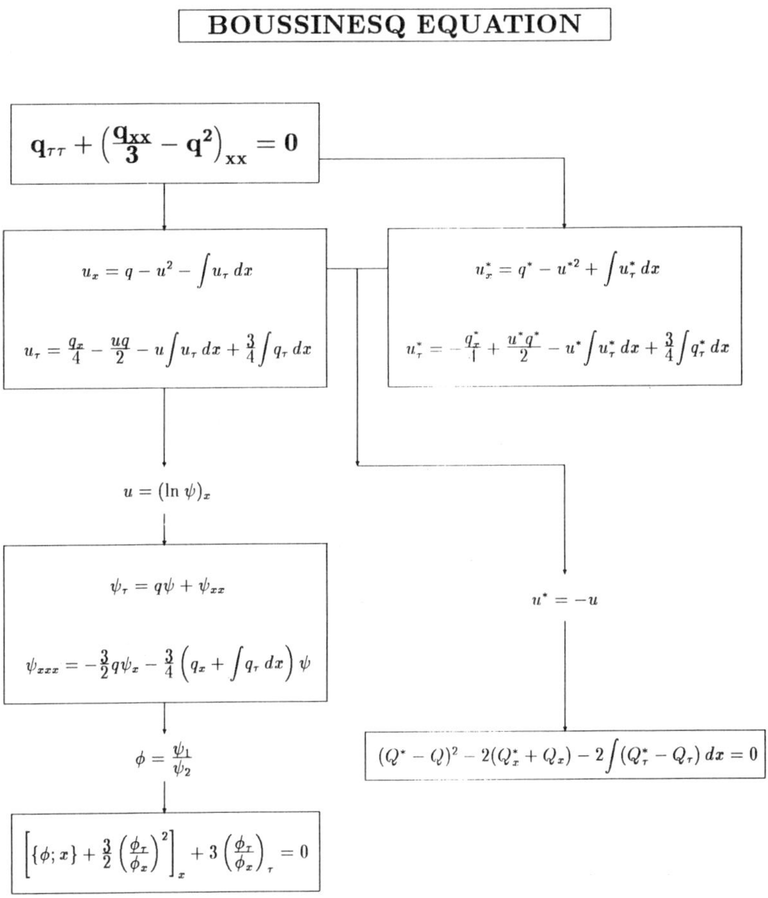

Figure 13 shows the results for the modified KP equation. Here the singularity manifold equation is given by the spectral function itself. Because of the small available space, we do not show the equation for u^*_t.

MODIFIED KADOMTSEV–PETVIASHVILI EQUATION

$$\mathbf{q_t = q_{xxx} - 6q^2 q_x + 3\int q_{yy}\,dx}$$

$$u_x = -2uq + u^2 - \int u_y\,dx$$

$$u_t = \left(-2uq_x - 2uq^2 - 4u_y - 4(q-u)\int u_y\,dx - 6u\int q_y\,dx\right)_x$$

$$u^*_x = -2u^* q^* + u^{*2} + \int u^*_y\,dx$$

$$u = -(\ln\psi)_x \qquad\qquad u^* = -u$$

$$\psi_{xx} = -2q\psi_x - \psi_y$$

$$\psi_t = -2\left(q_x + q^2 + 3\int q_y\,dx\right)\psi_x - 4\psi_{xy} - 4q\psi_y$$

$$\Updownarrow$$

$$\frac{\psi_t}{\psi_x} = \{\psi; x\} + 3\int\left(\frac{\psi_y}{\psi_x}\right)_y dx + \frac{3}{2}\left(\frac{\psi_y}{\psi_x}\right)^2$$

$$(Q^*_x - Q_x) + \exp[-(Q^* + Q)] + \exp(Q^* + Q)\int(Q^*_y + Q_y)\exp[-(Q^* + Q)]\,dx = 0$$

Figure 14 shows the results for the (2+1 dim.)-SK equation. Here the spectral function is θ (third-order scattering problem), which satisfies the singularity manifold equation. The auto-Bäcklund transform is given through the singularity manifold equation which is invariant under the Möbius group.

(2+1 DIM.)–SAWADA–KOTERA EQUATION

$$\mathbf{q_t = \left(q_{xxxx} + 5qq_{xx} + \frac{5}{3}q^3 + 5q_{xy}\right)_x - 5\int q_{yy}\,dx + 5qq_y + 5q_x\int q_y\,dx}$$

↓

$$u_x = q + u^2 - e^{\int u\,dx}\int\left(\int u_y\,dx\right)e^{-\int u\,dx}\,dx$$

$$u_t = \left(q_{xxx} + 2uq_{xx} + uq^2 + qq_x + 4q_y + 5u\int q_y\,dx - 3q\int u_y\,dx + 3q_x e^{\int u\,dx}\int\left(\int u_y\,dx\right)e^{-\int u\,dx}\,dx\right.$$
$$\left. -9e^{\int u\,dx}\int\left[\int u_{yy}\,dx - \left(\int u_y\,dx\right)^2\right]e^{-\int u\,dx}\,dx\right)_x$$

↓

$$u = -(\ln\theta_x)_x$$

↓

$$\theta_{xxx} = -q\theta_x - \theta_y$$

$$\theta_t = -3q_x\theta_{xx} - \left(q_{xx} - q^2 - 5\int q_y\,dx\right)\theta_x + 9\theta_{xxy} + 6q\theta_y$$

⇕

$$\left[\frac{\theta_t}{\theta_x} - \{\theta;x\}_{xx} - 4\{\theta;x\}^2 - 5\left(\frac{\theta_y}{\theta_x}\right)_{xx}\right]_x - 10\{\theta;x\}\left(\frac{\theta_y}{\theta_x}\right)_x + 5\left(\frac{\theta_y}{\theta_x}\right)_y = 0$$

$$q = -\frac{\theta_{xxx}}{\theta_x} - \frac{\theta_y}{\theta_x}$$

$$q^* = -\frac{\theta^*_{xxx}}{\theta^*_x} - \frac{\theta^*_y}{\theta^*_x}$$

↓

$$\theta^* = \theta^{-1}$$

↓

$$(Q^* - Q)\left[2(Q^* - Q)_{xx} + Q^*_x + Q_x\right] + \frac{1}{18}(Q^* - Q)^3 + 2\int(Q^*_y - Q_y)\,dx = 0$$

G. Baumann and T.F. Nonnenmacher

21. LIE-GROUP-SIMILARITY ANALYSIS OF COUPLED NONLINEAR SCHRÖDINGER EQUATIONS

1 Introduction

In recent years considerable interest has been attracted to the system of nonlinear Schrödinger equations [7,9,14]

$$
\begin{aligned}
i\hbar\, \varphi_{i_t} &= -\frac{\hbar^2}{2m}\, \varphi_{i_{xx}} + \kappa_i |\varphi_i|^2 \varphi_i + \lambda |\varphi_j|^2 \varphi_i \\
i\hbar\, \varphi^*_{i_t} &= -\frac{\hbar^2}{2m}\, \varphi^*_{i_{xx}} + \kappa_i |\varphi_i|^2 \varphi^*_i + \lambda |\varphi_j|^2 \varphi^*_i
\end{aligned}
\tag{1}
$$

$$i = 1,2; \qquad j = 1,2 \neq i$$

where φ_i denotes the complex fields for a particle with a property i; compare Yang [13] and Thacker [11] for a detailed discussion of the physical background. The cases studied so far are systems with equal coupling constants κ_i and λ which correspond to a homogeneous model [7,9,14] with respect to the coupling constants. System (1) shows a rich variety of behaviour for an inhomogeneous model. In most of the works the nonlinear system of partial differential equations has been investigated by the inverse scattering method [7]. In the present paper we study model (1) by applying the Lie-point-transformation method [4] which delivers a huge variety of similarity solutions. In a previous paper the authors [3] investigated a model with $\lambda = 0$ and $i = 1$; i.e. a single nonlinear Schrödinger equation. In this paper we study the system given above with the help of group theoretical methods reducing the system of partial differential equations with their continuous invariance groups to systems of ordinary differential equations (ODE). One of these systems will be considered in detail with the help of singular point analysis.

2 **Invariant Lie-point transformations**

In this section we give the symmetry groups under which the coupled nonlinear Schrödinger equations (1) are invariant. The introduction of real-valued and canonical conjugate fields S_i (phase) and ρ_i (density) as new variables related to the Schrödinger fields φ_i (and φ_i^*) by the canonical transformation

$$\varphi_i = \sqrt{\rho_i}\, e^{iS_i/\hbar} \qquad i = 1,2 \tag{2}$$

transforms (1) to

$$\begin{aligned}
H_1 &= \rho_{1t} + (\rho_1\sigma_{1x})_x = 0 \\
H_2 &= \rho_1^{\,2}\{\sigma_{1t} + \tfrac{1}{2}(\sigma_{1x})^2 + \Gamma_1\rho_1 + \mathrm{sgn}(\lambda)\rho_2 + \tfrac{1}{8}\rho_{1x}^2\} - \tfrac{1}{4}\rho_1\rho_{1xx} = 0 \\
H_3 &= \rho_{2t} + (\rho_2\sigma_{2x})_x = 0 \\
H_4 &= \rho_2^{\,2}\{\sigma_{2t} + \tfrac{1}{2}(\sigma_{2x})^2 + \Gamma_2\rho_2 + \mathrm{sgn}(\lambda)\rho_1 + \tfrac{1}{8}\rho_{2x}^2\} - \tfrac{1}{4}\rho_2\rho_{2xx} = 0
\end{aligned} \tag{3}$$

t and x are scaled time and space variables ($t \to \frac{|\lambda|}{\hbar}\, t$, $x \to \frac{(2m|\lambda|)^{1/2}}{\hbar}\, x$, $\sigma_i = \frac{S_i}{\hbar}$, the normalized phase, and $\Gamma_i = \frac{\kappa_i}{|\lambda|}$ $i = 1,2$ are the scaled coupling constants. Sgn(λ) denotes the corresponding sign of λ, (+) attraction, (-) repulsion between fields φ_i. The notion of group invariance that we shall use here can be found in Baumann & Nonnenmacher [3]. The following vector fields generating continuous symmetries of Eq. (3) are

$$X_2 = \partial t, \; X_4 = \partial_x \tag{4a}$$

$$X_5 = \partial_{\sigma_1}, \; X_6 = \partial_{\sigma_2} \tag{4b}$$

$$X_1 = 2t\,\partial_t + x\,\partial_x - 2\rho_1\,\partial_{\rho_1} - 2\rho_2\,\partial_{\rho_2} \tag{4c}$$

$$X_3 = x\,\partial_{\sigma_1} + x\,\partial_{\sigma_2} + t\partial_x \tag{4d}$$

The symmetries generated by the vector fields (4a) are responsible for translation invariance in time and space coordinates, (4b) allows translation invariance with respect to the phases σ_1 and σ_2. (4c) accounts for scaling in time, space, and for the densities ρ_1, ρ_2, (4d) guarantees Galilei invariance of solutions.

3 Similarity solutions

By using invariant groups we can reduce the number of independent variables. The resulting solutions are called similarity solutions.

The corresponding infinitesimal transformations of Eq. (4) are given by the relations

$$\begin{aligned}
\xi_1 &= 2c_1 t + c_2 \\
\xi_2 &= c_1 x + c_3 t + c_4 \\
\eta_1 &= - 2c_1 \rho_1 \\
\eta_2 &= c_3 x + c_5 \\
\eta_3 &= - 2c_1 \rho_2 \\
\eta_4 &= c_3 x + c_6 .
\end{aligned} \tag{5}$$

Integration of characteristic equations will deliver similarity solutions. These solutions can be classified by the arbitrary group-parameters c_i, $i = 1, \ldots, 6$. Our main interest in this paper is a subgroup of this 6 parameter group spanned by (5). Each choice of the c_i represents a different similarity solution of the characteristic equations

$$\frac{dt}{\xi_1} = \frac{dx}{\xi_2} = \frac{d\rho_1}{\eta_1} = \frac{d\rho_2}{\eta_3} = \frac{d\sigma_1}{\eta_2} = \frac{d\sigma_2}{\eta_4} . \tag{6}$$

If we set in (5) $c_1 = 0$, $c_i \neq 0$ for $i \neq 1$ and integrate (6), i.e. scaling in t, x, ρ_1, and ρ_2 are lost, the corresponding similarity solutions are

$$\rho_1(x,t) = F_1(\zeta), \qquad \rho_2(x,t) = F_3(\zeta)$$

$$\sigma_1(x,t) = \frac{c_3}{c_2}\{x - \frac{c_3}{3c_2} t^2 - \frac{c_4}{2c_2} t + \frac{c_5}{c_2}\}t + F_2(\zeta) \tag{7}$$

$$\sigma_2(x,t) = \frac{c_3}{c_2}\{x - \frac{c_3}{3c_2} t^2 - \frac{c_4}{2c_2} t + \frac{c_6}{c_2}\}t + F_4(\zeta)$$

where $\zeta = x - \frac{c_3}{2c_2} t^2 - \frac{c_4}{c_2} t$ is the so-called similarity variable. $F_i(\zeta)$ are up to now undetermined similarity functions. A straightforward integration of (6) with $c_1 = c_3 = 0$ reduces (7) to

$$\rho_1(x,t) = F_1(\zeta), \qquad \rho_2(x,t) = F_3(\zeta)$$
$$\sigma_1(x,t) = F_2(\zeta), \qquad \sigma_2(x,t) = F_4(\zeta) \tag{8}$$

with $\zeta = x - \frac{c_4}{c_2} t$ as moving wave variable. This simple subgroup $c_1 = c_3 = 0$ of (5) demonstrates that (3) allows solutions in the space-time domain with constant velocity of propagation $\frac{c_4}{c_2}$.

The general solution of Eq. (6) given by (7) for the subclass $c_1 = 0$ provides us with the general (functional) form of the similarity solutions. This form involves four undetermined functions depending upon one variable, the similarity variable. The reduction is achieved since the undetermined functions F_i are governed by a system of ordinary differential equations. When these equations are solved, we obtain similarity solutions by inserting F_i into the general form (7,8).

4 Similarity reduction

In order to determine the similarity solutions of the case $c_1 = c_3 = 0$ we derive the differential equations for the undetermined functions F_i, i=1,2,3,4 by inserting the general form (8) of the similarity solutions into Eq. (3). The result is an ODE for the F_i's which is called the reduced equation.

Inserting (8) into (3) with ζ specified according to a moving wave variable we obtain the four reduced equations

$$F_{2k-1}F''_{2k} + F'_{2k-1}(F'_{2k} - \frac{c_4}{c_2}) = 0 \tag{9a}$$

$$F_{2k-1}F''_{2k-1} - \frac{1}{2}(F'_{2k-1})^2 + F^2_{2k-1}\{4\frac{c_4}{c_2}F'_{2k} - 2(F'_{2k})^2 - 4\Gamma_k F_{2k-1} -4\,\mathrm{sgn}(\lambda)F_{\substack{2k+1\\ \mathrm{mod}(4)}} -4\frac{\delta_k}{c_2}\} = 0 \tag{9b}$$

$$\text{with } \delta_k = \begin{cases} c_5 \text{ for } k = 1 \\ c_6 \text{ for } k = 2 \end{cases} \quad k = 1,2$$

where primes denote derivatives with respect to ζ. This system contains besides the coupling parameters Γ_k four arbitrary constants c_i which determine via (5) the transformation properties of (3). The ratio $\frac{c_4}{c_2}$ represents the velocity of propagation of the solutions.

The first two equations (9a) integrate to implicit solutions of the F_{2k}'s

$$F_{2k}(\zeta) = \int \frac{I_{2k-1}}{F_{2k-1}(\zeta)} d\zeta + \frac{c_4}{c_2} \zeta + I_{2k} \qquad k=1,2. \tag{10}$$

The initial conditions $F'_{2k}(0) = \frac{c_4}{c_2}$, corresponding to constant velocity of propagation reduce Eq. (10) to

$$F_{2k}(\zeta) = \frac{c_4}{c_2} \zeta + I_{2k} \qquad k=1,2 \tag{11}$$

which, when inserted into Eqs. (9b), results in a reduction to

$$F_{2k-1}F''_{2k-1} - \frac{1}{2}(F'_{2k-1})^2 + F^2_{2k-1}\{\alpha_k - 4\Gamma_k F_{2k-1} - 4\,\mathrm{sgn}(\lambda) F_{\substack{2k+1 \\ \mathrm{mod}(4)}}\} = 0 \tag{12}$$

where α_k is given by

$$\alpha_k = \begin{cases} 2(\frac{c_4}{c_2})^2 - 4(\frac{c_5}{c_2}) & \text{for } k=1 \\ 2(\frac{c_4}{c_2})^2 - 4(\frac{c_6}{c_2}) & \text{for } k=2. \end{cases}$$

A substitution of the dependent variables $F_{2k-1} = q_k^2$ reduces system (12) to

$$\begin{aligned} q_1'' + \frac{\alpha_1}{2} q_1 - 2\Gamma_1 q_1^3 - 2\mathrm{sgn}(\lambda)\, q_1 q_2^2 = 0 \\ q_2'' + \frac{\alpha_2}{2} q_2 - 2\Gamma_2 q_2^3 - 2\mathrm{sgn}(\lambda)\, q_2 q_1^2 = 0. \end{aligned} \tag{13}$$

These equations describe a system of coupled non-linear oscillators. Closely related models to Eq. (13) have been studied by several authors [5,8,10] with respect to chaotic behaviour and quantum chaos. Our interest here relates to the question of integrability of these equations. A first step in this direction is a canonical formulation of equation (13). Using the Hamiltonian

$$H = \frac{1}{2}(p_1^2 + p_2^2) + \sum_{i=1}^{2} \frac{\alpha_i}{4} q_i^2 - \frac{\Gamma_i}{2} q_i^4 - \mathrm{sgn}(\lambda)\, q_1^2\, q_2^2 \tag{14}$$

and the classical Poisson-bracket for a two-dimensional system we can reconstruct Eq. (13), where the moments p_i are canonical conjugate to q_i. Our reduced system (13) is obviously equivalent to a two dimensional anharmonic oscillator with a quartic coupling.

By Arnold's theorem [2] the 2-dimensional dynamical system (13) is completely integrable if there exist a 2-dimensional set of integrals which are in involution. A first constant of motion is given by the Hamiltonian (14) because H does not depend explicitly on ζ. Looking for further integrals of involution, even for cases where we must specify some of the free parameters α_i and Γ_i in a specific way, we applied a singular point analysis to equation (13) in order to determine all possible parameter choices for integration.

5 Singular point analysis

We perform our singular-point analysis for the Painlevé test on the basis of the approach proposed by Ablowitz et al. [1] investigating the pole structure of the reduced system of ODE (13). An alternative approach developed by Weiss et al. [12] tests the Painlevé property of the original (unreduced) system of partial differential equations (PDE) (1) and has been applied recently by Sahadevan et al. [9] predicting only a single pair of resonances for the coupling constants $\Gamma_i = \pm 1$ as integrability condition. However, we will show that a singular-point analysis when applied to the reduced system of ODE's (13) will lead to additional resonances and to new, herertofore undiscovered similarity solutions of system (3).

Following the ideas of Ablowitz et al. [1] we calculated for system (13) the dominant behaviour with the ansatz $w_j \approx \beta_j(z-z_o)^{p_j}$. From this ansatz we got $p_j = -1$, $j = 1,2$. To determine the resonances we used the relation $w_j \approx \beta_j\tau^{-1} + w_j\tau^{-1}$, $\tau = z-z_o$. A straightforward calculation shows that r = -1 and r = 4 are two of the four possible resonances which occur independently of system parameters. Two further relations determine the resonances which depend on the coupling constants Γ_i. For the Painlevé property we need positive integer resonances. This condition leads to six pairs of values of the Γ_i's, first $\Gamma_i = \pm 1$ and $\Gamma_i = \pm 1/3$, $\Gamma_1 = -1/3$ and $\Gamma_2 = -8/3$ and finally $\Gamma_1 = -1/6$ and $\Gamma_2 = -8/3$. The determining equation of the resonance reduces to

$$\{(r-1)(r-2)\}^2 - 8(r-1)(r-2) + 12 = 0$$

for $\Gamma_i = \pm 1$ with solutions $r = \{-1,0,3,4\}$ and,

$$\{(r-1)(r-2)\}^2 - 6(r-1)(r-2) = 0$$

for $\Gamma_i = \pm 1/3$ with resonances at $r = \{-1,1,2,4\}$,

$$r(r+1)(r-2)(r-4) = 0$$

for ($\Gamma_1 = -1/3$, $\Gamma_2 = -8/3$) and ($\Gamma_1 = -1/6$, $\Gamma_2 = -8/3$) with resonances at $r = \{-1,0,2,4\}$.

Going through the third step, determining the constants of integration we have to assume the equality of $\alpha_1 = \alpha_2$ for the four first cases and $4\alpha_1 = \alpha_2$ with respect to the other two to get arbitrary coefficients in the expansions

$$w_j \approx \beta_j \tau^{-1} + \sum_{i=1}^{4} a_{ij} \tau^{i-1}.$$

The result of this analysis demonstrates that there exist six pairs of coupling constants which may lead to a second constant of motion.

6 Integrals of motion and solutions

As an example we calculate here the integrals of motion for the four parameter pairs $\Gamma_i = \pm 1$ and $\Gamma_i = \pm 1/3$. By introducing - in each case - a new set of coordinates a decoupling of q_1 and q_2 in (14) is achieved.

In case $\alpha_1 = \alpha_2 = \alpha$ with $\Gamma_1 = \Gamma_2 = \pm 1$ we introduce polar coordinates in the (q_1,q_2)- plane by $q_1 = r\cos(\theta)$ and $q_2 = r \sin(\theta)$ with $r = r(\zeta)$ and $\theta = \theta(\zeta)$. The resulting Hamiltonian has the form

$$H = \frac{1}{2}(p_r^2 + p_\theta^2 \frac{1}{r^2}) + \frac{\alpha}{4} r^2 \mp \frac{1}{2} r^4.$$

Because θ is a cyclic variable the conjugate momentum $p_\theta = \ell$ is conserved [6]. A transformation back to q_1 and q_2 coordinates gives a representation of the conserved quantity

$$\ell = q_1 p_2 - q_2 p_1 = I_1,$$

the "angular momentum".

For the second choice of parameters $\alpha_1 = \alpha_2 = \alpha$, $\Gamma_1 = \Gamma_2 = \pm 1/3$ we introduce the sum and difference ofcoordinates by $q_1 = \frac{1}{2}(x_1 + x_2)$ and $q_2 = \frac{1}{2}(x_1 - x_2)$. The transformed Hamiltonian decouples then into two equivalent systems

$$H_i = \frac{1}{4}\xi_i^2 + \frac{\alpha}{8}x_i^2 \mp \frac{1}{12}x_i^4, \qquad i = 1,2$$

with ξ_i as transformed momentum.

Each of these H_i is a constant of motion. The sum and difference of these integrals are also conserved quantities of motion. The sum of the H_i is the total Hamiltonian whereas the difference is a new invariant of motion. In the original coordinates this quantity reads:

$$I_2 = p_1p_2 + q_1q_2\left(\frac{\alpha}{2} \mp \frac{2}{3}(q_1^2 + q_2^2)\right).$$

A test of the involution condition $[I_k,H] = 0$, $k = 1,2$ is clearly satisfied.

The above considerations show that system (13) exhibits two groups of parameter choices in which a total integration of the equations can be achieved. The solutions for these cases are of elliptic and hyperbolic type. Special forms are given by

$$\rho_1 = \frac{3}{2}\alpha\,\mathrm{sech}^2((\alpha/2)^{1/2}(\zeta-\zeta_o))$$

$$\rho_2 = 0$$

for $\Gamma_i = \pm1/3$, and

$$\rho_1 = \frac{1}{2}\alpha\,\mathrm{sech}^2((\alpha/2)^{1/2}(\zeta-\zeta_o))\cos^2(I)$$

$$\rho_2 = \frac{1}{2}\alpha\,\mathrm{sech}^2((\alpha/2)^{1/2}(\zeta-\zeta_o))\sin^2(I)$$

for $\Gamma_i = \pm1$, where J and ζ_0 are constants of integration.

7 Discussion

Applying Lie-group methods to a system of coupled nonlinear Schrödinger equations we determined similarity variables and reduced the system of nonlinear PDE to a system of nonlinear ODE's of similarity functions $F_i(\zeta)$. An appropriate substitution $F_i(\zeta) = q_i^2(\zeta)$ transformed the system of ODE's into a system of ODE for $q_1(\zeta)$ and $q_2(\zeta)$. System (13) is equivalent to two coupled anharmonic oscillators. This system has been investigated by applying a singular-point analysis. In addition to the single pair of resonances with $\Gamma_i = \pm1$ (i=1,2) found earlier [14] we presented values of Γ_i and the corresponding resonances, for unknown classes of solutions to Eq. (3). For Γ_i's ($\Gamma_i = \pm1/3$, $\Gamma_i = \pm1$) we have calculated the integrals of motion and explicit forms of solutions.

The results presented here demonstrate that much more information about the coupled nonlinear Schrödinger system can be obtained when the singular-point analysis is applied to the reduced system of ODE rather than to the original system of nonlinear partial differential equations.

References

[1] Ablowitz, M.J., Ramani, A. & Segur, H. (1980). A connection between nonlinear evolution equations and ordinay differential equations of P-type I. *J. Math. Phys.* **21**, 715.

[2] Arnold, V.J. (1979). *Mathematical methods of classical mechanics.* Springer: New York.

[3] Baumann, G. & Nonnenmacher, T.F. (1987). Lie transformations, similarity reduction, and solutions for the nonlinear Madelung fluid equations with external potential. *J. Math. Phys.* **28**, 1250.

[4] Bluman, G.W. & Cole, J.D. (1974). *Similarity methods for differential equations.* Springer: New York.

[5] Carnegie, A. & Percival, J.C. (1984). Regular and chaotic motion in some quartic potentials. *J. Phys. A: Math. & Gen.* **17**, 801.

[6] Goldstein, H. (1951). *Classical mechanics.* Addison-Wesley: Reading, Massachusetts.

[7] Manakov, S.V. (1974). On the theory of two-dimensional stationary self-focusing of electromagnetic waves. *Sov. Phys. -JETP* **38**, 248.

[8] Pullen, R.A. & Edmonds, a.R. (1981). Comparison of classical and quantal spectra for a totally bound potential. *J. Phys. A: Math. & Gen.* **14**, L477.

[9] Sahadevan, R., Tamizhmani, K.M. & Lakshmanan, M. (1986). Painlevé analysis and integrability of coupled non-linear Schrödinger equations. *J. Phys. A: Math. & Gen.* **19**, 1783.

[10] Steeb, W.-H., Willet, C.M. & Kunick, A. (1985). Chaotic behaviour of a Hamiltonian with a quartic potential. *J. Phys. A: Math. & Gen.* **18**, 3269.

[11] Thacker, H.B. (1981). Exact integrability in quantum field theory and statistical systems. *Rev. Mod. Phys.* **53**, 253.

[12] Weiss, J., Tabor, M. & Carneval, G. (1983). The Painlevé property for partial differential equations. *J. Math. Phys.* **24**, 522.

[13] Yang, C.N. (1967). Some exact results for the many-body problem in one dimension with repulsive delta-function interaction. *Phys. Rev. Lett.* **23**, 1312.

[14] Zakharov, V.E. & Schulman, E.J. (1982). To the integrability of the system of two coupled nonlinear Schrödinger equations. *Physica* **4D**, 270.

P. Kaliappan, S. Andal, N. Rajeswari and S. Kalaiselvi.

22. PAINLEVÉ ANALYSIS AND SOLUTIONS OF THE NAGUMO EQUATION

The Nagumo Equation is analysed for Painlevé property, first by reducing it to an ordinary differential equation with the introduction of a wave variable. For specific values of the wave speed, the equation is of Painlevé type and exact travelling wave solutions are found for these specific wave speeds. Then the equation is analysed directly for Painlevé and the same set of solutions are found.

1 Introduction

To describe the shape and speed of pulses in the nerve of a squid, Nagumo has proposed the following equation [4]

$$\frac{\partial u}{\partial t} = \frac{\partial^2 u}{\partial x^2} - au + (a + 1)u^2 - u^3 \tag{1}$$

where $0 < a < 1$. Some aspects of wave solution for the above equation have been discussed by McKean [4]. Introducing the wave variable $z = x + ct$, eq.(1) reduces to ordinary differential equation (o.d.e.)

$$\frac{d^2u}{dz^2} - c\frac{du}{dz} - au + (a+1)u^2 - u^3 = 0 \tag{2}$$

In this paper we look for the Painlevé property of the Nagumo equation and find explicit travelling wave solutions with specific wave speeds. First we analyse the reduced o.d.e. (2) for Painlevé property by the method proposed by Ablowitz et al. [1]. (Actually this analysis was carried out by Airault & Kaliappan [2].) The equation has Painlevé property for three values of c and for these values we find three explicit wave solutions, by transforming the dependent variable bilinearly. Then we analyse the partial differential equation (p.d.e.) (1) for Painlevé property by the method proposed by Weiss et al. [6] and find the explicit travelling wave solutions which are the same as those found in the previous method. We also discuss the stability, boundedness and boundary values of the solutions.

2 Painlevé analysis of equation (2)

We perform a Painlevé test by using the algorithm found by Ablowitz et al. [1]. Assuming that the dominant behaviour of the solutions in a sufficiently small neighbourhood of a movable singularity z_0 is algebraic, let us put

$$u \sim \alpha(z - z_0)^p \qquad \text{as } z \to z_0 \text{ in (2)}$$

we find $p = -1$, $\alpha = \pm\sqrt{2}$. By taking $\alpha = +\sqrt{2}$, we find the resonances occur at $j = -1$ and $j = 4$. Substituting

$$u = \alpha(z - z_0)^p + \sum_{j=1}^{4} a_j(z - z_0)^{p + j} \text{ in (2)}$$

we get the recursion relationship as

$$Q(j)a_j - R_j\,(\alpha, a_1, a, \ldots a_{j-1}) = 0 \tag{3}$$

for $j = 1,2,3$, we find

$$a_1 = \frac{a + 1}{3} + \frac{c}{3\sqrt{2}} \tag{4}$$

$$a_2 = -\frac{\sqrt{2}a}{6} + \frac{\sqrt{2}a}{3}(a_1 + 1) - \frac{a_1^2}{\sqrt{2}} \tag{5}$$

$$a_3 = -\frac{\lambda}{4}a_2 - \frac{1}{4}aa_1 + \frac{1}{4}(a + 1)\,a_1^2 + \frac{1}{\sqrt{2}}(a + 1)\,a_2 - \frac{1}{4}a_1^3 - \frac{3}{\sqrt{2}}a_1a_2 \tag{6}$$

At the resonance $j = 4$, $Q(4) = 0$, then $R_4 = 0$ is a necessary condition for eq. (2) to have Painlevé property. This makes $a_3 = 0$. Then the compatibility condition of (4), (5), (6) gives

$$a_1 = \frac{1}{2}, \frac{a}{2}, \frac{a + 1}{2} \text{ with}$$

$$c = \frac{(1 - 2a)}{\sqrt{2}}, -\frac{a + 1}{\sqrt{2}}, -\frac{a - 2}{\sqrt{2}} \tag{7}$$

By taking $\alpha = -\sqrt{2}$, we obtain

$$c = -\frac{(1 - 2a)}{\sqrt{2}}, -\frac{(a + 1)}{\sqrt{2}}, -\frac{(a - 2)}{\sqrt{2}} \tag{8}$$

Thus we note that eq. (2) has Painlevé property if c takes the values given in (7) and (8).

3 **Solutions of (1)**

In eq. (1) let us transform the dependent variable by

$$u = \frac{\alpha\omega + \beta}{\nu\omega + \delta} \tag{9}$$

with

$$\alpha\delta - \nu\beta = K \neq 0 \tag{10}$$

Here α, β, ν, δ are arbitrary parameters restricted by (10). Finding u_t, u_{xx} from (9) and substituting these in (1) we get

$$\nu[\omega\omega_t - \omega\omega_x + 2\omega_x^2 + A_1\omega^2 + B_1\omega^3 + \delta C\omega_t - \omega_{xx} + A_2\omega + B_2] = 0 \tag{11}$$

where

$$KA_1 = a(2\alpha\nu\delta + \beta\nu^2) + (a + 1)(\alpha^2\delta + 2\alpha\nu\beta) + 3\alpha\beta^2 \tag{12}$$

$$KA_2 = a(2\nu\delta\beta + \alpha\delta^2) - (a + 1)(\beta\nu^2 + 2\alpha\beta\nu + 3\alpha^2\beta) \tag{13}$$

$$KB_1 = \alpha(\alpha^2 - (a + 1)\alpha\nu + a\nu^2) \tag{14}$$

$$KB_2 = B[\beta^2 - (a + 1)\delta\beta + a\delta^2] \tag{15}$$

All the terms inside the brackets near ν are non-linear with second degree except the term $B_1\omega^3$. All the terms in the brackets near δ are linear with first degree except the term B_2. Intentionally we shall make $B_1 = 0$ and $B_2 = 0$. This gives three values for α and β as follows

$$\alpha = 0,\ \nu,\ a\nu$$

$$\beta = 0,\ \delta,\ a\delta$$

From these values we can form six different combinations of the values of α and β.

(i)	$\alpha = 0,\ \beta = \delta$	(iv)	$\alpha = \nu,\ \beta = 0$
(ii)	$\alpha = 0,\ \beta = a\delta$	(v)	$\alpha = a\nu,\ \beta = 0$
(iii)	$\alpha = \nu,\ \beta = a\delta$	(vi)	$\alpha = a\nu,\ \beta = \delta$

Case (i) $\alpha = 0$, $\beta = \delta$.

Then equation (13) becomes

$$\nu[\omega\omega_t - \omega\omega_{xx} + 2\omega_x^2 - a\omega^2] = \delta[\omega_t - \omega_{xx} + (1-a)\omega] \tag{16}$$

Suppose we are interested in finding out travelling wave solutions. Take $\omega = e^{mz}$, where $z = x-ct$. Then eq. (16) becomes

$$e^{2mz}\nu(-m^2 + cm - a) = \delta(m^2 - cm + (1 + a))e^{mz}$$

Now make

$$-m^2 + cm - a = 0 \tag{17}$$

and

$$m^2 - cm + (1 + a) = 0 \tag{18}$$

These equations (17) and (18) give

$$m = \frac{1}{\surd 2} \qquad c = \frac{2a - 1}{\surd 2}$$

$$m = \frac{1}{\surd 2} \qquad c = -\frac{2a - 1}{\surd 2}$$

The solutions are

$$u = \frac{1}{1 + \nu e^{mz}} \tag{19}$$

with

$$m = \mp \frac{1}{\surd 2} \qquad c = \pm \frac{2a - 1}{\surd 2} \tag{20}$$

The combination (iv) also yields the same solution.

Case (ii)

The combinations (ii) and (iv) give the solutions as

$$u = \frac{-a}{1 + \nu e^{mz}} + a \tag{21}$$

with

$$m = \mp \frac{a}{\surd 2}, \qquad c = \equiv \frac{2 - a}{\surd 2} \tag{22}$$

Case (iii)

The combinations (iii) and (vi) give the solution as

$$u = \frac{a - 1}{1 + \nu e^{mz}} + 1 \tag{23}$$

with

$$m = \mp \frac{a - 1}{\sqrt{2}}, \qquad c = \pm \frac{a + 1}{\sqrt{2}}$$

5 Painlevé analysis and travelling wave solutions of equation (1)

We perform a Painlevé test for the p.d.e. (1) by the method as proposed by Weiss et al. [6]. A solution u of the p.d.e. (1) is taken in the form

$$u = \frac{1}{\emptyset^{\alpha}} - \sum_{j=0}^{\infty} u_j \emptyset^j \tag{24}$$

where $\emptyset$ is the analytic function whose vanishing defines a non-characteristic hyper surface. By inserting this expansion into the p.d.e., the values of α and the recursion relations for the functions u_j are found. Then the p.d.e. is said to have Painlevé property as proposed by Weiss et al. [6] provided that α is an integer, the resonance relation is consistent and that it contains the correct number of arbitrary functions.

By leading order analysis we find

$$\alpha = 1, \quad u_o = \sqrt{2}\, \emptyset_x \tag{25}$$

The recursion relations for u_j are found as

$$(j + 1)(j - 4)\emptyset_x^2 u_j = F(\emptyset_t, \emptyset_x, \ldots u_k; k<j) \tag{26}$$

we note that the truncated expansions of (24) can be taken as

$$u = \frac{\sqrt{2}\emptyset}{\emptyset} + u_1 \tag{27}$$

with the following conditions

$$\emptyset_t - 3\emptyset_{xx} + \sqrt{2}(a + 1)\emptyset_x - 3\sqrt{2}\, \emptyset_x u_1 = 0 \tag{28}$$

$$\emptyset_{xxx} - \emptyset_{xt} - a\emptyset_x + 2(a + 1)u_1\emptyset_x - 3\emptyset_x u_1^2 = 0 \tag{29}$$

$$(u_1)_{xx} - (u_1)_t - au_1 + (a + 1)u_1^2 - u_1^3 = 0 \tag{30}$$

Let us consider the constant solutions for u_1 of (30). The solutions are $u_1 = 0$, $u_1 = 1$, $u_1 = a$.

For $u_1 = 0$ eqs. (28) and (29) reduce to

$$\emptyset_t - 3\emptyset_{xx} + \sqrt{2}(a+1)\, \emptyset_x = 0 \tag{31}$$

$$\emptyset_{xxx} - \emptyset_{xt} - a\emptyset_x = 0 \tag{32}$$

Eqs. (31) and (32) are over determined to determine $\emptyset$, however we attempt to find specific solutions for the travelling waves of eq. (1).

Assuming $\emptyset = A + Be^{m(x-ct)}$ and substituting in (31) and (32) we find

$$m = \frac{-a}{\sqrt{2}}, \frac{+a}{\sqrt{2}} \tag{33}$$

$$c = + \frac{a - 2}{\sqrt{2}}, - \frac{a - 2}{\sqrt{2}} \tag{34}$$

Then by using (27), we get

$$u = \frac{a}{1 + De^{m(x + ct)}} \tag{35}$$

where

$$m = \mp \frac{a}{\sqrt{2}}, \quad c = \pm \frac{a - 2}{\sqrt{2}} \tag{36}$$

and

$$u = \frac{1}{1 + De^{m(x + ct)}} \tag{37}$$

where

$$m = \mp \frac{1}{\sqrt{2}}, \quad c = \pm \frac{1 - 2a}{\sqrt{2}} \tag{38}$$

Taking other values $u_1 = 1$, $u_1 = a$, we obtain (different from (34) and (37))

$$u = \frac{a - 1}{1 + De^{m(x + ct)}} + 1 \tag{39}$$

where

$$m = \mp \frac{1 - a}{\sqrt{2}}, \quad c = \pm \frac{a + 1}{\sqrt{2}} \tag{40}$$

5 Boundedness, boundary values and stability of solutions

In order that the solutions given by (19), (21) and (23) may be bounded, ν should be taken as greater than zero.

The solution given in (19) has the boundary values

$$\left.\begin{aligned} &u(-\infty) = 0,\ u(+\infty) = 1 \text{ for } m = -\frac{1}{\sqrt{2}}, \quad c = \frac{2a-1}{\sqrt{2}} \\ &u(-\infty) = 0,\ u(+\infty) = 0 \text{ for } m = \frac{1}{\sqrt{2}}, \quad c = \frac{1-2a}{\sqrt{2}} \end{aligned}\right\} \quad (41)$$

The solution given in (21) satisfies

$$\left.\begin{aligned} &u(-\infty)=a,\ u(+\infty)=0 \quad \text{for } m = \frac{a}{\sqrt{2}}, \quad c = \frac{2-a}{\sqrt{2}} \\ &u(-\infty)=0,\ u(+\infty)=a \quad \text{for } m = \frac{a}{\sqrt{2}}, \quad c = -\frac{(2-a)}{\sqrt{2}} \end{aligned}\right\} \quad (42)$$

and the solutions given in (23) satisfy

$$\left.\begin{aligned} &u(-\infty)=1,\ u(+\infty)=a \quad \text{for } m = -\frac{(1-a)}{\sqrt{2}}, \quad c = \frac{a+1}{\sqrt{2}} \\ &u(-\infty)=a,\ u(+\infty)=1 \quad \text{for } m = \frac{1-a}{\sqrt{2}}, \quad c = -\frac{(a+1)}{\sqrt{2}} \end{aligned}\right\} \quad (43)$$

We shall now discuss the stability of the solutions following the procedure adopted by Canosa [3]. Introducing the moving coordinate $z = x - ct$, eq. (1) becomes

$$u_t = u_{zz} + cu_z - au + (a + 1)u^2 - u^3 \quad (44)$$

Superimposing a small disturbance $v(z,t)$ on the travelling wave $u(z,t)$, that is

$$u(z,t) = u_o(z;c) + \varepsilon v(z,t) \quad (45)$$

where ε is a small parameter. We have the time dependent equation

$$v_t = v_{zz} + cv_z + (-a + 2(a+1)u_o - 3u_o^2)v \quad (46)$$

By letting $v = g(z)e^{-\lambda t}$ in (47)

$$\frac{d^2g}{dz^2} + c\frac{dg}{dz} + (-\lambda + 2(a + 1)u_o - eu_o^2 - a)g = 0 \quad (47)$$

If we restrict the perturbation to be non-zero only in finite domain then the eigenvalue problem for $g(z)$ has boundary condition $g(\pm L) = 0$.

By letting

$$g(z) = h(z)\, e^{-z/\sqrt{2}} \quad (48)$$

eq. (47) reduces to the eigenvalue problem

$$\frac{d^2h}{dz^2} + (\lambda - 3u_o^2 - 2(a + 1)u_o + a - \frac{\lambda^2}{4} + \frac{\lambda}{2})h = 0$$

$$h(\pm L) = 0$$

The expression $-3u_o^2 - 2(a + 1)u_o + a - \frac{\lambda^2}{4} + \frac{\lambda}{2}$ is positive for $\lambda = \frac{a - 2}{\sqrt{2}}$ and $\lambda = \frac{a + 1}{\sqrt{2}}$ and negative for $\lambda = \frac{1 - 2a}{\sqrt{2}}$ in the domain $-L \leq z \leq L$.

Now using the general theory of this kind of eigenvalue problem [5], we may conclude that all the eigenvalues of λ for the first two values of c, are positive and $v \longrightarrow 0$ as $t \longrightarrow \infty$. This means that the wave solutions (19) and (21) are stable to all small finite domain disturbances. For $c = \frac{1 - 2a}{\sqrt{2}}$, the eigenvalues of λ are negative and so $e^{-\lambda t} \rightarrow \infty$ as $t \rightarrow \infty$. Hence the wave solution (23) is not stable.

6 Conclusion

We wish to mention the following points. (i) The solutions given in (19), (21) and (23) are of biological interest. These solutions are Kink and Antikink type solutions. (ii) These solutions can be used to find new solutions (if any) from the equations (27), (28) and (29). One may attempt this.

Acknowledgements

The author is grateful for the financial assistance provided by the University Grants Commission, New Delhi, India and Department of Science and Technology, New Delhi, India, and to the Organizers of the Conference for the opportunity to attend to the International Conference held at Como, Italy.

References

[1] Ablowitz, M.J., Ramani, A. & Segur, H. (1980). *J. Math. Phys.* **7**, 15.

[2] Airault, H. & Kaliappan, P. (1984). Solutions Particulieres de l'equation u'' - u' = u(u-1)(u-a). *Comptes Rendus De l'Academie des Science*, Tome **299**, Serie 1, No. 8, 323.

[3] Canosa, J. (1973). *IBM J. Res. & Dev.* **17**, 307.

[4] McKean, H.P. (1975). Nagumo's Equation. *Comm. Pure & Appl. Math.* **28**, 323.

[5] Morse, P.M. & Feshback, H. (1953). *Methods of Theoretical Physics*. McGraw Hill: New York.

[6] Weiss, J., Tabor, M. & Carnevale, G. (1983). *J. Math. Phys.* **24**, 522.

Part 3
SPECIAL SOLUTIONS

M. Boiti, J.J.-P. Léon, L. Martina and F. Pempinelli

23. LOCALIZED SOLITONS IN THE PLANE[(0)]

1 Many of the nice properties of integrable nonlinear evolution equations in 1+1 dimensions have been successfully extended to 2+1 dimensions (see e.g. [1-9]). It is the case for the hamiltonian formalism and geometrical structure, the Bäcklund transformations and nonlinear superposition formula, the conservation laws and constants of the motion, the recursion operators and algebraic properties.

However one property is missing in the above list: the existence and scattering properties of two-dimensional solitons (exponentially localized spatial structures). Indeed, the only known localized objects in 2+1 dimensions are the so-called lump-solutions which decay algebraically at infinity and have no scattering properties.

In this paper (see also [13]) we give an explicit example of localized solitons interacting in the plane, which are solutions of the evolution equations related to the Zakharov-Shabat problem in 2+1 dimensions. They are derived in a standard way by using Bäcklund transformations and nonlinear superposition formulae.

The hyperbolic Davey-Stewartson (DSI) equation and a two-dimensional generalization of the modified KdV equation (2DMKdV) are considered in detail. In the first case solitons are pulses oscillating in a fixed place of the plane, while in the second case they are bell-shaped lumps moving with constant speed along the x-direction or along any direction in a right cone parallel to the y-axis.

Two-soliton solutions are also explicitly given.

For the 2DMKdV equation the most general case is considered, while for the DSI equation, for the sake of simplicity, only solitons pulsating at the same frequencies are considered. The initial position in the plane of the two solitons can be arbitarily chosen. When the soliton is sufficiently close to the other interacting soliton it generates at the expense of its size another shifted localized pulse or bump. The effect is proportional to the closeness of the two solitons, but differently along different directions in the plane. In the 2DMKdV case the only final result of this interacting pattern is to shift the position of the scattered solitons.

2 Let us consider the hyperbolic version of the 2x2 Zakharov-Shabat spectral problem in the plane

$$(\partial_x+\sigma_3\partial_y+Q)\psi=0, \qquad Q = \begin{pmatrix} 0 & q(x,y) \\ r(x,y) & 0 \end{pmatrix} \tag{1}$$

The complex spectral parameter k is introduced by seeking the eigenfunction $\psi(x,y,k)$ which satisfies the asymptotic behaviour

$$\psi(x,y,k)\ e^{-ik(\sigma_3 x-y)} \xrightarrow[k\to\infty]{} \mathbb{I} \tag{2}$$

and is bounded in the (x,y)-plane.

The 2x2 matrix spectral transform $R(k,l)$ of $Q(x,y)$ is defined as the measure of departure from analyticity of $\psi(x,y,k)$ as follows

$$\frac{\partial\psi}{\partial\bar{k}}(x,y,k) = \iint dl\wedge d\bar{l}\ \psi(x,y,l)R(k,l) \tag{3}$$

where the bar means complex conjugate. R in general results in being off-diagonal [10,11].

The time evolution of Q is fixed by introducing the 2x2 matrix dispersion relation $\Omega(k,t)$ and by requiring that the spectral transform $R(k,l,t)$ evolves as

$$R_t(k,l,t) = R(k,l,t)\ \Omega(k,t) - \Omega(l,t)\ R(k,l,t). \tag{4}$$

A very general choice for Ω and the related evolution equations has been given in [11] (see also the lecture by Léon and Pempinelli in this volume). Precisely one can choose

$$\Omega = \omega\ \sigma_3 \quad , \quad \omega = \omega_s + \omega_p \tag{5}$$

where $\omega_p = \omega_p(k,t)$ is a polynomial in k and $\omega_s = \omega_s(k,t)$ is any non-holomorphic function of k such that $\omega_s(k) \sim O(\frac{1}{k})$ as $k\to\infty$.

The simplest gauge transformation $B = B(Q',Q)$ which generates the Bäcklund transform Q' of Q (see [12]) is given by the equation

$$B(Q',Q) = \alpha\partial_y - \frac{1}{2}\sigma_3(Q'\alpha-\alpha Q) - \frac{1}{2}\sigma_3\alpha I(Q'^2-Q^2)+\beta \tag{6}$$

where

$$I = (\partial_x + \sigma_3\partial_y)^{-1} \tag{7}$$

with the boundary requirement that, in a given direction, lim IM = 0 for any diagonal (well behaved at infinity) matrix $M = M(x,y)$ and α, β are 2x2 constant diagonal matrices.

For $\alpha = \begin{pmatrix} 1 & 0 \\ 0 & 0 \end{pmatrix}$, $\beta = \begin{pmatrix} \lambda & 0 \\ 0 & 1 \end{pmatrix}$ we get the so-called elementary

Bäcklund gauge of first kind $B^{(I)}(Q',Q;\lambda)$, for $\alpha = \begin{pmatrix}00\\01\end{pmatrix}$, $\beta = \begin{pmatrix}10\\0\mu\end{pmatrix}$ the elementary Bäcklund gauge of second kind $B^{(II)}(Q',Q;\mu)$ and for $\alpha = \begin{pmatrix}10\\01\end{pmatrix}$, $\beta = \begin{pmatrix}\lambda 0\\0\mu\end{pmatrix}$ the Bäcklund gauge $B(Q',Q;\lambda,\mu)$ which generates the solitons and which we call, for brevity, the full Bäcklund gauge.

The most general one soliton solution Q has been derived in [12] and rederived in [11] for general ω by composing elementary Bäcklund gauges and by starting from the zero solution. The main property to be used is the commutativity of Bäcklund gauges which is given in this specific case by the formula

$$B(Q,0;\lambda,\mu) = B^{(II)}(Q,Q^{(I)};\mu)B^{(I)}(Q^{(I)},0;\lambda) = \\ =B^{(I)}(Q,Q^{(II)};\lambda)B^{(II)}(Q^{(II)},0;\mu). \tag{8}$$

It results that

$$Q^{(I)} = \begin{pmatrix}0 & 0\\ r^{(I)} & 0\end{pmatrix}, \quad Q^{(II)} = \begin{pmatrix}0 & q^{(II)}\\ 0 & 0\end{pmatrix}, \quad Q = \begin{pmatrix}0 & q\\ r & 0\end{pmatrix} \tag{9}$$

where

$$r^{(I)}=\rho(x+y,t)\ e^{-\lambda(x-y)}, \quad q^{(II)}=\eta(x-y,t)e^{\mu(x+y)} \tag{10}$$

$$q = \frac{q_y^{(II)}+ (\lambda-\mu)q^{(II)}}{1+ \frac{1}{4}\ r^{(I)}q^{(II)}}, \quad r = \frac{r_y^{(I)}- (\lambda-\mu)r^{(I)}}{1+ \frac{1}{4}\ r^{(I)}q^{(II)}} \tag{11}$$

with

$$\rho(x+y,t) = \int\int dl\wedge d\bar{l}\ e^{-il(x+y)}\ \tilde{\rho}\ (l,t)$$
$$\eta(x-y,t) = \int\int dl\wedge d\bar{l}\ e^{il(x-y)}\ \tilde{\eta}\ (l,t). \tag{12}$$

The time evolution is given by the equations

$$\tilde{\rho}(l,t) = \tilde{\rho}(l,0)\exp\{\int_0^t [\omega(l,t)+\omega(-i\lambda,t)]dt\}$$

$$\tilde{\eta}(l,t) = \tilde{\eta}(l,0)\exp\{-\int_0^t [\omega(I,t)+\omega(-i\mu,t)]dt\} \tag{13}$$

where $\tilde{\rho}(l,0)$ and $\tilde{\eta}(l,0)$ are arbitrary functions. The explicit form of $B(Q,0;\lambda,\mu)$

$$B(Q,0;\lambda,\mu) = \mathbb{I}\partial_y+B^{(1)}(Q,0;\lambda,\mu),$$

$$B^{(1)}(Q,0;\lambda,\mu) = \begin{pmatrix} \lambda - \frac{1}{4} r^{(I)} q & -\frac{1}{2} q \\ \frac{1}{2} r & \mu - \frac{1}{4} q^{(II)} r \end{pmatrix} \qquad (14)$$

will also be useful in the following. It provides in particular the eigenfunction solution of (1) [11]:

$$\psi(k,x,t) = \left\{ \mathbb{I} - \frac{i}{4} \begin{pmatrix} \frac{r^{(I)}q}{k+i\lambda} & \frac{2q}{k+i\mu} \\ \frac{-2r}{k+i\lambda} & \frac{q^{(II)}r}{k+i\mu} \end{pmatrix} \right\} \exp\,[ik(\sigma_3 x - y)] \qquad (15)$$

In [11] the spectral transform of Q is also computed and it results in

$$R(k,l) = i\pi \begin{pmatrix} 0 & \delta(k+i\mu)\tilde{\eta}(l,t)(l+i\lambda) \\ -\delta(k+i\lambda))\tilde{\rho}\ (l,t)\ (l+i\mu) & 0 \end{pmatrix} \qquad (16)$$

and, consequently, the solution Q is indeed related to discrete eigenvalues of the spectral problem (1).

In the following we choose

$$\tilde{\rho}(l,0) = \rho[\delta(l-i\mu)+\delta(l+i\mu)]$$

$$\tilde{\eta}(l,0) = \eta[\delta(l-i\lambda) + \delta(l+i\lambda)] \qquad (17)$$

with η and ρ arbitrary real constants.

Note that in this case both R(k,l) and $\psi(l)$ are distribution in l and one has to take into account that only their product is unambiguously defined via the equation (3).

Note also that the eigenfunction (15) does not represent a bound state of the spectral problem (1) for we may check that the residues of ψ in $k = -i\lambda$ or $k = -i\mu$ are not $L^2(\mathbb{R}^2)$ (they are asymptotically non-zero constants on the directions $x = \pm\, y$).

Decomposing ω as

$$\frac{\omega(k)}{ik} = c(k) + v(k) \qquad (18)$$

where c(k) (resp. v(k)) is odd (resp. even), we get for the one-soliton solution

$$q = \frac{\lambda\eta}{\Delta} e^{\alpha t} \qquad r = \frac{\mu\rho}{\Delta} e^{-\alpha t} \qquad (19)$$

$$\Delta = \gamma \cosh \xi_1 + \sqrt{\gamma(1+\gamma)} \cosh (\xi_2+\delta) \qquad (20)$$

$$\alpha = \lambda\, c\ (i\lambda) + \mu c\ (i\mu) \qquad (21)$$

$$\gamma = \frac{1}{4}\,\eta\rho \qquad \delta = \frac{1}{2}\,\ln\left(\frac{1+\gamma}{\gamma}\right) \tag{22}$$

$$\xi_1 = (\lambda+\mu)\,x - (\lambda-\mu)y-(\lambda v(i\lambda) + \mu v(i\mu))t$$

$$\xi_2 = (\lambda-\mu)\,x - (\lambda+\mu)y - (\lambda v(i\lambda)-\mu v(i\mu))t. \tag{23}$$

For v(ik) real, the above formula defines (up to the $e^{\pm\alpha t}$ in the numerator) a 2-dimensional bell-shaped solution, exponentially decreasing in all directions of the (x,y)-plane, moving without deformation along the direction defined by

$$[v(i\mu) + v(i\lambda)]y = [v(i\mu) - v(i\lambda)]x \tag{24}$$

with the velocity

$$v_s = \left[\frac{1}{2}\,(v^2(i\mu) + v^2(i\lambda))\right]^{1/2} \tag{25}$$

provided $\lambda\mu \neq 0$. If the factor $e^{\pm\alpha t}$, with c(ik) purely imaginary, is present the bell-shaped bump pulsates periodically in time. The initial position of the soliton can be moved arbitrarily by the translation

$$x \to x-x_0\,, \qquad y \to y-y_0\,, \qquad x_0,\ y_0 \in R. \tag{26}$$

It is worth noting that the eigenfunction

$$\nu = \psi\,\exp\,[-ik(\sigma_3 x-y)] \tag{27}$$

corresponding to Q in (19) is the solution of the following integral equation

$$\nu = \mathbb{I} + \frac{i}{2}\begin{pmatrix} -\dfrac{\alpha_1(x-y)}{k+i\lambda} & 0 \\ 0 & -\dfrac{\alpha_2(x+y)}{k+i\mu} \end{pmatrix} -$$

$$- \int_{\infty}^{x} dx'\exp[ik(x-x')\sigma_3] \begin{pmatrix} q\,\nu_{21} & q\,\nu_{22} \\ 0 & r\,\nu_{12} \end{pmatrix} (x',y-(x-x')\sigma_3)\cdot \tag{28}$$

$$\cdot\exp[-ik(x-x')\sigma_3] + \int_{x}^{\infty} dx'\exp[ik(x-x')\sigma_3] \begin{pmatrix} 0 & 0 \\ r\,\nu_{11} & 0 \end{pmatrix} (x',y-(x-x')\sigma_3)\cdot$$

$$\cdot\exp[-ik(x-x')\sigma_3]$$

for Im k > 0 and of an analogous one for Im k < 0. This integral equation differs from the integral equation given in [10] in having a different homogeneous term, which contains the functions

$$\alpha_1(x-y) = \frac{4\lambda\gamma}{\gamma+(\vartheta(\mu)+\gamma)\exp(\xi_1+\xi_2)} \tag{29}$$

$$\alpha_2(x+y) = \frac{4\mu\gamma}{\gamma+(\vartheta(-\lambda)+\gamma)\exp(\xi_2-\xi_1)} \,. \tag{30}$$

The first nonlinear equation in the hierarchy is known as the Davey-Stewartson eq. [10], it is obtained for $\omega(k) = ik^2$ and reads

$$iQ_t = -\frac{1}{2}\sigma_3(Q_{xx} + Q_{yy}) + \sigma_3 Q^3 + [Q,A], \tag{31}$$

$$(\partial_x+\sigma_3\partial_y)A = (Q^2)_y. \tag{32}$$

The auxiliary function A can be uniquely determined by using the auxiliary spectral equation [14]

$$i\psi_t + V\psi = ik^2\psi\sigma_3 \tag{33}$$

$$V = \sigma_3\partial^2_y + Q\partial_y - \frac{1}{2}\sigma_3 Q_x + \frac{1}{2}Q_y - \frac{1}{2}\sigma_3 Q^2 + A, \tag{34}$$

which fixes the time evolution of Q. One has to insert into (33) and (1) the asymptotic expansion of ν for large k in the form

$$\nu(x,y,k) = \mathbb{I} + \nu_1(x,y)/k + O\,(1/k^2). \tag{35}$$

One obtains for Q given in (19)

$$A = \frac{1}{2}(\partial_x - \sigma_3\partial_y)\partial_y \ln Q^2, \tag{36}$$

which corresponds with the following choice of the inverse operator of $(\partial_x + \sigma_3\partial_y)$ in (32)

$$(\partial_x + \partial_y)^{-1}f\,(x,y) = \int_\infty^x dx'f(x',y-x+x')+\alpha_1(x-y) \tag{37}$$

$$(\partial_x-\partial_y)^{-1}f(x,y) = \int_\infty^x dx'f(x',y+x-x')+\alpha_2(x+y) \tag{38}$$

The second equation in the hierarchy is the two-dimensional extension of the modified-Korteweg-deVries equation (2DMKdV) obtained for

$$\omega(k) = iak + ibk^3, \tag{39}$$

for which c(k) is zero in (18). This equation can be written

$$\frac{1}{b}(Q_t - aQ_x) = \frac{1}{4} Q_{xxx} + \frac{3}{4} Q_{xyy} - \frac{3}{2} Q^2 Q_x -$$

$$- \frac{3}{2} \sigma_3 [Q,A]_x - \frac{3}{4} \{Q_y, A\} + \frac{3}{4} [Q,B] \tag{40}$$

$$(\partial_x + \sigma_3 \partial_y) A = Q_y^2, \quad (\partial_x + \sigma_3 \partial_y) B = -\sigma_3 [Q_{xy}, Q] \tag{41}$$

The two generic simple solutions are those moving in the x- and y-directions. For $\mu = \lambda$, and $\eta\rho > 0$ we obtain a soliton moving with velocity $v=v(i\mu)$ along the x-direction denoted by

$$Q = Q(x-vt, y; \mu, x_0, y_0) \tag{42}$$

For $v(i\mu) = -v(i\lambda)$ and $\eta\rho > 0$ the soliton moves along the y-direction with velocity $v=v(i\mu)$, and is denoted by

$$Q = Q(x, y-vt; \lambda, \mu, x_0, y_0). \tag{43}$$

For $ab < 0$ and any v (or $ab > 0$ and $v^2 > a^2$) there are no real values μ and λ such that $v(i\mu) = -v(i\lambda)$ and solitons can move only along the x-direction.

For $ab > 0$ and $v^2 < a^2$ one gets solitons moving indifferently forward or backward both along the x- and y-directions, according to the choice of μ and λ.

In order to get the auxiliary functions A and B one can follow the same procedure used for the Davey-Stewartson equation by inserting the asymptotic expansion of ν for large k computed up to the second order

$$\nu(x,y,k) = \mathbb{I} + \nu_1(x,y)/k + \nu_2(x,y)/k^2 + O\left(\frac{1}{k^3}\right) \tag{44}$$

into the principal spectral problem (1) and into the following auxiliary spectral problem that fixes the time evolution of Q

$$\psi_t + V\psi = (iak + ibk^3)\, \psi\, \sigma_3 \tag{45}$$

with

$$V = \sigma_3 \partial_y^3 + Q^2 \partial_y^2 + V_1 \partial_y + V_0 \tag{46}$$

$$V_1 = -\frac{1}{2} \sigma_3 \partial_x + Q_y - \frac{1}{2} \sigma_3 Q^2 + \frac{3}{2} A$$

$$V_2 = \frac{1}{4} Q_{xx} + Q_{yy} - \frac{1}{4} \sigma_3 Q_{xy} - \frac{1}{2} Q^3 - \frac{3}{4} \sigma_3 [Q,A] -$$

$$- \frac{1}{4} [Q, Q_x] - \frac{1}{2} \sigma_3 Q_y Q + \frac{3}{4} A_y + \frac{3}{4} \sigma_3 [Q_y, Q] + \frac{3}{4} B.$$

After some algebra one gets for Q given in (19)

$$A = \frac{1}{2}\,(\partial_x - \sigma_3\partial_y)\,\partial_y \ln Q^2 \tag{47}$$

$$B = 0. \tag{48}$$

3 In order to get the solution Q_3 describing two interacting solitons Q_1 and Q_2 one can "superimpose" them by using the commutativity of the full Bäcklund gauges

$$\begin{aligned} &B(Q_3,Q_1;\lambda_2,\mu_2)\; B(Q_1,Q_0;\lambda_1,\mu_1) = \\ &= B(Q_3,Q_2;\lambda_1,\mu_1)\; B(Q_2,Q_0;\lambda_2,\mu_2) \end{aligned} \tag{49}$$

for $Q_0 = 0$.

The general full Bäcklund gauge is given in (6) for $\alpha = \mathbb{I}$ and $\beta = \begin{pmatrix}\lambda & 0\\ 0 & \mu\end{pmatrix}$ and in (14) for the special case $Q_0 = 0$.

From eq. (49) by equating the coefficients of the powers of ∂_y, after some algebra, one gets the following nonlinear superposition formula

$$q_3 = -\,\frac{4(\lambda_1^2-\lambda_2^2)}{D}\,(q_1 - q_2) \tag{50}$$

$$r_3 = -\,\frac{4(\mu_1^2-\mu_2^2)}{D}\,(r_1 - r_2) \tag{51}$$

where

$$\begin{aligned} D = (q_1-q_2)(r_1-r_2) - &\left[(\partial_x+\partial_y)\ln\left(\frac{q_2}{q_1}\right)\right]\cdot \\ \cdot&\left[(\partial_x-\partial_y)\ln\left(\frac{r_2}{r_1}\right)\right]. \end{aligned} \tag{52}$$

We are especially interested in the case $\alpha \neq 0$, $v = 0$, which corresponds to the DSI equation, and in the case $\alpha = 0$, $v \neq 0$, which corresponds to the 2DMKdV equation. It is easy to verify that, when $\lambda_1^2+\mu_1^2 = \lambda_2^2+\mu_2^2$, the solution of the 2DMKdV equation at time t_0, up to an overall factor $e^{\pm\alpha t}$, is equal to the solution of the DSI equation translated to a convenient point (x_0, y_0) of the plane. This means that one can get the two-soliton solution of the DSI equation by freezing the 2DMKdV solution at some time and by allowing it to pulsate periodically in time. Therefore, we consider in the following the time evolution of the 2DMKdV solution.

The two interacting solitons described by Q_3 can be different from the starting fields Q_1 and Q_2, not only because their relative position in the plane is shifted by the interaction, but also because their amplitude can be renormalized by the interaction. Therefore it is convenient to introduce the notation $\hat{Q}_1$ and $\hat{Q}_2$ for the two interacting solitons and to call them dressed solitons, in contrast with the bare solitons Q_1 and Q_2.

The dressed solitons $\hat{Q}_1$ and $\hat{Q}_2$ are determined by computing the asymptotic behaviour of Q_3 at $t=-\infty$ and the effect of the interaction by looking at $t=+\infty$.

Let us first consider two solitons moving along the x-direction with parameters μ_1 and μ_2.

If $\mu_1^2 - \mu_2^2 > 0$ in order to get **regular localized** dressed solitons $\hat{Q}_1$ and $\hat{Q}_2$ we choose as bare solitons

$$
\begin{aligned}
Q_1 &= Q(x-v_1t, y;\ \mu_1,\ x_{01} + i\pi/2\mu_1, y_{01}) \\
Q_2 &= Q(x-v_2t,\ y;\ \mu_2,\ x_{02},\ y_{02}).
\end{aligned}
\tag{53}
$$

The bare soliton Q_1 is chosen to be singular.

If $\mu_1^2-\mu_2^2 < 0$ we choose

$$
\begin{aligned}
Q_1 &= Q(x-v_1t, y;\ \mu_1,\ x_{01},\ y_{01}) \\
Q_2 &= Q(x-vt,\ y;\ \mu_2,\ x_{02}+i\pi/2\mu_2,\ y_{02}),
\end{aligned}
\tag{54}
$$

and the bare soliton Q_2 is singular.

In both cases the dressed solitons $\hat{Q}_1$ and $\hat{Q}_2$ behave as follows at $t = -\infty$

$$
\begin{aligned}
\hat{Q}_1 &\simeq -Q(x-v_1t, y;\ \mu_1, x_{01}-\delta_1, y_{01}) \\
\hat{Q}_2 &\simeq -Q(x-v_2t, y; \mu_2, x_{02}-\delta_2, y_{02})
\end{aligned}
\tag{55}
$$

and at $t = +\infty$

$$
\begin{aligned}
\hat{Q}_1 &\simeq -Q(x-v_1t,\ y;\ \mu_1,\ x_{01} + \delta_1,\ y_{01}) \\
\hat{Q}_2 &\simeq -Q(x-v_2t,\ y;\ \mu_2,\ x_{02} + \delta_2,\ y_{02}).
\end{aligned}
\tag{56}
$$

The only effect of their interaction is to shift their position along the x-direction.

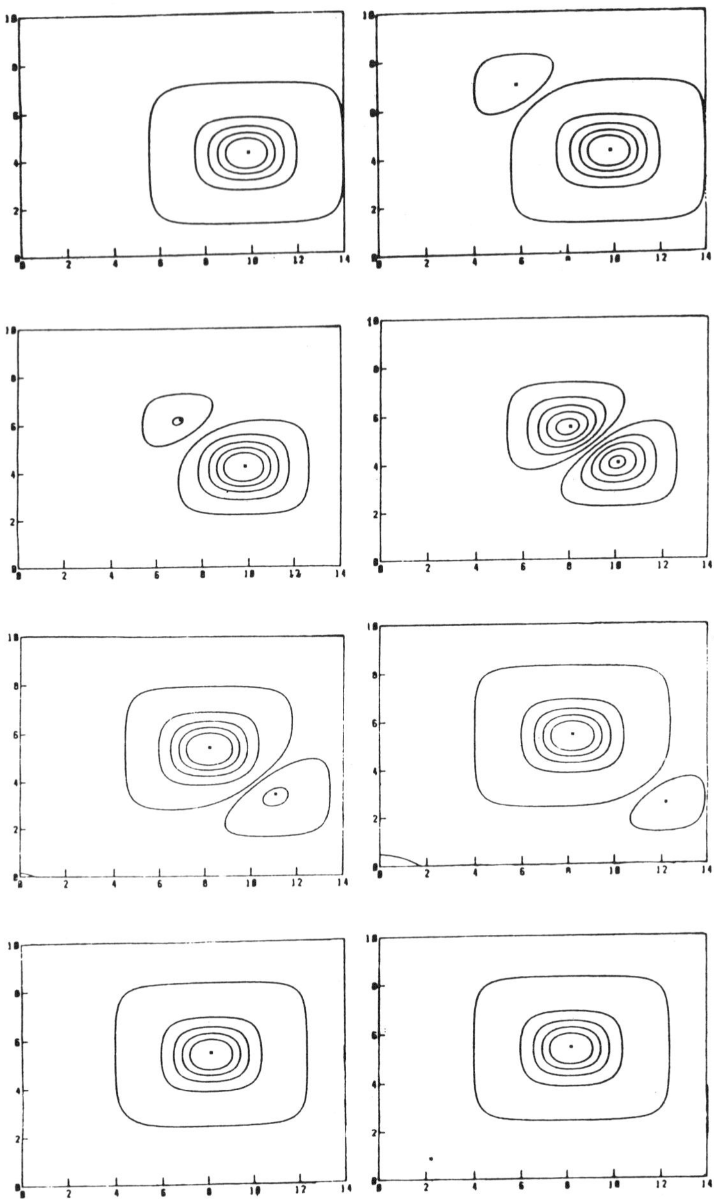

Figure 1 a and b. These two sets of plots show the patterns of the level contours at different times for the modulus of a x-soliton in its own reference frame, during the interaction with another, not visible, x-soliton. The parameters of the two solitons are:

$\mu_1 = \lambda_1 = .8 \quad v_1 = 0.0 \quad \rho_1 = \eta_1 = 4.0 \quad (x_{10}, y_{10}) = (7.0, 5.0)$

$\mu_2 = \lambda_2 = 1.0 \quad v_2 = 1.0 \quad \rho_2 = \eta_2 = 1.0 \quad (x_{20}, y_{20}) = (7.0, -3.0)$

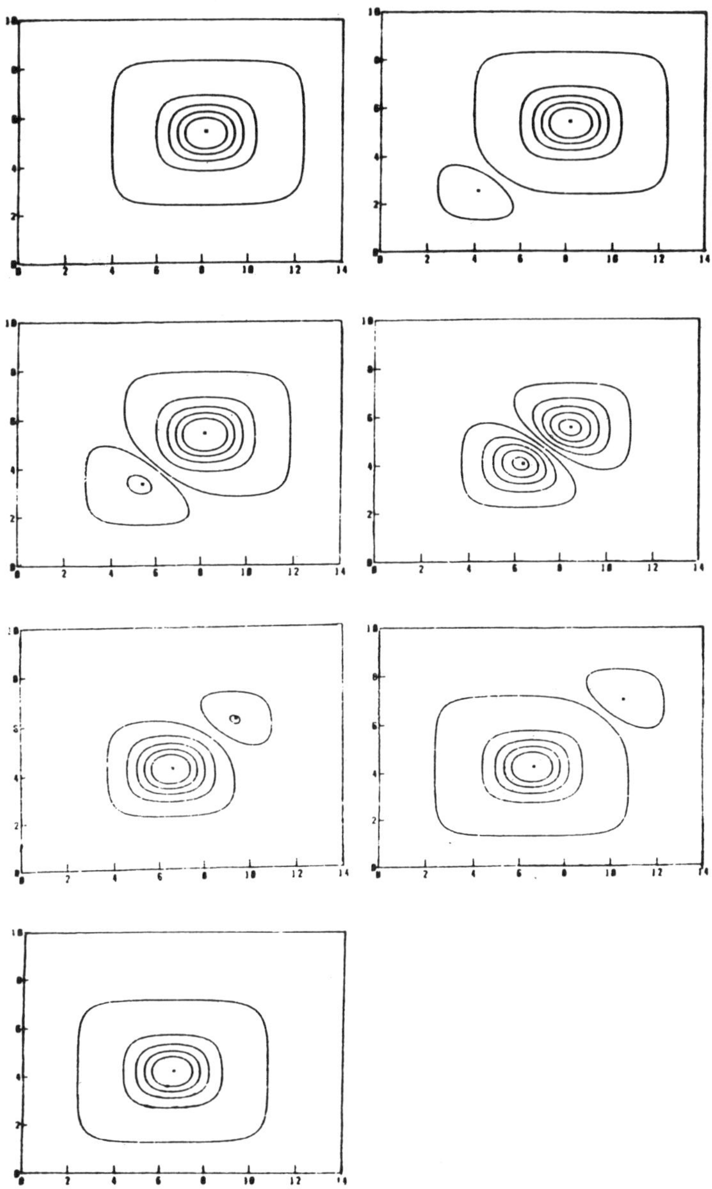

The former set of parameters corresponds to the soliton represented in the plots.

The time runs from -14 up to 14 ($\Delta t=2$) in our unity of time and the corresponding plots go from the left to the right and from the top to the bottom.

The lowest level plotted contour is 0.005 with steps of 0.065 in all plots, except ones at $t=\pm10.0$, ±8.0 where the correspondig values are 0.030 and 0.060 respectively, and at

$t=\pm 6$ where we have 0.010 and 0.060 respectively.
The star indicates the position of the relative maxima.
At $t=-6.0$ and -4.0 in the right bottom the second soliton appears.
Composing the plots at $t=\pm 14$ the reader gets an estimate of the x-shift undergone by the soliton.

The total real shift $2\delta_1$ in the position of the solitons $\hat{Q}_1$ is given by

$$\delta_1 = -\frac{1}{2\mu_1} \ln \left|\frac{\mu_1-\mu_2}{\mu_1+\mu_2}\right| \tag{57}$$

for $\mu_2(v_1-v_2) > 0$ and by

$$\delta_1 = \frac{1}{2\mu_1} \ln \left|\frac{\mu_1-\mu_2}{\mu_1+\mu_2}\right| \tag{58}$$

for $\mu_1(v_1-v_2) < 0$, while the total real shift $2\delta_2$ of $\hat{Q}_2$ is given by

$$\delta_2 = -\left|\frac{\mu_2}{\mu_1}\right| \delta_1 . \tag{59}$$

Let us now consider one soliton with parameter μ_1 moving along the x-direction and one soliton with parameters λ_2, μ_2 moving along y-direction.
We choose, for the sake of definiteness,

$$\begin{gathered} \mu_1^2 - \mu_2^2 > 0 , \quad \mu_1^2 - \lambda_2^2 > 0 \\ \mu_1(v_1+v_2) > \mu_1(v_1 -v_2) > 0 \\ \lambda_2(v_1+v_2) > \mu_2(v_1-v_2) > 0. \end{gathered} \tag{60}$$

Other cases can be easily examined in an analogous way.
The bare solitons have to be chosen as

$$\begin{aligned} Q_1 &= Q(x-v_1t,y;\ \mu_1,x_{01},y_{01}+i\pi/2\mu_1) \\ Q_2 &= Q(x,y-v_2t;\ \mu_2,\lambda_2,x_{02},y_{02}) \end{aligned} \tag{61}$$

and Q_1 is singular.
The dressed solitons $\hat{Q}_1$ and $\hat{Q}_2$ behave at $t = -\infty$ as

$$\hat{Q}_1 \simeq -A\, Q(x-v_1 t, y;\ \mu_1, x_{01}-\delta_1, y_{01}-\gamma_1)$$
$$\hat{Q}_2 \simeq -A\, Q(x, y-v_2 t;\ \mu_2, \lambda_2, x_{02}-\delta_2, y_{02}-\gamma_2) \tag{62}$$

and at t = +∞ as

$$\hat{Q}_1 \simeq -A\, Q(x-v_1 t, y;\ \mu_1, x_{01}+\delta_1, y_{01}+\gamma_1)$$
$$\hat{Q}_2 \simeq -A\, Q(x, y-v_2 t;\ \mu_2, \lambda_2, x_{02}+\delta_2, y_{02}+\gamma_2) \tag{63}$$

where the matrix A is given by

$$A = \begin{pmatrix} \sigma & 0 \\ 0 & 1/\sigma \end{pmatrix}, \quad \sigma = \left(\frac{\mu_1^2-\lambda_2^2}{\mu_1^2-\mu_2^2} \right)^{1/2} \tag{64}$$

and renormalizes the amplitude of both solitons.

$$\delta_1 + \gamma_1 = -\frac{1}{2\mu_1} \ln \left(\frac{\mu_1-\mu_2}{\mu_1+\mu_2} \right)$$
$$\delta_1 - \gamma_1 = -\frac{1}{2\mu_1} \ln \left(\frac{\mu_1-\lambda_2}{\mu_1+\lambda_2} \right) \tag{65}$$

are the shifts in the position of the first soliton $\hat{Q}_1$ and

$$\delta_2 + \gamma_2 = \frac{1}{2\mu_2} \ln \left(\frac{\mu_1-\mu_2}{\mu_1+\mu_2} \right)$$
$$\delta_2 - \gamma_2 = \frac{1}{2\lambda_2} \ln \left(\frac{\mu_1-\lambda_2}{\mu_1+\lambda_2} \right) \tag{66}$$

are the shifts in the position of the second soliton $\hat{Q}_2$.

Therefore the interaction of the two solitons $\hat{Q}_1$ and $\hat{Q}_2$ is an effective bidimensional scattering because the effect of the interaction is a shift in the position of the solitons **both** in the x- and y-directions. Moreover, it is worthwhile noticing that the total efects of the interaction between the two scattered solitons is independent on the initial position of the solitons that can be chosen in a completely arbitrary way. Consequently the interaction between solitons can be interpreted as representing an infinite range interaction, resulting from the nonlocal character common to all integrable equations in 2+1 dimensions.

Finally, let us note that if we make in the evolution equation (40) the change of variables

$$t' = bt \ , \quad x' = x+at \tag{67}$$

we get the evolution equation with dispersion relation $\omega(k)=ik^3$. Consequently, one gets solitons moving along the x'-axis and in any direction in a right cone with axis parallel to the y'-axis.

References

[1] Manakov, S.V. (1981). *Physica* **3D**, 420.
[2] Ablowitz, M.J., Bar Yaakov, D. & Fokas, A.S. (1983) *Studies Appl. Math.* **69**, 135.
[3] Fokas, A.S. & Ablowitz, M.J. (1983). *Studies Appl. Math* **69**, 211.
[4] Beals, R. & Coifman, R.R. (1985). *Proc. Symp. in Pure Math.* **43**, 45.
[5] Boiti, M., Léon, J.J.-P., Manna, M. & Pempinelli, F. (1986). *Inverse Problems* **2**, 271; (1987). **3**, 25.
[6] Boiti, M., Léon, J.J.-P. & Pempinelli, F. (1987). *Inverse Problems* **3**, 37 and 371.
[7] Santini, P.M. & Fokas, A.S. (1988). Recursion operators and bi-Hamiltonian structures in multidimensions I. *Comm. Math. Phys.* **115**, 375.
[8] Fokas, A.S. & Santini, P.M. (1988). Recursion operators and bi-Hamiltonian structures in multidimensions II. *Comm. Math. Phys.* **116**, 449.
[9] Boiti, M., Léon, J.J.-P. & Pempinelli, F. (1988). *Studies Appl. Math.* **78**, 1.
[10] Fokas, A.S. (1983). *Phys. Rev. Lett.* **51**, 3; Fokas, A.S. & Ablowitz, M.J. (1984). *J. Math. Phys.* **25**, 2494.
[11] Boiti, M., Léon, J.J.-P., Martina, L. & Pempinelli, F. Integrable nonlinear evolutions in 2+1 dimensions with non-analytic dispersion relations.
[12] Boiti, M., Konopelchenko, B.G. & Pempinelli, F. (1985). *Inverse Problems* **1**, 33.
[13] Boiti, M., Léon, J.J.-P., Martina, L. & Pempinelli, F. *Phys. Lett. A,* to be published.
[14] Schultz, C.L. Personal communication.

P.C. Sabatier

24. MODULATION OF TRAPPED WAVES GIVING APPROXIMATE TWO-DIMENSIONAL SOLITONS

In a previous paper, F. Calogero and the author studied the nonlinear modulation of dispersive trapped water-waves in a channel or along a beach (edge waves). They showed that the signal envelope (which is studied with convenient time scale and length scale) obeys the Nonlinear Schrödinger Equation. The result is extended here to water waves trapped by the bottom geometry of a basin which is infinitely extended into all horizontal directions. One can obtain along the guided path, envelope solitons of the Nonlinear Schrödinger Equation for a signal whose amplitude exponentially decreases in transverse directions. These bidimensional solitons are thus shown as asymptotic features for a large class of problems. They are in most cases unstable features, because of side-band instabilities. Nevertheless, the result suggests that the exact bidimensional solitons recently derived by Boiti et al. for special Equations (see chapter 23) are not an exceptional physical feature of two-dimensional evolutions.

This lecture, which is identical to that presented in the Workshop "Integrable Systems and Applications" M.Balabane, P.Lochak et al. Ed. (Springer Verlag, 1989) is not reproduced in the present volume.

V.K. Mel'nikov

25. CAPTURE AND CONFINEMENT OF SOLITONS IN NONLINEAR INTEGRABLE SYSTEMS

The goal of the present paper is nonlinear integrable systems having a rather unusual dynamics. In particular, solutions of these systems can describe capture and confinement of solitons. This means that the systems considered here have solutions describing solitons that come from infinity, then are captured into different oscillatory regimes (periodical, conditionally periodical, almost periodical, etc.) and remain in them during all the subsequent time intervals. This phenomenon turned out to be widespread among nonlinear integrable systems. However, in this paper we shall demonstrate this phenomenon by using only one rather interesting family of nonliner integrable systems as an example.

Thus, we use operators L and A of the form

$$L = \partial^2 + u, \quad A = \partial^{K_o+2} + \sum_{K=0}^{K_o} a_K \partial^K, \quad \partial = \frac{\partial}{\partial x}, \quad K_o \geq 0, \tag{1}$$

such that the operator

$$\Delta = [A,L] = A \cdot L - L \cdot A \tag{2}$$

has the zero order, i.e. it is the operator of multiplication by the function. As is known, this requirement determines the coefficients $a_o, \ldots, a_{K_o}$ up to the constants. We choose these constants from the requirement

$$a_o = \ldots = a_{K_o} \equiv 0 \text{ at } u \equiv 0. \tag{3}$$

Then, it is known that the coefficients $a_o, \ldots, a_{K_o}$ depend polynomially on u and its derivatives with respect to the space variable x of the corresponding order. Hence, the quantity Δ determined by (2) is the polynomial of u and its derivative with respect to x up to the (K_o+2)-th order.

Consider now the system of equations

$$f_o'' + (u-\lambda)\, f_o = 0, \quad f_m' = \psi_m f_o, \quad m = 1,\dots,N, \tag{4}$$

where ψ_m are certain (yet unknown) functions, m=1,...,N, and the primes denote the derivation with respect to x. Using the solutions f_o, $f_1,\dots,f_N$ of the system (4) we determine the quantities

$$g_o = \frac{\partial f_o}{\partial t} + cAf_o + \kappa \sum_{m=1}^{N} \varphi_m f_m, \tag{5}$$

$$g_m = \psi_m' f_o - \psi_m f_o' + (\lambda-\lambda_m) f_m, \; m = 1,\dots,N,$$

where c and κ are the constants and φ_m are yet unknown functions, m = 1,...,N. Let us find out what requirements are to be imposed on the functions u, $\varphi_1,\dots,\varphi_N$, $\psi_1,\dots,\psi_N$ so that the quantities g_o, $g_1,\dots,g_N$ determined by (5) should satisfy the conditions

$$(L-\lambda)g_o + \kappa \sum_{m=1}^{N} \varphi_m g_m = 0, \quad \frac{\partial g_m}{\partial x} \equiv 0, \; m=1,\dots,N, \tag{6}$$

under an arbitrary choice of the solutions f_o, $f_1,\dots,f_N$ of the system (4). By making simple calculations one can verify the validity, according to (1) and (4), of the equations

$$(L-\lambda)g_o + \kappa \sum_{m=1}^{N} \varphi_m g_m = \kappa \sum_{m=1}^{N} f_m (L-\lambda_m)\varphi_m -$$

$$- \{\frac{\partial L}{\partial t} + c[\,,L] - 2\kappa \sum_{m=1}^{N} \frac{\partial}{\partial x}(\varphi_m \psi_m)\}\, f_o,$$

$$\frac{\partial g_m}{\partial x} = f_o\, (L-\lambda_m)\psi_m, \quad m = 1,\dots,N.$$

Hence it follows that equations (6) must be valid under the following conditions:

$$\frac{\partial L}{\partial t} + c[A,L] = 2\kappa \sum_{m=1}^{N} \frac{\partial}{\partial x}(\varphi_m \psi_m),$$

$$(L-\lambda_m)\varphi_m = (L-\lambda_m)\psi_m = 0, \quad m = 1,\dots,N. \tag{7}$$

In the case when c and κ take real values, the system (7) has an invariant manifold $u = \bar{u}, \psi_m = \varepsilon_m \bar{\varphi}_m$, where $\varepsilon_m^2 = 1$ and the bar means complex conjugation. Motion on this manifold is determined by the system of equations

$$\frac{\partial L}{\partial t} + c[A,L] = 2\kappa \sum_{m=1}^{N} \varepsilon_m \frac{\partial}{\partial x} |\varphi_m|^2, \tag{8}$$

$$(L-\lambda_m)\varphi_m = 0, \quad m = 1,\dots,N.$$

The system (8) admits investigation by the inverse scattering method for the operator L of the form (1). Now let $u_o = u_o(x)$ be the real function of x with K_o+2 continuous derivatives satisfying the condition

$$\int_{-\infty}^{\infty} \{|xu_o(x)| + \sum_{K=0}^{K_o+2} |u_o^{(K)}(x)|\}dx < \infty. \tag{9}$$

Assume further that the Schrödinger equation

$$(L-\lambda)\varphi = 0 \tag{10}$$

has exactly N points of the discrete spectrum $\lambda = \lambda_m$, $m = 1,\dots,N$. Let finally $A_m(t)$ be arbitrary continuous functions of t, $m = 1,\dots,N$. Then, by using the inverse scattering method for the operator L one can find the solution $u = u(x,t)$, $\varphi_m = \varphi_m(x,t)$, $m = 1,\dots,N$, of the system (8) such that

$$u(x,o) = u_o(x),$$

$$\int_{-\infty}^{\infty} |\varphi_m(x,t)|^2\, dx = |A_m(t)|^2, \quad m = 1,\dots,N. \tag{11}$$

Indeed, let $\lambda = -S^2$, where $S\in(-\infty,\infty)$. We consider two solutions f_o^- and f_o^+, of eq. (10) meeting the requirements

$$f_o^- \sim \exp(-iSx), \quad x \to -\infty,$$

$$f_o^+ \sim \exp(iSx), \quad x \to \infty. \tag{12}$$

Then, allowing for (4) we assume at $m = 1,\dots,N$

$$f_m^- = \varepsilon_m \int_{-\infty}^{x} \bar{\varphi}_m(z) f_o^-(z,S)dz, \tag{13}$$

$$f_m^+ = -\varepsilon_m \int_x^\infty \bar{\varphi}_m(z)\ f_o^+(z,S)dz.$$

Substituting the functions f_o^-, f_o^+, f_1^-, $f_1^+, \ldots, f_N^-$, f_N^+ determined by (12) and (13) into equation (5), we get the quantities g_o^-, g_o^+, g_1^-, $g_1^+, \ldots, g_N^-$, g_N^+. Using (6) one can easily get that at $m = 1, \ldots, N$ the equation $g_m^- = g_m^+ \equiv 0$ is valid. Hence, by virtue of (3) it follows that

$$g_o^- = c(-iS)^{K_o+2}\ f_o^-,\ g_o^+ = c(iS)^{K_o+2}\ f_o^+. \tag{14}$$

At $S \neq 0$ the pair of the functions $f_o^-(x,S)$ and $f_o^-(x,-S)$ is linear independent. Consequently, the following equation holds:

$$f_o^+(x,S) = a(S)\ f_o^-(x,-S) + \mathcal{B}(S) f_o^-(x,S). \tag{15}$$

Now, in the equation

$$\frac{\partial f_o^+}{\partial t} + cAf_o^+ + \kappa \sum_{m=1}^{N} \varphi_m f_m^+ = c(iS)^{K_o+2}\ f_o^+ \tag{16}$$

following from (5) and (14) we change f_o^+ by the expression in the right-hand side of equation (15) and pass to the limit as $x \to -\infty$. After simple calculations we get

$$\frac{\partial a}{\partial t} \equiv 0,\ \frac{\partial \mathcal{B}}{\partial t} + c[(-iS)^{K_o+2} - (iS)^{K_o+2}]\mathcal{B} = 0. \tag{17}$$

Thus, evolution of the coefficients $a(S)$ and $\mathcal{B}(S)$ in our case is just the same as in the case of the equations of the KdV hierarchy [2].

Equation (15) results in the well known representation for the function $a(S)$, i.e.

$$a(S) = \frac{1}{2iS} \{\frac{\partial f_o^+}{\partial x} f_o^- - f_o^+ \frac{\partial f_o^-}{\partial x}\}.$$

Since the solutions f_o^- and f_o^+ admit analytic continuation in S into the upper half-plane, the function $a(S)$ also admits analytic continuation in S into the upper half-plane. In this case, the points of the discrete spectrum of the operator L correspond to zeroes $S = S_m$ of the function $a(S)$ in the upper half-plane. This means that at $S = S_m$ the following equation holds:

$$f_o^+(x,S_m) = B_m f_o^-(x,S_m), \tag{18}$$

where the quantities B_m are independent of x. According to the above-made assumptions, at m = 1,...,N the relation $\lambda_m + S_m^2 = 0$ is fulfilled. By virtue of (5) we have the equation

$$\varphi_m(x) = C_m f_o^+(x,S_m), \qquad (19)$$

where the quantities C_m are also independent of x, m = 1,...,N. Now, in equation (16) we substitute instead of f_o^+ the right-hand side of equation (18) and pass to the limit as $x \to -\infty$. According to (11), (18) and (19) we get

$$\frac{\partial B_m}{\partial t} + \{c[(-iS_m)^{K_o+2} - (iS_m)^{K_o+2}] - \kappa\epsilon_m |A_m(t)|^2\}B_m = 0. \qquad (20)$$

One can see that at $A_m(t) \equiv 0$ this equation coincides with an analogous relation for the equations of the KdV hierarchy.

Equations (17) and (20) fully determine the evolution of the scattering data for the operator L of the form (1). This enables one to investigate the system (8) by the inverse scattering method. In particular, these data determine the kernel of the integral Gel'fand-Levitan equation [1].

In the case when K_o is an odd integer number, using equation (20) we find that the phase velocity of the m-th soliton is proportional to the sum

$$\sigma_m = 2c(iS_m)^{K_o+2} + \kappa\varepsilon_m |A_m(t)|^2. \qquad (21)$$

At $c\kappa\epsilon_m < 0$ both the terms in the right-hand side of equation (21) have the same sign, and consequently, the phase velocity of the m-th soliton in this case has a constant sign, i.e. the m-th soliton moves all the time in the same direction. However, at $c\kappa\varepsilon_m > 0$ the terms in the right-hand side of equation (21) have different signs, and consequently, the phase velocity of the m-th soliton may change sign and the soliton itself may change its direction of motion. Thus, to an infinite number of changes of the sign of the quantity σ_m there corresponds the m-th soliton oscillatory motion. If there exists the time moment t_o such that at $t < t_o$ the quantity σ_m has a constant sign and at $t > t_o$ the quantity σ_m has an infinite number of changes of sign, then in this case the m-th soliton first moves in the constant direction and then is captured in the oscillatory regime.

Then, if K_o is an arbitrary even intetger number, the system (8) has the same form

$$\frac{\partial u}{\partial t} = 2\kappa \sum_{m=1}^{N} \varepsilon_m \frac{\partial}{\partial x} |\varphi_m|^2,$$

$$\frac{\partial^2 \varphi_m}{\partial x^2} + u\varphi_m = \lambda_m \varphi_m, \quad m = 1,\dots,N,$$

and equations (17) and (20) in this case become

$$\frac{\partial a}{\partial t} = \frac{\partial \mathscr{C}}{\partial t} \equiv 0, \quad \frac{\partial B_m}{\partial t} - \kappa\varepsilon_m |A_m(t)|^2 B_m = 0.$$

In conclusion, we should like to note that the validity of all the above limiting transitions is based on inequality (9) and can strictly be proved.

Acknowledgements

I would like to express my gratitude to F. Calogero and M. Kruskal for fruitful discussions of the above results.

References

[1] Gel'fand, I.M. & Levitan, B.M. (1951). On the determination of a differential equation from its spectral function. *Izv. Akad. Nauk SSSR, ser. Matem.*, **15(4)**, 309.

[2] Lax, P.D. (1968). Integral of nonlinear equations of evolution and solítary waves. *Commun. Pure and Appl. Math.* **21(5)**, 467.

F.A. Grünbaum

26. DARBOUX'S METHOD AND SOME "RANK TWO" EXPLICIT SOLUTIONS OF THE KP EQUATION

The purpose of this note is to show how certain explicit solutions of the KP equation, given in [2], can be obtained by use of Darboux's method as described, for instance, in [10].

1 Introduction

Among the many nonlinear partial differential equations that have attracted a great deal of attention, the Kadomtsev-Petviashvili equation [4] plays a fundamental role. We have

$$\frac{3}{4} U_{yy} = (U_t - \frac{1}{4} U_{xxx} - \frac{3}{4} UU_x)_x. \qquad (1)$$

The initial value problem, for sufficiently rapidly decaying initial data, can be linearized by an appropriate extension of the inverse scattering transform. Along these lines the equation is "integrable", a statement that should not be construed as meaning that lots of explicit solutions are readily available. The search for such explicit solutions is a separate job, and one that relies usually on deep algebraic-geometric structures. Indeed it is known since the work of I. Krichever [5] that given a compact Riemann surface one constructs explicit solutions of the KP equation by considering the Riemann theta function of such a surface; all the pure soliton solutions as well as the rational solutions are obtained out of these solutions by several degeneration procedures.

A search for new classes of explicit solutions was undertaken by I. Krichever & S. Novikov in [6,7,8]. These solutions are connected with "high rank vector bundles over algebraic curves" in the same fashion in which the previously known solutions were associated with "rank one bundles". The method of these papers, as made explicit in the case of "rank two" solutions consists of "reducing" the solution of a PDE in three variables to the solution of one PDE in two variables, namely x and t, to wit

$$c_t = \frac{1}{4} c_{xxx} + \frac{3}{8} \frac{1 - c_{xx}^2}{c_x} - \frac{3}{2} \wp(2c) c_x^3$$

This equation has been referred to as the Krichever-Novikov equation in [13] where the authors give a list of "integrable" equations of a certain type. $\wp$ denotes the Weierstrass p-function.

In [2] we give an independent derivation of this K-N equation, present three explicit solutions of it, and proceed to write down the corresponding "rank two" solutions of KP. These solutions are

$$\text{(a)} \quad U = 4 \frac{\partial^2}{\partial x^2} \log(x^5 - 20xy^2 + 20x_5)$$

$$\text{(b)} \quad U(x,y,t) = \frac{\partial^2}{\partial x^2} [4\log(27658a^2y^2 + 49152a^2x^2 - 20736a^2tx + 131072a^5x + 2187t^2 - 27648a^3t) + \frac{9x^2}{64a^2}]$$

with a any nonzero constant, and finally

$$\text{(c)} \quad u = - \frac{4x}{(3t+2)} + 4 \frac{\partial^2}{\partial x^2} \log(2y^2 - 3tx).$$

Of these three solutions the last two are new to this author, although more experienced workers in the field may have seen them already. Notice that the first one decays at infinity, the second one goes to a nonzero constant, and the third one grows linearly.

2 Rank two solutions of KP

Start with a formal pseudo-differential operator

$$L = D + \sum_{i=-\infty}^{-1} a_i(x_1, x_2, x_3, \ldots)D^i \qquad D \equiv \frac{d}{dx}$$

with coefficients that depend on the variables $x = x_1$, x_2, $t = x3$ (and possibly more variables x_4, x_5, ...).

Operators of this kind were already used by I. Schur [12]. They can be mulitiplied together usig Leibnitz's rule. In particular we denote with $(L^p)_+$ the result of taking the p-power of L and then retaining only the terms that have a nonnegative order in D = d/dx.

We say that we have a solution of the KP equations if

$$\frac{\partial L}{\partial x_2} = [L_+^2, L], \qquad \frac{\partial L}{\partial x_3} = [L_+^3, L].$$

This implies that $4a_1$ is a solution of the original KP equation

(1).

The solutions considered in [2] satisfy a pair of extra algebraic requirements, namely

(I) L^4 is a differential operator.

(II) The differential operator L_+^6 commutes with L^4 and the polynomial ring that they generate has rank two, i.e. the g.c.d. of the orders of such operators is two.

When these conditions are met we will say that we have a "rank two" solution of KP. There is something unsatisfactory about this definition and we address this issue now.

Given such a "rank two" ring of commuting differential operators, it is possible that a larger commutative ring might exist that contains it, and that its rank could be one. Given a concrete example it may be very hard to decide if this is the case. However as was seen in [2] this is indeed the case for the solution of the KP equation given in (a) above.

The implication of this is that example (a) could be equally well described as a "rank one" solution. More comments on this point, as well as a general picture that gives unity to this whole area can be found in the paper by E. Previato & G. Wilson [11].

For workers trying to find new explicit solutions of the KP equation, there are at least two pitfalls to avoid. One we have just referred to, the second one being that their "new" solution could also be obtained out of known ones by the appliction of some method like Darboux's, which we describe below.

This second pitfall may not be too easy to avoid. In [9], D. Levi, O. Ragnisco & A. Sym show the equivalence of the Darboux method with the so-called "dressing method". On the other hand, the construction in [11] exhibits the rank two solution of KP as an approprite dressing-up of solutions of the KdV equation.

3 The method of Darboux (and Moutard)

Readers intrigued by the reference to Moutard should look at Darboux [1].

Let L be a first order pseudo differential operator like the one considered above

$$L = D + \sum_{i=-\infty}^{-1} a_i(x_1, x_2, x_3, \ldots)D^i.$$

Let $\phi(x_1, x_2, x_3, \ldots; k)$ be a "wave function" for L, i.e. a nonzero function satisfying

$$L\phi = k\phi, \qquad \frac{\partial\phi}{\partial x_2} = (L^2)_+\phi, \qquad \frac{\partial\phi}{\partial x_3} = (L^3)_+\phi.$$

This makes $4a_1$ a solution of the KP equation (1).

Let $\tilde{L}$ be another such pseudo differential operator

$$\tilde{L} = D + \sum_{i=-\infty}^{-1} \tilde{a}_i(x_1 x_2 \ldots)D^i$$

We say that $\tilde{L}$ has been obtained from L by "transference of factors" or by one application of the "Darboux process" if

$$\tilde{L}(D - \frac{\psi_x}{\psi}) = (D - \frac{\psi_x}{\psi})L$$

where ψ is obtained from ϕ by the recipe

$$\psi(x_1 x_2 \ldots) \equiv \psi = \Sigma\, \alpha_j \left.\frac{\partial^j \phi}{\partial k^j}\right|_{k=k_0}.$$

This process can be iterated and leads to the situation of two first order pseudo differential operators L_1 and L_2 being obtained one from the other by p application of the "Darboux process".

One instance of this process converts the wave function ϕ and the "solution" a_1 of the KP equation (1) into a new such pair $\tilde{\phi}$ and $\tilde{a}_1$ according to the expression

$$\tilde{\phi} = (D - \frac{\psi_x}{\psi})\phi, \qquad \tilde{a}_1 = a_1 + \frac{\partial^2}{\partial x^2} \log \psi.$$

Recall that the solution of KP is actually $4a_1$. Another way of writing the equation gives $2a_1$ for its solution. Repeated application of this process with a sequence of choices $\psi_1, \psi_2, \ldots, \psi_p$ gives, after some work, the formulas

$$\tilde{\phi} = W(\psi_1, \ldots, \psi_p, \phi)/W(\psi_1 \ldots \psi_p)$$

$$\tilde{a}_1 = a_1 + \frac{\partial^2}{\partial x^2} \log W(\psi_1 \ldots \psi_p)$$

with

$$W(\psi_1, \psi_2, \ldots, \psi_p) \equiv \det(\psi_i^{(j-1)}).$$

We remark that each of the ψ_j is a linear combination of k derivatives of the original wave function ϕ. The proof of these formulas can be achieved by relying on one of Jacobi's identities for Wronskians, namely

$$W(f)W(f,g,h) = W(W(f,g), W(f,h)).$$

These expressions for $\tilde{\phi}$ and $\tilde{a}_1$ can be found, for instance, in [10].

In the next two sections we show how the method of Darboux can be used to produce two of the solutions of the KP equation given above.

Example (a). Take for L the operator

$$L = D.$$

The corresponding solution to the KP equation is $u = 0$. The corresponding wave function ϕ is given by

$$\phi(x_1 x_2 \ldots ;\ k) = e^{\Sigma k^j x_j} = \Sigma p_j(x_1 x_2 \ldots)k^j.$$

Here p_j denotes the so-called Schur polynomials, corresponding to partitions of the form

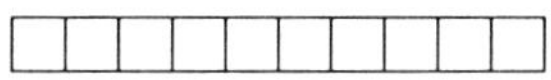

Since these p_j are (except for constants) the derivatives of ϕ at $k=0$, it is clear that one can pick p arbitrary linear combinations of the p_j, namely

$$\Sigma\gamma_j^{(1)}p_j,\quad \Sigma\gamma_j^{(2)}p_j,\quad \ldots\ ,\quad \Sigma\gamma_j^{(p)}p_j$$

and then form the Wronskian determinant

$$W = W(\Sigma\gamma_j^{(1)}p_j,\quad \ldots,\quad \Sigma\gamma_j^{(p)}p_j)$$

of these linear combinations to obtain a rational solution $U = 4\,\frac{\partial^2 \log}{\partial x^2}\,W$ of the KP equation (1). This fact seems to be due to Sato, and it is described in [3]. It is useful to recall, in the considerations above, that

$$\frac{\partial p_j}{\partial x_1} = p_{j-1}.$$

We believe that all rational solutions of the KP equation (1) that decay to zero as x grows are given by this recipe. We have never seen a proof of this fact. Nor have we ever seen, on the other hand, an expression for all the rational solutions of KP without any decay restrictions. Examples (b) and (c) are among the rational solutions that do not decay at infinity.

To obtain the example at issue, notice that

$$U = 4 \frac{\partial^2}{\partial x^2} \log(x^5 - 20y^2x + 20x_5)$$

$$x^5 - 20y^2x + 20x_5 = \det \begin{pmatrix} p_3 & p_4 & p_5 \\ p_0 & p_1 & p_2 \\ 0 & p_0 & p_1 \end{pmatrix} = W(p_5, p_2, p_1)$$

Example (b). To simplify matters we deal with the special case of $a = \frac{3}{4}$. The general case works exactly as below.

Take for L the operator

$$L = (D + \frac{1}{4})^{\frac{1}{2}}$$

The corresponding solution of the KP equation is

$$U = \frac{1}{2}.$$

Then the wave function is given by

$$\phi(x_1, x_2, \ldots; k) = e^{x_1 k + y(k^2 + \frac{1}{4}) + t(k^3 + \frac{3}{8} k) + \ldots}$$

Set

$$\phi^+(x,y,t) \equiv ((i+1)\phi + \frac{\partial\phi}{\partial k})(x,y,t, \tfrac{i}{2}) ,$$

$$\phi^-(x,y,t) = ((i+1)\phi + i \frac{\partial\phi}{\partial k})(x,y,t, -\tfrac{i}{2}).$$

With two applications of the Darboux method we get

$$\tilde{\phi} = W(\phi^+, \phi^-, \phi)/W(\phi^+, \phi^-)$$

$$U = \frac{1}{2} + 4 \frac{\partial}{\partial x^2} \log W(\phi^+, \phi^-)$$

This can be seen to be exactly the expression for the solution of example (b) in [2].

References

[1] Darboux, G. (1889). *Lecons sur la Theorie Generale des Surfaces,* 2eme part. Gauthiers-Villars: Paris.

[2] Grünbaum, F.A. (1989). The K-P equation: an alternative approach to the "rank two" solutions of Krichever and Novikov. *Phys. Lett. A* **139**, 3.

[3] Grünbaum, F.A. (1988). Differential equations in the spectral parameter: the higher order case. In *Inverse Problem: An Interdisciplinary Study*, p. 307. Sabatier, P. (ed.). Academic Press.

[4] Kadomtsev, B. & Petviashvili, P. (1970). On the stability of solitary waves in weakly dispersing media. *Sov. Phys. Dokl.* **15**(6), 539.

[5] Krichever, I.M. (1977). Methods of algebraic geometry in the theory of nonlinear equations. *Russ. Math. Surv.* **32**, 6, 183.

[6] Krichever, I.M. & Novikov, S.P. (1978). Holomorphic bundles over Riemann surfaces and the K-P equation. *Funct. Analysis and Appl.* **12**, 276.

[7] Krichever, I.M. & Novikov, S.P. (1979). Holomorphic bundles and nonlinear equations. Finite zone solutions of rank 2. *Sov. Math. Dokl.* **20**, 650.

[8] Krichever, I.M. & Novikov, S.P. (1980). Holomorphic bundles over algebraic curves and nonlinear equations. *Russ. Math. Surv.* **35**(6), 53.

[9] Levi, D., Ragnisco, O. & Sym, A. (1984). Dressing method vs. classical Darboux transformation. *Il Nuovo Cimento* **83**(1), 34.

[10] Matveev, V. (1979). Darboux transformation and explicit solutions of the KP equation, depending on functional parameters. *Lett. Math. Phys.* **3**, 213.

[11] Previato, E. & Wilson, G. (to appear). Vector bundles over curves and solutions of the KP equations. *AMS volume on "Theta functions"*, Maine 1986.

[12] Schur, I. (1905). Über vertauschbare lineare Differentialausdrücke. *Sitz. Berliner Math. Ges.* **4**, 2. See also (1973). *Ges. Abhandlungen*, I, 170. Springer: Berlin.

[13] Svinolupov, S., Sokolov, V. & Yamilov, R. (1983). On Bäcklund transformations for integrable equations. *Sov. Math. Dokl.* **28**, 165.

J.N. Elgin

27. COMMENTS ON THE PERIODIC NONLINEAR SCHRÖDINGER EQUATION

1 Introduction

Some aspects of quasiperiodic solutions of the nonlinear Schrödinger equation

$$iq_t + q_{xx} \pm q|q|^2 = 0 \qquad (1)$$

are discussed. Here, q is a complex amplitude, a suffix denotes partial differentiation, and the choice of signs describe either the focussing case (+), or the defocussing case (-). The technique used is based on the inverse spectral transform as developed for periodic problems. With the choice of minus sign in equation (1), the appropriate scattering problem is self-adjoint and the analysis is close to that for the KdV case, as described for example in Refs. (1,2). The approach adopted here is much in the spirit of Forest & McLaughlin (3), in their publication on the periodic sine-Gordon equation. A μ-representation is used throughout, with no mention at all of the theta-representation.

In §2, a single bilinear form of the scattering problem (essentially a density matrix) is introduced, and from that a most useful conserved quantity is immediately derived on which is imposed a suitable analytic structure (a cut Riemann surface).

Several authors have contributed to the periodic nonlinear Schrödinger problem (4,5). The main object here is to derive some of these results in a more succinct manner, and to emphasise the consequences of using a different scattering problem from that usually adopted in such studies (the Zakharov-Shabat system). In §3 some properties of the solutions are discussed.

2 Prelininaries

A bilinear form of the scattering problem is first introduced, then a set of equivalence transformations in defined and briefly discussed.

2.1 *Bilinear form of the scattering problem*

The scattering problem appropriate to the nls equation is

$$L\Phi = \zeta\Phi \tag{2}$$

Here, Φ is a solution matrix, ζ is an eigenvalue, and L is the operator

$$L = \begin{bmatrix} i\partial x & -iq \\ ir & -i\partial x \end{bmatrix} \tag{3}$$

with $r = \mp q^*$ corresponding to the choice of sign in equation (1). An equation governing the time evolution of Φ is introduced in a similar way - see equation (13) below. The adjoint operator L_A is now introduced such that

$$L_A\Psi = \beta\Psi \quad , \tag{4}$$

with the requirement

$$\Psi\sigma_3 L\Phi = (\Phi\sigma_3 L_A\Psi)^\dagger \tag{5}$$

Here, Ψ is a second solution matrix, σ_i are the Pauli matrices, and the 'dagger' superfix denotes Hermitian conjugate. It is then easily shown that, either

$$\beta = \zeta^*$$

or

$$\Psi^\dagger\sigma_3\Phi = 0 \tag{6}$$

With $\beta = \zeta^*$, the relationship between Ψ and Φ is

$$\Psi = -\frac{i}{2}\sigma_2\Phi^*\sigma_1 \tag{7}$$

where (*) denotes complex conjugate.

The required bilinear form is defined as

$$\rho = \Phi\Psi^\dagger = \frac{i}{2}\Phi\sigma_1\Phi^T\sigma_2 \tag{8}$$

where 'T' denotes transpose. With Φ given by

$$\Phi = \begin{bmatrix} \phi_1 & \bar{\phi}_1 \\ \phi_2 & \bar{\phi}_2 \end{bmatrix} ,$$

ρ has the structure

$$\rho = \begin{bmatrix} f & g \\ -h & -f \end{bmatrix} \tag{9}$$

where

$$f = -(\phi_1\bar{\phi}_2+\phi_2\bar{\phi}_1)/2$$
$$g = \phi_1\bar{\phi}_1$$
$$h = \phi_2\bar{\phi}_2 \tag{10}$$

From equation (2), and a similar equation governing the time evolution of Φ, the following evolution equations are found for ρ:

$$\rho_x = [M,\rho]$$
$$\rho_t = [B,\rho] \tag{11}$$

Here [,] is a commutator bracket,

$$M = \begin{bmatrix} -i\zeta & q \\ r & i\zeta \end{bmatrix} \tag{12}$$

corresponding to equation (2), and, for the nls equation, the appropriate form for B is

$$B = \begin{bmatrix} -iqr - 2i\zeta^2 & iq_x + 2\zeta q \\ -ir_x + 2\zeta r & iqr + 2i\zeta^2 \end{bmatrix} \tag{13}$$

with $r = \mp q^*$. Writing $\det\rho \equiv d = gh-f^2$, it follows from equation (9) that $\rho^2 = dI$. Since ρ^n (where n is any integer) obviously satisfies the same evolution equations (11) as ρ, it follows that $d_x = 0$ and $d_t = 0$; that is, d is a constant of the motion.

For reasons explained elsewhere (3), the constant is taken to be a simple polynomial in ζ; specifically, we take

$$f^2-gh = \prod_{k=1}^{2N+2} (\zeta-\zeta_k) \equiv P(\zeta)) \tag{14}$$

where the set of constants ζ_k (branch points on a cut Riemann surface) constitutes the fundamental scattering data which will be used in the reconstruction of the potential q.

To achieve the polynomial form for $P(\zeta)$, solutions for (f,g,h) are sought which are themselves polynomial in the eigenvalue ζ.

2.2 *Equivalence Transformations*

Clearly, the chosen form of the scattering problem (2) is not unique; one could choose to work instead with an equivalent system generated from (2) by any unitary transformation u: $\Phi \to \tilde{\Phi} = u\Phi$. The choice $u = (\sigma_3+\sigma_2)/\sqrt{2}$ transforms the Zakharov-Shabat (ZS) scattering problem (2) into one derived by Takhatajian & Feddeev (TF) [6], and used by Forest & McLaughlin in their study of the periodic sine-Gordon equation [3]. The point here is that the 'correct' choice of scattering problem can permit a solution to be found to the relevant problem when an inappropriate choice may not (as in the case of the sine-Gordon equation), or alternatively, the correct choice can lead to the simplest form for the algorithm used in the reconstruction of the potential q. For example, solving the mKdV equation using the TF scattering problem rather than the ZS gives a reconstruction formula for the potential q identical in form to that obtained in solutions to the periodic KdV problem (1,2) but with q now replaced by its Muira transform: $q \to q_x - q^2$.

For the nls problem, we first use the ZS scattering problem, then consider a second choice defined by taking $u = (\sigma_1+\sigma_3)/\sqrt{2}$. Then $\rho \to \bar{\rho} = -u\rho u$ and the components of $\bar{\rho}$ (in obvious notation) become

$$\bar{f} = \tfrac{1}{2}(h-g)$$

$$\bar{g} = -f + \tfrac{1}{2}(h+g)$$

$$\bar{h} = f + \tfrac{1}{2}(h+g) \qquad (15)$$

3 Periodic solutions of the NLS

The method by which periodic solutions to the nls are obtained is identical to that described elsewhere, and hence is only outlined here. Equations (11) are first written in component form, then solutions sought for (f,g,h) which are polynomial in the eigenvalue ζ; specifically, take

$$f = \sum_{n=0}^{N+1} \zeta^n f_n$$

$$g = \sum_{n=0}^{N} \zeta^n g_n = iq \prod_{k=0}^{N} (\zeta - \mu_k) \qquad (16)$$

where the structure of the latter form for g is inferred from inspection of the equation for g_x, and serves to introduce the improper spectral parameters μ_k. These are then substituted into the appropriate evolution equations and, on comparing like coefficients of ζ^n, relationships between the various coefficients f_n and the sums of products of the μ_i variables are obtained. A further set of relationships is found by making use of equation

(14). In this manner, it is found that

$$q_x/q = 2iS_1, \qquad q_t/q = 2i(S_2-S_1^2) \tag{17}$$

where

$$S_n = \sum_{i=1}^{N} \mu_i^n - \frac{1}{2} \sum_{i=1}^{N} \zeta_i^n \tag{18}$$

The potential q can then be reconstructed from a knowledge of the parameters μ_i, which are space-time dependent. Evolution equations for the latter are found from the equations for g_x and g_t evaluated at $\zeta = \mu_i$; thus,

$$\mu_{\ell x} = -\frac{2i\sigma_\ell\{\prod_{k=1}^{2N+3}(\mu_\ell-\zeta_k)\}^{\frac{1}{2}}}{\prod_{j\neq\ell}(\mu_\ell-\mu_j)}$$

$$\mu_{\ell t} = -2S'_{1,\ell}\,\mu_{\ell x} \tag{19}$$

where

$$S'_{n,\ell} = S_n-\mu_\ell^n \tag{20}$$

The algorithm outlined above differs in one important aspect from similar algorithms used in the solution of other nonlinear evolution equations: there is no additional constraint imposed on the parameters μ_i; they are free (at least in principle) to 'wander' anywhere at all on the cut Riemann surface. For the KdV or mKdV equations, the μ_i are constrained always to be real. For the sine-Gordon equation, the constraint

$$\prod_{k=1}^{N} |\mu_k|^2 = \text{constant}$$

applies - *i.e.*, the product is independent of both x and t. No such constraint applies for the focussing nls equation, nor indeed any other system isomorphic to the nls equation. Interestingly, the (continuum) Heisenberg chain - isomorphic to the nls - has a very similar algorithm to that outlined above; the same evolution equations for the μ_i are again found, only the reconstruction formula is slightly different. Specifically, q_x/q in the first of equations (17) is replaced by $-S_{z,x}+S_zS_{+,x}/S_+$ where $S_+=S_x+iS_y$, and S_z are the components of the spin vector.

There is an interesting link between multisoliton solutions and 'homoclinic' orbits of the μ_i parameters in the complex ζ-plane, as follows. In the limit of infinite period (in both x and t), branch points 'close' in pairs and a typical μ-trajectory links one such pair with a second pair; it is in this sense that we interpret the trajectory as homoclinic.

For the focussing nls equation, with N = 1, and $\zeta_1 = \zeta_2 = \zeta_3^* = \zeta_4^* = \xi + i\eta$, it is easily found that the solution for μ is

$$\mu(x,t) = \xi + i\mathrm{Tanh}(2\eta(x+4\xi t)) \tag{21}$$

leading, through equations (17), to the usual single soliton solution. Note that the real part of μ is always constant, and that the points $\xi + i\eta$ are approached in the limits t(or x) $\rightarrow \pm\infty$. By contrast the single soliton solution for the Heisenberg spin chain has the μ-variable moving from $\xi - i\eta$ to $\xi + i\eta$ along the longer arc of the circle passing through the three points $\xi \pm i\eta$ and the origin (details to be published elsewhere).

The above results all derive from the use of the ZS scattering problem. It might be argued that a disadvantage of this choice is that the final equations from which the potential q is obtained (equations (17)) are not particularly simple, and that perhaps an alternative choice might lend to simpler results. With the choice outlined at the end of section 2.2, the following results are obtained:

$$q_R = -\sum_{\ell=1}^{N} \mu_{\ell I}$$

$$q_I^2 = \sum_{\ell=1}^{N} (\mu_{\ell R}^2 - \mu_{\ell I}^2) - \frac{1}{2}\sum_{i=1}^{2N+2} \zeta_i^2 \tag{22}$$

$$q_{I,x} = 2\sum_{\ell=1}^{N} \mu_{\ell I}\mu_{\ell R}$$

where $q = q_R + iq_I$, $\mu_\ell = \mu_{\ell R} + i\mu_{\ell I}$. Clearly, these reconstruction formula are considerably simpler, but now the evolution equations for the μ-variables are a little more difficult:

$$\mu_{\ell x} = -\frac{2iS'_{1,\ell}\{\prod_{k=1}^{2N+2}(\mu_\ell - \zeta_k)^{\frac{1}{2}}\}}{\prod_{j\neq\ell}(\mu_\ell - \mu_j)}$$

$$\mu_{\ell t} = -2i \frac{(S'_{2,\ell}-S'^{2}_{1,\ell})\{\prod_{k=1}^{2N+2}(\mu_\ell-\zeta_k)\}^{\frac{1}{2}}}{\prod_{j\neq\ell}(\mu_\ell-\mu_j)} \tag{23}$$

Note that the first set of equations for $\mu_{\ell x}$ above are the same as the previous set for $\mu_{\ell t}$. As a final comment, it is noted that a set of 'dynamic' trace formula - similar to the set quoted by McKean & van Moerbeke [7] for the KdV equation (their corollary 12.1) - exist for the nls equation. Details of these will be published elsewhere.

References

[1] Novikov, S.P., (1974), *Funct. Anal. Appl.*, **8**, 236.

[2] Ablowitz, M.J. & Segur, H. (1981) "*Solitons and the Inverse Scattering Transform*", SIAM.

[3] Forest, M.G. & McLaughlin, D.W. (1988) *J. Math. Phys.*, **23**, 1248.

[4] Yan-Chow Ma & Ablowitz, M.J. (1981), SIAM, **65**, 113.

[5] Tracey, E.R., Chen, H.H. and Lee, Y.C. (1983) *Phys. Rev. Lett.* **53**, 218.

[6] Takhatajian, L.A. & Faddeev, L.D. (1974) *Theor. Math. Phys.*, **21**, 1046.

[7] McKean, H.P. & van Moerbeke, P. (1975) *Invent Math.*, **30**, 217.

C. Guha-Roy

28. SOME SPECIFIC SOLUTIONS OF COUPLED NONLINEAR EQUATIONS

We consider some coupled nonlinear evolution equations to look for exact solutions. By explicit analytical treatment, we show that coupled equations may have specific solutions in terms of either Jacobian elliptic function or Weierstrass elliptic function. Furthermore, solitary wave-like solutions may be found as simple limits of periodic solutions. Finally, a brief discussion and outlook is given to highlight our motivation.

1 Introduction and basic notions

Over the past several years, the investigation of exact solutions to nonlinear evolution equations (NLEE) applicable to a number of geophysical systems has been developed rapidly following the realization that those solutions play crucial roles in understanding fundamental properties of the system concerned. Nowadays, by virtue of recent advancement in mathematical physics, it has come to light that exact solutions which are found in terms of elliptic functions (Jacobian/Weierstrass) may yield in particular solitary wave profile. Thus an important criterion to characterize the NLEE seems to be their behaviour with respect to scattering of solitary waves. It is further noted that for dynamical equations governing planetary motions in the ocean and atmosphere, solitary wave solutions are likely to be nonlinear Rossby waves [18].

Since 1964, there exist many works [2,17-20] in connection with solitary Rossby waves. For instance, Long [17] and Benney [2] have observed long waves in a barotropic atmosphere with a horizontally sheared basic zonal flow and have derived the Korteweg-de Vries (KdV) equation for the nonlinear perturbations. Later on, Redekopp [19] has outlined the general theory of solitary waves in a zonal, planetary shear flow and has confined his attention to two specific atmospheric models. Besides, he has pointed out that the amplitude of long Rossby waves moving in a zonal shear flow is guided by the KdV equation or the modified KdV equation depending upon the distribution of the atmospheric density stratification. It has also been remarked [20] that the KdV equation is seen to be a valid asymptotic limit of the quasi-geostrophic forecast equation.

Recent trends have, however, been centred around a more complicated and sophisticated system which evolves as a coupled form of KdV equation. Indeed, Ito [12] has proposed a fascinating class of coupled KdV (cKdV) equations which have infinitely many symmetries and constants of motion. Employing Ito's effort, Kawamoto [14] has studied the different parametric representations of Ito's equation and has obtained cusp like soliton solutions. It may be mentioned that Antonowicz & Fordy [1] have currently derived Ito's equation in the context of the novel spectral problem. One can strand other papers (for example: [11], [8]) too, which typically deal with the cKdV equation as primitive equation. In contrast, quite a few publications may appear that glimpse the coupled version of KdV equation with higher nonlinear terms. Owing to these interests, we think it would be worthwhile to consider model equations which may be an extension of the cKdV equation. The coupled equations we propose to examine are

$$p_t = \alpha p^n p_x + \beta q^n q_x + \gamma p_{3x} \tag{1}$$

$$q_t = \delta q q_x + \mu (pq)_x \tag{2}$$

which represent the cKdV equation and its generalization systems (including modified form) depending on the positive integer n. Here the subscripts "t" and "x" stand for the partial differentiations w.r.t. the corresponding variables, p and q being real wave fields and the parameters α, β, γ, δ and μ are chosen arbitrarily.

The aim of this presentation, based on my two recent papers [5,7], is to review some interesting results in context with equations (1) and (2). In the first part, i.e. in section 2, we exploit an analytical treatment to work out the mathematical properties of our basic model equations and report some so-called specific solutions with special reference to solitary wave solutions. However, interest is also directed toward a more exact solution which can be obtained in terms of Weierstrass or Jacobian elliptic function provided we treat the equations in a different manner. Furthermore, solitary wave like solutions may be found as simple limits of periodic solutions. This is actually done in section 3. The final section is devoted to a brief discussion and outlook with the applicability of the solutions in physical situations.

2 **Mathematical framework and specific solutions**

When attempting to find out a solitary wavetrain, one generally seeks a localized solution that depends on the travelling wave variable, $\sigma = x \pm \eta t$ (η being the signal speed) thereby reducing the partial differential equation of two variables into an ordinary differential equation in σ. By imposing a stream function into some coupled nonlinear wave systems, we have established [9] that if one of the solutions is of the travelling wave form then the other must assume the same dependence of variable. Inspired by this desire, we cast (for the moment) the set of transformations as

$$p = p(\sigma \equiv x+\eta t), \quad q = q(\sigma \equiv x+\eta t) \tag{3}$$

To start our discussion, it is convenient to let n = 2. It is because one may see subsequently that the case n = 1 is incorporated into our analysis.

Applying (3) into eqs. (1) and (2) [for n = 2], we find

$$\eta p_{\sigma} = \alpha(p^3/3)_{\sigma} + \beta(q^3/3)_{\sigma} + \gamma p_{3\sigma} \tag{4}$$

$$\eta q_{\sigma} = \delta(q^2/2)_{\sigma} + \mu(pq)_{\sigma} \tag{5}$$

From (5), one can easily obtain a relation between p and q as

$$\delta q^2 + 2\mu pq - 2\eta q - 2\nu = 0 \tag{6}$$

where ν is the constant of integration. With no loss in generality we can assert [5,7] ν to be zero, and as a consequence (6) gives

$$q = \frac{2\mu}{\delta}\left(\frac{\eta}{\mu} - p\right) \tag{7}$$

which, in turn, mathematically implies that we require p to be regular everywhere.

To proceed further, we eliminate q from (4) and (7) and integrate the equation for p twice. The straightforward manipulations, on account of the boundary conditions that p and its derivatives with respect to σ vanish as $|\sigma|$ tends to infinity, give rise to the following expression:

$$(p')^2 = \varepsilon_2 p^2 + \varepsilon_3 p^3 + \varepsilon_4 p^4 \equiv \sum_{j=2}^{4} \varepsilon_j p^j \tag{8}$$

where a prime indicates differentiation w.r.t. σ and

$$\varepsilon_2 = (\eta/\gamma)(1+8\beta\mu\eta/\delta^3), \tag{9a}$$

$$\varepsilon_3 = (-1)(\eta/3\gamma)(16\beta\mu^2/\delta^3), \tag{9b}$$

$$\varepsilon_4 = (1/6\gamma)(8\beta\mu^3/\delta^3-\alpha). \tag{9c}$$

It may be remarked that one can readily select values for the constants ε_2, ε_3, ε_4 appearing in eq. (8) for evaluating numerical solutions, but here we mainly present some explicit solutions of (8) from purely analytical considerations.

Denoting $\varepsilon = (\varepsilon_3^2-4\varepsilon_2\varepsilon_4)$, one can write the solitary wave solution of (8) as

$$p(\sigma) = \frac{4\varepsilon_2}{[\sqrt{\varepsilon}\exp(\sqrt{\varepsilon_2}(\sigma+\sigma_o))-2\varepsilon_3+\sqrt{\varepsilon}\exp(-\sqrt{\varepsilon_2}(\sigma+\sigma_o))]} \tag{10}$$

provided $\varepsilon_2 > 0$, $\varepsilon_3 < 0$ and $\varepsilon > 0$.

It is observed that if $\varepsilon_3 = -2\sqrt{\varepsilon_2\varepsilon_4}$, (8) yields kink (+) and antikink (-) solutions which are given by

$$p(\sigma) = (\varepsilon_2/\varepsilon_3)[1\pm\tanh\{\sqrt{\varepsilon_2}(\sigma+\sigma_o)/2\}] \tag{11}$$

Besides, there may be many other solutions too [10], such as:

$$p(\sigma) = \frac{2\varepsilon_2}{[d\sqrt{(-\varepsilon)}\sinh\{\sqrt{\varepsilon_2}(\sigma+\sigma_o)\}-\varepsilon_3]}, \quad \text{for } \varepsilon_2 > 0,\ \varepsilon < 0 \tag{12}$$

$$p(\sigma) = \frac{2\varepsilon_2}{[d\cdot\exp\{\pm\sqrt{\varepsilon_2}(\sigma+\sigma_o)\}-\varepsilon_3]}, \quad \text{for } \varepsilon_2 > 0,\ \varepsilon = 0 \tag{13}$$

$$p(\sigma) = \frac{2\varepsilon_2}{[d\sqrt{\varepsilon}\sin\{\sqrt{(-\varepsilon_2)}(\sigma+\sigma_o)\}-\varepsilon_3]}, \quad \text{for } \varepsilon_2 < 0,\ \varepsilon > 0 \tag{14}$$

where $d = \pm 1$.

It is interesting to note that a remarkable aspect emerges from (8) when one notices that $(\alpha/\beta) = (2\mu/\delta)^3$. Subject to this condition, the expression for p takes the form

$$(p')^2 = \sum_{j=2}^{3} \varepsilon_j\, p^j \tag{15}$$

For $\varepsilon_2 > 0$ and $\varepsilon_3 < 0$, we can again determine the solitary wave solution as

$$p(\sigma) = -(\varepsilon_2/\varepsilon_3)\,\mathrm{sech}^2[\sqrt{\varepsilon_2}(\sigma+\sigma_o)/2] \tag{16}$$

provided that $0 \le p \le -(\varepsilon_2/\varepsilon_3)$.

It is further instructive to point out that for n=1, our coupled equations can be transformed into the form (15) that has the solution (16).

A completely analogous treatment can also be performed to our coupled systems with n = 3,4,... and the possible solitary wave solutions (see for instance [4],[6 and references quoted therein]) to eqs. (1) and (2) may then be found for suitable parameteric values ε_j.

Before we switch over to the next section for seeking exact solutions, it is also worth remarking that the systems (1) and (2) are reduced to Ito-type equations if we pose $\delta = 0$ and n = 1. For such a case, travelling wave solutions can be obtained in the form of cusp-soliton solutions [14] under certain conditions.

3 Exact solutions

In order to extract exact solutions of eqs. (1) and (2), we follow the procedure as outlined in Guha-Roy [7] which is based on Kano & Nakayama's [13] method. For the purpose, as before, let us consider n = 2, and rewrite eq. (8) as follows:

$$p'' = \sum_{i=1}^{3} \lambda_i p^i \tag{17}$$

where

$$\lambda_1 = \varepsilon_2,\ \lambda_2 = (3/2)\varepsilon_3 \text{ and } \lambda_3 = 2\varepsilon_4 \tag{18}$$

We primarily turn our attention to look for [15,16] a solution of the type

$$p(\sigma) = \frac{k\mathbf{P}(\sigma)}{[1+l\mathbf{P}(\sigma)]} \tag{19}$$

where k and l are arbitrary parameters.

In (19), $\mathbf{P}(\sigma)$ denotes the Weierstrass **P**-function and fulfils the condition

$$(\mathbf{P}')^2 = 4\mathbf{P}^3 - g_2\mathbf{P} - g_3 \tag{20}$$

provided g_2 and g_3 are real constants obeying $g_2^3 > 27g_3^2$.

Now by inserting (19) in (17) and then equating the coefficients of the powers of **P** from both sides, we get

$$\lambda_1 l^2 + \lambda_2 kl + \lambda_3 k^2 = -2l \tag{21a}$$

$$2\lambda_1 l + \lambda_2 k = 6 \tag{21b}$$

$$\lambda_1 = (3/2) l g_2 \tag{21c}$$

$$0 = 2lg_3 - (1/2)g_2 \tag{21d}$$

Combining (21a) and (21b), we have

$$5\lambda_1 l^2 + 4\lambda_2 kl + 3\lambda_3 k^2 = 0 \tag{22}$$

which may also be written as

$$(al + bk)^2 = 0 \tag{23a}$$

where

$$a = \sqrt{5\lambda_1}, \ b = \sqrt{3\lambda_3} \text{ and } \lambda_2 = \tfrac{1}{2}\sqrt{15\lambda_1\lambda_3} \tag{23b}$$

Now on the application of (23) in eqs. (21a) and (21b), one finds k and l uniquely, so that

$$k = 4(15/\lambda_1\lambda_3)^{1/2}, \quad l = -(6/\lambda_2)(15\lambda_3/\lambda_1)^{1/2} \tag{24}$$

Using (9) and (18) we can express k and l in (24) as

$$k = (12\sqrt{5}\gamma\delta^3)/\sqrt{\eta F}, \quad l = ((3\sqrt{5}\gamma\delta^3)/(4\beta\mu^2\eta))\sqrt{G/\eta} \tag{25}$$

where

$$F = \{(8\beta\mu\eta+\delta^3)(8\beta\mu^3-\alpha\delta^3)\} \text{ and } G = \{(8\beta\mu^3-\alpha\delta^3)/(8\beta\mu\eta+\delta^3)\} \tag{26}$$

In the same way, g_2 and g_3 are obtained from eqs. (21c) and (21d), namely,

$$g_2 = (2\lambda_1/3l), \quad g_3 = (g_2/4l) \tag{27}$$

Hence the exact periodic solution is given by

$$p(\sigma) = \frac{k\mathbf{P}(\sigma+\rho; g_2, g_3)}{[1+l\mathbf{P}(\sigma+\rho; g_2, g_3)]} \tag{28}$$

where k, l, g_2 and g_3 are determined from eqs. (25) and (27) and ρ is a constant of integration of (20).

It is now clear that one may find the exact bounded periodic solution as

$$p(\sigma) = \frac{k[e_3+(e_2-e_3)\mathrm{sn}^2\{(e_1-e_3)^{1/2}\sigma+\rho_o\}]}{[1+l(e_3+(e_2-e_3)\mathrm{sn}^2\{(e_1-e_3)^{1/2}\sigma+\rho_o\})]} \tag{29}$$

where sn is the Jacobian elliptic function having modulus, $m = \frac{e_2-e_3}{e_1-e_3}$ and e_1, e_2, e_3 are real roots of $4z^3-g_2z-g_3 = 0$ so that $e_1>e_2>e_3$.

Since for a solitary wave $m \sim 1$, we have $e_2 \sim e_1$. Hence in the limiting case the solitary wave profile may be expressed in the following form

$$p(\sigma) = \frac{k[e_1-(e_1-e_3)\mathrm{sech}^2\{(e_1-e_3)^{1/2}\sigma+\rho_o\}]}{[1+l(e_1-(e_1-e_3)\mathrm{sech}^2\{(e_1-e_3)^{1/2}\sigma+\rho_o\})]} \tag{30}$$

Similar arguments can be advocated for the systems (1) and (2) corresponding to n = 3,4 and so on.

Meanwhile, one should note that once p is known then q can easily obtain through the connection (7).

4 Discussion and outlook

In the preceding sections, the nonlinear coupled equations (1) and (2) which we have encountered appear to be noteworthy from the mathematical as well as physical point of view. This is because some of the present results provide a close similarity to the kinds of flow patterns exhibited by a restricted class of solutions concerning large amplitude Rossby waves. In a paper [3], certain physically important solutions of such a class of waves have been reported when one relaxes the weak nonlinearity restriction. It has been found that finite amplitude long wave equations can be derived into the following forms:

$$a_f^2 = \tau^2 a^2 (1 - \frac{a}{a_o})^2, \quad a_f^2 = \tau^2 a^2 (1 - \frac{a}{a_o}) \tag{31}$$

provided strongly nonlinear counterparts are considered. In (31), $\tau > 0$ and a_o is a constant. The relevant solutions representing the two waves are given by

$$a = \frac{a_o}{2}\{1-\tanh(\frac{1}{2}\tau f)\}, \quad a = a_o\{\mathrm{sech}^2(\frac{1}{2}\tau f)\} \tag{32}$$

which are analogous to the forms as given in (11) and (16). As emphasized in the latter case, there was no curvature in the undisturbed velocity profile.

To conclude, it can be stated that our coupled systems may exhibit soiltary wave solutions depending on the precise choice of ε_j/λ_i, which may be useful to disentangle complex behaviour of nonlinear phenomena and its qualitative physical features.

Acknowledgements

The work reported herein is supported by a grant from the Council of Scientific & Industrial Research, Govt. of India. The author is pleasurably indebted to Prof. D.K. Sinha (Pro-Vice Chancellor, Calcutta University) and Dr. B. Bagchi (Vidyasagar University) for their keen interest and valuable advice. He is also thankful to the organisers for warm hospitality at Como (Italy) during the workshop, and grateful to the Tata Institute of Fundamental Research (N.B.H.M.) for providing partial travel support.

References

[1] Antonowicz, M. & Fordy, A.P. (1988). Coupled KdV equations associated with a novel Schrödinger spectral problem. In *Nonlinear Evolutions, Proc. 6th Workshop on NEEDS,* pp. 145-160, Léon, J.J-P. (ed.). World Scientific: Singapore.

[2] Benney, D.J. (1966). Long nonlinear waves in fluid flows. *J. Math. & Phys.* **45**, 52.

[3] Benney, D.J. (1979). Large amplitude Rossby waves. *Stud. in Appl. Math.* **60**, 1.

[4] Guha-Roy, C. (1986). A note on the cnoidal wave solution of a coupled nonlinear equation. In *Mathematical Aspects of Nonlinear Studies in Oceanic Sciences,* pp. 47-53, Sinha, D.K. (ed.). South Asian Publishers: New Delhi.

[5] Guha-Roy, C. (1987). Solitary wave solutions of a system of coupled nonlinear equations. *J. Math. Phys.* **28**, 2087.

[6] Guha-Roy, C. (1987). A note on stationary solitary wave solutions. *J. Phys. Soc. Jpn.* **56**, 1625.

[7] Guha-Roy, C. (1988). Exact solutions to a coupled nonlinear equation. *Int. J. Theor. Phys.* **27**, 447.

[8] Guha-Roy, C., Bagchi, B. & Sinha, D.K. (1985). Solitary wave solution of a new type of a coupled K-dV equation. JU/DM/3/85, preprint.

[9] Guha-Roy, C., Bagchi, B. & Sinha, D.K. (1986). Traveling-wave solutions and the coupled Korteweg-de Vries equation. *J. Math. Phys.* **27**, 2558.

[10] Guha-Roy, C. & Sinha, D.K. (1986). On the solitary wave solutions of certain coupled nonlinear evolution equations. In *Third Indian Conference on Ocean Engineering* Vol. I, pp. A85-A89.

[11] Hirota, R. & Satsuma, J. (1981). Soliton solutions of a coupled Korteweg-de Vries equation. *Phys. Lett.* **85A**, 407.

[12] Ito, M. (1982). Symmetries and conservation laws of a coupled nonlinear wave equation. *Phys. Lett.* **91A**, 335.

[13] Kano, K. & Nakayama, T. (1981). An exact solution of the wave equation $u_t+uu_x-u_{5x} = 0$. *J. Phys. Soc. Jpn.* **50**, 361.

[14] Kawamoto, S. (1984). Cusp soliton solutions of the Ito-type coupled nonlinear wave equation. *J. Phys. Soc. Jpn.* **53**, 1203.

[15] Krishnan, E.V. (1982). An exact solution to the classical Boussinesq equation. *J. Phys. Soc. Jpn.* **51**, 2391.

[16] Krishnan, E.V. (1986). On the Ito-type coupled nonlinear wave equation. *J. Phys. Soc. Jpn.* **55**, 3753.

[17] Long, R.R. (1964). Solitary waves in the Westerlies. *J. Atmos. Sci.* **21**, 197.

[18] Malanotte-Rizzoli, P. (1982). Planetary solitary waves in Geophysical flows. In *Advances in Geophysics,* vol. 24, pp. 147-224, Saltzman, B. (ed.). Academic Press: Inc.

[19] Redekopp, L.G. (1977). On the theory of solitary Rossby waves. *J. Fluid Mech.* **82**, 725.

[20] Redekopp, L.G. & Weidman, P.D. (1978). Solitary Rossby waves in zonal shear flows and their interactions. *J. Atmos. Sci.* **35**, 790.

J. Cieśliński

29. GEOMETRIC METHOD OF INTEGRATION OF THE INHOMOGENEOUS HEISENBERG CHAIN

1 Introduction

In the present paper we consider the following nonlinear system:

$$S_{,2} = S \wedge (fS_{,1})_{,1}, \tag{1a}$$

$$S \cdot S = 1, \tag{1b}$$

where $S=S(x_1,x_2)$ is $\mathbb{E}^3$-valued function of two real variables x_1 and x_2, the comma means differentiation, ^ denotes skew (or vector) product in $\mathbb{E}^3$, $f=f(x_1,x_2)$ is a given real function, and the centre dot denotes the scalar product in $\mathbb{E}^3$. The above system has an important physical application as equations of motions of the classical 1-dimensional continuous inhomogeneous Heisenberg ferromagnet. The case $f \equiv$ const corresponds to the homogeneous Heisenberg model.

There will be presented a geometric approach proposed by A.Sym [14,13] and some new solutions of the equation (1) obtained that way [4,5].

2 Physical motivation

Consider the mechanical system defined on the Cartesian product of N unit spheres by the following hamiltonian and Poisson brackets [1,5]:

$$H = - \sum_{k=1}^{N-1} f_k(t)\, S_k \cdot S_{k+1}, \tag{2}$$

$$\{S_k^A, S_j^B\} = \sum_{c=1}^{3} \epsilon^{ABC} S_k^C\, \delta_{jk} \qquad \text{for } j,k = 1,2,\ldots,N, \tag{3}$$

where $S_k=(S_k^1,S_k^2,S_k^3)$ are confined to the unit sphere and $f_k(t)$ are given functions. This system describes inhomogeneous 1-dimensional

Heisenberg ferromagnetic spin model. Equations of motion read as follows:

$$S_{k,t} = \{S_k, H\} = S_k \wedge [f_k(t)S_{k+1} + f_{k-1}(t)S_{k-1}]. \quad (4)$$

Performing the standard 'continuous limit' procedure we arrive at the inhomogeneous Heisenberg chain (IHC) equation (1).

3 Integrability of the model

Lakshmanan & Bullough [8] in the case of f linear in x_1 and Balakrishnan [1] in the case of an arbitrary f have shown that IHC equation (1) can be reduced to the so-called inhomogeneous nonlinear Schrödinger equation (INSE):

$$iq_{,2} + (fq)_{,11} + 2qR = 0, \quad (5a)$$

$$R_{,1} = (fqq^*)_{,1} + f_{,1}qq^*, \quad (5b)$$

where $i^2=-1$, $q=q(x_1,x_2)$ is $\mathbb{C}$-valued function, $R=R(x_1,x_2)$ is real function and the star denotes the complex conjugate. Sometimes INSE is written in the integro-differential form (after elimination of R).

INSE was proved to be integrable in the case of f linear in x_1 by Calogero & Degasperis [3]. However it is very difficult to reconstruct S-solution from the knowledge of the corresponding q-solution. This problem is equivalent to the reconstruction of a curve from the knowledge of its curvature and torsion which are given in terms of the q-solution. The advantage of our approach consists in obtaining S-solutions in an explicit way.

In the integrable case $(f_{,11}=0)$ we constructed the Darboux-Bäcklund transformation (DBT) [4]. Thus we are able to obtain N-soliton S-solutions. It should be stressed that the spectral parameter is x_2('time')-dependent.

Different integrability tests indicate that for other functions f INSE seems to be non-integrable [7,5]. Nevertheless we are able to find particular solutions S for certain functions f even in that case [14,5].

4 Geometry of inhomogeneous Heisenberg chain

Consider an arbitrary surface Σ embedded in $\mathbb{E}^3$ and described by a position vector $r(x_1,x_2)\in\Sigma$, where x_1,x_2 are some coordinates on Σ.

Following objects can be defined in each point of Σ: tangent vectors $r_{,1}$ and $r_{,2}$, the normal vector n, coefficients of the I fundamental form $g_{ij} = r_{,i}\cdot r_{,j}$ and coefficients of the II fundamental form $b_{ij} = n\cdot r_{,ij}$.

It is always possible to choose (at least locally) such

coordinates that $r,_1 \cdot r,_1 = 1$ and $r,_1 \cdot r,_2 = 0$. They are called orthogonal semi-geodesic coordinates because parametric lines x_2=const are geodesics. Any one-parameter family of geodesics defines such coordinates.

Theorem 1 (A. Sym): Let Σ be an arbitrary (smooth enough) surface in $\mathbb{E}^3$ endowed with orthogonal semi-geodesic coordinates x_1, x_2 such that $b_{11} \neq 0$. Then:

$$S,_2 = S \wedge (fS,_1),_1 \quad \text{and} \quad S \cdot S = 1, \tag{6}$$

$$iq,_2 + (fq),_{11} + 2qR = 0, \tag{7a}$$

$$R,_1 = (fqq^*),_1 + f,_1 qq^*, \tag{7b}$$

where: $S = r,_1$, $f = -\gamma/b_{11}$, $q = (b_{11}/2)\exp(i\alpha)$, $\gamma^2 = g_{22}$, $\alpha,_1 = b_{12}/\gamma$, $2R = \alpha,_2 - b_{22}/\gamma$.

Remark 1: Equations (6) and (7) are identical with (1) and (5).

Remark 2: The explicit knowledge of fundamental forms g_{ij}, b_{ij} for some orthogonal semi-geodesic coordinates allows us to obtain one particular solution S and one particular solution q,R for some coupling function f defined by these coordinates.

Proof of the theorem: n, $r,_1$ and $r,_2$ form an orthogonal frame for any x_1, x_2. It implies that $r,_2$ is proportional to $r,_1 \wedge n$. Moreover: $r,_1 \cdot r,_{11} = r,_2 \cdot r,_{11} = 0$ (because $r,_1$ is unit vector orthogonal to $r,_2$). Thus: $r,_{11} = b_{11} \cdot n$, $r,_2 = \gamma n \wedge r,_1$, and, finally:

$$r,_2 = -(\gamma/b_{11})\, r,_1 \wedge r,_{11}, \tag{8}$$

which after differentiation yields the equation (6), where $S = r,_1$ and $f = -\gamma/b_{11}$. By the way we obtained another interesting nonlinear system (8). To complete the proof it is enough to notice that the system (7) is satisfied by virtue of Gauss-Mainardi-Codazzi equations for the surface Σ.

5 New solutions of IHC equation

We applied results of the previous section to the case of surfaces of revolution. In this case it is possible to calculate (in quadratures at least) all one-parameter families of geodesics. Without entering into technical details we present the final result [5]:

$$S = [1-(A/y)^2]^{1/2}([1-\dot{y}^2]^{1/2},\ \dot{y}\cos\phi,\ \dot{y}\sin\phi) +$$

$$+ [A/y](0, -\sin\phi, \cos\phi) \qquad (9)$$

is a solution of the equation (1) for the function f given by

$$f = y^3[A\beta'+A'\sigma,_A][(y^2-A^2)(1-\dot{y}^2)]^{1/2}/[y\ddot{y}(y^2-A^2)-(1-\dot{y}^2)A^2] \qquad (10)$$

where $y=y(s)$, $A=A(x_2)$, $\beta=\beta(x_2)$ are arbitrary functions satisfying (for any s and x_2) constraints: $\dot{y}^2\le 1$ and $A^2\le y^2$, dots and primes denote the ordinary differentiation (with respect to s and x_2 respectively), $\sigma=\sigma(s,A)$ is an auxiliary function defined below ($\sigma,_A$ is its partial derivative):

$$\sigma = \int^s (1-[A/y(\xi)]^2)^{-1/2} d\xi, \qquad (11a)$$

and depending on x_1,x_2 in the following way:

$$\sigma = x_1 - \int^{x_2} A(t)\beta'(t)dt. \qquad (11b)$$

The above formulas allow us (after elimination of σ) to express the parameter s by x_1 and x_2 (in an involved form at least). Finally:

$$\phi = \beta(x_2) + A(x_2)\int^s [y^2(\xi)-A^2(x_2)]^{-1/2} y^{-1}(\xi)d\xi. \qquad (12)$$

To have explicit expressions for $f(x_1,x_2)$ and $S(x_1,x_2)$ it is necessary to carry out the integration in formulas (11a) and (12). It was done in a number of cases [5]. The application of formulas (9), (10) to the unit sphere ($y(s)=\cos s$), cylinder ($y(s)=1$), and conus ($y(s)=s\cdot\sin\delta$, δ=const) reproduces results of Sym & Wesselius [14].

An interesting subclass of solutions is obtained when we take $A(x_2)\equiv$const, $\beta(x_2)\equiv x_2$. It is not difficult to notice that in this case f is a function of $\sigma=x_1-Ax_2$ only (the travelling wave solution). In the simplest case $A(x_2)\equiv 0$, $\beta(x_2)\equiv x_2$ (the corresponding family of geodesics consist of 'meridians') formulas (11), (12) after a trivial integration yield (for an arbitrary function y(s)): $s=\sigma=x_1$ and $\phi=x_2$, and, finally, we obtain:

$$f = y[1-\dot{y}^2]^{1/2}/\ddot{y}, \qquad (13)$$

$$S = ([1-\dot{y}^2]^{1/2}, \dot{y}\cos x_2, \dot{y}\sin x_2), \qquad (14)$$

where $y=y(x_1)$ is an arbitrary function such that $\dot{y}^2\le 1$.

6 Integrable case ($f_{,11} \equiv 0$)

An integrable nonlinear equation can be usually represented as integrability condition

$$U_{,2} - V_{,1} + [U,V] = 0 \qquad (15a)$$

for a system of two matrix linear equations:

$$\Phi_{,1} = U\Phi \text{ and } \Phi_{,2} = V\Phi, \qquad (15b)$$

where Φ,U and V depend on x_1, x_2 and on an additional ('spectral') parameter λ in such a way that λ does not enter the nonlinear system (15a) [10]. The linear system (15b) is called the linear problem (or Lax pair) corresponding to the nonlinear system (15a).

The form of the linear problem for INSE (5) in the case of f linear in x_1 is well known [6]:

$$U = i\lambda\sigma_3 + q\sigma_+ - q^*\sigma_- \qquad (16a)$$

$$V = i[R-2\lambda^2 f]\sigma_3+[i(qf)_{,1}-2\lambda qf]\sigma_+ +[i(q^*f)_{,1}+2\lambda q^*f]\sigma_- \qquad (16b)$$

where: $\sigma_3 = \begin{pmatrix}1 & 0\\ 0 & -1\end{pmatrix}$, $\sigma_+ = \begin{pmatrix}0 & 1\\ 0 & 0\end{pmatrix}$, $\sigma_- = \begin{pmatrix}0 & 0\\ 1 & 0\end{pmatrix}$,

$f=f(x_1,x_2)=x_1\mu(x_2)+\nu(x_2)$, μ and ν are arbitrary functions, and

$$\lambda_{,2} = -2\mu(x_2)\lambda^2. \qquad (17a)$$

We can easily express λ as an explicit function of x_2 and of a constant of integration, λ_o:

$$\lambda(x_2) = \lambda_o/[1 + 2\lambda_o \int^{x_2}\mu(\tau)d\tau]. \qquad (17b)$$

In fact λ_o is a genuine spectral parameter but it is convenient and more natural to use x_2-dependent λ. The system (15a) for U,V given by (16) is identical with (5).

Theorem 2: Let $\mathring{\Phi} \equiv \Phi|_{\lambda=0} = \Phi(x_1,x_2,0)$ is a solution of the system (15b) in which U,V are given by (16) with λ=0. Then

$$S = (i/2)\mathring{\Phi}^{-1}\sigma_3\mathring{\Phi}, \qquad (18)$$

is a solution of the following nonlinear equation:

$$S,_2 = [S,(fS,_1),_1], \tag{19}$$

where square brackets denote the commutator ($[A,B]\equiv AB-BA$).

Proof: It is enough to calculate both sides of (19) taking into account (15b) and (16) where we substitute $\lambda\equiv 0$ (i.e. $\lambda_0=0$).

Remark 1: S is an element of the Lie algebra su(2) (spanned by $\sigma_+-\sigma_-$, $i\sigma_3$ and $i\sigma_++i\sigma_-$) which can be identified with 3-dimensional Euclidean vector space $\mathbb{E}^3$. The scalar and vector products of two elements a,b of the algebra su(2) are defined as follows:

$$a \cdot b = -2\mathrm{Tr}(ab) \ , \tag{20a}$$

$$a \wedge b = [a,b] \ . \tag{20b}$$

Corollary: S given by (18) solve IHC (1) because $S\cdot S=1$ and (19) holds.

Remark 2: One can prove that S defined by (18) is equal to $r,_1$, where

$$r = (1/2)(\Phi^{-1}\Phi,_\lambda)|_{\lambda=0} \tag{21}$$

is su(2) $\cong \mathbb{E}^3$-valued function describing the soliton surface [12] for INSE (5). In a special case $\lambda\equiv$const Sym [11] noticed that $r,_1$ solves the homogeneous Heisenberg chain.

7 Darboux-Bäcklund transformation (DBT) for IHC ($f,_{11} = 0$)

The following conditions are sufficient for matrices U and and V to form the Lax pair (16) for INSE [4]:

$$U = U_0 + \lambda U_1, \tag{22a}$$

$$V = V_0 + \lambda V_1 + \lambda^2 V, \tag{22b}$$

$$U_k = U_k(x_1,x_2) \in su(2), \quad V_k = V_k(x_1,x_2) \in su(2), \tag{22c}$$

$$U,_2 - V,_1 + [U,V] = 0 \text{ for each value of } \lambda, \tag{22d}$$

$$U_1 \cdot U_0 = U_1 \cdot V_1 = 0 \quad \text{and} \quad U_1 \equiv i\sigma_3, \tag{22e}$$

$$\lambda,_2 = -2f,_1\lambda^2 \quad \text{and} \quad \lambda,_1 = 0, \tag{22f}$$

where f,q and R (entering (16)) are defined by:

$$f = (i/8)V_2 \cdot \sigma_3 , \tag{23a}$$

$$q = -(1/2)U_0 \cdot \sigma_- , \tag{23b}$$

$$R = (i/4)V_0 \cdot \sigma_3 , \tag{23c}$$

and the scalar product is given by (20a).

Suppose that we know a particular solution q,R of INSE (5) defining matrices U,V (16). Then any transformation of these matrices preserving equations (22) leads (through (23)) to some new solution of INSE. DBT discussed below is an example of such transformation.

Consider the following gauge transformation of the system (15b) given by a matrix $D=D(x_1,x_2,\lambda)$:

$$\tilde{\Phi} = D\Phi, \tag{24a}$$

$$\tilde{U} = D_{,1}D^{-1} + DUD^{-1} , \tag{24b}$$

$$\tilde{V} = D_{,2}D^{-1} + DVD^{-1} . \tag{24c}$$

(24a) implies following transformations of r,S defined by (18), (21) (we denote $D|_{\lambda=0}=\mathring{D}$, $(D_{,\lambda})|_{\lambda=0}=\mathring{D}_{,\lambda}$, $\Phi|_{\lambda=0}=\mathring{\Phi}$):

$$\tilde{r} = r + (1/2)\mathring{\Phi}^{-1}(\mathring{D}^{-1}\mathring{D}_{,\lambda})\mathring{\Phi} , \tag{24d}$$

$$\tilde{S} = (\mathring{\Phi}^{-1}\mathring{D}\mathring{\Phi})^{-1}S(\mathring{\Phi}^{-1}\mathring{D}\mathring{\Phi}) . \tag{24e}$$

Formulas (24bc) preserve (for any D) the condition (22d). Moreover we assume the form of λ given by (22f). To satisfy other conditions (22) we postulate the following form of D (in close analogy with the well known case $\lambda\equiv$const, described for example by Manakov et al. [10] and Levi et al. [9]):

$$D = [(\lambda-\lambda_1)/(\lambda-\mu_1)]^{1/2}\{I + [(\lambda_1-\mu_1)/(\lambda-\lambda_1)]P\} \tag{25}$$

where λ_1 and μ_1 are some complex functions, I is an identity 2x2 matrix, and $P=P(x_1,x_2)$ is an arbitrary projector ($P^2\equiv P$). For D given by (25) equations (24bc) are equivalent to:

$$\begin{aligned}\tilde{U} = U &+ [(\lambda_1-\mu_1)/(\lambda-\lambda_1)]P(U-\partial_1)(I-P) + \\ &+ [(\mu_1-\lambda_1)/(\lambda-\mu_1)](I-P)(U-\partial_1)P + \\ &+ (1/2-P)\partial_1\log[(\lambda-\lambda_1)/(\lambda-\mu_1)] ,\end{aligned} \tag{26a}$$

$$\tilde{V} = V + [(\lambda_1-\mu_1)/(\lambda-\lambda_1)]P(V-\partial_2)(I-P) +$$

$$+ [(\mu_1-\lambda_1)/(\lambda-\mu_1)](I-P)(V-\partial_2)P + \tag{26b}$$

$$+ (1/2-P)\partial_2\log[(\lambda-\lambda_1)/(\lambda-\mu_1)] ,$$

where ∂_k denotes the differentiation with respect to x_k.

Now we demand that $\tilde{U}$ and $\tilde{V}$ defined by (26) should satisfy conditions (22abc). It implies (like in the case λ=const) that: $\mu_1 = \lambda_1^*$, P is hermitean ($P \equiv P^\dagger$) and $\ker P = \Phi(x_1, x_2, \lambda_1)W$, where W is any 1-dimensional vector subspace of $\mathbb{C}^2$. The last from conditions sufficient for $\tilde{U}, \tilde{V}$ to have the form (22ab) is that λ_1 should satisfy (21f). Finally we obtain:

$$\tilde{U}_1 = U_1 , \tag{27a}$$

$$\tilde{U}_0 = U_0 + \mathrm{Im}\lambda_1[U_1, \mathcal{A}] , \tag{27b}$$

$$\tilde{V}_2 = V_2 , \tag{27c}$$

$$\tilde{V}_1 = V_1 + \mathrm{Im}\lambda_1[V_2, \mathcal{A}] , \tag{27d}$$

$$\tilde{V}_0 = V_0 + \mathrm{Im}\lambda_1[V_1, \mathcal{A}] + \mathrm{Im}\lambda_1\mathrm{Re}\lambda_1[V_2, \mathcal{A}] +$$

$$+ (1/2)(\mathrm{Im}\lambda_1)^2[[V_2, \mathcal{A}], \mathcal{A}] - 2\mu\mathrm{Im}\lambda_1\mathcal{A}, \tag{27e}$$

where $su(2) \ni \mathcal{A} \equiv i(I-2P)$ is given by:

$$\mathcal{A} = i[(\xi\xi^*-1)\sigma_3 + 2\xi\sigma_+ + 2\xi^*\sigma_-]/[1+\xi\xi^*] , \tag{28}$$

and $\xi = \xi_1/\xi_2$, where $\mathbb{C}^2 \ni (\xi_1, \xi_2) = \Phi(x_1, x_2, \lambda_1)c$, and, finally c is an arbitrary constant element of $\mathbb{C}^2$. The crucial point is that $\mathcal{A}$ and P are expressed by an 'old' wave function Φ.

It is not difficult to see that the transformation (27) is compatible with conditions (22e). Thus all conditions (22) are preserved by the transformation (27). Moreover the function f does not change (compare (23a) with (27c)). We constructed then DBT for INSE. The final formula for D read:

$$D = \cos\vartheta(\lambda) + \mathcal{A}\sin\vartheta(\lambda) \tag{29}$$

where $\vartheta(\lambda)$ is an argument of $\lambda_1-\lambda$ ($\lambda_1-\lambda = |\lambda_1-\lambda|\exp[i\vartheta(\lambda)]$). Now we are able to write down explicit formulas for DBT of the objects defined before (see (18),(21),(23) and (24)):

$$\tilde{q} = q - 4\mathrm{Im}\lambda_1 \xi/[1+\xi\xi^*] \ , \qquad (30a)$$

$$\tilde{r} = r + [\mathrm{Im}\lambda_1/(2\lambda_1^*\lambda_1)]\overset{\circ}{\Phi}^{-1}\mathcal{A}\overset{\circ}{\Phi} \ , \qquad (30b)$$

$$\tilde{S} = (\cos\overset{\circ}{\vartheta} - \overset{\circ}{\Phi}^{-1}\mathcal{A}\overset{\circ}{\Phi}\sin\overset{\circ}{\vartheta})S(\cos\overset{\circ}{\vartheta} + \overset{\circ}{\Phi}^{-1}\mathcal{A}\overset{\circ}{\Phi}\sin\overset{\circ}{\vartheta}) \ , \qquad (30c)$$

where $\overset{\circ}{\vartheta}=\vartheta(0)$. A complicated expression for $\tilde{R}$ is omitted.

Example: Let $f(x_1,x_2) = x_1$. Applying DBT to the trivial solution ($q \equiv 0$, $R \equiv 0$, $\Phi \equiv \exp(i\sigma_3 x_1\lambda)$, $S = (i/2)\sigma_3$) we obtain one soliton solution of IHC (1):

$$S = (-\sin B \sin 2\vartheta \ \mathrm{sech} Q + 2\sin^2\vartheta \ \mathrm{th} Q \ \mathrm{sech} Q \cos B,$$
$$-\cos B \sin 2\vartheta \ \mathrm{sech} Q - 2\sin^2\vartheta \ \mathrm{th} Q \ \mathrm{sech} Q \sin B,$$
$$1 - 2\sin^2\vartheta \ \mathrm{sech}^2 Q), \qquad (31)$$

where $B = 2x_1\mathrm{Re}\lambda_1 + B_0$, $Q = -2x_1\mathrm{Im}\lambda_1 + Q_0$, ϑ is an argument of λ_1, $\lambda_1 = \lambda_{10}/[1+2\lambda_{10}x_2]$, λ_{10} is a complex constant, B_0 and Q_0 are real constants.

8 Conclusions

The paper can be divided in two distinct parts. In sections 4 and 5 there is presented a geometric method of finding particular solutions of inhomogeneous Heisenberg chain equation (1) by constructing one-parameter families of geodesics on surfaces. This method was applied to surfaces of rotation. The resulting coupling functions f have usually complicated dependence on x_1,x_2. Nevertheless we can find among them some interesting examples, listed by Cieśliński et al. [5]. It should be stressed that IHC is in general nonintegrable.

In sections 6 and 7 we consider the integrable subcase of IHC (the coupling function f is linear in x_1). The spectral parameter is x_2-dependent in this case. We constructed the Darboux-Bäcklund transformation using the method which can be easily generalized. We are able to construct DBT for some other systems with the spectral parameter λ dependent on x_1 and x_2 ($\lambda_{,1}$ and $\lambda_{,2}$ should be polynomials in λ). Interesting examples of such systems and corresponding Lax pairs are listed for example by Burtsev et al. [2]. The work in this direction is in progress.

After completing this work I was told by Prof. M. Lakshmanan about his paper [15] in which many interesting results on the IHC can be found. There was obtained for example the 1-soliton solution of the IHC by the inverse scattering method.

Acknowledgements

I would like to express my gratitude to Prof. Antoni Sym whose ideas underlie all this work, to Prof. Decio Levi for numerous discussions from which I benefitted greatly, and to Dr. Willem Wesselius for the cooperation and hospitality during my visit in Enschede.

This work was done in part in the framework of the Rome-Warsaw exchange agreement and was partially supported by the Polish Ministry of National Education (Research Project CPBP-01-03).

References

[1] Balakrishnan, R. (1982). On the inhomogeneous Heisenberg chain. *J. Physics C* **15**, L1305.

[2] Burtsev, S.P., Zakharov, V.E. & Mikhailov, A.V. (1987). Inverse Scattering Method with variable spectral parameter. (In Russian). *Theoret. & Math. Phys.* **70**, 323.

[3] Calogero, F. & Degasperis, A. (1978). Exact solution via the spectral transform of a generalisation with linearly x-dependent coefficients of the nonlinear Schrödinger equation. *Lett. al Nuovo Cimento* **22**, 420.

[4] Cieśliński, J. (1987). The Darboux-Bäcklund transformation for inhomogeneous Heisenberg ferromagnet. In *Proceedings of the IV Symposium on Classical Analysis.* Kazimierz Dolny.

[5] Cieśliński, J., Sym, A. & Wesselius, W. (1989). On geometry of inhomogeneous Heisenberg ferromagnet. Twente University Memorandum No. 789 to be published in *Nonlinearity*.

[6] Gupta, M.R. (1979). Exact inverse scattering solution of a nonlinear evolution equation in a non-uniform medium. *Phys. Lett. A* **72**, 420.

[7] Kaup, D.J. (1986). The nonintegrability of the 'generalized nonlinear Schrödinger equation'. *Clarkson University preprint* INS #58 (February 1986).

[8] Lakshmanan, M. & Bullough, R.K. (1980). Geometry of generalized nonlinear Schrödinger and Heisenberg ferromagnetic spin equations with linearly x-dependent coefficients. *Phys. Lett. A* **80**, 287.

[9] Levi, D., Ragnisco, O. & Bruschi, M. (1980). Extension of the Zakharov-Shabat generalized inverse method to solve differential-difference equatins. *Nuovo Cimento A* **58**, 56.

[10] Manakov, S.V., Novikov, S.P., Pitaievsky, L.P. & Zakharov, V.E. (1980). *Theory of Solitons.* Plenum, New York.

[11] Sym, A. (1983). Soliton surfaces II. Geometric unification of solvable nonlinearities. *Lett. al Nuovo Cimento* **36**, 307.

[12] Sym, A. (1985). Soliton surfaces and their applications. In *Geometric Aspects of the Einstein Equations and Integrable Systems,* p. 154, Martini, R. (ed.). Springer-Verlag: New York.

[13] Sym, A. (1989). *Geometry of Solitons*, Vol. 1, to be published by Reidel Publishing Company, Dordrecht.

[14] Sym, A. & Wesselius, W. (1987). Geodesic congruences and

inhomogeneous Heisenberg ferromagnet. *Phys. Lett. A* **120**, 183.
[15] Lakshmanan, M. & Ganesan, S. (1985). Geometrical and gauge equivalence of the generalized Hirota, Heisenberg and WKIS equations with linear inhomogeneities. *Physica A* **132**, 117.

J. Hietarinta

30. RECENT RESULTS FROM THE SEARCH FOR BILINEAR EQUATIONS HAVING THREE-SOLITON SOLUTIONS

We report the results obtained recently in the search for bilinear equations having three-soliton solutions. Such equations are likely to be integrable, possibly even base equations for integrable hierarchies of equations. The complete results for Korteweg-de Vries and nonlinear Schrödinger type bilinear equations are presented here together with their nonlinear forms. These include an interesting combination of Davey-Stewartson and Kadomtsev-Petrivashivili equations.

1 Introduction

To put our approach in a proper perspective let us recall the known properties of soliton systems, which can be listed e.g. as follows:

(i) Existence of special solutions: (a) travelling wave solutions, (b) multisoliton solutions. In fact solitons should go through anything suffering only phase shifts.

(ii) Algebraic properties: (a) There is an infinite sequence of conserved quantities and corresponding hierarchy of equations. (b) The equation is a consistency condition for a pair of linear equations, the Lax pair.

(iii) The nonlinear initial value problem can be solved by a sequence of linear problems, the inverse scattering method.

In order to generalize known integrable systems and obtain new ones it is possible to start with any one of the above properties, as other contributions to these proceedings show. When the search for generalizations is made at the multisoliton level the assumptions on the underlying structure are different, and most likely less demanding, than at other levels. For this reason we believe that new interesting types of integrable equations can be best found when one searches for equations with multisoliton solutions.

2 Hirota's bilinear formalism

A particularly suitable formalism for the study of multisoliton solutions is the bilinear formalism of Hirota [11,12]. Originally it was used just for the construction of various solutions, but recently the bilinear formalism has been found to have rather deep mathematical meaning in the fundamental theory of solitons. this has been studied extensively by the Kyoto school [1,14].

As an example of Hirota's formalism let us consider the Korteweg-de Vries (KdV) equation in the potential form

$$v_{xxx} - 3v_x^{\;2} + v_t = 0. \tag{1}$$

Let us introduce a new dependent variable F by

$$u = -2\partial_x \log F \tag{2}$$

(Recall that Gel'fand-Levitan-Marchenko equation gives $v = -2\partial_x \log \det M$.) For F we get from (1) the ,equation [8]

$$F_{xxxx}F - 4F_{xxx}F_x + 3F_{xx}^{\;2} + Ff_{xt} - F_xF_t = 0 \tag{3}$$

This can be written in a more compact form if we define new operators $D_x, D_t, \ldots$ by [11,12]

$$P(D_x,\ldots)F\cdot F = P(\partial_x - \partial'_{x'},\ldots)F(x,\ldots)F(x',\ldots)\Big|_{x'=x,\ldots}, \tag{4}$$

so that eq. (3) becomes

$$(D_x^4 + D_xD_t)F\cdot F = 0. \tag{5}$$

All integrable NEE's have been written in the bilinear form, mostly by Hirota. There is no universal method for this and each case must be analysed separately. In the above KdV case only one equation and one new dependent variable is needed, but in many other cases one needs several independent and dependent variables. For the modified KdV equation (mKdv)

$$v_{xxx} + 24v^2v_x + v_t = 0 \tag{6}$$

one introduces **two** dependent variables f and g by $v = \partial_x \arctan(g/f)$, which yields [9]

$$\begin{cases} (D_x^3 - D_t)g\cdot f = 0 \\ D_x^2(f\cdot f + g\cdot g) = 0. \end{cases} \tag{7}$$

For the nonlinear Schrödinger equation (nlS)

$$y_{xx} + iy_t + |y|^2 = 0 \tag{8}$$

one uses y = G/F, where F is real, and then (8) is equivalent to [10]

$$\begin{cases} (D_x^2 + iD_t)G\cdot F = 0, \\ D_x^2 F\cdot F = |G|^2. \end{cases} \tag{9}$$

Higher members of a hierarchy are obtained by adjoining new "time"-variables.

3 The multisoliton solutions

The bilinear form has the special property that equations in that form automatically have one- and two-soliton solutions. Consider the general KdV type equation [11,12]

$$P(D_x, D_t, \ldots)F\cdot F = 0 \tag{10}$$

First of all for the vacuum or zero-soliton solution (0SS) we take F = 1, which by (2) yields u = 0 as expected. (This implies the condition $P(0,\ldots) = 0$.) For the 1SS we take

$$F = 1 + e^{\eta}, \tag{11}$$

where $\eta = px + \Omega t + \ldots + \eta^0$, and the soliton **parameters** p, Ω, ... are restricted by the **dispersion relation** $P(p,\Omega,\ldots) = 0$. Then F solves (10) and the quantity

$$\partial_x^2 \log F = p^2[\cosh(\eta/2)]^{-2}$$

derived from it looks like a soliton.

For a 2SS we take a linear combination of 1SS's with an interaction term:

$$F = 1 + e^{\eta_1} + e^{\eta_2} + B_{12}e^{\eta_1+\eta_2}, \tag{12}$$

where, as before,

$$\eta_i = p_i x + \Omega_i t + \ldots + \eta_i^0, \tag{13}$$

and the parameters are restricted by

$$P(p_i, \Omega_i, \ldots) = 0, \forall i. \tag{14}$$

This two soliton ansatz (12) with (13)and (14) solves (10) if

$$B_{12} = -\frac{P(p_1 - p_2, \Omega_1 - \Omega_2, \ldots)}{P(p_1 + p_2, \Omega_1 + \Omega_2, \ldots)} . \tag{15}$$

This exercise shows that 2SS's are automatic for a bilinear equation of type (10), however, the existence of higher soliton solutions is a much more complicaated issue. One can try to construct multisoliton solution using the expansion

$$F = 1 + \sum_{k=1}^{M} \epsilon^k f_k(x, \ldots), \tag{16}$$

which, when substituted to (10), yields f_i recursively; for an N-soliton solution we should start (16) with

$$f_1 = \sum_{i=1}^{N} \exp(\eta_i). \tag{17}$$

For known integrable systems the expansion (16) with (17) yields

$$F = \sum_{\mu_i \in \{0,1\}} \exp(\Sigma_{0<i<j\le N} A_{ij}\mu_i\mu_j + \Sigma_{i=1}^{N} \eta_i\mu_i), \tag{18}$$

where $\exp A_{ij} = B_{ij}$ is given in (15). In particular expansion (16) therefore truncates at $k = N$. If one attempts to use the truncating ansatz (18) for the general equation (10) one finds for the existence of an N-soliton solution the conditions

$$\sum_{\sigma_i = \pm 1} P(\Sigma_{i=1}^{k} \sigma_i p_i, \ldots) \prod_{0<i<j\le k} [P(\sigma_i p_i - \sigma_j p_j, \ldots)\sigma_i\sigma_j] \doteq 0, \tag{19}$$

for $k = 3, \ldots, N$. Here $\doteq$ means that the equation must hold whenever $p_i, \ldots$ are such that (14) holds. Eq. (19) should be taken as a condition for P and not for the parameters, which are only subject to the dispersion relation (14).

The above construction was done for the bilinear equation (10), but similar ansatzes have been constructed for many other types of equations by Hirota and others.

Since already the existence of three-soliton solutions is so demanding one should find out what these special equations are that satisfy the three-soliton condition. This suggests the following

PROGRAM. Find all bilinear equations having three- (and four-) soliton solutions.

So far we have studied the following types of bilinear equations

(i) KdV type: $PF \cdot F = 0$, [3].

(ii) Modified-KdV type: $\begin{cases} B_0 F \cdot G = 0 \\ A(F \cdot F + G \cdot G) = 0 \end{cases}$, A quadratic in [4].

(iii) Sine-Gordon type: $\begin{cases} B_e F\cdot G = 0 \\ A(F\cdot F + G\cdot G) = 0 \end{cases}$, A quadratic in [5].

(iv) nlS type: $\begin{cases} B_c G\cdot F = 0 \\ AF\cdot F = |G|^2 \end{cases}$, A quadratic in [6].

(v) Benjamin-Ono type: $B_c G\cdot G^* = 0$, [6].

Here, P, A, B_e are even real polynomials, B_o an odd real polynomial, and B_c a complex polynomial s.t. B(iX,...) is real. It is quite demanding to check the relevant three-soliton condition so computer algebra is an absolute necessity to do the computations. We used REDUCE [2] on IBM 3081 and IBM 3033.

4 The KdV-type results

The KdV-type results were given in [3] in bilinear form. Here we would like to give the results together with the nonlinear form, which is obtained with the substitution $F = e^u$. Usually one still has to take derivatives and make changes of variables before a bona-fide nonlinear evolution equation is obtained. This last step is not made here, because no universally applicable rules seem to exist.

To the results below one can make linear changes of variables and projections to a fewer number of variables. Modulo such changes the list below is **complete** among polynomial bilinear equations of type (10). The results are as follows:

(a) Kadomtsev-Petviashivili equation [20]:

$$(D_x^4 - 4D_xD_t + 3D_y^2)F\cdot F = 0 \tag{20}$$

and in its nonlinear form

$$u_{xxxx} + 6u_{xx}^{\ 2} - 4u_{xt} + 3u_{yy} = 0 \tag{21}$$

This equation contains the well known special cases of Korteweg-de Vries and Boussinesq equations.

(b) Ito equation [13]:

$$(D_x^3D_t + aD_x^2 + D_tD_y)F\cdot F = 0 \tag{22}$$

$$u_{txxx} + 6u_{tx}u_{xx} + au_{xx} + u_{ty} = 0 \tag{23}$$

This is also a 2+1-dimensional equation, note however the different leading monomial.

(c) A new 1+1 dimensional combination of (a) and (b):

$$(D_x^4 - D_xD_t^3 + aD_x^2 + bD_xD_t + cD_t^2)F\cdot F = 0 \tag{24}$$

$$u_{xxxx} + 6u_{xx}^2 - (u_{tttx} + 6u_{tx}u_{tt}) + au_{xx} + bu_{xt} + cu_{tt} = 0 \tag{25}$$

This is a new result which suggests a combination of KdV- and Ito-equations. It can also be given in a completely x - t-symmetric way: $(D_xD_t(D_x^2 + \sqrt{3}D_xD_t + D_t^2) + \ldots)F\cdot F = 0$.

(d) Ramani equation [18]:

$$(D_x^6 + 5D_x^3D_t - 5D_t^2 + D_xD_y)F\cdot F = 0 \tag{26}$$

$$u_6 + 30u_4u_{xx} + 60u_{xx}^3 + 5u_{txxx} + 30u_{tx}u_{xx} - 5u_{tt} + u_{xy} = 0 \tag{27}$$

($u_4 = u_{xxxx}$, etc.). As a subcase it contains the Sawada-Kotera equation, obtained by $t \longrightarrow 0$, $y \longrightarrow t$.

The previously known results (a), (b), and (d) fit very well in the Kyoto classification [1,14]. For example one can assign nonzero weights for the variables x,t, and y so that all monomials in the bilinear equation have the same weight. For case (c) this is not possible. For cases (a) and (d) one gets the standard nonlinear form by taking two derivatives of the equation and defining $w = 2u_{xx}$. For (b) and (c) this does not work so well, because the fourth order terms contain also t-derivatives.

In addition to the above we have obtained also several equations with multiple factors. For example $(D_x^2-1)^N D_x^M D_t^P F\cdot F = 0$ has 3SS whenever M and P are odd. The dispersion relation $(p^2 - 1)p\Omega = 0$ gives a union of linear subspaces, and the physical meaning of solutions with such a parameter-space is open.

5 The nlS-type results

In our previously published searches [4,5] we assumed that the auxiliary polynomial A is quadratic. Here we would like to report on the new and complete results of nonlinear Schrödinger type without such restrictions [7]. (We have now almost completed the search for similar generalizations for the mKdV and sG type equations [7].) The general form of nlS-type equations is

$$\begin{cases} B(D_x, \ldots)G\cdot F = 0 \\ A(D_x, \ldots)F\cdot F = |G|^2. \end{cases} \tag{28}$$

It is important to realize that there can now be more than one kind of soliton for the system (28) even when we start from the vacuum soliton $F = 1$, $G = 0$. We have the standard nlS-type solitons (dispersion relation given by $B(p, \ldots) = 0$)

$$G = e^{\eta_B}, \quad F = 1 + Ke^{\eta_B+\eta_B^*}, \tag{29}$$

and also KdV type solitons (dispersion relation given by A(p,...) = 0)

$$G = 0, \quad F = 1 + Le^{\eta_A}. \tag{30}$$

The known integrable systems have the property that their 1SS's can be combined to make multisoliton solutions. If the dispersion relation given by A is nonlinear enough to support soliton solutions by itself then we may require that combinations of the two kinds of solutions above must be allowed. (If the polynomial A (or B) factorizes then it is possible to have solitons with the same structure but different dispersion relations. This possibility was included already in [3-6].)

The construction of a proper ansatz, with the two kinds of solitons above, is not too difficult following the ideas given in (16) and (17). For a combination of N nlS-type solitons and M KdV-type solitons the result has the usual general form

$$\begin{aligned} F &= \sum_{\mu}^{(0)} \exp(\Sigma_{i<j} A_{ij}\mu_i\mu_j + \Sigma_i \eta_i\mu_i), \\ G &= \sum_{\mu}^{(1)} \exp(\Sigma_{i<j} A_{ij}\mu_i\mu_j + \Sigma_i \eta_i\mu_i). \end{aligned} \tag{31}$$

Here $\mu = (\mu_1,\ldots,\mu_{2N+M})$, $\mu_i \in \{0,1\}$ and the index a in $\Sigma_{\mu}^{(a)}$ means that the μ-summation is further restricted by

$$\sum_{i=1}^{N} \mu_i - \sum_{i=N+1}^{2N} \mu_i = a. \tag{32}$$

The function η has the usual form (13) but the dispersion relation for its parameters depends on the index i. In fact for i = 1,...,N we associate η_i with a nlS-type soliton, $\eta_{i+N} = \eta_i^*$, and finally for i = 2N + 1,...,2N + M η_i is associated with a KdV-type soliton. Therefore we get the dispersion relations

$$\begin{cases} B(p_i,\ldots) = 0, & \text{for } i = 1,\ldots,N \\ B(-p_i,\ldots) = 0, & \text{for } i = N+1,\ldots,2N \\ A(p_i,\ldots) = 0, & \text{for } i = 2N+1,\ldots,2N+M. \end{cases} \tag{33}$$

[The middle equation is automatic whenever $[B(p_i,\ldots)]^* = \pm B(-p_i^*,\ldots)$, which is the assumption here.] For the coefficients A_{ij} we have [7]

$$e^{A_{ij}} = \begin{cases} 2A(p_i - p_j,\dots), & \text{for } 1\le i<j\le N, \\ 1/[2A(p_i + p^*_{j-N},\dots)], & \text{for } 1\le i\le N<j\le 2N, \\ 2A(p^*_{i-N} - p^*_{j-N},\dots), & \text{for } N<i<j\le 2N, \\ -B(p_i - p_j,\dots)/B(p_i + p_j,\dots), & \text{for } 1\le i\le N, 2N+1\le j\le 2N+M, \\ -A(p_i - p_j,\dots)/A(p_i + p_j,\dots), & \text{for } 2N\le i<j\le 2N+M, \\ -B(-p^*_{i-N} + p_j,\dots)/B(-p^*_{i-N} - p_j,\dots), & \text{for } N<i\le 2N<j\le 2N+M. \end{cases}$$

Using the ansatz (31) with (32-34) one can derive a condition, corresponding to (19), for the existence of multi-soliton solutions [7]. In our search program we checked the conditions corresponding to 3B, 2B+A, B+2A, 4B, 3B+A, 2B+2A, and B+3A solitons. [Note, however, that here one complex soliton counts as two.] We assumed that the dispersion conditions (33) had no multiple factors. The conditions for 3A and 4A solitons were studied in [3] so the A-polynomial is fixed by those results.

For the nonlinear Schrödinger equation there is a canonical first step in the transformation from the bilinear form to the nonlinear form:

$$F = e^u,\ G = ve^u,\ u \text{ real}, \tag{35}$$

that is, $v = G/F$ and $u = \log F$. Often one has to make further substitutions, derivations and integrations to obtain a conventional form. We will use (35) to exhibit our results also in this more familiar form.

The **complete** results can be expressed in the form of the following three equations (various transformations and projections are allowed, as usual);

(a) Melnikov's equation:

$$\begin{cases} (D_x^2 + iD_y + c)G\cdot F = 0, \\ (a(D_x^4 - 3D_y^2) + D_xD_t)F\cdot F = |G|^2. \end{cases} \tag{36}$$

The transformation (35) yields now

$$\begin{cases} v_{xx} + iv_y + cv + 2u_{xx}v = 0, \\ a(2u_{xxxx} + 12u_{xx}^2 - 6u_{yy}) + 2u_{xt} = |v|^2 \end{cases} \tag{37}$$

Next we operate by ∂_x^2 on the second equation, and put $w = 2u_{xx}$ to obtain

$$\begin{cases} v_{xx} + iv_y + cv + wv = 0, \\ a[w_{xxxx} + 3(w^2)_{xx} - 3w_{yy}] + w_{xt} = (|v|^2)_{xx}. \end{cases} \tag{38}$$

This equation has been studied by Melnikov in a number of articles [15,16,17]. The special case with a = 0 was obtained in [6], it goes to nlS if $t \to x$, $y \to t$, for then we can satisfy the second equation with $w = |v|^2$. For $v = 0$ the second equation reduces to the KP-equation (the sign of the yy-term is fixed!), so the result is a combination of nlS and KP. Note that the first equation has only two independent variables, x and y, while the second has x, t, and y.

(b) A combination of KP and DS:

$$\begin{cases} (i\alpha D_x^3 + 3D_xD_y - 2iD_t + c)G\cdot F = 0, \\ [a(\alpha^2D_x^4 - 3D_y^2 + 4\alpha D_xD_t) + bD_x^2]F\cdot F = |G|^2. \end{cases} \tag{39}$$

The special case a = 0 was included in [6]. The substitution (35) gives

$$\begin{cases} i\alpha(v_{xxx} + 6v_xu_{xx}) + 3(v_{xy} + 2u_{xy}v) - 2iv_t + cv = 0, \\ a[2\alpha^2(u_{xxxx} + 6u_{xx}{}^2) - 6u_{yy} + 8\alpha u_{xt}] + 2bu_{xx} = |v|^2. \end{cases} \tag{40}$$

There are now **two** different limits yielding equations that have appeared in the literature before:

(i) If we put $v = 0$, $\alpha = 1$, $a = 1$, $b = 0$, operate on the second equation by ∂_x^2 and define $w = 2u_{xx}$ then this equation becomes the KP-equation.

(ii) If we take $\alpha = 0$, $c = 0$, operate on the second equation by $\partial_x\partial_y$ and define $w = u_{xy}$ then we obtain

$$\begin{cases} 3v_{xy} - 2iv_t + 6wv = 0, \\ -6aw_{yy} + bw_{xx} = (|v|^2)_{xy}. \end{cases} \tag{41}$$

For $b = -6a = 1/\delta$, $w = -z + \frac{1}{2}\delta|v|^2$ and after a 45° rotation in the (x,y)-plane this equation gets the familiar form of the Davey-Stewartson (DS) equation for v and z.

In contrast to (38) both equations in (40) have 3 variables. Thus this equation is a true 2+1 combination of KP and DS equations. Note however that we had to use different reductions to obtain these equations. The two different limits have also different associated natural scalings. For these reasons one cannot see the connection in the standard form of representation for these equations, but in the Hirota formalism a combination can be written down. This emphasizes again the fundamental nature of Hirota's formalism.

(c) Generalization of Ito's equation:

$$\begin{cases} (i\alpha D_x^3 + 3cD_x^2 + i(bD_x - 2dD_t) + g)G\cdot F = 0, \\ (\alpha D_x^3 D_t + aD_x^2 + (b + 3c^2)D_x D_t + dD_t^2)F\cdot F = |G|^2 \end{cases} \quad (42)$$

This is a combination of Ito's equation (22) and Hirota's generalization of nlS [10]. Transformation (35) yields now

$$\begin{cases} i\alpha(v_{xxx} + 6u_{xx}v_x + 3c(v_{xx} + 2u_{xx}) + i(bv_x - 2dv_t) + gv = 0, \\ 2\alpha(u_{xxxt} + 6u_{xx}u_{xt}) + 2au_{xx} + 2(b + 3c^2)u_{xt} + du_{tt} = |v|^2. \end{cases} \quad (43)$$

6 Conclusions

In previous papers [3-6] we have initiated the search for bilinear equations having three-soliton solutions. Here we have presented the complete results for KdV- and nlS-type equations, together with their nonlinear forms.

It is quite surprising how few equations pass the three-soliton condition. This suggests that the equations found by this condition are most likely to be integrable. As further evidence in that direction we have recently found that these equations do pass the Painlevé test [19].

There are still many properties that must be checked before the new equations can be shown to be integrable, but the results so far are positive.

References

[1] Date, E., Jimbo, M., Kashiwara, M. & Miwa, T. (1982). *Publ. RIMS, Kyoto Univ.* **18**, 1077.
[2] Hearn, A.C. (1985). *REDUCE User's Manual Version 3.2.* Publ. CP78, rev 4/85. Rand: Santa Monica.
[3] Hietarinta, J. (1987). *J. Math. Phys.* **28**, 1732.
[4] Hietarinta, J. (1987). *J. Math. Phys.* **28**, 2094.
[5] Hietarinta, J. (1987). *J. Math. Phys.* **28**, 2586.
[6] Hietarinta, J. (1988). *J. Math. Phys.* **29**, 628.
[7] Hietarinta, J. (1989). Work in progress.
[8] Hirota, R. (1971). *Phys. Rev. Lett.* **27**, 1192.
[9] Hirota, R. (1972). *J. Phys. Soc. Jpn.* **33**, 1456.
[10] Hirota, R. (1973). *J. Math. Phys.* **14**, 805.
[11] Hirota, R. (1976). Direct method of finding exact solutions of nonlinear evolution equations. In *Bäcklund Transformations, the Inverse Scattering Method, and Their Applications*, pp. 40-68, Miura, R.M. (ed.). Springer-Verlag: Berlin.
[12] Hirota, R. (1980). Direct methods in soliton theory. In *Solitons*, pp. 157-176, Bullough, R.K. & Caudrey, P.J. (eds.). Springer-Verlag: Berlin.
[13] Ito, M. (1980). *J. Phys. Soc. Jpn.* **49**, 771.

[14] Jimbo, M. & Miwa, T. (1983). *Publ. RIMS, Kyoto Univ.* **19**, 943.

[15] Melnikov, V.K. (1986). *Phys. Lett.* **118A**, 22.

[16] Melnikov, V.K. (1987). *J. Math. Phys.* **28**, 2603.

[17] Melnikov, V.K. (1987). *Commun. Math. Phys.* **112**, 639.

[18] Ramani, A. (1981). In *Fourth International Conference on Collective Phenomena,* p. 54, Lebowitz, J. (ed.). New York Academy of Sciences: New York.

[19] Grammaticos, B., Ramani, A. & Hietarinta, J. (1989). Work in progress.

[20] Satsuma, J. (1976). *J. Phys. Soc. Jpn.* **40**, 286.

Part 4

ALGEBRAIC METHODS AND HAMILTONIAN THEORY

P.M. Santini

31. ON THE OPERATOR STRUCTURES UNDERLYING INTEGRABILITY

A review of results on the algebraic properties of integrable systems in an arbitrary number of dimensions is presented. We first show that physically relevant evolutions like the Korteweg-de Vries, the nonlinear Schrödinger, the Benjamin-Ono, the Chiral fields equations and their integrable mulidimensional analogues are generated by different realisations of the same operator structures. Then we summarise the theory recently developed by Fokas and the author, associated with the recursion operators and with the bi-Hamiltonian formulation of integrable systems in multidimensions. Finally, we discuss a natural application of this theory, the "deformations approach", in which the richness of the algebraic structure of a given integrable system is explored, with the goal of constructing integrable generalisations in multidimensions.

1 Introduction

In this review we discuss the algebraic properties of integrable nonlinear evolution equations in multidimensions, often referring to the following celebrated examples: the Korteweg-de Vries (KdV)

$$u_t = u_{xxx} + 6uu_x, \tag{1.1}$$

the nonlinear Schrödinger (NLS)

$$iu_t + \frac{1}{2}u_{xx} - u|u|^2 = 0 \tag{1.2}$$

the intermediate long wave (ILW)

$$u_t = \hat{T}u_{xx} + 2uu_x, (\hat{T}f)(x) := (\eta)^{-1}P\int_{-\infty}^{\infty}\coth((\pi/\eta)(\xi - x))f(\xi)d\xi, \tag{1.3a}$$

and its $\eta \longrightarrow \infty$ limit, the Benjamin-Ono (BO)

$$u_t = \hat{H}u_{xx} + 2uu_x, \quad (\hat{H}f)(x) := \pi^{-1}P\int_{-\infty}^{+\infty}(\xi - x)^{-1}f(\xi)d\xi, \tag{1.3b}$$

the Chiral fields (CF)

$$(g^{-1}g_t)_x = \alpha(g^{-1}g_x)_t \tag{1.4}$$

equations and their 2+1 dimensional analogues: the Kadomtesev-Petviashvili (KP)

$$u_t=u_{xxx}+6uu_x+\alpha^2\partial_x^{-1}u_{yy}, \tag{1.5}$$

the Davey-Stewartson (DS)

$$iu_t+\frac{1}{2}(u_{x_1x_1}+\alpha^2u_{x_2x_2}) = uv \tag{1.6a}$$

$$v_{x_1x_1}-\alpha^2v_{x_2x_2} = (|u|^2)_{x_1x_1}+\alpha^2(|u|^2)_{x_2x_2}, \tag{1.6b}$$

the 2+1 dimensional version of the ILW and BO

$$u_t = \hat{T}D^2u+2uDu, \ D :== \alpha\partial_y, \tag{1.7a}$$

$$u_t = \hat{H}D^2u+2uDu, \tag{1.7b}$$

and the self-dual Yang-Mills

$$(g^{-1}g_t)_x = \alpha(g^{-1}g_y)_z. \tag{1.8}$$

In particular:

i) we emphasize that equations (1.1-8) are nothing but *different* realisations of the *same* operator structure [1] and then they are algebraically *equivalent*! This unified perspective allows a more profound understanding of the basic algebraic properties underlying integrability (§2);

ii) we outline the theory developed by Fokas and the author [3,4,5] associated with the recursion operators and the bi-Hamitonian structures of integrable systems. This theory which generalises to a multidimensional context results by Magri, Gel'fand, Diki, Dorfman, Fokas and Fuchssteiner[2)], describes integrable systems regardless of the number of dimensions, degenerates into the previous one in all cases in which the previous one is adequate, incorporates very naturally the notion of Bäcklund transformations and the associated Calogero's results [15,16] into the scheme and leads to the existence of infinitely many symmetries and constants of motion in involution under distinct Poisson brackets (§3);

iii) we review the dimensional deformations [6], an approach used to explore the richness of the operator structure associated with a given integrable system, with the goal of generating integrable generalisations (deformations) in higher dimensions (§4). This approach is a natural application of the theory discussed in §3 and finds a natural collocation in the unified presentation of §2.

2 A common operator structure underlying integrability

2.1 *Bilocal Representation of Integrable Systems.*

Nonlinear integrable systems like equations (1.1-8) are elements of hierarchies of integrable equations that can be represented in the following *bilocal* form [9,6,3,4,5,10]

$$\delta(x-x')u'_t = \beta_n\delta(x-x')K^{(n)}(u,u'), \quad u = u(x,t), u' := u(x',t), \tag{2.1a}$$

$$K^{(n)}(u,u') := \Phi^n\hat{K}^0\cdot 1, \tag{2.1b}$$

which obviously implies

$$u_t=\beta_n\int_{\mathcal{R}^\upsilon} dx'\delta(x-x')K^{(n)}(u,u') = \beta_n K^{(n)}(u,u') =: \beta_n k^{(n)}(u). \tag{2.1c}$$

where x,x' are υ -dimensional vectors of components x_i,x'_i, $i=1,..,\upsilon$,

$$\delta(x-x') = \prod_{i=1}^{\upsilon}\delta(x_i-x'_i)$$

(δ is the Dirac function), $K^{(n)}$ belongs to suitable space of S of polynomials in $u=u(x,t)$, $u'=u(x',t)$ and their integrals and derivatives, and the "recursion operator" Φ and the "starting operator" $\hat{K}^0$ are operator valued functionson S. Through this paper m and n are nonnegative integers.

2.2 *The Operator Structures Underlying Integrability*

Integrable systems admit in general *more than one* bilocal representation (2.1), corresponding to more than one pair $\Phi,\hat{K}^0$ [8,9], but in this presentation we shall focus only on a specific one, which is the operator structure *common* to all the equations (1.1-8).

Precisely it turns out that the integrable systems (1.1-8) can be viewed as *different bilocal representations* of the *same operator structure* [1], given by

$$\Phi = q^-(d^-)^{-1}, \tag{2.2a}$$

$$\hat{K}^0 = q^-C, \quad d^-C = 0, \tag{2.2b}$$

where the basic operators q^-,d^- are defined in the following way

$$\left(\begin{pmatrix} q^- \\ d^- \end{pmatrix} f\right)(x,x') := \int_{\mathcal{R}^{\upsilon}} dx''\left(\begin{pmatrix} q(x,x'') \\ d(x,x'') \end{pmatrix} f(x'',x') - \right.$$

$$\left. -f(x,x'')\begin{pmatrix} q(x'',x') \\ d(x'',x') \end{pmatrix}\right) \qquad (2.3)$$

in terms of their kernels $q(x,x')$and $d(x,x')$, that can be thought as NxN matrix functions.

If $d=\delta(x-x')\sigma$, where σ is a constant NxN diagonal matrix, then d^-reduces to the usual commutation operator between matrices

$$d^- f = \hat{\sigma}f: = \sigma f - f\sigma \ , \qquad (2.4)$$

and we can derive the following interesting "off-diagonal" reduction of the operator structure (2.2) [1]

$$\Phi = (q_D^- + (1-\Pi)q_0^- - q_o^-(q_D^-)^{-1}\Pi q_0^-)\hat{\sigma}^{\ -1}, \qquad (2.5a)$$

$$\hat{K}^0 = q_0^- C_D, \qquad (2.5b)$$

where $q_0$and q_D are the off-diagonal and diagonal parts of q, Π f is the diagonal part of f and C_D is a diagonal matrix.

2.3 *Different Representation*

Different choices of the kernels q and d give rise to different bilocal representations of the nonlocal operators q^- and d^-; in particular *differential* equations like (1.1-8) are generated by kernels q and d proportional to the Dirac function and its derivatives; in the following we list the different bilocal realisations which generate equations (1.1-8).

Examples 2.1

1) If $d=\delta'(x_1-x_1'),q=\delta(x_1-x_1')u'+\delta''(x_1-x_1')$ and u is a scalar; then $d^-=\partial_{+1},q^-=u-u'+\partial_{+1}\partial_{-1}$, where

$$\partial^j_{i\pm} := (\partial_{xi}\pm\partial_{xi'})^j \ , \qquad (2.6)$$

and (2.2) reduces to the bilocal representation of the KdV class; the KdV equation (1.1) is the fourth member ($n=3,\beta_3=1/2$) of the class (2.1) [8,5].

2) If $d=\delta^1_1$, $q=\delta u'+\delta^2_1+\alpha\delta^1_2$ where

$$\delta_i^j := \delta^j(x_i - x_i') \prod_{s\neq i} \delta(x_s - x_s'), \tag{2.6a}$$

$$\delta^j(x_i - x_i') := \partial^j \delta(x_i - x_i')/\partial x_i^j, \tag{2.6b}$$

then $d^- = \partial_{1+}$, $q^- = u - u' + \partial_{1+}\partial_{1-} + \alpha\partial_{2+}$ and (2.2) reduces to the bilocal representation of the KP class (2.1); the KP equation (1.5) is its fourth member ($n=3$, $\beta_3 = 1/2$) [8,5].

3) If $d = -(1/2)\delta(x_1 + i\eta - x_1')$, $q = \delta(x_1 - x_1')u' + i\delta'(x_1 - x_1')$, then

$$d^- = (1 - ee')(2e')^{-1}, \quad q^- = u - u' + i\partial_{1+},$$

$$(ef)(x,x') := f(x + i\eta, x'), \quad (e'f)(x,x') := f(x, x' + i\eta)$$

and (2.2) reduces to the bilocal representation of the ILW class; the ILW equation (1.3a) is its third member ($n=2$, $\beta_2 = 1/4i$) [11,5].

4) If $d = (-1/2)\delta(x_1 + i\eta - x_1')$, $q = \delta u' + \alpha\delta_2^1$, then $d^- = (1 - ee')(2e')^{-1}$, $q^- = u - u' + \alpha\partial_{2+}$ and (2.2) becomes the bilocal representation of the 2+1 dimensional ILW equation (1.7a), corresponding to $n=2$, $\beta_2 = 1/4i$ in (2.1) [5].

5) If $d = \delta'(x_1 - x_1')$, $q = \delta(x_1 - x_1')u'$, then $d^- = \partial_{1+}$, $q^- f = uf - fu'$ and we obtain the bilocal representation of the CF algebra,; if $d = \delta'(x_1 - x_1')\delta(x_2 - x_2')$, $q = \delta u' + \delta(x_1 - x_1')\delta'(x_2 - x_2')$, then $d^- = \partial_{1+}$, $q^- f = uf - fu' + \alpha\partial_{2+} f$ and we obtain the bilocal representation of the SDYM algebra. The CF and SDYM equations are generated by these algebras via the mechanism discussed in §4.2 (see also [6]).

6) If $N=2$, $d = \delta\sigma$, $\sigma = \mathrm{diag}(1 -1)$, $q_0 = \delta u_0'$, $q_D = I\delta'(x_1 - x_1')$, then $d^- f = \hat{\sigma} f$, $q_D^- = \partial_{1+}$, $q_0^- f = uf - fu'$ and (2.5) reduces to the bilocal represenation of the NLS class; the NLS equation comes from (21) for $n=2$, $\beta_2 = 1/4i$, $\bar{u}_{12} = u_{21} = u$ [3,6].

7) If $N=2$, $d = \delta\sigma$, $\sigma = \mathrm{diag}(1,-1)$, $q_0 = \delta u_0'$, $q_D = \delta_1^1 I + \alpha\delta_2^1\sigma$, then $q_D^- f = \partial_{1+} f + \alpha(\sigma f_{x2} + f_{x2'}\sigma)$ and (2.4) reduces to the representation of the DS class; the DS equation coincides with (2.1) for $n=2$, $\beta_2 = 1/4i$ [3,6].

All the equations (1.1-8), generated by the structures (2.2,5), are described by the theory summarised in the next section.

3 **Algebraic properties of integrable systems**

In this section we review the basic results of the theory developed in [3,4,5,6] associated with recursion operators of integrable systems. We shall closely follow the terse and concise presentation given in [6], §2. For the sake of concreteness we shall always refer to the explicit Examples 2.1 and to their operator structures (2.2), although the results presented have a more general validity (see for example [6]).

3.1 *Operator Structure and Canonical Lie Algebra*

The operator structure of a given integrable system is completely described when we assign

(i) the recursion operator Φ;

(ii) the starting operator(s) $\hat{K}^0$;

(iii) a Lie algebra $\tilde{H}$ of suitable functions $H(x,x')$ endowed with the bracket

$$[H^{(1)},H^{(2)}]_I := \int_{\mathcal{R}^\upsilon} dx''(H^{(1)}(x,x'')H^{(2)}(x'',x')-$$

$$-H^{(2)}(x,x'')H^{(1)}(x'',x')), \tag{3.1}$$

and containing an *abelian* subalgebra $\tilde{h}$ of functions $h=h(x-x')$;

In the Examples 2.1 Φ and $\hat{K}^0$ depend on the basic operators q^-, d^- defined in (2.3). Definition (2.3) implies the following operation

$$q^-[g(x,x')]f(x,x') = \int_{\mathcal{R}^\upsilon} dx''(g(x,x'')f(x'',x')-$$

$$-f(x,x'')g(x'',x')) =: (g^-f)(x,x'), \tag{3.2}$$

where $q^{-\prime}[g]$ denotes the Frechet derivative of the operator q^- in the direction of the intergral operators g^- and, consequently,

$$(L(q^-))'[g^-] = \partial_\varepsilon L(q^-+\varepsilon g^-)|_{\varepsilon=0}. \tag{3.3}$$

Moreover we assume that d^- is functionally constant, namely that $(d^-)'[g]=0$. The following properties hold:

Properties 3.1

i) The recursion operator Φ is a hereditary or Nijenhuis operator, namely

$$\Phi'[\phi f]g-\Phi\Phi'[f]g \text{ is symmetric w.r.t. } f \text{ and } g, \qquad (3.4)$$

ii) The starting operators $\hat{K}^0$, acting on elements of $\tilde{H}$, satisfy the equation

$$[\hat{K}^0H^{(1)},\hat{K}^0H^{(2)}]_d = c\Phi^i\hat{K}^0[H^{(1)},H^{(2)}]_I \ ,H^{(1)},H^{(2)} \in \tilde{H}, \qquad (3.5)$$

for some nonnegative integer i and constant c, in terms of the Lie bracket $[,]_d$ defined by

$$[X^{(1)},X^{(2)}]_d := X^{(1)'}[X^{(2)}]-X^{(2)'}[X^{(1)}]. \qquad (3.6)$$

iii) Φ is a strong symmetry for $\hat{K}^0H$ or, equivalently, the Lie derivative of Φ in the direction $\hat{K}^0H$ is zero, namely

$$\Phi'[\hat{K}^0H]+\Phi(\hat{K}^0H)' - (\hat{K}^0H)'\Phi = 0 \qquad (3.7)$$

iv) Φ admits a factorization $\Phi=\Theta^{(2)}(\Theta^{(1)})^{-1}$ in terms of the two compatible Hamiltonian operators $\Theta^{(1)}=d^-,\Theta^{(2)}=q^-$, namely if $\Theta:=\Theta^{(1)}+\xi\Theta^{(2)}$, then for any constant ξ, we have that

a)

$$\Theta^* = -\Theta, \qquad (3.8a)$$

where L* denotes the adjoint of L with respect to the symmetric bilinear form

$$\langle g,f\rangle := \mathrm{tr}\int_{\mathcal{R}^{2\upsilon}} dxdx'g(x',x)f(x,x') \qquad (3.8b)$$

(the trace operation is dropped if g and f are scalars);

b) Θ satisfies the Jacobi identity w.r.t. the bracket

$$\{a,b,c\} := \langle a,\Theta'[b]c\rangle; \qquad (3.9)$$

v) $\hat{\Gamma}^0H :=(\Theta^{(1)})^{-1}\hat{K}^0H$ is a gradient, namely

$$(\hat{\Gamma}^0H)' := (\hat{\Gamma}^0H)'^*. \qquad (3.10)$$

vi) The eigenfunctions of $\Phi^*= (d^-)^{-1}q^-$ are the "bilocal squared eigenfunctions" of the associated linear problem, namely if

$$\Phi^{*}w = \lambda w, w = w(x,x') \tag{3.11}$$

then

$$w = v(x)\tilde{v}(x'), \tag{3.12}$$

where v and $\tilde{v}$ are the right and left eigenfunctions of the associated spectral problem

$$d_L v = \lambda q_L v,\ d_R\tilde{v} = \lambda q_R\tilde{v}, \tag{3.13}$$

$$(\begin{pmatrix} q_L \\ d_L \end{pmatrix} f)(x) := \int dx'' \begin{pmatrix} q \\ d \end{pmatrix}(x,x'')f(x''), \tag{3.14a}$$

$$(\begin{pmatrix} q_R \\ d_R \end{pmatrix} f)(x) := \int dx'' f(x'') \begin{pmatrix} q \\ d \end{pmatrix}(x'',x), \tag{3.14b}$$

$$q_L - q_R = q^{-}; \tag{3.14c}$$

Remark 3.1

i) The factorisation $\Phi=\Theta^{(2)}(\Theta^{(1)})^{-1}$ in terms of the two Hamiltonian operators implies the following "well-coupling" condition $\Phi\Theta^{(1)}=\Theta^{(2)}\Phi^{*}$.

ii) Properties 3.1 are not independent, for example:

a) the factorisation of a compatible pair of Hamiltonian operators gives rise to a hereditary operator Φ;

b) there exists a one to one correspondence between spectral problems and their associated recursion operators [3,8]; in particular spectral problems yield hereditary operators Φ [3] (see [12] for the derivation of (2.2) from the abstract spectral problem (3.14));

c) the recursion operator Φ algorithmically implies $\hat{K}^{0}H$, furthermore if Φ is hereditary, it is also a strong symmetry for $\hat{K}^{0}H$ (see §4 of [3]).

Properties 3.1 characterize the Lie algebra underlying integrability; namely we have the following

Theorem 3.1. Let μ be the linear hull of the operators $\hat{K}^{(m)}:=\Phi^{m}\hat{K}^{0}$, namely

$$\mu := \{\hat{K}\colon\ \hat{K} = \sum_{j} c_j \hat{K}^{(m_j)}, m_j \geq 0, c_j \in C\}, \tag{3.15}$$

then Properties 2.1, i-iii) imply that the vector space

$$\kappa_{\tilde{H}} := \{X : X = \hat{K}H,\ H(x,x') \in \tilde{H}, \hat{K} \in \mu\} = :\mu(\tilde{H}) \qquad (3.16)$$

is a Lie algebra endowed with the Lie bracket$[,]_d$ and

$$\mu(c_1H^{(1)}+c_2H^{(2)}) = c_1\mu(H^{(1)})+c_2\mu(H^{(2)}), \qquad (3.17a)$$

$$[\mu(H^{(1)}),\mu(H^{(2)})]_d \subset \mu([H^{(1)},H^{(2)}]_I),H^{(i)} \in \tilde{H}, i = 1,2, \qquad (3.17b)$$

where $\mu(H)$ is the hull of vectors $\hat{K}$ $H,\hat{K} \in \mu$.

Remark 3.2

i) Equations (3.17) imply that $\kappa_{\tilde{h}}$ $:\{X:X=\hat{K}h, h=h(x-x')\in\tilde{h},\hat{K}\in\mu\}$ is an abelian subalgebra of $\kappa_{\tilde{H}}$.

ii) While the abelian algebra $\kappa_{\tilde{h}}$ generates commuting (time independent) symmetries and constants of motion in involution, the nonabelian algebra $\kappa_{\tilde{H}}$ generates time dependent symmetries [4] and is strictly related to the mastersymmetries theory[13,14].

Definition 3.1. The vector space $\kappa_{\tilde{H}}$ satisfying the set of properties illustrated above, is called a *canonical* Lie algebra.

3.2 *Distributional projection and directional derivative*

As we have seen §2, the evolution equations (1.1-8) are obtained choosing q and d to be generalised functions; correspondingly the basic operator q^- becomes a bilocal operator u^-, and Φ and $\hat{K}^0$ assume a bilocal structure involving the function u evaluated at two different points x and x' (see **Examples 2.1**).

Since $\kappa_{\tilde{H}}$ is canonical with respect to the Frechet derivative of the nonlocal operator q^-, any particular choice (like the one given in **Examples 2.1**) of $q(x,x')$ preserves the canonicity of the obtained bilocal algebra with respect to the so called *directional d-derivative*

$$u^-_d[g(x,x')] := q^{-\prime}[g(x,x')] = g^-. \qquad (3.18)$$

of the bilocal operator u^-. For this reason any bilocal realisation of an operator canonical Lie algebra $\kappa_{\tilde{H}}$ is hereafter called canonical too.

The d-derivative of the bilocal operator u^- satisfies the following *projective* formula

$$u^-_d[\delta(x-x')g(x,x')]f(x,x') = g(x,x)f(x,x') -f(x,x')g(x',x') =$$

$$=: u^{-}_{f}[g]f(x,x'), \tag{3.19}$$

where the subscript f denotes the usual Frechet derivative with respect to u, namely

$$L_f(u,u')[g] = \partial_\epsilon L(u+\epsilon g(x,x),u'+\epsilon g(x',x'))\big|_{\epsilon=0} \tag{3.20}$$

The d-derivative and the usual Frechet derivative are the main tool of this theory.

3.3 *Integrable evolution equations*

Now we are ready to state the main theorem for integrable systems.

Theorem 3.2. **Let** $\Theta^{(1)}+\xi\Theta^{(2)}$ be a bilocal Hamiltonian operator for all constants ξ. Define $\Phi := \Theta^{(2)}(\Theta^{(1)})^{-1}, \hat{\Gamma}^0 := (\Theta^{(1)})^{-1}\hat{K}^0$ and assume that

i) Φ is a strong symmetry for $\hat{K}^0 H$;

ii) the starting symmetries $\hat{K}^0 H$ satisfy equation (3.5);

then

1) Φ is a hereditary or Nijenhuis operator;

2) $\Phi^m\Theta^{(1)}$ are Hamiltonian operators;

3) $\kappa_{\tilde{H}}$ is a canonical Lie algebra endowed with the Lie bracket $[,]_d$.

Moreover if, in addition,

iii) $\hat{\Gamma}^0 H$ is an extended gradient,

iv) the distrubution $\delta(x-x')K^{(n)}(u,u')$ belongs to the abelian sub-algebra $\kappa_{\tilde{h}}$,

then

4) $\Sigma^{(m)} := \Phi^m\hat{K}^0\cdot 1$ and $\Gamma^{(m)} := (\Theta^{(1)})^{-1}\Sigma^{(m)} = (\Phi^*)^m\hat{\Gamma}^0\cdot 1$ are extended symmetries and extended gradients of the conserved quantities I_n in involution respectively, for equation (2.1), namely

$$\Sigma^{(m)}{}_f[K^{(n)}(u,u)] = (\delta K^{(n)})_d[\Sigma^{(m)}], \tag{3.21a}$$

$$\Gamma^{(m)}{}_f[K^{(n)}(u,u)] = -(\delta K^{(n)})_d^{*}[\Gamma^{(m)}], \tag{3.21b}$$

$$(\Gamma^{(m)})_d = (\Gamma^{(m)})_d^{*} \Leftrightarrow I_{md}[f] = \langle\Gamma^{(m)},f\rangle, \tag{3.21c}$$

$$\{I_m,I_n\}_i := \langle\delta\Gamma^{(m)},\Theta^{(i)}\Gamma^{(n)}\rangle, i = 1,2; \tag{3.21d}$$

5) equations (2.1a,b) are *extended bi-Hamiltonian systems*,

since they can be written in the following two extended Hamiltonian forms

$$\delta(x-x')u'_t = \delta(x-x')\Theta^{(1)}\Gamma^{(n)} = \delta(x-x')\Theta^{(2)}\Gamma^{(n-1)}; \tag{3.22}$$

6) $\sigma^{(m)}=\sigma^{(m)}(u):=\Sigma^{(m)}(u,u)$ and $\gamma^{(m)}=\gamma^{(m)}(u):=\Gamma^{(m)}(u,u)$ are commuting symmetries and gradients of the conserved quantities I_m involution respectively for equations (2.1), namely

$$\sigma^{(m)}{}_f[k^{(n)}] = k_f^{(n)}[\sigma^{(m)}], \tag{3.23a}$$

$$\gamma^{(m)}{}_f[k^{(n)}] = -k^{(n)}{}^+_f[\gamma^{(m)}], \tag{3.23b}$$

$$\gamma^{(m)}{}_f = \gamma^{(m)}{}^+_f \Leftrightarrow I_{mf}[f] = (\gamma^{(m)},f) := \int_{\mathcal{R}^\nu} dx\gamma^{(m)}f; \tag{3.23c}$$

7) the equations $\Sigma^{(m)}=\Sigma^{(m)}(u,u')=0$ are auto BT's equations (2.1)

Remark 3.3

i) Theorem 3.2 shows that symmetries and BT's of an integrable system originate from the same entity: the *extended symmetry*, defined in equation (2.21a). In this equation *two different* operations on bilocal structures appear: the usual Frechet derivative and the novel d-derivative, which takes account of the nonlocal origin of bilocal operators.

ii) If the the hypothesis iv) of Theorem 3.2 is satisfied, namely if

$$\delta(x-x')K^{(n)}(u,u') = \sum_{l=1}^{n} b_{n,l}\Phi^{n-l}\hat{K}^0\cdot h^{(l)},\ h^{(l)}\in \tilde{h}, \tag{3.24}$$

for some constants $b_{n,l}$, then the operators $(\delta K^{(n)})_d$, appearing in the equation (2.36), are well defined and read

$$(\delta K^{(n)})_d = \sum_{l=1}^{n} b_{n,l}(\Phi^{n-l}\hat{K}^0\cdot h^{(l)})d. \tag{3.25}$$

iii) In the BT $\Sigma(u,u')=0$, the functions u,u′ are now viewed as two different solutions $u=u(x)$, $u'=u'(x)$ of $u_t=k(u)$ and, correspondingly, in equations (3.21) the even derivative ∂_{i+} of the operator $(\delta K^{(n)})_d$ is replaced by ∂x_i; the odd derivative ∂_{i-} is absent (see theorem 4.1 of [3]).

Theorem 3.2 provides a nonambiguous definition of integrability.

Definition 3.2. A nonlinear evolutionary system $u_t=k(u)$ is integrable if it is associated with a canonical Lie algebra $\kappa_{\tilde{H}}$ in the way prescribed by **Theorem 3.2**, namely the following *two* conditions must be simultaneously satisfied:

A) there exists an underlying canonical Lie algebra $\kappa_{\tilde{H}}$ generated by the bilocal operators Φ and $\hat{K}^0$;

B) the evolution equation takes the distributional form (2.1a), and $\delta K^{(n)}(u,u')$ belongs to the canonical algebra $\kappa_{\tilde{H}}$, namely equation (3.24) holds.

Conditions A and B are at the basis of the deformation approach presented in the following section.

4 Deformations and integrable systems

In the deformation theory[6] we propose an elementary algebraic approach to multidimensions showing that, given an integrable system, say in 1+1 dimensions, it is possible to generate its integrable multidimensional extensions in a very straightforward manner, deforming in a simple way the associated operator structures, and using the theoretical results reviewed in §3.

Operating at the level of simple operator structures, this approach has allowed to derive, with essentially *no* calculations involved, all the known examples of integrable multidimensional systems and some novel ones. Moreover the simplicity of the operations involved has also allowed to isolate and focus the technical and conceptual reasons for the almost total absence of nonlinear integrable systems in more than 2+1 dimensions.

The deformation approach has been the way in which the conceptual unity of the theory (briefly illustrated in §2) has become apparent, leading to the identification of elementary structures underlying integrability (the algebraic "roots" of integrability) and, at the same time, to the construction of solvable algebraic systems[1].

In the light of the above unified point of view, the deformation approach

i) finds a very natural collocation, indeed: two different bilocal operator structures connected via a deformation are nothing but *different* bilocal realisations of the *same* nonlocal operator structure!

ii) answers the following basic question connected to the abstract structures (2.2): does *any* choice of the kernels q and d of the nonlocal operators q^-, d^- give rise to integrable systems and, if not, what are the necessary constraints? Equivalently, given a couple of basic operators q^- and d^- which give rise to an integrable system, how can one *deforme* them in new operators $q^- \longrightarrow q^- + \chi^-$, in order to generate other integrable systems, eventually in higher or lower dimensions?

These and other questions are exhaustively answered by the "deformation approach" which we summarise in this section in the way it was originally formulated in [6].

The theory reviewed in §3 is based on two facts:

i) the existence of a *canonical* Lie algebra,

ii) the associated evolution equations belong to this algebra.

Consequently the dimensional deformation approach is based on the following two steps:

1) we introduce a suitable dimensional deformation of the bilocal Lie algebra of the integrable system under consideration, in order to generate a higher dimensional generalisation; this can always be done in a straightforward manner, once the operator structure of the integrable system is known;

2) we check whether the obtained multidimensional Lie algebra can be used to generate integrable multidimensional systems. This check consists of computing a simple commutator relation and it is also very straightforward.

4.1 *The deformations approach*

A deformation of the Lie algebra $\kappa_{\tilde{H}}$ is obtained through a suitable deformation of the basic operators q^-, d^-. This is achieved replacing their kernels q and/or d by $q+\chi$ and/or $d+\chi, \chi=\chi(x,x')$; then, correspondingly,

$$q^- \longrightarrow q^- + \chi^- . \tag{4.1}$$

Of particular interest for us are deformations of the type (4.1) which are associated with an increase of the number of dimensions of the vectors x and x′ and, correspondingly, give rise to higher dimensional systems; they are called *dimensional deformations.*

An example of *dimensional deformation* from, say, a one dimensional canonical algebra to a three dimensional one is given by

$$\chi^- = \alpha(\partial_{x_2} - \partial_{x'_2}) + \beta(\partial^2_{x_3} - \partial^2_{x'_3}), \tag{4.2a}$$

corresponding, via the mechanism (2.3), to the kernel

$$\chi = \alpha\delta^1_2 + \beta\delta^2_3. \tag{4.2b}$$

Remark 4.1. With respect to the d-derivative

$$(q^- + \chi^-)_d[g] := g^- \tag{4.3}$$

the obtained algebra is still canonical (see also §3.2) then deformations of a canonical Lie algebra allow to construct very easily multidimensional canonical Lie algebras satisfying condition A of §3. On the other hand it is also very easy to check whether or not the associated multidimensional system (2.1a,b) belongs to this algebra (condition B of §3). In order to check condition B or, equivalently, equation (3.24), it is indeed enough to compute the commutator $[\Phi,h]$, $[\hat{K}^0,h]$. It is perhaps convenient to illustrate this in two explicit examples.

4.2 *Examples of dimensional deformations*

1) If we consider the bilocal algebra generating the NxN NLS

class, described in **Examples 2.1,4**), and we look for deformations of χ of $q_D = I\delta_1'$ of the type

$$\chi = \sum_{i=2}^{\nu} J_i \delta_i^1, \tag{4.4a}$$

then the "new" operator $\bar{q}_D$ reads

$$\bar{q}_D f = \sum_{i=1}^{\nu} (J_i f_{x_i} + f_{x_i'} J_i), \quad J_1 = I \tag{4.4b}$$

and

$$[\Phi, \delta] = \left(\sum_{i=2}^{\nu} \delta_i^1 \hat{J}_i \right) (\hat{a})^{-1}, \quad [\hat{K}^0, \delta] = 0 \tag{4.5}$$

Condition B is satisfied if

$$\hat{J}_i = b_i \hat{a} \Leftrightarrow J_i = e_i I + b_i a, \tag{4.6}$$

since it is straightforward to check that $K^{(n)}$ satisfy equation (3.24) with $b_{n,l} = (-1)^l$ and $h^{(l)} := (\sum_{i=2}^{\upsilon} b_i \partial_{xi})^l \delta(x-x')$. The N-wave interaction in $\upsilon+1$ dimensions

$$u_{ij_t} = (a_i - a_j)^{-1} \sum_{i=1}^{\nu} ((e_l + b_l a_i) c_j - (e_l + b_l a_j) c_i) u_{ij_{x_l}} -$$

$$- \sum_{k=1}^{N} \left(\frac{c_i - c_k}{a_i - a_k} - \frac{c_k - c_j}{a_k - a_j} \right) u_{ik} u_{kj} \tag{4.7}$$

is the second member of the hierarchy (2.1) for n=1, $\beta_1 = 1$.

2) It is very easy to convince oneself that, for a generic deformation of the canonical Lie algebra, condition B is *not* satisfied. Then the associated multidimensional systems are *not* integrable, according to **Definition 3.2**, and *none* of the beautiful properties stated in **Theorem 3.2** are enjoyed by the associated class of evolution equations.

For example, if $\upsilon=3$, $q=\delta u'$, $d=\delta_1^1$, then the deformation $\chi = \alpha\delta_2^1 + \beta\delta_3^2$ of the operator algebra (2.2) implies that

$$[\Phi, h] = 2\beta h' \partial_{3+} \partial_{1+}^{-1}, \quad [\hat{K}^0, h] = 2\beta h' \partial_{3+}; \tag{4.8}$$

consequently it is easy to see that $\delta \Phi^n \hat{K}^0 \cdot 1 \notin \kappa_{\tilde{h}}$ and the associated

3+1 dimensional equations are not integrable.

The problem of the existence of nontrivial integrable structures in more than 2+1 dimensions is open and it is, in our opinion, one of the most outstanding one. This approach, in its simplicity, allows to focus concretely the associated theoretical and technical difficulties.

Technically speaking, the almost total absence of integrable systems in multidimensions (more than 2+1 dimensions) is related to the difficulty of satisfying condition B for a generic deformation of the *available* canonical algebras.

4.3 *Elementary deformations and integrable systems with an arbitrary number of time variables.*

The simplest possible deformation of a canonical Lie algebra is of course achieved

i) increasing the number of dimensions of the vectors x,x′,

ii) keeping unchanged the structure of q and d, namely choosing $\chi=0$ in formula (4.1).

We refer to it as an *elementary* dimensional deformation.

This apparently trivial deformation gives rise to interesting integrable multidimensional systems containing an *arbitrary* number of variables that can not be distinguished from the time variable t. Their derivation relies upon the existence of *ghost* classes (see [6]), a common feature of the bilocal realisations of the algebras (2.2.5) for which

$$[\Phi,h] = [\hat{K}^0,h] = 0,\ \hat{K}^0 = u^-,\ u^-f = uf-fu', \tag{4.9}$$

and it consists of the following two steps:

i) we introduce an elementary deformation, making the vectors x and x′ υ-dimensional;

ii) we make use of the following distributional identity

$$\delta_i^1 \Phi^{n_i} u^- \cdot 1 = -\delta(x-x')\Phi^{n_i} u'_{x'_i}, \tag{4.10}$$

which follows directly from equation (4.9). Then the distributional evolution equation

$$\delta u' = \delta K^{(n)}(u,u'), \tag{4.11}$$

$$K^{(n)}(u,u') = \sum_{i=1}^{\nu} c_i \Phi^{n_i} u'_{x'_i} + \beta_n \Phi^n \hat{M}\cdot 1, \tag{4.12}$$

is integrable, since the identity (4.10) implies

$$\delta K^{(n)}(u,u') = -\sum_{i=1}^{\nu} c_i \Phi^{n_i} u^- \cdot \delta_i^1 + \beta_n \Phi^n \hat{M}\cdot\delta \in \kappa_{\tilde{h}}, \tag{4.13}$$

namely condition B is satisfied.

Then the results of **Theorem 3.2** apply to the class of evolution equations (4.11,12), whose projection gives

$$u_t = \sum_{i=1}^{\nu} c_i \phi^{n_i} u_{x_i} + \beta_n \phi^n \hat{m}\cdot 1 =: k^{(n)}, \qquad (4.14a)$$

$$\phi := (\hat{u}+\alpha\partial_{x_2})(\partial_{x_1})^{-1}, \qquad (4.14b)$$

$$\hat{m} := \hat{u}+\alpha\partial_{x_2}. \qquad (4.14c)$$

In particular $\sum(u,u')=a_1\Phi^{m_1}\bar{u}\cdot 1+a_2\Phi^{m_2}\hat{M}\cdot 1$ and $\Gamma(u,u')=(\Theta^{(1)})^{-1}\sum(u,u')$ are extended symmetries (BT's) and extended gradients of conserved quantities in involution for the evolution equations (4.14).

Remark 4.2

i) The evolution equation (4.14) shows that the variables $x_i, 2\le i\le\upsilon$, introduced via the deformation approach described in this section, play the *same* role as t; for this reason they will be called *time variables.* In this sense, it is worthwhile to remark again that integrable *evolution* equations should be correctly viewed as *elementary dimensional* deformations (corresponding to the introduction of the variables $x_2=t$) of their *stationary* analogues.

ii) Calogero was the first to derive examples of integrable equations with an arbitrary number of time variables [15,16].

Examples 4.2

i) In order to derive the SDYM equation (1.8) we notice that the bilocal algebra of **Examples (2.1,5)** satisfies conditions (4.9) for c=1. Then the distributional identity (4.10) holds and the class of integrable multidimensional systems (4.14) can be constructed. The SDYM equation (1.8) is equivalent, with the positions $u=g^{-1}g_{x_2}, u_{x_3}=(g^{-1}g_t)_{x_1}$, to the equation [18]

$$u_t = \alpha\partial_{x_1}^{-1}u_{x_2x_3}+u\_\partial_{x_1}^{-1}u_{x_3}, \qquad (4.15)$$

obtained from (4.14a) for $c_i=0, i\neq 3, c_3=1, n_3=1$. The CF equation is obtained from (1.8) in the reduction $x_2=x_1$, $x_3=t$ [17,18].

ii) Elementary deformations can also be used to generate the following novel (and peculiar) class of multidimensional equations

$$v_t = \beta_n \prod_{i=1}^{n+1} v_{x_i} =: \beta_n k^{(n)}. \qquad (4.16)$$

If we consider the canonical algebra (2.2) with $q=\delta u', d=\delta'(x_1-x_1')$

and u scalar, the identity (4.19) generalises to

$$\left(\prod_{i=1}^{n+1} \delta^{1}(x_i - x'_i)\right) \Sigma^{(n)}(u,u') = (-1)^{n+1}\delta(x-x')\left(\prod_{i=1}^{n+1} \partial^{-1}_{x'_1}\partial_{x'_i}u'\right)_{x'_1}, \tag{4.17a}$$

$$\Sigma^{(n)}(u,u') := \Phi^{n}u^{-}\cdot 1 = ((u-u')\partial^{-1}_{1+})(u-u') =$$

$$\frac{1}{(n+1)!}\,(\partial_{1+}(\partial^{-1}_{1+}(u-u'))^{n}). \tag{4.17b}$$

Then the evolution equation

$$\delta(x-x')u'_t = \beta_n\delta(x-x')\left(\prod_{i=1}^{n+1} \partial^{-1}_{x'_1}\partial_{x'_i}u'\right)_{x'_1} \tag{4.18}$$

belongs to the canonical algebra $\kappa_{\tilde{H}}$ and reduces to equation (4.16) via the change of variable $v_{x_1}=u$.

Remark 4.4

i) $\Sigma^{(m)}=((m+1)!)^{-1}(\partial_{1+}((v-v')^{m+1}))$ are extended symmetries of the evolution equations (4.18), namely equations (3.36a) are satisfied *but* the corresponding BT's $\Sigma^{(m)}(v,v')=0$ are trivial (they imply $v'=v$), and the corresponding commuting symmetries are zero.

ii) Analogously $\Gamma^{(m)}=\partial^{-1}_{1+}\Sigma^{(m)}$ are extended gradients of conserved quantities for equations (4.18), but the projected gradients $\gamma^{(m)}$ are zero and then the corresponding constants of motion are trivial (they do not depend on v).

The only result of **Theorem 2.2** which does not trivilise is

$$k_f^{(n)}[k^{(m)}] = k_f^{(m)}[k^{(n)}], \tag{4.19}$$

which follows from **Th.4**. of [2]. Then $k^{(n)}$ form an infinite dimensional abelian Lie algebra or, in other words, the evolution equations (4.16) describe commuting flows.

iii) It turns out that equations (4.16) can be integrated using the methods of the characteristics [19].

References

[1] P.M.Santini, The algebraic structures underlying integrability and solvable algebraic systems, Preprint PM.88/55, Montpellier 1988. Inverse Problems (in press).

[2] F.Magri, *J.Math.Phys.* **19**,1156 (1978); I.M.Gel'fand and I.Y.Dorfman, *Func.,Anal.Appl.* **13**,13 (1979); B.Fuchssteiner

and A.S. Fokas, *Physica* **4D**,47k (1981).
[3] P.M.Santini and A.S.Fokas, *Comm.Math.Phys.* **115**,375 (1988).
[4] A.S.Fokas and P.M.Santini, *Comm.math.Phys.* **116**,449 (1988).
[5] P.M.Santini and A.S.Fokas, The bi-Hamiltonian formulations of integrable evolution equations in multidimensions, in Nonlinear Evolutions, Proceedings of the IV Workshop on Nonlinear Evolution Equations and Dynamical Systems; edited by J.Leon, World Scientific Publishing Company, Singapore (1988).
[6] P.M.Santini, Inverse Problems **5**, 67 (1989).
[7] A.S.Fokas and P.M.Santini, *Stud.Appl.Math.* **75**,179 (1986).
[8] M.Bruschi, private communication.
[9] A.S.Fokas and P.M.Santini, *J.Math.Phys.* **29**,604 (1988).
[10] B.G.Konopelchenko, Recursion and group theoretical structures of the integrable equations in 1+1 and 2+1 dimensions. Bilocal approach, Preprint INP Novosibirsk, USSR, 1987.
[11] P.M.Santini, Bi-Hamiltonian formulatiions of the Intermediate Long Wave equation, Preprint INS 80, Clarkson University, 1987; Inverse Problems (submitted to).
[12] A.S.Fokas and P.M.Santini, A unified approach to recursion operators, Preprint INS 101, Clarkson University, 1988.
[13] A.S.Fokas and B.Fuchssteiner, *Phys.Lett.* **86A**,341 (1981).
[14] W.Oevel and B.Fuchssteiner, *Phys.Lett.* **88A**,323 (1982). H.H.Chen, Y.C.Lee and J.E.Lin, *Physica* **9D**,439 (1983). B.Fuchssteiner, *Progr.Theoret.Phys.* **70**,150 (1983).
[15] F.Calogero, *Lett.Nuovo Cimento* **14**,443 (1975).
[16] F.Calogero, *Lett.Nuovo Cimento* **14**,537 (1975); F.Calogero and A.Degasperis, *Nuovo Cimento* **32B**,201 (1976); *Nuovo Cimento* **39B**,1 (1977).
[17] M.Bruschi, D.Levi and O.Ragnisco, *Lett.Nuova Cimento* **33**,263 (1982).
[18] M.Bruschi, D.Levi and O.Ragnisco, *Phys.Lett.* **88A**,379 (1982).
[19] We thank W.Craig for showing us that equation (4.16) can be integrated using the method of the characteristics.

B.G. Konopelchenko

32. OPERATOR REPRESENTATION OF THE INTEGRABLE EQUATIONS

1 **Introduction**

A starting and central point of the inverse spectral transform method (see e.g. [1-4]) is a correspondence between a given nonlinear differential equation and a certain set of the auxiliary linear problems

$$L_i\Psi = 0 \qquad i = 1,\dots,d \tag{1}$$

where L_i are linear (usually, differential) operators. Namely, the condition of compatibility of the linear system (1) is equivalent to the system of nonlinear equations for the coefficients of the operators L_i, which is integrable with the help of the problems (1) [1-4].

In the present paper we will discuss the different operator forms of the compatibility condition for the linear system (1). We will demonstrate that the generic operator representation of these compatibility conditions is the system of the operator equations

$$[L_i,L_k] = \sum_{l=1}^{d} \gamma_{ikl}\, L_l \qquad (i,k = 1,\dots,d) \tag{2}$$

where $[L_i,L_k] = L_iL_k - L_kL_i$ and γ_{ikl} are some operators.

The famous Lax represenation $\frac{\partial L}{\partial t} = [L,A]$ for the Korteweg-de Vries equation was the first form of the operator representation of the integrable equations [5]. Then the more general operator form of the compatibility conditions for the integrable equations has been introduced by Zakharov & Shabat [6]. This is the commutativity condition

$$[L_1,L_2] = 0 \tag{3}$$

of two linear operators L_1 and L_2. A great number of the integrable equations in 1+1 dimensions have such an operator representation [1-4]. The well-known nonlinear Schrödinger equation, sine-Gordon equation, N-waves equations and Landau-Lifshitz equation are among them [1-4].

The commutativity operator representation (3) is generalizable to the 2+1-dimensional case (two spatial dimensions). The operators L_1 and L_2 which lead to the nontrivial nonlinear equations in 2+1 dimensions are of the form [6]

$$L_1 = \partial_y + \sum_{n=0}^{N} U_n(x,y,t)\, \partial_x^n$$

$$L_2 = \partial_t + \sum_{m=0}^{M} V_m(x,y,t)\, \partial_x^m \qquad (4)$$

where U_n and V_m are matrix-valued functions of the independent variables x, y and t.

The main feature of such operators L_1 and L_2 is that they contain only the first order derivatives over the two variables y and t. The class of corresponding integrable equations is rather restricted. Nevertheless it contains the Kadomtsev-Petviashvili equation, Davey-Stewartson equation, N-waves equations and some other integrable equations in 2+1 dimensions [1-4,6].

2 Triad operator representation

The commutativity of two operators L_1 and L_2 is a very strong and useful condition. It guarantees the existence of the complete set of the common eigenfunctions for these operators. On the other hand the condition (3) is very restrictive. Only for the operators L_1, L_2 of the form (4) it gives nonlinear equations for the coefficients U_n and V_m in 2+1 dimensions. The condition (3), for example, for the operator $L_1=\partial_x^2+\partial_y^2+U(x,y,t)$ immediately leads to linear equations for U(x,y,t) [7,6.4].

A way for passing over this obstacle has been found by Manakov [8]. Based on the properties of the two-dimensional scattering problems he has proposed to use instead of Lax representation, the triad operator representation $\frac{\partial L}{\partial t} = LA-BL$ or, equivalently,

$$[L_1, L_2] = B_1 L_1 \qquad (5)$$

where B_1 is a differential operator.

Manakov has presented also the first, rather complicated, example of the nonlinear system representable in the form (5).

A simple 2+1-dimensional equation which has the operator representation (5) has been found by Nizhnik [9] and Veselov &

Novikov [10]. It is of the form

$$U_t + k_1 U_{\xi\xi\xi} + k_2 U_{\eta\eta\eta} - 3k_1(U\partial^{-1}U_\eta)_\eta - 3k_2(U\partial_\eta^{-1}U_\xi)_\xi = 0 \qquad (6)$$

where U(x,y,t) is a scalar function, $\partial_\xi = \partial_x - \sigma\partial_y$, $\partial_\eta = \partial_x + \sigma\partial_y$, $\sigma^2 = \pm 1$ and k_1, k_2 are arbitrary constants. The operators L_1, L_2, B_1 for the Nizhnik-Veselov-Novikov (NVN) equation (6) are [9,10]

$$L_1 = -\left(\partial_x{}^2 - \sigma^2\partial_y{}^2\right) + U(x,y,t),$$

$$L_2 = \partial_t + k_1\partial_\xi{}^3 + k_2\partial_\eta{}^2 - 3k_1\partial_\xi{}^{-1}U_\eta \cdot \partial_\eta - 3k_2\partial_\eta^{-1}U_\xi \cdot \partial_\xi,$$

$$B_1 = -3k_1\partial_\xi^{-1}U_{\eta\eta} - 3k_2\partial_\eta^{-1}U_{\xi\xi} \qquad (7)$$

The system [11]

$$q_t = \Delta q + \alpha(q_\xi)^2 - \beta(q_\eta)^2 + 2\alpha\partial_\eta^{-1}U_\xi - 2\beta\partial_\xi^{-1}U_\eta,$$

$$U_t = -\Delta U + 2\alpha(Uq_\xi)_\xi - 2\beta(Uq_\eta)_\eta \qquad (8)$$

is another example of the nonlinear system which has the operator representation (5). Here q(x,y,t) and u(x,y,t) are scalar functions, $\Delta = -\alpha\partial_\xi^2 + \beta\,\partial_\eta^2$ and α,β are arbitrary constants. The corresponding operators L_1, L_2, B_1 are

$$L_1 = \partial_\eta\partial_\xi + q_\xi\partial_\eta + U,$$

$$L_2 = \partial_t + d\partial_\xi^2 + \beta\partial_\eta^2 + 2\beta\, q_\eta\partial_\eta + 2\beta\partial_\eta^{-1}U_\xi,$$

$$B_1 = 2\Delta q.$$

At $\beta=0$ the system (8) is reduced to the 2+1-dimensional integrable generalization of the dispersive long-waves equations found in [12]. The papers [11-14] contain other examples of the 2+1-dimensional integrable equations which possess the triad operator representation (5).

There is a certain arbitrariness in the choice of the operators L_2 and B_1 for a given nonlinear equation. Zakharov [15] has demonstrated that for the triad operator representation (5) the operators L_2 and B_1 are defined up to the transformation

$$L_2 \to L_2 = L_2 + CL_1,$$

$$B_1 \to B_1 = B_1 + [L_1, C] \qquad (9)$$

where C is an arbitrary differential operator.

This arbitrariness allows us to choose the operators L_2 and B_1 in the most convenient form. On the one hand starting from the commutativity representation (3) for a given nonlinear equation (e.g. Kadomtsev-Petviashvili equation) one can construct an infinite set of the triad representation for this equation using the transformation (9). On the other hand starting from given triad representation one can try to find the corresponding commutativity representation. It is not difficult to show that it is impossible to convert the triad representation (5) for the NVN equation (6) into the commutativity representation if one is restricted by the scalar operators. But this becomes possible if one combines the transformation (9) with the transition to the 2x2 matrix form of the operators L_1, L_2 and B_1 [11]. For example, for the Veselov-Novikov equation (NVN equation (6) with $\sigma = i$ and $k_1=k_2=1$) one obtains the commutativity representation $[L_1^M, L_2^M] = 0$ with the operators [11]

$$L_1^M = \partial_y - i\begin{pmatrix}1 & 0\\ 0 & -1\end{pmatrix}\partial_x + i\begin{pmatrix}0 & 1\\ U & 0\end{pmatrix},$$

$$L_2^M = \partial_t - 8\partial_x^3 - \begin{pmatrix}6U + 6(\bar{\partial}^{-1}\partial U), & 0\\ 6(\partial U) + 6(\bar{\partial}U), & 6U+6(\partial^{-1}\bar{\partial}U)\end{pmatrix}\partial_x + \tag{10}$$

$$+ \begin{pmatrix}6U_x , & 3\bar{\partial}^{-1}\partial U - 3\partial^{-1}\bar{\partial}U\\ 6\bar{\partial}U_x + 3U(\bar{\partial}^{-1}\partial U-\partial^{-1}\bar{\partial}U) , & 6U_x + 3\partial U-(\partial^{-1}\bar{\partial}^2U)\end{pmatrix}$$

where $\partial_\xi = \partial$, $\partial_\eta = \bar{\partial}$.

The corresponding 2x2 matrix operators L_1^M and L_2^M for the system (8) with $\alpha = 1$, $\beta = 0$ are [11]

$$L_1^M = \partial_y + \frac{1}{\sigma}\begin{pmatrix}1 & 0\\ 0 & -1\end{pmatrix}\partial_x - \frac{1}{\sigma}\begin{pmatrix}0 & 1\\ U & q_\xi\end{pmatrix},$$

$$L_2^M = \partial_t + \begin{pmatrix}4 & 0\\ 0 & 0\end{pmatrix}\partial_x^2 - 2\begin{pmatrix}0 & 1\\ U & 0\end{pmatrix}\partial_x \tag{11}$$

$$+ \begin{pmatrix}-U + 2\partial_\eta^{-1}U_\xi , & -q_\xi\\ 2U_\xi + U_\eta - Uq_\xi , & q_{\xi\xi} - (q_\xi)^2 - U + 2\partial_\eta^{-1}U_\xi\end{pmatrix}.$$

Matrix commutativity representation can also be constructed for other integrable equations which possess a scalar triad representation [11].

Note, however, an essential difference between the equations which have the scalar commutativity representation and those which possess the matrix commutativity representation (e.g. between KP and NVN equations). For the equations of the second type the commutativity representation is a reducible one. The irreducible operator representation for such a type of equation (e.g. NVN equation) is a scalar triad operator representation.

One can also consider the 1+1-dimensional nonlinear equations which possess the triad operator representation. This gives some new integrable systems in 1+1-dimensions [14,11].

At last, it is not difficult to see that the triad operator equation (5) is equivalent to the system of equations

$$L_1\Psi = 0$$

$$[L_1, L_2]\ \Psi = 0. \qquad (12)$$

The system (12) presents the weak commutativity condition proposed in [13,12]. Therefore the approach to 2+1-dimensional equations based on the weak commutativity condition is equivalent to the triad operator representation approach.

3 Quartet operator representation

The triad operator representation can be generalized in a rather obvious way. Indeed, the more general operator form of the compatibility condition of the linear system $L_1\Psi = 0$, $L_2\Psi = 0$ is the following [16]

$$[L_1,\ L] = B_1L_1 + B_2L_2 \qquad (13)$$

where B_1 and B_2 are differential operators.

The class of operators L_1 and L_2 which demands the consideration of more general operator equations than (5) has been pointed out first by Zakharov in the paper [17]. These are the operators of the form

$$L_1 = A_1(\partial_x)\partial_y + B_1(\partial_x),$$

$$L_1 = A_2(\partial_x)\partial_t + B_2(\partial_x) \qquad (14)$$

where A_1, A_2, B_1, B_2 are matrix-valued differential over x operators, the coefficients of which depend on x, y and t. The corresponding compatibility condition has been presented in [17] in the form of a rather complicated operator system. In fact this system is equivalent to equation (13) with the operators B_1 and B_2 of the form $B_1=C_1(\partial_x)\partial_t+D_1(\partial_x)$, $B_2=C_2(\partial_x)\partial_y + D_2(\partial_x)$ where C_1,C_2,D_1,D_2 are differential over x operators.

The simplest known nonlinear system in 2+1-dimensions which has the quartet representation is the following system [18]

$$U_t + U_{\xi\xi\xi} + U_{\eta\eta\eta} - 3(U\partial_\xi^{-1}U_\eta)_\eta - 3(U\partial_\eta^{-1}U_\xi)_\xi - (\varphi^2)_{\xi\eta} = 0,$$

$$-\varphi_{\xi\eta} + U\varphi = 0 \tag{15}$$

where $U(x,y,t)$, $\varphi(x,y,t)$ are scalar functions, $\partial_\xi = \partial_x - \sigma\partial_y$, $\partial_\eta = \partial_x + \sigma\partial_y$ and $\sigma^2 = \pm 1$. The corresponding operators L_1, L_2, B_1, B_2 are

$$L_1 = \begin{pmatrix} -\partial_\xi\partial_\eta + U, & 0 \\ -2\varphi\partial_\xi \quad , & -\partial_\xi \end{pmatrix},$$

$$L_2 = \begin{pmatrix} \partial_t + \partial_\xi^3 + \partial_\eta^3 - 3\partial_\xi^{-1}U_\eta \cdot \partial_\eta - 3\partial_\eta^{-1}U_\xi \cdot \partial_\xi \ , & \varphi \\ (\varphi^2)\eta \quad , & -\varphi\partial_\eta \end{pmatrix}, \tag{16}$$

$$B_1 = \begin{pmatrix} -e\partial_\xi^{-1}U_{\eta\eta} - 3\partial_\eta^{-1}U_{\xi\xi} \ , & -\varphi + \varphi_\eta \\ -(\varphi^2)_\eta \quad , & \varphi\partial_\eta \end{pmatrix}.$$

Similar to the triad representation the operators L_1, L_2, B_1, B_2 in the quartet representation for a given nonlinear equation are defined nonuniquely but up to the transformations [16]

$$L_i \to L_i' = \sum_{k=1}^{2} Q_{ik}L_k, \qquad N_i \to N_i' = \sum_{k=1}^{2} N_k\hat{Q}_{ki} \tag{17}$$

where $N_1 = B_1 + L_2$, $N_2 = B_2 - L_1$ and Q_{ik} and $\hat{Q}_{ki}$ are arbitrary differential operators which obey the constraints $\sum_{k=1}^{2} \hat{Q}_{1k}Q_{ki} = \partial_{il}$. In the particular case $Q_{ii} = \hat{Q}_{ii} = 1$ $(i=1,2)$, $Q_{12} = \hat{Q}_{12} = 0$ and $Q_{21} = -\hat{Q}_{21} = C$ the transformation (17) is reduced to the transformation

$$L_1 \to L_1' = L_1, \qquad L_2 \to L_2' = L_2 + CL_1, \tag{18}$$

$$B_1 \to B_1' = B_1 + [L_1, C] - B_2C. \qquad B_2 \to B_2' = B_2.$$

At $B_2 = 0$ (triad representation) this transformation coincides with the transformation (9).

4 Generic operator representation of the integrable equations

Nonlinear systems in 2+1 dimensions which are equivalent to the compatibility condition of the two linear equations $L_1\Psi = 0$ and

$L_2\Psi = 0$ have one common and important feature. Their integrability is hereditarily a 1+1-dimensional one. Indeed, two-dimensional differential operators L_1, L_2 can be obtaied from the one-dimensional NxN matrix operators with the special matrix structure in the limit $N \to \infty$. For the KP equation this has been demonstrated in [19]. Hence the procedure of the integration of the corresponding 2+1-dimensional equations inherits the main feature of the 1+1-dimensional procedure. Furthermore, such integrable equations always contain one detached independent variable. The general operator representation for such a type of equation in the quartet representation (13).

An essentially new feature appears when one proceeds to the consideration of the nonlinear systems which contain all three independent variables x_1, x_2, x_3 on an equal footing and more or less symmetrically. It may be in this case that it is not sufficient to associate the nonlinear equation with only two auxiliary linear equations. This can be easily seen on the simplest example. Let us consider the two linear problems

$$L_1\Psi = \left(\partial_{x_1}\partial_{x_2} + U_1\partial_{x_1} + U_2\partial_{x_2}\right)\Psi = 0,$$

$$L_2\Psi = \left(\partial_{x_1}\partial_{x_3} + V_1\partial_{x_1} + V_3\partial_{x_3}\right)\Psi = 0. \qquad (19)$$

The compatibility condition of this system (the equality $\partial_{x_3}\partial_{x_1}\partial_{x_2}\Psi = \partial_{x_2}\partial_{x_1}\partial_{x_3}\Psi$) gives $U_2 = V_3 = \phi$, $V_1 = \dfrac{\partial\varphi}{\partial x_3}$, and $U_1 = -\dfrac{\partial\varphi}{\partial x_2}\cdot\phi^{-1}$. So we have not obtained any nontrivial nonlinear system.

The situation changes radically if one demands that the equation

$$L_3\Psi = \left(\partial_{x_2}\partial_{x_3} + W_2\partial_{x_2} + W_3\partial_{x_3}\right) \qquad (20)$$

holds in addition to the system (19). The compatibiliy condition of all three auxiliary equations (19) and (20) now gives the nontrivial nonlinear system. This system is of the form

$$\partial_{x_i}\partial_{x_k}g_l + \partial_{x_i}g_l\cdot\frac{A_k}{\lambda_l-\lambda_i} + \partial_{x_k}g_l\cdot\frac{A_i}{\lambda_l-\lambda_k} +$$

$$g_l\,\frac{A_iA_k}{(\lambda_l-\lambda_i)(\lambda_l-\lambda_k)} - \left(\partial_{x_i}g_k + \frac{g_kA_i}{\lambda_k-\lambda_i}\right)g_k^{-1}\left(\partial_{x_k}g_l + \frac{g_lA_k}{\lambda_l-\lambda_k}\right) \qquad (21)$$

$$-\left(\partial_{x_k} g_i + \frac{g_i A_k}{\lambda_i - \lambda_k}\right) g_i^{-1} \left(\partial_{x_i} g_l + \frac{g_l A_i}{\lambda_l - \lambda_i}\right) = 0 \qquad \begin{pmatrix} i,k,k = 1,2,3 \\ i \neq k \neq l \end{pmatrix}$$

where $g_i(x_1,x_2,x_3)$ are NxN matrices, $[A_i,A_k] = 0$ $(i,k=1,2,3)$, $\lambda_i (i=1,2,3)$ are arbitrary constants and there is no summation over repeated indices. This system has been found by Zakharov & Manakov (by another approach) in [20].

The system (21) is the three-dimensional integrable generalization of the well-known principal chiral fields equations in 1+1 dimensions.

The operator representation of the system (21) is rather new. Namely, the system (21) is equivalent to the system of the three operator equations

$$[L_{ik}, L_{nk}] = \alpha_{ink} L_{in} + \beta_{ink} L_{ik} + \gamma_{ink} L_{nk} \tag{22}$$

$$\begin{pmatrix} i,k,n = 1,2,3 \\ i \neq n \neq k, \ i \neq k \end{pmatrix}$$

where

$$L_{ik} = \partial_{x_i} \partial_{x_k} + U^i_{ik} \partial_{x_i} + U^k_{ik} \partial_{x_i}, \quad (i \neq k) \tag{23}$$

and

$$U^i_{ik} \equiv \left(-\frac{\partial g_i}{\partial x_k} + g_i \frac{A_k}{\lambda_i - \lambda_k} \right) g_i^{-1}.$$

The coefficients α_{ink}, β_{ink} and γ_{ink} are

$$\alpha_{ink} = \frac{\partial U^n_{nk}}{\partial x^k} - \frac{\partial U^i_{ik}}{\partial x^k} - \left[U^n_{nk}, U^i_{ik}\right],$$

$$\beta_{ink} = \frac{\partial U^n_{nk}}{\partial x^i} - \frac{\partial U^i_{ik}}{\partial x^n} - \left[U^k_{nk}, U^i_{ik}\right], \tag{24}$$

$$\gamma_{ink} = \frac{\partial U^n_{nk}}{\partial x^i} - \frac{\partial U^k_{ik}}{\partial x^k} - \left[U^n_{nk}, U^k_{ik}\right].$$

There is no summation over repeated indices in the formulae (22)-(24).

It is easy to see that the operator system (22) is nothing but the operator system (2) for d=3 rewritten in the double-indices notations.

The Zakharov-Manakov system (21) is the first representative of the class of the generic integrable three-dimensional nonlinear systems. The generic three-dimensional integrable nonlinear systems seem to be equivalent to the compatibility condition of the three auxiliary linear problems (1) and their operator representations are given by the operator equations (2) [21]. The main feature of the integration procedure for the generic three-dimensional integrable equations is that in this case one should solve simultaneously all three linear problems $L_i\Psi=0$ (i=1,2,3) in contrast to the hereditary 1+1-dimensional equations.

The construction proposed can be easily generalized to the case of the arbitrary number of the independent variables. For the two-dimensional case d=2 the generic integrable systems have the operator representation $[L_1,L_2] = \gamma_1 L_1+\gamma_2 L_2$. The equation [21]

$$\Delta\varphi_y-4\,\alpha\,\varphi_{xy} + 6\alpha^2(\varphi\cdot\Delta^{-1}(\varphi^2)_{xy})_x + 6\alpha\beta\varphi_y \cdot \Delta^{-1}(\varphi^2)_{yy} = 0 \qquad (25)$$

is the representative of the class of such generic two-dimensional integrable equations. Here $\varphi(x,y)$ is a scalar function, $\Delta=\alpha\partial_x^2-\beta\partial_y^2$ and α,β are arbitrary constants.

Equation (25) has the operator representation $[L_1,L_2] = \gamma_1 L_1+\gamma_2 L_2$ where

$$L_1 = \partial_x\partial_y + \frac{1}{2}\varphi\partial_y - \frac{1}{4}\varphi_y - \frac{3}{4}\alpha\Delta^{-1}(\varphi^2)_{xy},$$

$$L_2 = \alpha\partial_x^2 + \beta\partial_y^2 - \alpha\varphi\partial_x + \frac{\alpha}{2}\varphi_x + \frac{3}{4}\alpha\varphi^2 - \frac{3}{2}\alpha\Delta^{-1}I\varphi^2)_{xx}, \qquad (26)$$

$$\gamma_1 = -2\alpha\varphi_x, \qquad \gamma_2 = -\varphi_y.$$

The nonlocal equation (25) can be easily rewritten as a local system by introducing the function Ψ such that $\Delta\Psi = (\varphi^2)_y$.

At $\beta = 0$ equation (25) is reduced to the very simple equation

$$\varphi_{xy} - 2\varphi\partial_x^{-1}(\varphi^2)_y = 0. \qquad (27)$$

Note that the noncommutative operator representation of the two-dimensional equations different from the quartet representation has been considered in [22].

Multidimensional (d>3) integrable systems which have been constructed in the papers [23-25] have the commutative operator representations (2) (all $\gamma_{ikl} = 0$).

5 Conclusion

The operator representation (2) for the integrable equations represents the logical completion of the idea of the initial Lax and commutativity operator representations. Such operator representation reveals the deep interrelation between the nonlinear integrable systems and the closed algebras of the differential operators.

The closed operator algebras defined by the equations (2) can be considered as the certain generalization of the Lie algebras to the case when the structure constants become the differential operators too. As far as the theory of the Lie algebras is very rich in content and applications then it seems the theory of the closed algebras of the differential operators of the type (2) should play a fundamental role in the theory of the integrable systems.

Acknowledgements

I am very grateful to the Physics Department of the Lecce University and especially to M. Boiti and F. Pempinelli for their kind hospitality.

References

[1] Zakharov, V.E., Manakov, S.V., Novikov, S.P. & Pitaevsky, L.P. (1980). *Soliton Theory. The Method of Inverse Problem.* Plenum (1984): Nauka, Moskwa.

[2] Ablowitz, M.J. & Segur, H. (1981). *Solitons and Inverse Scattering Transform.* SIAM: Philadelphia.

[3] Calogero, F. & Degasperis, A. (1982). *Spectral Transform and Solitons I.* North Holland: Amsterdam.

[4] Newell, A.C. (1985). *Solitons in Mathematics and Physics.* SIAM: Philadelphia.

[5] Lax, P.D. (1968). *Commun. Pure Appl. Math.* **21**, 467.

[6] Zakharov, V.E. & Shabat, A.B. (1974). *Funct. Analysis and its Appl.* **8**, 23.

[7] Perelomov, A.M. (1976). *Lett. Math. Phys.* **1**, 175.

[8] Manakov, S.V. (1976). *Uspekhi Matem. Nauk.* **31**, 245.

[9] Nizhnik, L.P. (1980). *Doklady Acad. Nauk SSSR* **254**, 332.

[10] Veselov, A.P. & Novikov, S.P. (1984). *Doklady Acad. Nauk SSSR* **279**, 20.

[11] Konopelchenko, B.G. (1988). *Inverse Problems* **4**, 151.

[12] Boiti, M., Leon, J.-J.P. & Pempinelli, F. (1987). *Inverse Problems* **3**, 37.

[13] Boiti, M., Leon, J.-J.P., Manna, M. & Pempinelli, F. (1986). *Inverse Problems* **2**, 271.

[14] Mel'nikov, V.K. (1983). *Lett. Math. Phys.* **7**, 129.

[15] Zakharov, V.E. (1980). The inverse scattering method. In *Solitons,* p. 243. Bullough, R. & Caudrey, P. (eds.). Springer-Verlag: Berlin.

[16] Konopelchenko, B.G. (1988). *J. Phys. A: Math. Gen.* **20**, L1057.

[17] Zakharov, V.E. (1987). Integrable systems in multidimensional spaces. *Lecture Notes in Physics*, vol. 153, p. 190. Springer-Verlag: Berlin.

[18] Dubrovsky, V.G. & Konopelchenko, B.G. (1988), to be published.

[19] Caudrey, P. (1986). *Inverse Problems* **2**, 281.

[20] Zakharov, V.E. & Manakov, S.V. (1985). *Funct. Analysis and its Appl.* **19**, 11.
[21] Konopelchenko, B.G. (1987). On operator representation of integrable equations, preprint 87-164 of Inst. Nucl. Phys., Novosibirsk.
[22] Novikov, S.P. & Veselov, A.P. (1986). *Physica* **18D**, 267.
[23] Saveliev, M.V. (1986). *Theor. Math. Phys.* **69**, 1234.
[24] Ablowitz, M.J., Beals, R. & Tenenblat, K. (1986). *Studies in Appl. Math.* **74**, 177.
[25] Ablowitz, M.J., Costa, D.G. & Tenenblat, K. (1987). *Studies in Appl. Math.* **77**, 37.

S. Carillo and B. Fuchssteiner

33. NON COMMUTATIVE SYMMETRIES AND NEW SOLUTIONS OF THE HARRY DYM EQUATION

The well known links between the KdV, the mKdV and the Harry Dym equation, whose extensions to the whole hierarchies have been obtained by Rogers-Nucci, will be reconsidered in this paper. We introduce a Bäcklund chart involving also the KdV-singularity hierarchy as well as the KdV interacting soliton structure; the study of the links depicted therein will allow us to obtain the recursion operator, together with its implectic-symplectic factorization. Furthermore this chart will reveal the very rich symmetry structure of the Harry Dym equation. It turns out that there are four different hierarchies of non-linear evolution equations enjoying the property to be symmetry generators for the Harry Dym equation. These hierarchies do not commute. An additional fifth hierarchy of symmetries is then found by reciprocal transformation of the trivial symmetries of the KdV "zero-soliton" equation. It is remarkable that the symmetry structure exhibited by the Harry Dym equation is isomorphic to that of the Kawamoto equation. This result enforces our belief that all the integrable non-linear systems in 1+1-dimensions are, more or less, related to each other. As a last result we will point out how the links in the Bäcklund chart indicate a way to construct from the KdV soliton structure explicit solutions for the Harry Dym equation.

1 The Bäcklund chart involving KdV, mKdV, KdV singularity, the KdV "zero soliton" and the Harry Dym hierarchy

It is well known that the Harry Dym and the KdV-mKdV equations are related [1] as well as it is known that such a link holds for the whole hierarchies [20]. In [10] we reconsidered these links in order to involve the KdV-singularity manifold equation and the interacting soliton equation. Indeed both play a fundamental role in our considerations. This insofar as out of invariances of the KdV-singularity we will obtain new invariances of the Harry Dym hierarchy, and from solutions of the KdV-interacting equation we obtain explicit solutions for the Harry Dym equation.

We consider the following Bäcklund chart:

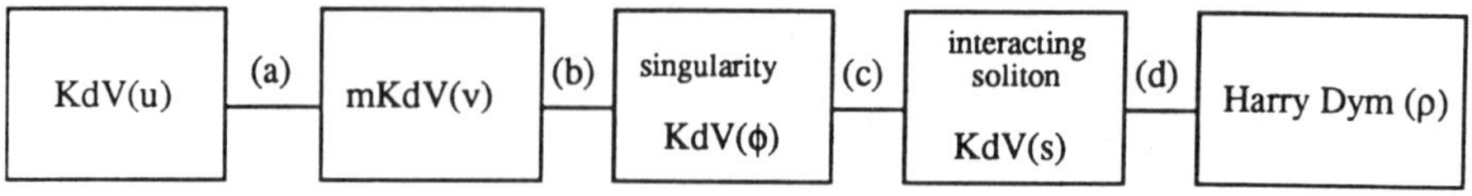

where
(a) is the usual Miura transformation,
(b) is a Cole-Hopf transformation,
(c) is the introduction of a potential,
(d) is the reciprocal link.

A detailed presentation of all the links exhibited in this Bäcklund chart is given below.

The KdV hierarchy can be written in the form

$$u_t = \Phi(u)^n u_x, \qquad n = 0,1,2,... \tag{1}$$

where

$$\Phi(u) = D^2+2DuD^{-1}+2u \tag{2}$$

is its hereditary recursion operator [3,5], whose implectic-symplectic factorization [4] is

$$\Phi(u) = \Theta(u)J(u) \tag{3}$$

where

$$\Theta(u) = D^3+2Du+2uD \tag{4}$$

$$J(u) = D^{-1} \tag{5}$$

This factorization gives a compatible bi-hamiltonian structure [17,18,12,13,14]. The first member in the hierarchy is the KdV equation

$$u_t = u_{xxx}+6uu_x. \tag{6}$$

The Miura transformation

$$u+v_x+v^2 = 0 \tag{a}$$

applied to the KdV equation transforms it into the mKdV equation

$$v_t = v_{xxx}-6v^2v_x, \tag{7}$$

and transforms $\Phi(u)$ into the hereditary recursion operator [4] for the mKdV

$$\Phi_2(v) = D^2-4DvD^{-1}v \tag{8}$$

so that the whole hierarchy (1) goes over into the mKdV hierarchy

$$v_t = \Phi_2(v)^n v_x, \qquad n - 0,1,2,... \tag{9}$$

Successively, the application of the Cole-Hopf type transformation

$$v = \frac{1}{2}\frac{\phi_{xx}}{\phi_x} \tag{b}$$

relates the mKdV hierarchy to the KdV-singularity hierarchy

$$\phi_t = \phi_x B_n(\{\phi;x\}) \tag{10}$$

where $\{\phi;x\}$ denotes the Schwarzian derivative and where

$$B_0(\{\phi x\}) = 1 \tag{11}$$

$$B_1(\{\phi;x\}) = \{\phi;x\} \tag{12}$$

and

$$B_{n+1}(\{\phi;x\}) = D^{-1}(D^3+D\{\phi;x\}+\{\phi;x\}D)B_n(\{\phi;x\}) \tag{13}$$

The base member of this hierarchy is

$$\phi_t = \phi_x(\phi;x\}. \tag{14}$$

Now the introduction of a "bona fide" potential

$$s = \phi_x \tag{c}$$

transforms the KdV-singularity equation into

$$s^2 s_t = s^2 s_{xxx} - 3ss_x s_{xx} + \frac{3}{2}s_x^3 \tag{15}$$

which turns out to be the interacting soliton equation of the KdV [8] for spectral parameter λ equal to zero. We will address this equation as the **zero-soliton** equation for the KdV and we will call the hierarchy

$$s_t = [sB_n(S)]_x \tag{16}$$

with

$$S = \left(\frac{s_x}{s}\right)_x - \frac{1}{2}\left(\frac{s_x}{s}\right)^2 \tag{17}$$

the **zero-soliton hierarchy**. The relation between the singularity manifold and the zero-soliton equation is not at all coincidental but rather the result of structural reasons [11]. This relation provides a better understanding of the evolution of singularities [10].

Further application of the reciprocal transformation

$$\bar{x} := D^{-1} s(x) \qquad \text{(d)}$$

$$\rho(\bar{x}) = \rho(\bar{x}(s,x)) := s(x)$$

between the s-manifold on which the flow (16) is defined and the manifold of functions $\rho=\rho(\bar{x})$ in the new independent variable $\bar{x}$, produces the Harry Dym hierarchy

$$\rho_t = \rho^2 B_n(\bar{S})_{\bar{x}} \qquad (18)$$

where

$$\bar{S} = \rho\rho_{\bar{x}\bar{x}} - \frac{1}{2}\rho^2_{\bar{x}} \qquad (19)$$

whose base member is the Harry Dym equation

$$\rho_t = \rho^3 \rho_{\bar{x}\bar{x}\bar{x}} . \qquad (20)$$

All the transformations which were involved in the chain from the KdV to the Harry Dym equation can be regarded as Bäcklund transformations since the transformations (a),(b),(c) are well known Bäcklund transformations and because the reciprocal transformation (d) can be regarded as such if, instead of the manifold consisting of the dependent variable alone, we consider the manifold given by pairs formed by the dependent and the independent variables. The transformation (d) defines, at least locally [2,4], a Bäcklund transformation:

$$T_{(s,x)} \longrightarrow T_{(\rho,\bar{x})}$$

between the two respective tangent spaces. Therefore it is possible to transfer the infinitesimal structure using the transformation formulae [4,10]; only the computations become more difficult.

In the following we show how to get the recursion operator for the Harry Dym directly out of the corresponding one for the KdV. First we combine the Bäcklund transformations (a),(b) and (c) into

$$u = \frac{1}{2}\left(\frac{s_x}{s}\right)_x - \frac{1}{4}\left(\frac{s_x}{s}\right)^2 = -\sqrt{s}\left(\frac{1}{\sqrt{s}}\right)_{xx} \tag{21}$$

giving a Bäcklund transformation which relates directly the KdV hierarchy to its zero soliton hierarchy. Now the operator

$$\Pi: T_u \longrightarrow T_s$$

going from the tangent space of the u-manifold (KdV vector field) to the tangent space of the s-manifold (zero-soliton vector field) is delivered by explicit computations [2] to be

$$\Pi = 2Ds\Theta(u)^{-1} \tag{22}$$

where $\Theta(u)$ has to be expressed in terms of the s variable by means of (21). Knowing Π, then the recursion operator for the KdV zero-soliton equation as well as its implectic-symplectic factorization are given by

$$\Phi_3(s)=\Pi\Phi(u)\Pi^{-1} \text{ (hereditary operator)} \tag{23}$$

$$\Theta_3(s)=\Pi\Theta(u)\Pi^{+} \text{ (implectic operator)} \tag{24}$$

$$J_3(s)=(\Pi^{+})^{-1}J(u)\Pi^{-1} \text{ (symplectic operator)} \tag{25}$$

in particular the recursion operator can be expressed in the form

$$\Phi_3(s)=D_s\{D^2+s+D^{-1}SD\}s^{-1}D^{-1}. \tag{26}$$

Furthermore, the operator

$$\bar{\Pi}: T_{(s,x)} \longrightarrow T_{(\rho,\bar{x})}$$

mapping the (s,x)-vector fields into the $(\rho,\bar{x})$ ones, is shown [2] to be

$$\bar{\Pi} = (I-\rho_{\bar{x}}D_x^{-1}) = (I-s_x/s\, D_x^{-1}) = sD_x(1/s)D_x^{-1} = \rho^2D_{\bar{x}}(1/\rho)D_{\bar{x}}^{-1}1/\rho. \tag{27}$$

There, in order to avoid confusion, we adopted the notation

$$D_x = \frac{d}{dx}, \qquad D_{\bar{x}} = \frac{d}{d\bar{x}}. \tag{28}$$

Then it follows

$$D_x = \rho D_{\bar{x}}, \qquad (D_{\bar{x}}\rho_{\bar{x}})\rho = \rho^2 D_{\bar{x}} \tag{29}$$

and the composition of the operator $\hat{\Pi}$ mapping the u-manifold into the ρ-manifold can be easily obtained by the composition of the two operators Π and $\bar{\Pi}$

$$\hat{\Pi} = \bar{\Pi}\cdot\Pi. \tag{30}$$

Finally the application of the formulae (23)-(25) provides the Harry Dym recursion operator

$$\hat{\Phi}(\rho) = \hat{\Pi}\Phi(u)\hat{\Pi}^{-1} = \rho^3 D_{\bar{x}}^{\,3}\rho\, D_{\bar{x}}^{\,-1}\rho^{-2} \tag{31}$$

whose factorization follows to be

$$\hat{\Phi}(\rho) = \hat{\Theta}(\rho)\hat{J}(\rho) \tag{32}$$

where

$$\hat{\Theta}(\rho) = \hat{\Pi}\Theta(u)\hat{\Pi}^{+} = 4\rho^2 D_{\bar{x}}\rho^{-1}D_{\bar{x}}^{\,-2} \tag{33}$$

$$\hat{J}(\rho) = (\hat{\Pi}^{+})^{-1}J(u)\hat{\Pi}^{-1} = \tfrac{1}{4}D_{\bar{x}}^{\,2}\rho D_{\bar{x}}^{\,-1}\rho D_{\bar{x}}^{\,3}\rho D_{\bar{x}}^{\,-1}\rho^{-2}. \tag{34}$$

It is remarkable that although the hereditary recursion operator we obtained for the Harry Dym equation is exactly the well known one [19,16], this is not the case for the factorization. This is a consequence of the reciprocal link which does not preserve the usual bilinear form between tangent and cotangent bundle [2,10].

Now, using known invariances of the non-linear evolution equations involved in the Bäcklund chart, new invariances can be deduced for other equations therein. For instance, the invariance under the Möbius group of transformations, characteristic of the singularity equations, implies for the Harry Dym equation the invariance under the transformations

$$\rho(\bar{x}) \longrightarrow \hat{\rho}\ (\hat{x}) = \frac{(c\bar{x}+d)^2}{ad-bc}\,\rho(\bar{x}) \tag{35}$$

$$\bar{x} \longrightarrow \hat{x} = \frac{a\bar{x}+b}{c\bar{x}+d} \tag{36}$$

The same invariance was noticed for the 2+1 dimensional Harry Dym equation [23] and the Kawamoto hierarchy [22]. The Kawamoto equation plays a role similar to that of the Harry Dym equation in a Bäcklund chart involving the CDGSK instead of the KdV equation.

2 The symmetry structure of the Harry Dym equation

The observation that all the transformations we have given so far involve differential equations is crucial in order to retrieve the Harry Dym symmetry structure. Indeed, since in general different operators have non empty kernels, usually, these transformations are not injective. In particular, going backward, through the Bäcklund chart, from the Harry Dym to the KdV equation, we find that the vector fields

$$T_0(\rho) = \rho_{\bar{x}} \tag{37}$$

$$R_0(\rho) = \bar{x}\rho_{\bar{x}} - \rho \tag{38}$$

$$N_0(\rho) = \frac{1}{2}\bar{x}^2\rho_{\bar{x}} - \bar{x}\rho \tag{39}$$

do all belong to the kernel of the transformation going from the Harry Dym to the KdV equation, i.e. they are all mapped to zero by the transformations under consideration [11]. Furthermore, letting

$$K_0(\rho) = \rho^3\rho_{\overline{xxx}} \tag{40}$$

we notice that

$$[K_0, T_0] = [K_0, R_0] = [K_0, N_0] = 0. \tag{41}$$

Now, since $\hat{\Phi}(\rho)$ is a hereditary operator having vanishing Lie-derivative in the direction of K_0, i.e.

$$\hat{\Phi}[K, A] = [K, \hat{\Phi}, A] \text{ for all } A \tag{42}$$

we find that when A is one of the fields (37)-(39), let them be denoted as V, it follows

$$[K, \Phi, V] = 0. \tag{43}$$

And now successive application of hereditariness of the operator $\Phi(\rho)$ implies that the vector field defined by the Harry Dym equation commutes with all members of the following hierarchies

$$K_n = \hat{\Phi}^n K_0, \quad n \in \mathbb{Z} \tag{44}$$

$$T_n = \hat{\Phi}^n T_0, \quad n \in \mathbb{Z} \tag{45}$$

$$N_n = \hat{\Phi}^n N_0, \quad n \in \mathbb{Z} \tag{46}$$

$$R_n = \hat{\Phi}^n R_0, \quad n \in \mathbb{Z} \tag{47}$$

This means that the Harry Dym equation admits four hierarchies of completely integrable equations as symmetry generators. Since not all these fields commute among each other, the Harry Dym equation has a huge non-commutative symmetry group which is of larger size than the symmetry group of the KdV or the mKdV equation. However we must remark that all these additional hierarchies of symmetry generators for the Harry Dym equation, being related to the kernel of the transformation operator mapping the KdV zero soliton into the Harry Dym equation itself, are mapped into zero (modulo known invariances) by going backward from the Harry Dym to the KdV zero-soliton field. This is in close analogy to the Kawamoto equation [15,2]. Indeed the symmetry groups of Harry Dym equation and Kawamoto equation are isomorphic [10]: in both cases these non-abelian symmetry groups have a semisimple part isomorphic to the affine Lie algebra $A_1^{(1)}$ associated to $sl(2,\mathbb{C})$. Moreover, the hierarchies of symmetry generators for the Harry Dym and the Kawamoto equations do share their base members. However this does not carry over to the whole hierarchies since these are generated by different hereditary operators, a fact which is obvious by looking at the explicit form of the recursion operator of the Kawamoto equation:

$$\Phi(\rho) = \hat{\Theta}(\rho)\hat{J}(\rho) \tag{48}$$

where

$$\Theta(\rho) = \rho^2 D\Theta_1(u,\rho)^{-1}\rho^2 D \tag{49}$$

and

$$\hat{J}(\rho) = D^{-1}\rho^{-2}\Theta_1(u,\rho)J_1(u,\rho)\Theta_1(u,\rho)D^{-1}\rho^{-2}. \tag{50}$$

Here u was given by

$$u = \rho\rho_{xx} - \frac{1}{2}\rho_x^{\,2} \tag{51}$$

and $\Theta_1(u,,\rho)$ and $J_1(u,\rho)$ are:

$$\Theta_1(u,\rho) = (\rho D)^3+u\rho D+\rho Du \tag{52}$$

$$J_1(u,\rho) = (\rho D)^3+2((\rho D)^2u(\rho D)^{-1}+(\rho D)^{-1}u(\rho D)^2). \tag{53}$$

We would like to know whether all the equations given by (44)-(47) are hamiltonian. Since these equations are obtained by application of a hereditary operator out of the fields Q_0 (where Q=K,T,N,R), and since this operator is invariant by the Q_0 itself, it suffices [4] to show that the fields Q_0 are hamiltonian, then the hamiltonian structure carries over to all of the Q_n. In fact this is easily shown. We recall ([4] or [5]) that, having fixed some implectic operator $J(\rho)$ and some duality, then a vector field $Q(\rho)$ is said to be **hamiltonian** if

$$J(\rho)Q(\rho) = \nabla H(\rho), \quad H(\rho) \text{ scalar field} \tag{54}$$

Here ∇ denotes the gradient with respect to the duality between tangent bundle and cotangent bundle:

$$<\nabla H(\rho),\nu> = \frac{\partial}{\partial\varepsilon}_{|\varepsilon=0|} H(\rho+\varepsilon\nu). \tag{55}$$

Now, taking $J(\rho)$ to be a suitable implectic operator for the Harry Dym equation, for example[1]

$$J = \rho^{-1}D^{-1}\rho^{-2} \tag{56}$$

and fixing the duality between tangent field $f(\rho)$ and a cotangent field $\gamma(\rho)$ as the one which is derived (out of the usual one for the KdV) by application of the transformations given in our Bäcklund chart [2,4]:

$$<\gamma(\rho)f(\rho)> = \int_{-\infty}^{+\infty} \gamma(\rho)f(\rho)\frac{d\bar{x}}{\rho} . \tag{57}$$

Then the hamiltonian structure is given by the observation that

$$JK_0 = \nabla H_K,\ JN_0 = \nabla H_N,\ JR_0 = \nabla H_R,\ JT_0 = \nabla H_T \tag{58}$$

where formally the conservation laws H are given as

$$H_K = -\frac{1}{2}\int_{-\infty}^{+\infty} \frac{\rho_{\bar{x}}\rho_{\bar{x}}}{\rho}\, d\bar{x} \tag{59}$$

$$H_N = \frac{1}{4}\int_{-\infty}^{+\infty} \frac{\bar{x}^2}{\rho^2}\, d\bar{x} \tag{60}$$

$$H_R = \frac{1}{2}\int_{-\infty}^{+\infty} \frac{\bar{x}}{\rho^2}\, d\bar{x} \tag{61}$$

---

[1] At this point for the hamiltonian formulation of the Harry Dym equation we have used instead of (34) an equivalent operator (see [19]). This change of hamiltonian representation was done for the sake of facilitation of the computations necessary in order to check the hamiltonian structure.

$$H_T = \frac{1}{2} \int_{-\infty}^{+\infty} \frac{1}{\rho^2} d\bar{x} \qquad (62)$$

Of course, as scalar quantities these conservation laws only make sense under suitable boundary conditions. However, their Lie algebra meaning does not depend on such boundary conditions, because it always is possible to work with density valued scalar fields instead of ℝ-valued fields (see [6]). Thus their formal gradients are closed one-forms.

When considering the symmetry group generators of those dynamical systems where the flow is given by any of the vector fields (37)-(39), we observe that they do not possess symmetry groups which are isomorphic to that of the Harry Dym equation (exactly as it was observed in the study on the Kawamoto equation). Such a behaviour may seem to contradict the belief that all symmetry groups, related to non-linear evolutions equations obtained via reciprocal links, should have, more or less, the same structure. Unfortunately, all the symmetry groups generated by the additional hierarchies indeed turn out to be isomorphic to each other when the time-dependent symmetry generators (master symmetries) are taken into account. The complete analogy obtained in the study of the Harry Dym and the Kawamoto equation does not surprise us at all, but was expected since these two equations play exactly the same role in two similar Bäcklund charts involving the KdV-mKdV and the CDGSK-Fordy hierarchies [22], respectively.

Furthermore, the KdV Bäcklund chart leads to the discovery of additional symmetries for the Harry Dym equation. These can be obtained as a result of known invariances exhibited by equations appearing therein; in particular (see [10]) the trivial symmetry

$$s \longrightarrow 1/s \qquad (63)$$

for the KdV zero-soliton equation gives that the completely integrable integro-differential equation

$$\rho_t = \hat{G}(\rho) \qquad (64)$$

where

$$\hat{G}(\rho) = \rho^2 D_{\bar{x}} 1/\rho D_{\bar{x}}^{-2} 1/\rho^2 \qquad (65)$$

is an infinitesimal symmetry generator for the Harry Dym equation.

3 Explicit solutions and soliton structure

The intimate connection between the interacting soliton [8] and the singularity structure (see [10,11]) for completely integrable systems plays a fundamental role. This connection is important not only for getting a better knowledge of structural properties, but also because it provides a useful tool in the construction of explicit solutions. How this leads to explicit solutions for the

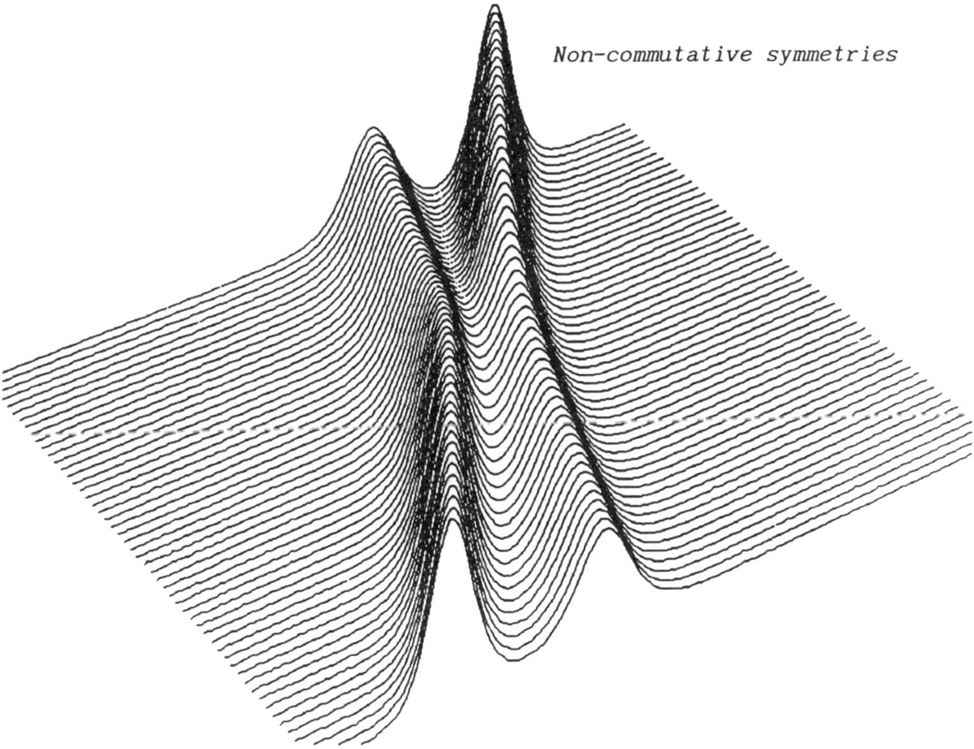

Figure 2. Two-soliton solution of the KdV equation (c1=1.44, c2=2.56).

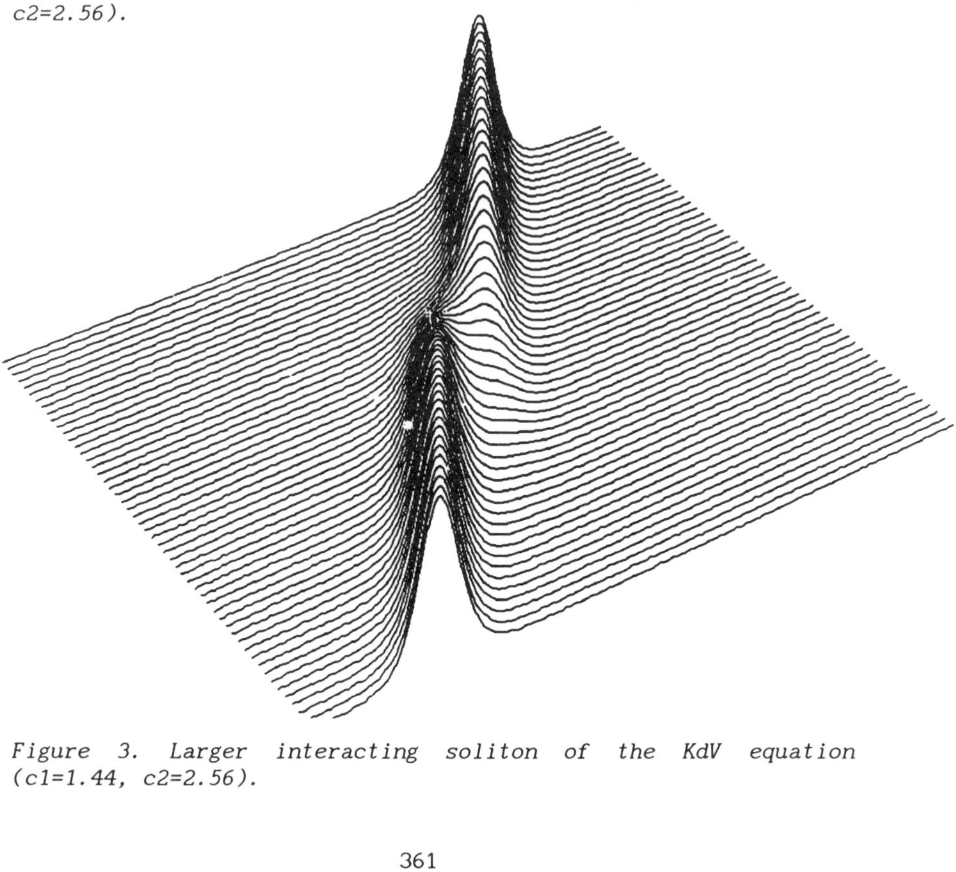

Figure 3. Larger interacting soliton of the KdV equation (c1=1.44, c2=2.56).

Harry Dym equation will be briefly explained in this section.

The KdV interacting soliton equation

$$s^2 s_t = s^2 s_{xxx} - 3ss_x s_{xx} + \frac{3}{2}s_x^{\ 3} + \frac{3}{2}\lambda\ s^2 s_x \tag{66}$$

describes the dynamics of the λ-eigenvector of the recursion operator of the KdV equation [8,9]. For $\lambda=0$ and for ϕ and s related via

$$s = \phi_x \tag{67}$$

this equation produces the KdV singularity equation. Hence the KdV zero-soliton equation can be regarded as a "potential" singularity KdV equation.

For general values of λ (66) becomes

$$\phi_t = \phi_x\{\phi;x\}+3/2\lambda\phi_x \tag{68}$$

which differs from the KdV singularity equation just for the last term representing the generator of translation invariance. Now, since the translation invariance generator lives in the kernel of

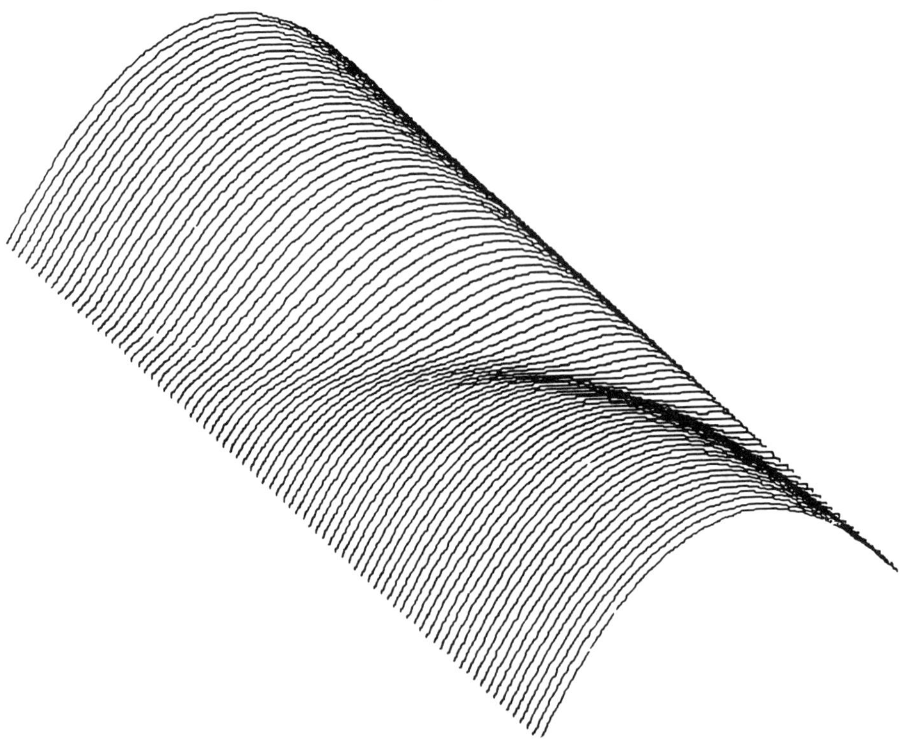

Figure 4. Solution of the Harry Dym equation coming from the larger interacting soliton of a two-soliton solution of the KdV equation (c1=1.44, c2=2.56).

the reciprocal transformation, we have a direct way to relate solutions of the KdV interacting soliton to solutions of the Harry Dym equation:

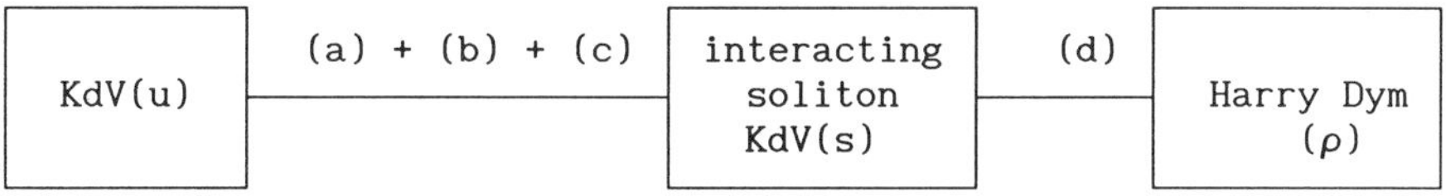

Figure 5.

The above picture shows how, via the transformation operators in section 1, solutions for the KdV can produce solutions for the Harry Dym equation via the interacting soliton equation: starting with a known solution of the KdV equation, by means of successive reparametrizations the corresponding solution of the Harry Dym equation is generated.

An outline of the procedure is the following:

(a) Choose a multilsoliton solution of the KdV equation, for example the two soliton which is plotted in Fig. 2.

(b) In case a two soliton is chosen define s by

$$s_x = (\phi-\lambda_1)u_x$$

where λ_1 is the asymptotic speed of one of the two solitons. The factor $\phi-\lambda_1$ is chosen in order to annihilate all but one of the solitons. Of course, if n solitons occur asymptotically then n-1 such factors have to be applied [7,8,9]. The quantity s then evolves according to the interacting soliton equation. In Fig. 3 this solution is plotted for the kdV solution given in Fig. 2. Thus a solution is obtained for the interacting soliton equation.

(c) The last step consists of the application of the reciprocal transformation which is nothing but a reparametrization of the soliton solution involving both the dependent and independent variable. The result of this step is plotted in Fig. 4.

It should be remarked that this procedure provides a tool for getting solutions of the Harry Dym equation, not only restricted to a segment, but also on half line. These solutions come up when the procedure is applied to singular solutions, for example a degenerate two soliton in case $\lambda_1=\lambda_2$. In fact some solutions of this kind have already been obtained [21] for the Harry Dym equation, but there a critical dependency on the initial data was exhibited.

References

[1] Calogero, F. & Degasperis, A. (1977). Special solutions of coupled nonlinear evolution equations with bumps that behave

as interacting particles. *Lett. Nuovo Cimento* **19**, 525.

[2] Carillo, S. & Fuchssteiner, B. (1987). The abundant symmetry structure of nonlinear equations obtained by reciprocal links, *J. Math. Phys.*

[3] Fuchssteiner, B. (1979). Aplication of Hereditary Symmetries to Nonlinear Evolution equations. *Nonlinear Analysis TMA* **3**, 849.

[4] Fuchssteiner, B. & Fokas, A.S. (1981). Symplectic Structures, Their Bäcklund Trasformations and Hereditary Symmetries. *Physica* **4D**, 47.

[5] Fuchssteiner, B. (1981). The Lie algebra structure of Nonlinear Evolution Equations admitting Infinite Dimensional Abelian Symmetry Groups. *Progr. Theor. Phys.* **65**, 861.

[6] Fuchssteiner, B. (1982). The Lie algebra structure of Degenerate Hamiltonian and Bi-Hamiltonian systems. *Prog. Theor. P hys.* **68**, 1082.

[7] Fuchssteiner, B. (1983) Mastersymmetries, Higher-order Time-dependent symmetries and conserved Densities of Nonlinear Evolution Equations. *Progr. Theor. Phys.* **70**, 1508.

[8] Fuchssteiner, B. & Aiyer, R.N. (1987). Multisolitons, or the discrete Eigenfunctions of the Recursion Operator of non-linear Evolution Equations: II. Background. *J. Phys.* **A20**, 375.

[9] Fuchssteiner, B. (1988). The dynamical behaviour of interacting solitons. In *Proceedings of the workshop of nonlinear Evolution equations and Dynamical Systems,* p. 13. World Scientific Publ.: Singapore.

[10] Fuchssteiner, B. & Carillo, S. (1988). Soliton Structure versus Singularity Analysis: Third-order Completely Integrable Nonlinear Differential Equations in 1+1-dimensions. *Physica A.*

[11] Fuchssteiner, B. & Carillo, S. (1988). The Soliton Singularity Transform, in this volume.

[12] Gelfand, I.M. & Dorfmann, I.Y. (1974). Hamiltonian Operators and Algebraic Structures related to them. *Funktsional'nyi Analiz i Ego Prilozheniya* **13**, 13.

[13] Gelfand, I.M. & Dorfmann, I.Y. (1980). The Schouten bracket and Hamiltonian operators. *Funktsional'nyi Analiz i Ego Prilozheniya* **14**, 71.

[14] Gelfand, I.M. & Dorfmann, I.Y. (1981). Hamiltonian Operators and Infinite-Dimensional Lie-Algebras. *Funktsional'nyi Analiz i Ego Prilozheniya* **15**, 23.

[15] Kawamoto, S. (1985). An exact Transformation from the Harry Dym equation to the modified KdV equation. *J. Phys. Soc. Japan* **54**, 2055.

[16] Leo, M., Soliani, R.A. & Solombrino, L. (1983). On the isospectral-eigenvalue Problem and the recursion operator of the Harry Dym equation. *L. al Nuovo Cimento* **38**, 45.

[17] Magri, F. (1978). A simple model of the integrable Hamiltonian equation. *J. Math. Phys.* **19**, 1156.

[18] Magri, F. (1980). A Geometrical Approach to the Nonlinear Solvable equations. In *Nonlinear Evolution equations and Dynamical Systems*, p. 233, Boiti, M., Pempinelli, F. & Soliani, G. (eds.). Springer Verlag.

[19] Oevel, W. (1984). *Rekursionsmechanismen fur Symmetrien und Erhaltungssatze in integrablen Systemen.* Disstertation: Paderborn.
[20] Rogers, C. & Nucci, M.C. (1986). On Reciprocal Bäcklund transformations and the Korteweg-de Vries Hierarchy. *Physica Scripta* **33**, 289.
[21] Rogers, C. & Guo, H. (1987). The solution of the initial Value problem of the Harry Dym equation, preprint.
[22] Rogers, C. & Carillo, S. (1987). On Reciprocal Properties of the Caudrey-Dodd-Gibbon and Kaup-Kupershmidt Hierarchies. *Physica Scripta* **36**, 865.
[23] Rogers, C. (1987). The Harry Dym equation in 2+1 dimensions: A Reciprocal Link with the Kadomtsev Petviashvili equation. *Phys. Lett.* **120**, 15.

V.S. Gerdjikov

34. Z_N-REDUCTIONS AND NEW INTEGRABLE VERSIONS OF DERIVATIVE NONLINEAR SCHRÖDINGER EQUATIONS

1 Our aim here is to present natural generalizations of the derivative nonlinear Schrödinger (NLS) equation [5] and their gauge equivalent Heisenberg ferromagnet (HF) type equations, related to the simple Lie algebras $\mathfrak{sl}(N,\mathbb{C})$. To do this we shall employ the group of reductions, introduced by A.V. Mikhailov [8]. This notion has led to the discovery of a whole class of 2-dimensional integrable models, generalizing the sine-Gordon model (see the review paper by Drinfeld & Sokolov [1], and the references therein). We consider only the simplest possible case, when the underlying algebra is $\mathfrak{sl}(N,\mathbb{C})$ and the group of reduction is $\mathbb{Z}_N$, $\mathbb{D}_N$ and $\mathbb{Z}_N\otimes\mathbb{Z}_2$. In this case the generalizations of the derivative NLS equations have the form:

$$i\psi_{k,t} + \gamma\left(\operatorname{cotg}\frac{\pi k}{N}\cdot\psi_{k,x} + i\sum_{p=1}^{N-1}\psi_p\,\psi_{k-p}\right)_x = 0,\quad k = 1,2,\dots,N-1 \tag{1}$$

where γ is a constant and the index k-p should be understood modulus N, $\psi_o = \psi_N = 0$. The system (1) allows also the involutions:

$$\psi_k = -\,\psi_k^*,\qquad\qquad \gamma = -\,\gamma^* \tag{2a}$$

$$\psi_k = \psi_{N-k}^*,\qquad\qquad \phi = \gamma^* \tag{2b}$$

The HF type equations, gauge equivalent to (1) have the form:

$$S_t - \gamma\,\frac{d}{dx}\,\mathrm{ad}_S^{-1}\,(S^2(x,t))_x = 0,\quad \mathrm{ad}_S\,X = [S,X], \tag{3}$$

with the additional restriction $S^N(x,t) = 1$. This last condition generalizes the relation $S^2(x,t) = 1$, characteristic for the HF equation. The analogue of the involutions (2) for the NLEE (3) are:

$$S^{\dagger}(x,t) = -\omega^{-1} S_1 S(x,t) S_1, \quad S_1 = \sum_{j=1}^{N-1} E_{j,N-j+1} \tag{4a}$$

$$S^{*}(x,t) = \omega^{-1} S_1 S(x,t) S_1 \tag{4b}$$

One can check that $(S^2)_x$ lies in the image of ad_S, where ad_S^{-1} may be expressed as a polynomial of ad_S.

The particular cases of (1), (2) with N = 3 have been reported by Fordy & Gibbons [2] and Mikhailov [8]. After this paper was completed the author was informed that the systems (1) for N > 3 have also been known to A.V. Mikhailov and J. Gibbons.

2 Let us consider a Lax pair of the form:

$$\mathcal{L}\chi(x,t,\lambda) = (\frac{d}{dx} + U(x,t,\lambda))\, \chi(x,t,\lambda) = 0, \tag{5}$$

$$\mathcal{M}\chi(x,t,\lambda) = (\frac{d}{dt} + V(x,t,\lambda))\, \chi(x,t,\lambda) - \lambda^2 \chi(x,t,\lambda)\, V_2 = 0, \tag{6}$$

where the potentials $U(x,t,\lambda)$ and $V(x,t,\lambda)$ are polynomials of λ:

$$U(x,t,\lambda) = U_o(x,t) + \lambda\, U_1 \tag{7}$$

$$V(x,t,\lambda) = V_o(x,t) + \lambda\, V_1(x,t) + \lambda^2\, V_2 \tag{8}$$

We impose on (5) and (6) a $\mathbb{Z}_N$-reduction by [8]:

$$U(x,t,\lambda) = K^{-1}\, U(x,t,\lambda\omega)\, K\,, \quad V(x,t,\lambda) = K^{-1}\, V(x,t,\lambda\omega)\, K \tag{9}$$

where $K^N = 1$ and $\omega = \exp(2\pi i/N)$. We choose K to be:

$$K = \sum_{j=1}^{N} E_{j,j+1}\,, \quad (E_{jk})_{lm} = \delta_{jl}\, \delta_{km} \tag{10}$$

here and below all indices are understood modulus N, so that the last term with j = N in (10) equals $E_{N,1}$.

The automorphism Ad_K defines a grading in the Lie algebra $s\ell(N,\mathbb{C}) = \bigoplus_{k=0}^{N-1} g^{(k)}$, where $g^{(k)}$ is the eigenspace of Ad_K corresponding to its eigenvalue ω^{-k}, $k = 0,1,\ldots,N-1$. The calculations are much simpler if we introduce a convenient basis in $g^{(k)}$, namely:

$$J_s^{(k)} = \sum_{j=1}^{N} \omega^{kj} E_{j,j+s}, \quad K^{-1} J_s^{(k)} K = \omega^{-k} J_s^{(k)} \tag{11}$$

Obviously, $J_s^{(k)}$ satisfies the commutation relations:

$$[J_s^{(k)}, J_l^{(m)}] = (\omega^{ms} - \omega^{kl}) J_{s+l}^{[k+m]} \tag{12}$$

and (9) is identically satisfied if U_k, $V_k \in g^{(k)}$. Let us put:

$$U_o(x) = \sum_{j=1}^{N-1} \psi_j(x,t) J_j^{(o)}, \quad U_1 = - a J_o^{(1)}, \quad V_2 = - b J_o^{(2)} \tag{13}$$

Then the requirement that (5) and (6) are compatible for all values of λ allows us to express $V_o(x,t)$ and $V_1(x,t)$ in terms of $\psi_j(x,t)$ and finally to obtain the NLEE (1).

The Lax pair allows also one of the two involutions:

$$U^{\dagger}(x,t,\lambda) = - S_1 U(x,t,\varepsilon\lambda^*) S_1, \tag{14a}$$

or

$$U^*(x,t,\lambda) = S_1 U(x,t,\varepsilon\lambda^*) S_1, \tag{14b}$$

where S_1 is defined in (4a), and analogously for $V(x,t,\lambda)$. The relations (14a,b) give directly (2a,b). If (2b) holds, then for odd $N = 2N_o+1$ the NLS (1) is reduced to a system of N_o equations for N_o complex-valued functions $\psi_j(x,t)$, $j=1,\dots,N_o$. For even $N = 2N_o$ the system (1) consists of N_o equations for N_o-1 complex-valued functions $\psi_j(x,t)$, $j=1,\dots,N_o-1$ and one real-valued function $\psi_{N_o}(x,t)$. Note, that in the equation for $\psi_{N_o}(x,t)$ the term with the second x-derivative is missing, since $\mathrm{cotg}(\pi/2) = 0$.

It is easy to check that $S_1 K^{\dagger} S_1 = K$ and $S_1 K^* S_1 = K^{-1}$, i.e. the reduction (9) commutes with the involution (14a) but does not commute with (14b). This reflects the fact that the reduction group of the NLEE (1), (2a) is $\mathbb{Z}_N \otimes \mathbb{Z}_2$, while (1), (2b) possess $\mathbb{D}_N$ as reduction group.

3 The Lax pair (5), (6) is gauge equivalent to:

$$\bar{\mathcal{L}}\xi(x,t,\lambda) + (\frac{d}{dx} - \lambda S(x,t)) \xi(x,t,\lambda) = 0, \tag{15}$$

$$\bar{\mathcal{M}}\xi(x,t,\lambda) = (\frac{d}{dt} + \lambda\bar{V}_1(x,t))\xi(x,t,\lambda) - \lambda^2[\xi(x,t,\lambda), S^2(x,t)] = 0 \tag{16}$$

where $\xi(x,t,\lambda) = \chi_o^{-1}(x,t)\ \xi(x,t,\lambda)$ and bar means a similarity transformation with the solution $\chi_o(x,t)$ of (5), (6) with $\lambda = 0$, so that:

$$S(x,t) = \chi_o^{-1}(x,t)\, J_o^{(1)}\, \chi_o(x,t) \qquad (17)$$

Then the compatibility condition of (15) and (16) gives the NLEE (3). This equation contains the operator ad_S^{-1}, which is the inverse of $\mathrm{ad}_S X = [S,X]$.

Let us consider the operator $\mathcal{D}_1 X = X - J_o^{(N-1)} X J_o^{(1)}$; obviously $\overline{\mathcal{D}}_1 \overline{X} = S^{-1}\mathrm{ad}_S\, \overline{X}$. Using (11), (12) one checks that all basic elements $J_p^{(k)}$ are eigenvectors of $\mathcal{D}_1$ with eigenvalues $\alpha_p = 1-\omega^p$. Following [4] we project out the zero eigenvectors and analyse the corresponding characteristic equation using its gauge invariance. Finally we get:

$$\mathrm{ad}_S^{-1} X = \sum_{q=0}^{N-2} c_q (\mathrm{ad}_S)^q\, (S^{-q-1}X) \qquad (18)$$

where c_q are some constants; i.e. the equations (3) may be written down in polynomial in S form.

4 The direct and the inverse scattering problems for the system (5) have been solved by Mikhailov [8]. The continuous spectrum Γ of (5) fills up the following lines in the complex λ-plane:

$$\Gamma:\ \arg\lambda = -\beta\,\frac{2\pi k}{N}\,,\quad \beta = \arg\alpha \pm \frac{\pi}{2}\,,\quad k = 1,2,\ldots,N. \qquad (19)$$

The reductions imply also symmetries between the discrete eigenvalues (d.e.) of (5). Thus (9) implies that if λ_1 is a d.e. of (5), then so are $\lambda_1\omega^j$ for all $j = 1,\ldots,N-1$.

The transfer matrix $T(\lambda,t) = \psi^{-1}(x,t,\lambda)\, \phi(x,t,\lambda)$; $\lambda \in \Gamma$ is as usual the ratio of the Jost solutions of (5), defined by:

$$\lim_{x\to\infty} \psi(x,t,\lambda)\, e^{\lambda U_1 x} = \lim_{x\to-\infty} \phi(x,t,\lambda)\, e^{\lambda U_1 x} = 1, \qquad (20)$$

for all values of t. The inverse scattering problem for (5) is equivalent to a Riemann-Hilbert problem which allows one to calculate the reflectionless potentials of (5) [8]. Then to find the corresponding soliton solutions of equations (1) it is enough to know the time dependence of $T(\lambda,t)$, which is given by:

$$\frac{dT}{d\lambda} + \lambda^2\, [V_2,\, T(\lambda,t)] = 0. \qquad (21)$$

Since V_2 is a diagonal matrix, then the diagonal elements of $T(\lambda,t)$ are time-independent and may be considered as generating functionals of the integrals of motion. The first two of them A_1 and A_2, obtained as coefficients in the asymptotic expansion: $\ln T_{11}(\lambda) = (\lambda\omega)^{-1} A_1 + (\lambda\omega)^{-2} A_2 + \mathcal{O}(\lambda^{-3})$, have the form:

$$A_1 = \frac{1}{2\alpha} \int_{-\infty}^{\infty} dx \sum_{p=1}^{N-1} \psi_p \, \psi_{N-p}(x,t) \tag{22}$$

$$A_2 = \frac{1}{2\alpha^2} \int_{-\infty}^{\infty} dx \left\{ \sum_{p=1}^{N-1} i \operatorname{ctg} \frac{\pi p}{N} (\psi_{p,x} \psi_{N-p} - \psi_{N-p,x} \psi_p)(x,t) - \right.$$
$$\left. - \frac{2}{3} \sum_{p+n+m=N} \psi_p \, \psi_n \psi_m \, (x,t) \right\} \tag{23}$$

where $p+n+m = 0 \pmod N$. Note that under the involution (3b) both A_1 and A_2 become real. The density of A_1 may be interpreted as the number of particles density, while the kinetic term in A_2 acquires the meaning of the momentum density.

5 The system of equations (1) allows a hierarchy of Hamiltonian structures. Here we will use the approach of Kulish & Reiman [6], developed for generic (non-restricted) systems. In cases which involve reductions one should: (i) calculate the generic Poisson brackets; (ii) impose the reduction as constraints on these brackets; (iii) calculate the corresponding Dirac brackets.

However, if we define $\psi_j(x)$ as linear functionals of $U(x,\lambda)$ by: $\psi_j(x) = \frac{1}{N} \operatorname{tr} U(x,\lambda) \, J^{(0)}_{N-j}$, then the results of Kulish & Reiman [6] surprisingly give us directly the correct answer [7]:

$$\{\psi_j(x) \, , \, \psi_k(y)\} = \delta_{k+j-N} \, \delta'(x-y) \tag{24}$$

Obviously, the Poisson brackets (24) and the Hamiltonian $H = b\, A_2$ (23) lead to (1).

6 Let us make a few concluding remarks.

The generic integrable NLEE possesses a hierarchy of Hamiltonian structures, generated by a certain Λ-operator. Its spectral expansions and applications to the theory of the corresponding NLEE are well known, see e.g. [3]. However, the reduced case has some peculiarities [2] which must be taken care of when studying the properties of the corresponding Λ-operator. Done in a gauge covariant way [4] this would enable one to construct the $\tilde{\Lambda}$-operator, generating the HF type equations (3), and then to find the interrelations between the Hamiltonian structures of (1) and (3).

The leading power of λ in $\mathcal{M}$ (6) should be an exponent of the corresponding algebra [1]. Since 2 is an exponent only for A_n, there would be no new examples of $\mathbb{Z}_n$-reduced NLS type equations. However, 3 is an exponent for all the algebras from the classical series. Choosing $\mathcal{M}$ to be cubic in λ one will get series of KdV type equations; the simplest of them are known [1].

The limit $N \longrightarrow \infty$ leads to 2+1 dimensional models. Indeed, if

$$u(x,t,y) = \sum_{j=1}^{N-1} \psi_j(x,t) \exp(2\pi i j y/N) \tag{25}$$

then (2b) means that $u(x,t,y)$ is real-valued. Retaining only the leading term of $\text{cotg}(\pi k/N)$ and rescaling $x = \xi/N$, $\tau = t/N$ in (1) we get:

$$u_{y\tau} + \gamma\, u_{\xi\xi} + \gamma(u^2)_{\xi y} = 0 \tag{26}$$

which resembles the Kadomtsev-Petviashvilli equation without the dispersive term.

Acknowledgements

I am grateful to Dr. S.V. Manakov, Dr. A.V. Mikhailov, Dr. J. Gibbons and Prof. G. Wilson for useful discussions. It is my pleasure to thank the organizers: Prof. A. Degasperis, Dr. A.P. Fordy and Prof. M. Lakshmanan for giving me the opportunity to participate in the present Workshop.

References

[1] Drinfeld, V.G. & Sokolov, V.V. (1984). Lie algebras and equations of Korteweg-de Vries type. In *Contemporary Problems of Mathematics. Latest Achievements,* 24, pp. 81-180. VINITI: Moscow.

[2] Fordy, A.P. & Gibbons, J. (1981). Factorization of operators. II. *J. Math. Phys.* **22**, 1170.

[3] Gerdjikov, V.S. & Kulish, P.P. (1981). The generating operator for the non linear system. *Physica* **3D**, 549.

[4] Gerdjikov, V.S. & Yanovski, A.B. (1985). Gauge covariant formulation of the generating operator. 2. Systems on homogeneous spaces. *Phys. Lett.* **110A**, 53.

[5] Kaup, D.J. & Newell, A.C. (1978). An exact solution of a derivative nonlinear Schrödinger equation. *J. Math. Phys.* **19**, 798.

[6] Kulish, P.P. & Reiman, A.G. (1983). *Scientific Notes of LOMI seminars* **123**, 67 (in Russian).

[7] Kupershmidt, B. & Wilson, G. (1981). Conservation laws and symmetries of generalized sine-Gordon equations. *Comm. Math. Phys.* **81**, 189.

[8] Mikhailov, A.V. (1981). The reduction problem and the inverse scattering method. *Physica* **3D**, 73.

V.S. Gerdjikov and D.J. Kaup

35. REDUCTION OF 3 x 3 POLYNOMIAL BUNDLES AND NEW TYPES OF INTEGRABLE 3-WAVE INTERACTIONS

1. Our aim is to show that the group of reductions proposed by Mikhailov [2] can be effectively used in the construction of new versions of integrable nonlinear evolution equations (NLEE) in 1+1 dimensions. We illustrate this by two examples, which lead to new integrable versions of the well-known 3-wave interaction [1,3].

Let us consider a matrix Lax pair, first order in $\frac{d}{dx}$ and $\frac{d}{dt}$ of the form:

$$\left(i \frac{d}{dx} + U(x,t,\lambda)\right) \psi (x,t,\lambda) = 0 \qquad (1)$$

$$\left(i \frac{d}{dt} + V(x,t,\lambda)\right) \psi (x,t,\lambda) = 0 \qquad (2)$$

Following [2] we will say that it possesses a $\mathbb{Z}_N$ group of reductions if U and V satisfy the relations:

$$K^{-1} U(x,t,\lambda) K = U(x,t,\lambda\omega), \quad K^{-1} V(x,t,\lambda) K = V(x,t,\lambda\omega) \qquad (3)$$

where K is a constant matrix such that $K^N = I$ and $\omega = \exp(2\pi i/N)$. In what follows we shall also impose the involution (or $\mathbb{Z}_2$-reduction):

$$U^{\dagger} (x,t,\lambda) = B_1^{-1} U(x,\varepsilon\lambda) b_1, \qquad (4)$$

and the same for $V(x,t,\lambda)$, where $B_1^{\dagger} B_1^{-1} = I$ and $\varepsilon = \pm 1$. We limit ourselves to the simplest possible case, when U and V are 3 x 3 matrix-valued functions, depending polynomially in λ.

2. As first example we shall consider U and V to be quadratic on λ:

$$U(x,t,\lambda) = \sum_{k=0}^{2} U_k(t,x)\lambda^k, \quad V(x,t,\lambda) = \sum_{k=0}^{2} V_k(x,t)\lambda^k \tag{5}$$

where

$$U_0(x,t) = \begin{bmatrix} u_{11} & u_{22} & 0 \\ u_{21} & u_{22} & 0 \\ 0 & 0 & u_{33} \end{bmatrix} \qquad U_1(x,t) = \begin{bmatrix} 0 & 0 & u_{13} \\ 0 & 0 & u_{23} \\ u_{31} & u_{32} & 0 \end{bmatrix} \tag{6}$$

$$V_0(x,t) = \begin{bmatrix} v_{11} & v_{12} & 0 \\ v_{21} & v_{22} & 0 \\ 0 & 0 & v_{33} \end{bmatrix} \qquad V_1(x,t) = \begin{bmatrix} 0 & 0 & v_{13} \\ 0 & 0 & v_{23} \\ v_{31} & v_{32} & 0 \end{bmatrix} \tag{7}$$

and

$$-U_2 = \mathrm{diag}\,(a_1, a_2, a_3), \qquad -V_2 = \mathrm{diag}\,(b_1, b_2, b_3) \tag{8}$$

are constant diagonal matrices, whose entries are pairwise different: $a_1 \neq a_2 \neq a_3 \neq a_1$ and $b_1 \neq b_2 \neq b_3 \neq b_1$. Obviously, U and V defined by (5)-(8) satisfy (3) with N = 2 and K = diag (1,1,-1). Imposing the involution (4) with $B_1 = \mathrm{diag}\,(1, \gamma, \alpha)$, α, γ - real constants, leads to:

$$\begin{aligned} u_{21} &= \gamma u_{12}^*, \quad u_{31} = \alpha u_{13}^*, \quad u_{32} = \frac{\alpha}{\gamma} u_{23}^* \\ v_{21} &= \gamma v_{12}^*, \quad v_{31} = \alpha v_{13}^*, \quad v_{32}^* = \frac{\alpha}{\gamma} v_{23}^* \\ u_{jj} &= u_{jj}^*, \quad v_{jj} = v_{jj}^*, \quad a_j^* = a_j, \quad b_j^* = b_j \end{aligned} \tag{9}$$

After some calculations we find that the compatibility condition for (1), (2) gives us the following expressions for v_{ij} in terms of u_{ij}:

$$\begin{aligned} v_{13} &= \eta_{13} u_{13}, \quad v_{23} = \eta_{23} u_{23} \\ v_{12} &= \eta_{12} u_{12} + \frac{\alpha\kappa}{\gamma} u_{13} u_{23}^* \end{aligned} \tag{10}$$

where

$$\eta_{2j} = \frac{b_i - b_j}{a_i - a_j}, \quad \kappa = \frac{\eta_{23} - \eta_{13}}{a_1 - a_2} \tag{11}$$

and the following NLEE for u_{ij}:

$$
\begin{aligned}
i(\partial_t - \eta_{13}\partial_x) u_{13} &= (\eta_{23} - \eta_{12}) u_{12} u_{23} - \frac{\alpha\kappa}{\gamma} u_{13}|u_{23}|^2 \\
&\quad + u_{13}[\eta_{13}(u_{11} - u_{22}) + v_{33} - v_{11}] \\
i(\partial_t - \eta_{23}\partial_x) u_{23} &= \gamma(\eta_{13} - \eta_{12}) u_{13} u_{12}^* - \kappa\alpha|u_{13}|^2 u_{23} \\
&\quad + u_{23}[\eta_{23}(u_{22} - u_{33}) + v_{33} - v_{22}] \\
i(\partial_t - \eta_{12}\partial_x) u_{12} &= \frac{\alpha\kappa}{\gamma}[i\partial_x(u_{13} u_{23}^*) + u_{13} u_{23}^*(u_{11} - u_{22})] \\
&\quad + u_{12}[\eta_{12}(u_{11} - u_{22}) + v_{22} - v_{11}]
\end{aligned} \tag{12}
$$

In (10) ((12)) we have written down only the relations (NLEE's) for v_{ij} (u_{ij}) with $i < j$; the corresponding ones with $i > j$ are obtained from them by complex conjugation and the use of (9). As regards the diagonal elements u_{ij} and v_{ij}, they satisfy:

$$
\begin{aligned}
& i(v_{jj,x} - u_{jj,t}) + F_{jj}(x,t) = 0 \\
& F_{11} = -F_{22} = \alpha\kappa(u_{12} u_{13}^* u_{23} - u_{12}^* u_{13} u_{23}^*); \; F_{33} = 0
\end{aligned} \tag{13}
$$

We can fix up the diagonal terms u_{jj} and v_{jj} by choosing the gauge of the Lax pair. There are many possibilities to do this:

$$
\begin{aligned}
u_{jj} &= \mu_j|u_{13}|^2 + \gamma\nu_j|u_{23}|^2 \\
v_{jj} &= \mu_j\eta_{13}|u_{13}|^2 + \gamma\nu_j\,\eta_{23}|u_{23}|^2 \qquad j = 1,2,3
\end{aligned} \tag{14}
$$

Then equations (13) are a direct consequence of (12), if the constants μ_j, ν_j are related by:

$$
\mu_j = \frac{a_2 - a_3}{a_1 - a_3}\nu_j + \theta_j, \quad \theta_1 = -\theta_2 = -\frac{\alpha}{\gamma(a_1 - a_3)}, \quad \theta_3 = 0 \tag{15}
$$

Note that only differences of μ_j and ν_j occur in (12), and by a phase transformation on u_{ij} and v_{ij}, one can also transform some of the constants in (15) to zero. For example, one may easily phase transform all three μ_j's to be equal or all three ν_j's to be equal. But because of (13), with F_{jj} nonzero in general, one may never phase transform all three μ_j's or all three ν_j's to be zero.

Thus the first two equations in (12) contain in addition to the usual bilinear in u_{ij} nonlinearities, also cubic terms. In the third equation in (12) the usual bilinear in u_{ij} term appears under an x-derivative; here we also have cubic terms.

Note that both reductions on U and V commute between themselves, so the total reduction group is $\mathbb{Z}_2 \otimes \mathbb{Z}_2$.

As an example, let us choose $a_3 = 0$, $a_2 = -a_1$, $b_1 = b_2 = 0$, $b_3 = -ca_1$, $a_1 = -c/\kappa$, $u_{13} = E$, $u_{23} = F$, $u_{12} = N$, $\gamma = -1$, $\nu_2 = \nu_1 + 1/c$, $\nu_3 = \nu_2 - 3/(2c)$, and $\alpha\kappa = 1$, then (12) reduces to

$$i\,(\partial_t - c\,\partial_x)\,E = -\,cNF$$

$$i\,(\partial_t + c\partial_x)\,F = -\,cN^*E \tag{16}$$

$$i\partial_t N = -\,i\,\partial_x(F^*E) + F^*E\,(|E|^2 - |F|^2)\,\frac{1}{c} + N\,(|E|^2 + |F|^2)$$

These equations remind one of two counterpropagating electromagnetic beams, E and F, interacting via some density modulation, N; and these are another example of integrable interactions between long waves and short waves [4,5].

3. Now consider U and V to be cubic in λ:

$$U\,(x,t,\lambda) = \sum_{k=0}^{3} U_k\,(x,t)\lambda^k\ , \quad V\,(x,t,\lambda) = \sum_{k=0}^{3} V_k\,(x,t)\lambda^k \tag{17}$$

and such that U_0, U_3, V_0 and V_3 are diagonal:

$$U_0 = \mathrm{diag}\,(u_{11}, u_{22}, u_{33}), \quad -U_3 = +\,\mathrm{diag}\,(a_1, a_2, a_3)$$

$$V_0 = \mathrm{diag}\,(v_{11}, v_{22}, v_{33}), \quad -V_3 = \mathrm{diag}\,(b_1, b_2, b_3) \tag{18}$$

where a_j, b_j are the same constants as in (8) above. The matrices U_1, U_2, V_1, V_2 are given by:

$$U_1\,(x,t) = \begin{bmatrix} 0 & u_{12} & 0 \\ 0 & 0 & u_{23} \\ u_{31} & 0 & 0 \end{bmatrix}, \quad U_2\,(x,t) = \begin{bmatrix} 0 & 0 & u_{13} \\ u_{21} & 0 & 0 \\ 0 & u_{32} & 0 \end{bmatrix} \tag{19}$$

$$V_1\,(x,t) = \begin{bmatrix} 0 & v_{12} & 0 \\ 0 & 0 & v_{23} \\ v_{31} & 0 & 0 \end{bmatrix}, \quad V_2\,(x,t) = \begin{bmatrix} 0 & 0 & v_{13} \\ v_{21} & 0 & 0 \\ 0 & v_{32} & 0 \end{bmatrix} \tag{20}$$

With this choice U and V automatically satisfy the reduction condition (3) with N = 3 and K = diag $(1, \omega, \omega^2), \omega = \exp(2\pi i/3)$. We can also impose the involution (4)

with $\varepsilon = -1$ and $B_1 = \begin{bmatrix} 0 & 0 & \varepsilon_1 \\ 0 & 1 & 0 \\ \varepsilon_1 & 0 & 0 \end{bmatrix}$, $\varepsilon_1 = \pm 1$, which gives:

$$\begin{array}{lll} u_{12} = -\varepsilon_1 u^*_{23}, & u^*_{31} = -u_{31}, & u^*_{13} = u_{13} \\ u_{21} = \varepsilon_1 u^*_{32}, & u^*_{11} = -u_{33}, & u^*_{22} = -u_{22} \\ & a^*_3 = -a_1, \quad a^*_2 = -a_2 & \end{array} \tag{21}$$

and analogous relations for v_{ij} and b_j. The compatibility condition now gives:

$$\begin{array}{ll} v_{12} = \eta_{12} u_{12} + \kappa\varepsilon_1 u_{13} u^*_{21}; & v_{13} = \eta_{13} u_{13} \\ v_{31} = \eta_{13} u_{31} + \kappa\varepsilon_1 |u_{21}|^2; & v_{21} = \eta_{12} u_{21} \end{array} \tag{22}$$

where η_{ij} and κ are expressed through a_j, b_j as in (11). The corresponding NLEE have the form:

$$\begin{aligned} i(\partial_t - \eta_{13}\partial_x)u_{13} &= \varepsilon_1 (\eta_{12} - \eta_{23})|u_{12}|^2 + \kappa u_{13} u_{12} u_{21} + u^*_{12} u^*_{21}) \\ &\quad + u_{13} [\eta_{13} (u_{11} - u_{33}) + v_{33} - v_{11}] \\ i(\partial_t - \eta_{12}\partial_x)u_{21} &= -\varepsilon_1 (\eta_{13} - \eta_{23}) u^*_{12} u_{31} \\ &\quad - \kappa u_{21} (u^*_{12} u^*_{21} + u_{13} u_{31}) \\ &\quad + u_{21} [\eta_{12} (u_{22} - u_{11}) + v_{11} - v_{22}] \\ i(\partial_t - \eta_{12}\partial_x) u_{12} &= i\kappa\varepsilon_1 (u_{13} u^*_{21})_x + \kappa\varepsilon_2 u_{13} u^*_{21} (u_{11} - u_{22}) \\ &\quad + u_{12} [\eta_{12} (u_{11} - u_{22}) + v_{22} - v_{11}] \\ i(\partial_t - \eta_{13}\partial_x) u_{31} &= i\kappa\varepsilon_1 (|u_{21}|^2)_x + \kappa\varepsilon_1 |u_{21}|^2 (u_{33} - u_{11}) \\ &\quad + u_{31} [\eta_{13} (u_{33} - u_{11}) + v_{11} - v_{33}] \end{aligned} \tag{23}$$

For the diagonal elements u_{jj}, v_{jj} we get

$$u_{jj,t} - v_{jj,x} = 0 \ , \ j = 1,2,3 \tag{24}$$

If we fix the gauge in the simplest possible way, choosing $v_{jj} = u_{jj} = 0$, then what we get is a modification of the 3-wave equations with additional cubic nonlinearities in the first two equations in (23); the nonlinearities in the two last equations of (23) just acquire additional x-derivative.

However, we can choose another gauge by requiring that the corresponding linear problems (1), (2) become equivalent to a Riemann-Hilbert problem with canonical normalization. This leads to:

$$u_{11} = \frac{u_{12}\, u_{21}}{a_1 - a_2} + \frac{u_{13}\, u_{31}}{a_1 - a_3} - \frac{\varepsilon_1\, u_{13}\, |u_{21}|^2}{(a_1 - a_3)(a_1 - a_2)}$$

$$u_{22} = - \frac{u_{12}^{*}\, u_{21}^{*}}{a_2 - a_3} - \frac{u_{12}\, u_{21}}{a_1 - a_2} + \frac{\varepsilon_1\, u_{13}\, |u_{21}|^2}{(a_1 - a_2)(a_2 - a_3)} \tag{25}$$

and

$$v_{11} = \frac{\eta_{12}\, u_{12}\, u_{21}}{a_1 - a_2} + \frac{\eta_{13}\, u_{13}\, u_{31}}{a_1 - a_3} + \theta_{11}\, \varepsilon_1\, u_{13}|u_{21}|^2$$

$$v_{22} = - \frac{\eta_{23}\, u_{12}^{*}\, u_{21}^{*}}{a_2 - a_3} - \frac{\eta_{12}\, u_{12}\, u_{21}}{a_1 - a_2} + \theta_{22}\, \varepsilon_1\, u_{13}|u_{21}|^2$$

$$\theta_{11} = \frac{3a_1\, \kappa - \eta_{23}}{(a_1-a_2)(a_1-a_3)} \ , \quad \theta_{22} = \frac{-3a_2\, \kappa + \eta_{13}}{(a_1-a_2)(a_2-a_3)} \tag{26}$$

In deriving the last line of (26) we have used also that $\sum_{j=1}^{3} a_j = 0$; u_{33} and v_{33} are defined by $\sum_{j=1}^{3} u_{jj} = \sum_{j=1}^{3} v_{jj} = 0$. From (25),(26) we see that such choice of the gauge leads to additional quartic and quintic nonlinearities in the NLEE (23).

At the end we give the explicit form of the NLEE (23) in the simplest possible case, when $u_{jj} = v_{jj} = 0$ and moreover $\eta_{12} = \eta_{23} = c$ = real and $\eta_{13} = 0$. Denoting

$$u_{12} = E(x,t) \ , \quad u_{13} = n(x,t) \ , \quad u_{23} = - \varepsilon_1\, E^{*}(x,t)$$

$$u_{21} = F(x,t) \ , \quad u_{31} = iN(x,t) \ , \quad u_{32} = \varepsilon_1\, F^{*}(x,t) \tag{27}$$

we get the following system for the two real n(x,t), N(x,t) and the two complex-valued functions E(x,t) and F(x,t):

$$\partial_t n = g\, n\, (E^* F^* + EF)$$

$$\partial_t N = g\, \varepsilon_1\, \partial_x\, (F^* F)$$

$$\partial_t F = -g\, E^*\, (F^* F) - i\, g\, N\, n\, F$$

$$\partial_t E = i\, \varepsilon_1\, g\, \partial_x\, (n\, F^*) \qquad (28)$$

Here $g = -i\kappa$ is a real-valued constant.

The question for possible physical applications of these equations is open.

Acknowledgements

One of us (V.S.G.) is grateful to Profs. A. Degasperis, A. Fordy and M. Lakshmanan for giving him the opportunity to participate in the Workshop, and the other (D.J.K.) wishes to acknowledge that this research has been supported in part by the Office of Naval Research, through Grant #N00014-88-K-0153, and the National Science Foundation through Grant #DMS-8501325.

References

[1] Kaup, D.J. (1976). The three wave interaction - a non-dispersive phenomenon. *Stud. in Appl. Math.* **55**, 9.

[2] Mikhailov, A.V. (1981). The reduction problem and the inverse scattering method. *Physica* **3D**, 73.

[3] Zakharov, V.E. & Manakov, S.V. (1975). On the theory of resonant interaction of wave packets in nonlinear media. *J. Exp. Theor. Phys. (USSR)* **69**, 1654.

[4] Benny, D.J. (1977). A General Theory for Interactions between Short and Long Waves. *Stud. in Appl. Math.* **56**, 81.

[5] Newell, A.C. (1978). Long Waves-Short Waves; a Solvable Model. *SIAM J. Appl. Math.* **35**, 650.

A. Csizmazia

36. SYLVESTER'S LAW IN INFINITE DIMENSIONS

A summary is given of some of the author's results concerning the "CS equations" (C:Congruent; S: Schroedinger operators). The CS equations are strongly 2+1-D (two spatial dimensions in addition to time) generalizations of the cornerstone (higher order) Korteweg-de Vries equations of soliton theory. The CS equations are obtained from a radical generalization, the "C equation" (C:Congruence), of the Heisenberg-Lax equation of quantum mechanics and soliton theory. The C equation is the time derived form of pairwise invertible congruence among the members of a time dependent family of self adjoint linear operators. In the finite dimensional case Sylvester's law indicates that the general C equation has as its only invariants the rank and the signature of the evolving self adjoint operator. The C equation for operators on an infinite dimensional Hilbert space can behave differently. In particular, the CS equations share an infinite number of independent "time invariants". Topics for further research are included.

The **Korteweg-de Vries (KdV) equation** is $\dot{u} = 2(-u_{xx}+u^2)_x$. x,t (t:time) are real variables. u is a real valued function of x,t. It is required that on the line u and each of its spatial derivatives vanishes rapidly at infinity. The case with u periodic in x is also of interest. The KdV equation is the simplest nonlinear equation in the hierarchy of "higher order KdV equations", KdV_1, KdV_2 (=KdV)...KdV_j,... The **$KdV_{\ell+1}$ equation** has the form $\dot{u} = -2\delta_x[T_\ell(u)]$. $T_\ell = P\partial_x^{-1}$ with P the differential operator part of $(\partial_x-\partial_x^{-1}\circ u)^{2\ell+1}$.

The KdV_j equation has the infinite sequence of independent time invariants $I_{\ell+1} = (2\ell+1)^{-1}\int T_\ell(u)dx$ for ℓ a natural number. The KdV equations and their invariants issue from the **Heisenberg-Lax (H-L) equation** $\dot{L}_t = [Q_t,L_t]$, $Q_t^* = -Q_t$. The H-L equation plays a fundamental role in quantum mechanics and in soliton theory. The H-L equation is the time derived form of unitary equivalence.

$U_t^* L_t U_t = L_0$, $U_t^* = U_t^{-1}$; $\dot{U}_t = Q_t U_t$, $U_0 = 1$. Thus the spectrum of L_t is independent of t. In the H-L equation take L_t to be the 1-D Schroedinger operator $-\partial_x^2 + u(x,t)$. Require Q_t to be a differential operator. The $KdV_{\ell+1}$ equation is obtained with Q the differential operator of $2A^{2\ell+1}$. Here $A = -\partial_x + \partial_x^{-1} \circ \frac{u}{4} + \frac{u}{4}\partial_x^{-1} + \ldots = -A^*$ is a square root of $\partial_x^2 - u$. The above invariants of the $KdV_{\ell+1}$ equation can be expressed in terms of the discrete spectrum of L_t.

Let M be the plane or the torus with standard coordinates x,y. Define $d = \partial_x + i\partial_y$. Then $d^* = -\partial_x + i\partial_y$, $\Delta = \partial_x^2 + \partial_y^2 = -d^* d = -dd$. Let b be in $C^\infty(M)$. Take $b_{\underline{k}} = (d^*)^k(b)$. At each instant in time let u be a real valued function defined on M. When M is the plane require that u and each of its spatial derivatives vanish rapidly at infinity. Let $s = \omega(u)$ with $\omega = d^{-1}d^*$. When M is the plane s is required to vanish at infinity. On the torus s is taken so as to have 0 as its mean value. The **CS equation** is $\dot{u} = 2\mathrm{Re}\, d^*(u_{\underline{2}} + 3su)$ [2].

The CS equation is a **strongly** 2+1-D generalization of the KdV equation in the sense of Theorem 1 below. Say M is the plane. Let x', y' be obtained from x, y by rotation through the angle θ. Assume $f(x', t_1)$ [respectively $g(y', t_2)$] is a solution of the KdV equation in the coordinates x', t_1 (respectively y', t_2). Set $t_1 = t\cos(3\theta)$, $t_2 = t(-\sin 3\theta)$.

Theorem 1: The (nonlinear) CS equation in x,y,t coordinates has, for each θ, the solution $u(x,y,t) = f(x',t_1) + g(y',t_2)$.

Note: This u is not required to vanish at infinity.

The CS equation has the independent invariants

$$I_1 = \int_M u, \quad I_2 = \frac{1}{3}\int su, \quad I_3 = \frac{1}{5}\int u(2s^2 + s_{\underline{2}}).$$

An infinite number of independent possible invariants are described below.

[The formal equation $\dot{u} = (\Delta u_x)u_y - (\Delta u_y)u_x$ is invariant under a rotation of the x,y coordinates. This equation has $\int_M (u_x)^2 + (u_y)^2$ and $\int F(u)$ as invariants. The equation does not appear to have any nontrivial 1+1-D specializations in x,y.]

We now discuss the origin of the CS equation. Set $(L,H) = LH + H^*L$. This "C product" (C:Congruence) is a simultaneous generalization of the Lie product ($H^* = -H$) and of the Jordan product ($H^* = H$). The **C equation** is $\dot{L}_t = -(L_t, H_t)$. The C equation is the time derived form of the congruences $C_t^* L_t C_t = L_0$, with C_t invertible. $t \longrightarrow C_t$ is "generated" by $t \longrightarrow H_t$: $\dot{C}_t = H_t C_t$; $C_0 = 1$. The "special" C equation has L_t equal to the 2-D Schroedinger operator $-\Delta+u$. H_t is required to be a differential operator with real valued coefficients [2,3].

Research Topic: What is the significance in quantum mechanics of the C equation?

Indeterminacy principle. Assume $L^* = L$. Then $(L,H) = (L,H')$ if $H = H' + SL$ with $S^* = -S$. Each solution of the special C equation can be reduced, via a judicious choice of S, to the canonical form

$$H = Td^* - T^*d \ , \quad T = T_\ell = \Sigma_n (d^*)^{\ell-n} \circ a_n (d^*)^{\ell-n} \ , \quad a_0 = 1.$$

Research Topics: What interpretations does the CS equation have in physics? On the plane does the CS equation have soliton-like solutions vanishing rapidly at infinity? Make computer studies of this equation.

The higher order CS equations will now be speficied. Define $T_q[\alpha] = \sum_{j=0}^{q} [\binom{q}{j} + \binom{q-1}{j}] \binom{\alpha}{k} (d^*)^j \circ u_{\underline{k}}$ with $k = 2q-1-j$. Set $T_{p,q} = T_q[\alpha]$ with $\alpha = p+q-1$. Put $p = \ell+1-n$. The "central recursion equation" is $\mathbf{d(a_n) = \Sigma\, T_{p,q}(a_{n-q})}$, $q = 1,2,\dots,n$. Take $a_0 = 1$. Note that for $n \geq 1$, a_n has a parametric dependence on ℓ: $a_n = a_n\langle\ell\rangle$. Momentarily assume that the coefficients of T_ℓ in the canonical form of H satisfy this central recursion equation. The "$CS_{\ell+1}$ equation" is $\mathbf{\dot{u} = 2Re\ d^*[T_\ell(u)]}$. a_n exists when the right hand side of the central recursion equation is a divergence.

Theorem 2: The C equation, with H_t in canonical form, holds if and only if the coefficients a_n of H_t satisfy the central recursion equation and u evolves according to the $CS_{\ell+1}$ equation.

Allow $\ell = 0,1,2,\dots$ Set $r = 2\ell+1$ and $a_n' = r^{-1}a_n$. Take $I_{\ell+1} = \frac{1}{r}\int T_\ell(u) = \int u a_\ell'\langle\ell\rangle$.

Conjecture: Each CS_k equation has the infinite sequence of (independent) invariants $I_1, I_2, I_3, \dots$

Research Topic: How does the scattering data for $-\Delta$+u evolve as u varies according to the $CS_{\ell+1}$ equation?

Set $L' = (d^*)^{-1} \circ (-\Delta+u) = d+(d^*)^{-1} \circ u$. Say $K = Td^*$ with T a pseudodifferential operator (ψdo) of the form $1 + (d^*)^{-1} \circ a_1 (d^*)^{-1} + (d^*)^{-2} \circ a_2 (d^*)^{-2} + \dots$ Note that K is a d^*-operator, i.e. K involves only d^* derivatives or their inverses. Assume: K **commutes with L'**. Say r is an odd natural number. Take $H\langle r\rangle$ to be the total part $H(r)^+$ of order ≥ 0 of $H(r) = 2\mathrm{Re}(K^r)$.

Theorem 3: The $H\langle r\rangle$ are the canonical solutions of the C equation.

The requirement that K be a d^*-operator eliminates the possibility of taking K = L. **K commutes with L′ precisely when the a_n satisfy the central recursion equation with $\ell = 0$.**

Research Topics: Prove K exists. Give an explicit expression for $a_n\langle\ell\rangle$ in terms of the coefficients of K.

Ghost operator conundrum. We do not claim that there exist C_t generated by $H\langle r\rangle$ for $r > 1$. These $H\langle r\rangle$ are unbounded and not skew adjoint for general u. However H(r) is skew adjoint relative to the inner product $\langle f,g\rangle_L = \langle Lf,g\rangle$ with $\langle a,b\rangle = \int_M a\bar{b}$. Momentarily assume L is invertible. Then $L^{1/2}H(r)L^{-1/2}$ is skew adjoint relative to $\langle\ ,\ \rangle$. **For each t with $0 \leq t \leq T$ let H_t be a ψdo of order ≤ 0. Assume $t \longrightarrow H_t$ is continuous in the operator norm topology. Require that each C_t be a ψ_{do} of order ≤ 0. Then there is a unique family $t \longrightarrow C_t$ generated by $t \longrightarrow H_t$. Moreover each C_t is invertible [2].**

The KdV hierarchy can be derived from the C equation in 1-D. We simply cut our results down from two spatial dimensions to the case of dependence only on x (and t). Take $L_t = -\partial_x^{\,2}+u(x,t)$. In the C equation, H_t is to be in the "canonical form" $-2T\partial_x$, with T a symmetric differential operator (with real valued coefficients). Take $K = \partial_x^{\,-1}L_t$. Set $H\langle r\rangle = 2(K^r)^+$ with $r = 2\ell+1$. **The $H\langle r\rangle$ are the only canonical solutions of the C equation.** The $\mathrm{KdV}_{\ell+1}$ equation arises from the C equation on $H\langle r\rangle$, since $H\langle r\rangle = Q+SL_t$ with: $S = -S^*$; and $Q = -Q^*$ the above solution of the Lax equation. This yields the form of the $\mathrm{KdV}_{\ell+1}$ equation given in the review.

We return to 2+1-D. Say $\dot{I}_{\ell+1} = \int_M \beta_\ell \dot{u}$. The function β_ℓ is the "gradient" of $I_{\ell+1}$. Next we make a conjecture on how to obtain the gradients. Allow $h = -1,0,1,\ldots$ Take $r = 2(h+1)$. Let $T'_h = K^r(d^*)^{-1}$. Then $T'_n = \sum_{n=1}^{\infty} (d^*)^{h-n}\circ[g_n d^* + \frac{1}{2}(g_n)_{\underline{1}}](d^*)^{h-n}$ for some functions $g_n\langle h\rangle(x,y)$. $I_{\ell+1}$ is "normal" means the gradient of $I_{\ell+1}$ is $g_\ell\langle\ell-1\rangle$.

Conjecture: Each I_k is normal.

Fix h. The $g_n\langle h\rangle$ satisfy the "corecursion equation" presented next. Define $T'_q[\alpha] = \sum_{j=0}^{q}[\binom{q}{j}\binom{\alpha-1}{k}+\binom{q-1}{j}\binom{\alpha}{k}](d^*)^j \circ u_{\underline{k}}$ with $k = 2q-1-j$. Put $T'_{p,q} = T'_q[\alpha]$ with $\alpha = p+q-1$. $g_0 = 1$. $d(g_n) = \sum_{q=1}^{n} T'_{p,q}(g_{n-q})$ with $p = h+2-n$.

Theorem 4: The CS equation has each normal I_k as an invariant.

Lemma 1 involves the remainders defined next. Each differential operator on M has a unique representation $d^*\circ Fd + d^*\circ R' + L'd + r$. Here F, R', L' are differential operators. R', L' are "thin", i.e. of degree 0 in d, d^* respectively. The "remainder" r is a function. This yields the representations $d^*\circ R + Ld + r$ with: $R = Fd + R'$, $L = L'$; and $R = R'$, $L = d^*F + L'$. Let w,k,n, be integers $\geq 0,2,1$ respectively. Set $q = w+k-n$. Require $q \geq 1$. Set $h = q-n$. Below we determine the "elite" differential operators $R_{n,h}[w,k]$, $L_{-h,q}[w,k]$. Abbreviations: $R_{n,h}[w] = R_{n,h}\langle k\rangle = R_{n,h}[w,k]$ and similarly for $L_{-h,q}[w,k]$. Require $L_{-h,q}\langle 2j\rangle$, $R_{n,h}\langle 2j+1\rangle$ to be thin. Put $R_{n,h}[w,k] = 0$, $L_{-h,q}[w,k] = 0$ when $|h| > w$, $|h-1| > w$ respectively. Let $u_{i,j}$ be the function $(d^*)^i d^j(u)$. The elite operators corresponding to the initial value $k = 2$ are uniquely determined from $u_{0,q-1}(-d)^{q-1}T'_{1,n} = d^*\circ R_{n,h}\langle 2\rangle + L_{-h,q}\langle 2\rangle d$. Determine the other elite operators via the following recursion over w,k.

$$\mathbf{d^*\circ R_{n,h}[w,k+1] + L_{-h,q}[w,k+1]d + r =}$$

$$\mathbf{\Sigma_p\, T_p[q-1]^*\, R_{n,h-p}[w+1-p,k] + \Sigma_p\, L_{-h-p,q}[w+1-p,k]T'_p[n]}$$

for some remainder function $r = r_{n,q}[w] = r_{n,q}\langle k\rangle$.

Lemma 1: Each CS_j equation has every normal I_v as an invariant, if the remainders $r_{n,q}\langle k\rangle$ vanish for $k \geq 3$.

For $w = 0,1$ the remainders vanish. In fact $L_{-1,n+1}[0] = u^{2n+1}$, $R_{n,0}[0] = u^{2n}$. $R_{n+1,0}[1] = c\theta u^{2n-1}$ with $c = 2\binom{n+2}{3}$ and $\theta = uu_{1,1} - u_{1,0}u_{0,1}$. $L_{-1,n+1}[1] = c'\theta u^{2(n-1)}$. Here $c' = a(2n+1)/3$ with $a = \binom{n+1}{2}$. Let v,h be $2^{-1}[b(u_{1,0})^2 - uu_{2,0}]$ with $b = 1,3$ respectively. v',h' are defined similarly using the d derivative.

$$R_{n+1,-1}[1] = a[d^{*}\circ u_{1,0}u - v - 3^{-1}(4n-1)h]\circ u^{2(n-1)}.$$

$$L_{0,n+1}[1] = (ad^{*}\circ u_{1,0}u - c'2h)\circ u^{2n-1}.$$

$$R_{n,1}[1] = u^{2(n-1)}a[-uu_{0,1}d + v' - 3^{-1}(4n-1)h'].$$

$$L_{-2,n+2}[1] = u^{2n-1}(-auu_{0,1}d - c2h').$$

Example: Fix $x',y' > 0$. let $u(x,y) = e^{-(x'x+y'y)}$ for $x,y > 0$. Each $r_{n,q}\langle k\rangle$, $L_{-h,q}\langle 2j\rangle$, $R_{n,h}\langle 2j+1\rangle$ vanishes. Set $z = x' + iy'$. For $k \geq 2$ let $R\langle k\rangle = \Sigma_{n,q}\ \varepsilon^{n}\lambda^{q}R_{n,h}\langle k\rangle$. Define $L\langle k\rangle$ similarly. Say $k = 2j$ or $k = 2j + 1$. Allow $c = 1,2,\ldots,j$. Take $R'\langle k\rangle$ to be the ψdo $\varepsilon^{j}(\Pi_{c}\{1-\varepsilon\bar{z}c[d^{*}-(k-c)\bar{z}]\})^{-1}\circ u^{k}\lambda^{j}\Pi_{c}\{1-\lambda zc[d-(k-c)z]\}^{-1}$. Then: $R\langle 2j\rangle = R'\langle 2j\rangle$. Also: $L\langle 2j+1\rangle = R'\langle 2j+1\rangle[1-\lambda z(j+1)(d-jz)]^{-1}$.

The proofs of Theorem 4 and Lemma 1 given in Csizmazia [3] have the virtue of avoiding the use of nonlocal operators.

Research Topics: Express $g_n\langle h\rangle$ in terms of the coefficients of K. Prove the conjectures on the invariants and their gradients. Can a proof of invariance be based on the generalization of the adjoint operator arising from the C product?

An exotic (nonlocal) "continued fraction" appears below. For $h \neq -1$ set $g'_n\langle h\rangle = [2(h+1)]^{-1}g_n\langle h\rangle$. "HOT" stands for: highest order (degree) terms. HOT a'_n (HOT g'_n) consists of the terms of a'_n (respectively g'_n) which are nonlinear in u of order n. Explicit expressions for HOT $a'_\ell\langle\ell\rangle$ and HOT $g'_{h+1}\langle h\rangle$ result from the "continued fraction" defined next and the theorem below. Let J be the formal power series $j_1\lambda+\ldots+j_n(x,y)\lambda^{n}+\ldots$ Determine the j_n recursively via $J = F(J)$ with $F(J) = \lambda\omega[u(1-J)^{-1}]$. Take $J_0 = 0$. Define J_n for $n \geq 1$ using $J_{n+1} = F(J_n)$. In J_n the coefficient of λ^{k} is j_k for $k = 1,\ldots,n$. In the limit

$$J = \lambda\omega\left(\cfrac{u}{1-\lambda\omega\left(\cfrac{u}{1-\cdot_{\cdot_{\cdot}}}\right)}\right)$$

Note the novel role of the nonlocal operator ω.

Theorem 5: $\sum_{\ell\geq 1}\lambda^{\ell}\mathrm{HOT}\ a'_\ell\langle\ell\rangle = (1-J)^{-1}$. $\sum_{h\geq 0}\lambda^{h+1}\mathrm{HOT}\ g'_{h+1}\langle h\rangle = -\ln(1-J)$. **More generally:**

$$\sum_{n=0}^{\infty} \lambda^n \text{HOT } a'_n\langle p-1+n\rangle = (2p-1)^{-1}(1-J)^{-(2p-1)}.$$

$$\sum_{n\geq 1} \lambda^n \text{HOT } g'_n\langle p-2+n\rangle = [2(p-1)]^{-1}[-1+(1-J)^{-2(p-1)}].$$

[3].

Research Topic: Extend the results of Theorem 5 to formulas for $a'_n\langle \ell\rangle$, $g'_n\langle h\rangle$.

A foreshadowing of defeat is dispelled as we now relate. We are employing the C equation involving linear operators on the infinite dimensional pre-Hilbert space of C^∞, complex-valued functions with compact support defined on the space manifold M. We seek nonlinear evolution equations with an infinite number of time invariants. We bet on obtaining such equations from the C equation. Analogy with the C equation involving linear operators on the finitie dimensional space $\mathbb{R}^n$ warns us that we are gambling at high risk. The latter C equation amounts to pairwise invertible congruence as above. (L_0 is assumed to be self adjoint.) Sylvester's law of inertia states that self-adjoint operators L_t, L_0 on $\mathbb{R}^n$ have the same rank and signature (inertial index) if and only if L_t is invertibly congruent with L_0. Thus the general C equation in finite dimensions has as its only invariants the rank and the signature of the evolving self-adjoint operator. Nonetheless the infinite dimensional case considered apparently yields an infinite number of invariants.

Research Topic: Explain these invariants using the C equation and operator theory. Can this lead to a new fundamental principle for congruences which are more general than unitary equivalences and yet conserve quantities in addition to the (analogues) of rank and signature?

We now turn our attention to analogues in infinite dimensions of Sylvester's law [2]. There is an infinite dimensional analogue of the signature of a self-adjoint operator. Let L be a ψdo which is elliptic, self-adjoint and of positive order on sections of a vector bundle over a compact oriented Riemannian manifold. L is said to be a "special operator". L has pure point spectrum λ_j. The function

$$\eta_L(s) = \sum_{\lambda_j\neq 0} (\text{sign } \lambda_j)|\lambda_j|^{-s} = \text{tr}[(L^2)^{-\frac{s+1}{2}} L]$$

turns out to be holomorphic for s with real part >> 0. It has a meromorphic extension to the whole plane with s = 0 not a pole. $\eta_L(0)$ is a natural definition for the signature of L [1].

On the 2-D torus the signature of the Schroedinger operator L =

$-\Delta+u$ is $\eta_L(0) = -\frac{1}{4\pi}\int_M u$. This follows from

$$\eta_L(0) = \frac{1}{(2\pi)^2}\int_M d(x)d(y)\int_{s\in\mathbb{R}^2-\{0\}}(ds)\frac{1}{2\pi i}\int_\gamma (d\lambda)e^{-\lambda}r_{-4}$$

with r_{-4} and γ defined as follows. The symbol r of $(\lambda-L)^{-1}$ has an asymptotic expansion $\sum_{j\le -2} i^j r_j(x,s,\lambda)$ where $r_j(x,ps,p^2\lambda) = p^j r_j(x,s,\lambda)$ for any $p > 0$. γ is a circle of small radius traversed in the counterclockwise sense about the origin. $\int u$ is an invariant of the $CS_{\ell+1}$ equation. Thus the associated C equation has $\eta_L(0)$ as an invariant [3]. This can also be proven directly from the C equation using the formula for $\dot{\eta}_{L_t}(z)$ presented below.

Research Topic: Can the nonlocal "invariants" $I_2, I_3, \ldots$ be expressed in terms of spectral information arising from $-\Delta + u$ and K?

We pass to the general point of view. Let $t \longrightarrow L_t$ be a C^∞ one-parameter family of special operators on the Riemannian manifold M of dimension N. Since $\eta_{L_t}(z) = \mathrm{tr}[(L_t^2)^{-\frac{1}{2}(z+1)} \cdot L_t]$ it is not hard to show [1] that $\dot{\eta}_{L_t}(z) = -z\ \mathrm{tr}[(L_t^2)^{-\frac{1}{2}(z+1)}\dot{L}_t]$ for Re z >> 0. [Use: $\mathrm{tr}(AB) = \mathrm{tr}(BA)$ if A and B are of trace class; and $(L^2)^{-z/2}L$ is of trace class for Re(z) >> 0.] Thus $\eta_{L_t}(0)$ varies with t precisely when $\mathrm{tr}[(L_t^2)^{-\frac{1}{2}(z+1)}\dot{L}_t]$ has a pole at $z = 0$. Let us now assume that L_t evolves according to the C equation. Let B_t be the "residual" operator $\frac{1}{2}(H_t+H_t^*)$. Set $E = \mathrm{tr}[(L_t^2)^{-z/2}(\frac{L_t}{|L_t|})B_t]$. Then $\dot{\eta}_{L_t}(z) = 2zE$. Set $\eta\langle t\rangle = \eta_{L_t}(0)$. Thus $\dot{\eta}\langle t\rangle = 2r_t$ with r_t the residue at $z = 0$ of E.

We illustrate with the C equation involving H<3>.

$$\dot{\eta}\langle t\rangle = \frac{1}{2\pi^2}\int_M d(x)d(y)\int_{\mathbb{R}^2-\{0\}}(ds)\frac{1}{2\pi i}\int_\gamma (d\lambda)e^{-\lambda}\lambda T_{-4}$$

with γ as above and T_{-4} defined as follows. $\ell = 1$ so B_t is a function $B_t = B(x,y,t)$. The symbol of $(\frac{1}{\lambda-L_t})\circ B_t$, namely $r\circ B_t$, has an asymptotic expansion $\sum_{h\le -2} i^h T_h$ where $T_h(x,ps,p^2\lambda) = p^h T_h(x,s,\lambda)$ for any $p > 0$. Calculating T_{-4} leads to $\dot{\eta}\langle t\rangle = 0$. For the details for this and the material below see Csizmazia [2].

We now give a counterxample showing that Sylvester's law cannot be naively generalized to infinite dimensions. Let M be the 3-D torus. We consider ψdo's acting on 4-tuples of complex valued functions on M. We specify a self-adjoint square root D of $-\Delta$.

$$D = \begin{pmatrix} 0 & \text{div} \\ -\text{grad} & \text{curl} \end{pmatrix} = \begin{pmatrix} 0 & \partial x_1 & \partial x_2 & \partial x_3 \\ -\partial x_1 & 0 & -\partial x_3 & \partial x_2 \\ -\partial x_2 & \partial x_3 & 0 & -\partial x_1 \\ -\partial x_3 & -\partial x_2 & \partial x_1 & 0 \end{pmatrix}$$

Set $J_t = 1-tF$ with $F = \Delta^{-1}D\Delta^{-1}$. Take $L_t = J_t^* D J_t$. Then

$L_t = D + t(2\Delta^{-1}) + t^2(-\Delta^{-3}D)$.

Theorem 6: $\dot{\eta}\langle t\rangle \neq 0$, even though $L_t = J_t^* D J_t$ with each J_t invertible for small $|t|$.

Proof. Set $\eta\langle t\rangle = \eta_{L_t}(0)$. We have $\eta\langle t\rangle = \frac{1}{(2\pi)^2}\int_M (dx) \int (ds)(I^+ - I^-)$ with $x = (x_1,x_2,x_3) \in M$, s the nonzero covariable in $\mathbb{R}^3$ and for $\sigma = \pm$, $I^\sigma = \frac{1}{2\pi i}\int_{\gamma_\sigma} (d\lambda)e^{-\sigma\gamma}\operatorname{tr} r_{-4}$. γ_+,γ_- are paths akin to the respective branches $x > 0$, $x < 0$ of $x^2 - y^2 = \varepsilon$ with ε a small positive number. The matrix r_{-4} is obtained as follows. Regard $\theta = (s,\lambda)$ as a new covariable. The symbol r of $(\lambda-L)^{-1}$ has an asymptotic expansion $\sum_{k\le -1} i^k r_k$ with $r_k(ps,p\lambda) = p^k r_k(s,\lambda)$ for $p > 0$. A calculation shows $\eta\langle t\rangle = \mu t$ with $\mu = \frac{ab}{\pi^3}\int_0^\infty e^{-\rho}d\rho$. Here $a = \int_M 1$ is the volume of M. b is the surface area of the unit sphere in $\mathbb{R}^3$. $\mu \neq 0$. So $\dot{\eta}\langle t\rangle \neq 0$.

A more general context illuminates the preceding counterexample. Suppose L_0 is of special type. Assume f is a real valued function of a real variable such that $f(L_0)$ has order < 0. Set $L_t = J_t^* L_0 J_t$. $\eta\langle 0\rangle = 2r_0$ with r_0 the residue defined above. r_0 depends on L_0 and B_0. $B_0 = -f(L_0)$. Assume $L_0^2 = -\Delta$ + lower order terms. Set $f(\theta) = -\frac{\theta}{|\theta|}(\theta^2)^{-N/2}$ with N the dimension of M. Then r_0 is the residue at

$z = 0$ of $\mathrm{tr}(L_0^2)^{-\frac{1}{2}(z+N)}$. This residue is the volume of M multiplied by a nonzero constant $\left[= \frac{n}{2^{n-1}\Gamma(\frac{N}{2})\pi^{N/2}}\right.$ if L_t acts on n-tuples of complex valued functions] according to a celebrated theorem of Weyl. Note that B_0 is of order $-N$. So B_0 is not of trace class.

$\dot{\eta} = 2r_t$ has another consequence. Suppose the special operators L_t evolve according to the C equation.

Theorem 7: The signature of L_t is independent of t, if each residual operator is of trace class.

Proof. It suffices to prove the following.

Lemma 2: Say L is a special operator. Assume T is of trace class. Then $\mathrm{tr}[(L^2)^{-s}T]$ has an analytic extension from Re s >> 0 to Re s > 0. It has a continuous extension to Re s ≥ 0. Also assume T is a ψdo. Then there is a meromorphic extension to the whole plane.

Proof of Lemma 2. L^2 has an orthonormal basis of eigenvectors $\phi_0, \ldots \phi_n, \ldots$ with $L^2(\phi_n) = \lambda_n\phi_n$. The required extension to Re s ≥ 0 is $\Sigma\, \lambda_n^{-s}\langle T\phi_n, \phi_n\rangle$. (Use: $\lambda_n \geq 1$ for large n.)

References

[1] Atiyah, M.F., Patodi, V.K. & Singer, I.M. (1976). Spectral asymmetry and Riemannian geometry, III. *Math. Proc. Camb. Phil. Soc.* **79**, 71.

[2] Csizmazia, A. (1984). *Invariants under Congruence for Infinite Dimensional Operators.* Thesis, University of California, Berkeley.

[3] Csizmazia, A. (1988). The Heisenberg-Lax equation generalized. In *Some Topics on Inverse Problems. Proceedings of the XVIth Workshop on the Interdisciplinary Study of Inverse Problems (Montpellier, France, Nov. 30 - Dec. 4, 1987).* World Scientific Publishing Co.: Singapore.

M. Antonowicz and A.P. Fordy

37. HAMILTONIAN OPERATORS AND MIURA MAPS FROM LAX OPERATORS: ENERGY DEPENDENT SCHRODINGER OPERATORS AND THEIR SUPER EXTENSIONS

1 Introduction

In a number of recent papers [1-4] we discussed the two generalised Schrödinger equations :

$$\mathbb{L}_1\psi \equiv \left(\sum_0^{N-1}\lambda^i(\varepsilon_i\partial^2 + u_i)\right)\psi = \lambda^N\psi \tag{1.1a}$$

$$\mathbb{L}_2\psi \equiv \left(\partial^2 + \sum_1^{N} u_i\lambda^i\right)\psi = a^2\psi \tag{1.1b}$$

where a is a constant. We have shown that the isospectral flows of each of these spectral problems possess (N+1) compatible Hamiltonian structures $\mathbf{B}_0,\ldots,\mathbf{B}_N$. When N=1 these spectral problems give rise respectively to the KdV and Harry Dym hierarchies, both bi-Hamiltonian. When N=2, (1.1a) includes the dispersive water waves (DWW) and Ito hierarchies, both tri-Hamiltonian. More recently [5] we have derived a sequence of Miura maps and modified equations associated with these spectral problems and given a super extension of all our results [6]. Particular cases of our construction are the sKdV, sDWW and super-Harry Dym equations and their modifications. It is the purpose of this paper to give a brief and introductory review of our results so far.

In section 2 we give a brief synopsis of the Hamiltonian theory of NLEEs, referring to [7,8] for further details.

In section 3 we use a <u>single construction</u> to derive :

(a) the isospectral flows of (1.1) and corresponding time evolutions of eigenfunctions ψ.

(b) an infinite hierarchy of constants of motion.

(c) the locally defined Hamiltonian structures associated with (1.1).

In section 4 we introduce a generalisation of the factorisation method [9-11] and construct Miura maps and modified systems.

In section 5 we introduce a super-extension of the energy dependent Schrödinger operators (1.1) and extend the above constructions to this case.

2 Hamiltonian Property.

In this paper we are concerned with systems of NLEEs (in (1+1)-dimensions) which can be written in Hamiltonian form :

$$\mathbf{u}_t = \mathbf{B}\delta\mathcal{H} , \tag{2.1}$$

where $\mathbf{B}$ is a (matrix differential) Hamiltonian operator. $\delta\mathcal{H}$ is the variational derivative of functional $\mathcal{H}$ and generally a vector: $\delta \equiv \delta_u \equiv (\delta_0, \cdots, \delta_{N-1})^T$, $\delta_i = \dfrac{\delta}{\delta u_i}$. In the context of analysis and physics one would then deal with constants of motion and Poisson brackets in their integral form, respectively :

$$\mathbb{H} = \int \mathcal{H}dx \ , \ \{\mathbb{K},\mathbb{H}\} = \int \delta\mathcal{K}\mathbf{B}\delta\mathcal{H}dx \tag{2.2}$$

which would involve particular boundary conditions on the functions $u_i(x)$. To avoid any such considerations it is customary to work within the framework of differential algebras.

Roughly speaking, when calculating conserved densities and Poisson brackets we work modulo exact x-derivatives. The justification for this is that such exact derivatives contribute only boundary terms in the above definitions of $\mathbb{H}$ and $\{\mathbb{K},\mathbb{H}\}$. Thus, with appropriate boundary conditions, these terms vanish. Furthermore, the total x-derivatives constitute the kernel of the variational derivative operator. Thus, we may consider two functions of u_i and their x-derivatives to be equivalent if they differ by a total x-derivative. The corresponding equivalence class (and, by an abuse of notation, any convenient representative) is written $\mathcal{H}$, as it was in (2.1) and (2.2).

Conservation Laws

Let $K[u_i]$ be a function on phase space. Then :

$$\frac{dK}{dt} = \mathbf{K}'\mathbf{u}_t = (\delta K)^T\mathbf{u}_t + \frac{dF}{dx} \tag{2.3a}$$

where $\mathbf{K}'$ denotes the Fréchet derivative (operator) defined by $\mathbf{K}'[\mathbf{u}]\mathbf{v} = \dfrac{d}{d\varepsilon} K[\mathbf{u}+\varepsilon\mathbf{v}]\Big|_{\varepsilon=0}$. The second step corresponds to an integration by parts. Adding an exact derivative J_x to K does not change δK but does change F to $\tilde{F} = F+J_t$. Thus we may choose any convenient representative of the equivalence class $\mathcal{K}$ of K. Using

(2.1) we find :

$$\mathcal{K}_t = \frac{\delta\mathcal{K}}{\delta u_i} B_{ij} \frac{\delta\mathcal{H}}{\delta u_j} + \tilde{F}_x . \tag{2.3b}$$

Defining the quadratic form $\{\mathcal{K},\mathcal{H}\}$ by :

$$\{\mathcal{K},\mathcal{H}\} = \frac{\delta\mathcal{K}}{\delta u_i} B_{ij} \frac{\delta\mathcal{H}}{\delta u_j} \tag{2.4}$$

it follows from (2.3) that whenever $\{\mathcal{K},\mathcal{H}\} = 0 \pmod{\mathrm{Im}\partial}$, we have the local conservation law :

$$\mathcal{K}_t = \mathcal{F}_x \tag{2.3c}$$

where $\mathcal{F}$ is the flux corresponding to $\mathcal{K}$.

Poisson Brackets and Hamiltonian Operators

The quadratic form (2.4) defines a Poisson bracket if and only if it is skew-symmetric and satisfies the Jacobi identity :ie. if for any 3 functions $\mathcal{H},\mathcal{J},\mathcal{K}$:

(i) $\{\mathcal{H},\mathcal{J}\}+\{\mathcal{J},\mathcal{H}\} \in \mathrm{Im}\,\partial$ (2.5a)

(ii) $\{\{\mathcal{H},\mathcal{J}\},\mathcal{K}\}+\{\{\mathcal{K},\mathcal{H}\},\mathcal{J}\}+\{\{\mathcal{J},\mathcal{K}\},\mathcal{H}\} \in \mathrm{Im}\,\partial$. (2.5b)

When (2.4) satisfies conditions (2.5a,b) then **B** is called a Hamiltonian operator (or Hamiltonian structure). Property (2.5a) is guaranteed by choosing operator **B** to be skew adjoint, while the Jacobi identity (2.5b) is a much stronger (and much more complicated) constraint (see [7,8] for details). Thus, most skew symmetric operators are <u>not</u> Hamiltonian.

Remark

For two Hamiltonians to Poisson commute *wrt* (2.4) means that the *rhs* of (2.4) is an exact derivative. In the analytic context the *rhs* is a boundary term which can only be 'thrown away' with an appropriate choice of boundary condition.

A system of evolution equations is said to be bi-Hamiltonian if there exist two Hamiltonian operators $\mathbf{B}_0$ and $\mathbf{B}_1$ and two Hamiltonians $\mathcal{G}$ and $\mathcal{H}$ such that

$$\mathbf{u}_t = \mathbf{B}_0\delta\mathcal{G} = \mathbf{B}_1\delta\mathcal{H} \quad . \tag{2.6}$$

It is particularly interesting if the operator $\mathbf{B}_0 + \mathbf{B}_1$ is also Hamiltonian, in which case $\mathbf{B}_0$ and $\mathbf{B}_1$ are said to be compatible (in general the sum of the Poisson brackets would fail to satisfy the Jacobi identity). The importance of compatibility is that it enables us (under certain conditions) to construct an infinite hierarchy of (Poisson commuting) Hamiltonians. This important condition was first noticed by Magri [12]. It is now possible to state a useful lemma (see [7] for a proof).

Lemma

If $\mathbf{B}_0$ and $\mathbf{B}_1$ are compatible Hamiltonian operators, with $\mathbf{B}_0$ non-degenerate, and

$$\mathbf{B}_1 \delta\mathcal{G} = \mathbf{B}_0 \delta\mathcal{H}\ , \quad \mathbf{B}_1 \delta\mathcal{H} = \mathbf{B}_0 \mathbf{K} \tag{2.7a}$$

then there exists a function $\mathcal{K}$ s.t. $\mathbf{K} = \delta\mathcal{K}$.

To prove the existence of an infinite hierarchy of Hamiltonians, $\mathcal{H}_n$, related to compatible Hamiltonian operators $\mathbf{B}_0, \mathbf{B}_1$, we need to check that two conditions hold :

(i) $\exists$ an infinite sequence of vector functions $\mathbf{K}_0, \mathbf{K}_1, \ldots$ satisfying

$$\mathbf{B}_1 \mathbf{K}_n = \mathbf{B}_0 \mathbf{K}_{n+1} \tag{2.7b}$$

(ii) $\exists$ two function(al)s $\mathcal{H}_0$ and $\mathcal{H}_1$ s.t.

$$\mathbf{K}_0 = \delta\mathcal{H}_0, \quad \mathbf{K}_1 = \delta\mathcal{H}_1$$

It then follows from the Lemma that there exist function(al)s $\mathcal{H}_n$ s.t.

$$\mathbf{K}_n = \delta\mathcal{H}_n \quad \forall\ n \geq 0\ . \tag{2.7c}$$

Remark

Condition (i) is not always easy to check, although it is for our systems. Indeed, it may not even be satisfied, as shown by an example of Kupershmidt [13].

For this construction, it is of no advantage for a system to be <u>more</u> than bi-Hamiltonian. However, the existence of multi-Hamiltonian stuctures does lead to a rich supply of (multi-)Hamiltonian modifications.

Recursion Operator

Suppose we use (2.7b) to define an evolution parameter t_n by :

$$\mathbf{u}_{t_n} = \mathbf{B}_0 \mathbf{K}_n = \mathbf{G}_n \tag{2.8}$$

We define an integro-differential operator $\mathbf{R}$ by formally inverting $\mathbf{B}_0$:

$$\mathrm{R} = \mathrm{B}_1 \mathrm{B}_0^{-1} \tag{2.9}$$

Then

$$\mathbf{R}\mathbf{u}_{t_n} = (\mathbf{B}_1 \mathbf{B}_0^{-1}) \mathbf{B}_0 \mathbf{K}_n = \mathbf{B}_1 \mathbf{K}_n = \mathbf{B}_0 \mathbf{K}_{n+1} = \mathbf{u}_{t_{n+1}}$$

Thus $\mathbf{R}$ maps flows onto flows. $\mathbf{R}$ is called the recursion operator since it can be used to generate the infinite sequence of flows

(2.8) once we have the first. It is known [7] that $\mathbf{R}^{\dagger}$ satisfies the Lax type equation :

$$\mathbf{R}^{\dagger}_{t_n} = [-\left(\mathbf{G}'_n\right)^{\dagger}, \mathbf{R}^{\dagger}] \tag{2.10a}$$

where $\mathbf{G}'_n$ is the Fréchet derivative of the *rhs* of (2.8). This is the integrability condition of the spectral problem (squared eigenfunctions) :

$$\mathbf{R}^{\dagger}\Phi = \lambda\Phi \tag{2.10b}$$

and the linear evolution :

$$\Phi_t = -\left(\mathbf{G}'_n\right)^{\dagger}\Phi \tag{2.10c}$$

which is just the adjoint of the linearisation of equation (2.8). Thus the bi-Hamiltonian system (2.8) has the Lax representation (2.10).

3 Hamiltonian Operators from Lax Equations.

In this paper we consider a fairly general second order scalar spectral problem :

$$\mathbb{L}\psi \equiv (\varepsilon\partial^2 + u)\psi \equiv \sum_{0}^{N} \lambda^i(\varepsilon_i\partial^2 + u_i)\psi = 0 \tag{3.1a}$$

with ε_i being constant and u_i functions of x.

We look for time evolutions of the wave function ψ of the form :

$$\psi_t = \mathbb{P}\psi \equiv (\tfrac{1}{2}P\partial + Q)\psi \tag{3.1b}$$

where P and Q are functions of u_i and their x-derivatives, and of the spectral parameter λ. A simple calculation leads to

$$\mathbb{L}_t - [\mathbb{P},\mathbb{L}] = u_t + \varepsilon Q_{xx} - \tfrac{1}{2}Pu_x + \tfrac{1}{2}\varepsilon(P_{xx} + 4Q_x)\partial + \varepsilon P_x\partial^2 . \tag{3.2a}$$

Evidently, we cannot expect the usual Lax equation to hold. However, the integrability conditions of (3.1a,b) imply that $(\mathbb{L}_t - [\mathbb{P},\mathbb{L}])\psi = 0$ for eigenfunctions of (3.1a). To match the coefficient of ∂^2 we must take :

$$\mathbb{L}_t - [\mathbb{P},\mathbb{L}] = P_x\mathbb{L} . \tag{3.2b}$$

This further implies that $P_{xx} + 4Q_x = 0$, so that (3.2b) takes the remarkably simple form :

$$u_t = \left(\tfrac{1}{4}\varepsilon\partial^3 + \tfrac{1}{2}(u\partial + \partial u)\right)P \equiv JP . \tag{3.2c}$$

Remark

On the phase space defined by just one function u, the operator J, defined by (3.2c) is Hamiltonian, being (when $\varepsilon = 1$) just the second Hamiltonian structure of the KdV equation. The operator J is the basic unit out of which all our Hamiltonian operators are built.

With ε and u defined by (3.1a), the operator J takes the form :

$$J = \sum_{k=0}^{N} \lambda^k J_k \equiv \sum_{k=0}^{N} \lambda^k \left(\frac{1}{4}\varepsilon_k \partial^3 + \frac{1}{2}(u_k \partial + \partial u_k) \right) . \tag{3.3a}$$

Equation (3.2c) then takes the form

$$\sum_{k=0}^{N} \lambda^k u_{kt} = \left(\sum_{k=0}^{N} \lambda^k J_k \right) P . \tag{3.3b}$$

In [5] we continued the general development to include both KdV and Harry Dym type equations. Here we just consider the KdV reduction. This simplifies some of the formulae and statements :

KdV Case $u_N = -1$, $\varepsilon_N = 0$, so that $J_N = -\partial$.

To construct the 'polynomial' time evolutions we first seek a solution of :

$$J\mathcal{P} = 0 \ , \ \mathcal{P} = \sum_{k=0}^{\infty} P_k \lambda^{-k} \tag{3.4a}$$

written explicitly as :

$$J_0 P_{k-N} + J_1 P_{k-N+1} + \ldots + J_N P_k = 0, \qquad \forall k \geq 0. \tag{3.4b}$$

A polynomial expansion $P_{(m)}$ is then defined by :

$$P_{(m)} = (\lambda^m \mathcal{P})_+ = \sum_{k=0}^{m} P_{m-k} \lambda^k . \tag{3.4c}$$

Upon substitution of $P_{(m)}$ into (3.3c) (with t_m parametrising the corresponding evolution) the coefficients of λ^k, $k \geq N$, are identically zero, whilst the remaining ones give the equations of motion for $u_0, \ldots, u_{N-1}$:

$$\begin{pmatrix} u_0 \\ \cdot \\ \cdot \\ \cdot \\ u_{N-1} \end{pmatrix}_{t_m} = \begin{pmatrix} 0 & & & J_0 \\ & & \cdot^{\cdot^{\cdot}} & \vdots \\ & \cdot^{\cdot^{\cdot}} & & \vdots \\ J_0 & \cdots & \cdots & J_{N-1} \end{pmatrix} \begin{pmatrix} P_{m-N+1} \\ \cdot \\ \cdot \\ \cdot \\ P_m \end{pmatrix} . \tag{3.4d}$$

It is a remarkable fact that the scalar recursion relation (3.4b) can be written as an $N \times N$ matrix equation in exactly (N+1) different ways :

$$\mathbf{B}_n\mathbf{P}^{(k-1)} = \mathbf{B}_{n-1}\mathbf{P}^{(k)} , \qquad n = 1,\dots,N, \tag{3.5a}$$

where $\mathbf{P}^{(k)} = (P_{k-N+1},\dots,P_k)^T$ and the matrix differential operators $\mathbf{B}_n$ are determined by the following requirement : $\mathbf{B}_n$ is skew adjoint and the nth row of each matrix equation (3.5a) is just (3.4b), the remaining ones being identities. Explicitly, $\mathbf{B}_n$ are :

$$\mathbf{B}_n = \left(\begin{array}{cccc|cccc} 0 & & & J_0 & & & & \\ & & \cdot & \vdots & & 0 & & \\ & \cdot & & \vdots & & & & \\ J_0 & \cdots & \cdots & J_{n-1} & & & & \\ \hline & & & & -J_{n+1} & \cdots & \cdots & -J_N \\ & & & & \vdots & & \cdot & \\ & 0 & & & \vdots & \cdot & & \\ & & & & -J_N & & & 0 \end{array}\right) \tag{3.5b}$$

and satisfy the formal relation $\mathbf{B}_n = \mathbf{R}\mathbf{B}_{n-1}$ where :

$$\mathbf{R} = \mathbf{B}_1\mathbf{B}_0^{-1} = \left(\begin{array}{cccc|c} 0 & \cdots & \cdots & 0 & -J_0J_N^{-1} \\ \hline 1 & & & & -J_1J_N^{-1} \\ & \ddots & & 0 & \vdots \\ & & \ddots & & \vdots \\ 0 & & & 1 & -J_{N-1}J_N^{-1} \end{array}\right) . \tag{3.5c}$$

In [5] we prove 3 basic facts :

(1) The operators $\mathbf{B}_n$ are each Hamiltonian and, furthermore, are mutually compatible.
(2) The recursion relation (3.4b) can be solved for all k, subject to the condition $\varepsilon_N = 0$.
(3) The vectors $\mathbf{P}^{(k)}$ given by (3.5a) are variational derivatives of a sequence of function(al)s $\mathcal{H}_k$ (the Hamiltonians).

Then it follows from (3.5a) that the equations of motion (3.4d) can be written in Hamiltonian form in (N+1) distinct ways :

$$\mathbf{u}_{t_m} = \mathbf{B}_N\delta\mathcal{H}_m = \dots = \mathbf{B}_0\delta\mathcal{H}_{m+N} . \tag{3.6}$$

We refer to [5] for the details.

4 Miura Maps

Miura presented his famous transformation 20 years ago [14]. He showed that if :

$$u = -v_x - v^2 \tag{4.1a}$$

and v satisfies the MKdV equation (4.1b) then u satisfies the KdV equation (4.1c). The property of most interest for this paper is that (4.1a) can be used to construct the second Hamiltonian structure of the KdV equation out of the single <u>local</u> Hamiltonian structure of the MKdV equation (see [10,15]). The MKdV equation can be written in Hamiltonian form

$$v_t = v_{xxx} - 6v^2 v_x = (-\partial)\frac{\delta\tilde{\mathcal{H}}}{\delta v} \quad , \quad \tilde{\mathcal{H}} = \frac{1}{2}(v_x^2 + v^4) \ . \tag{4.1b}$$

If we denote (4.1a) by $u = M[v]$, the Fréchet derivative M' of M is given by $(-\partial-2v)$. Given any functional $\mathcal{H}[u]$ we define $\tilde{\mathcal{H}}[v]$ by $\tilde{\mathcal{H}}[v] = \mathcal{H}\circ M[v]$ (mod $\mathrm{Im}\partial$). It is then an easy matter to show that, as a consequence of (4.1b),

$$u_t = M'(-\partial)(M')^{\dagger}\delta_u\mathcal{H} = (\partial^3 + 4u\partial + 2u_x)\delta_u\mathcal{H} = u_{xxx} + 6uu_x \tag{4.1c}$$

when $\mathcal{H} = \frac{1}{2}u^2$. For an arbitrary differential mapping $u = M[v]$ this process would take us out of the differential algebra setting, since the differential operator $M'(-\partial)(M')^{\dagger}$, which has coefficients given in terms of v and its derivatives, would not normally be <u>locally</u> <u>defined</u> in terms of just u and its derivatives.

Remark

This remarkable property enables us to deduce that (4.1a) is a map between hierarchies rather than just between the KdV and MKdV equations.

Remark

The Hamiltonian nature of the third order differential operator (4.1c) follows from that of $(-\partial)$ through the formula $M'(-\partial)(M')^{\dagger}$.

In a more general algebraic setting, let $\mathbf{u} = (u_0,\ldots,u_{N-1})^T$ and $\mathbf{v} = (v_0,\ldots,v_{N-1})^T$ be the (respectively) unmodified and modified variables . Then

Definition

The mapping $\mathbf{u} = M[\mathbf{v}]$, is a Miura map for Hamiltonian operator $\tilde{\mathbf{B}}$ (acting in the v space) if :

(i) M is not invertible ,

(ii) $\mathbf{B} = \mathbf{M}'\tilde{\mathbf{B}}(\mathbf{M}')^{\dagger}\Big|_{u=M[v]}$ is locally defined in terms of u and its derivatives.

This definition is adopted from [15].

Remark

The Hamiltonian nature of the operator **B** follows from that of $\tilde{\mathbf{B}}$ provided Miura map $\mathbf{u} = M[\mathbf{v}]$ is nondegenerate (injective) (see [15] for a more detailed discussion of this).

Factorisation of Differential Operators

The relationship of Miura maps to the factorisation of differential operators is discussed in [9,10,11,15]. The map (4.1a) can be obtained from the Schrödinger operator by the following identification :

$$\mathbb{L} = \partial^2 + u = (\partial + v)(\partial - v) \tag{4.3a}$$

The spectral problem $\mathbb{L}\psi_1 = \lambda\psi_1$ for the KdV equation can be used to define that for the MKdV equation :

$$(\partial - v)\psi_1 = \lambda\psi_2 , \qquad (\partial + v)\psi_2 = \psi_1 . \tag{4.3b}$$

This notion is easily extended to higher order Lax operators [10,11,15].

We now generalise the factorisation approach described above to the case of the energy dependent Schrödinger operator (3.1). It is not enough to just choose v to be polynomial in λ in order to obtain u as a polynomial. We replace the factorisation (4.3a) by a quadratic form. The N modifications mentioned in the introduction arise from a sequence of N such quadratic forms.

We denote the modified variables by $\mathbf{v} = (v_0,\dots,v_N)^T$.

Define

$$\ell_k = \alpha_k\partial + v_k , \quad \alpha_k \text{ constants}, \quad k = 0,\dots,N, \; \ell = (\ell_0,\dots,\ell_N) \tag{4.4}$$

Let Λ be any constant, λ-dependent, $(N+1)\times(N+1)$ matrix, and use this to define a λ-dependent second order differential operator by the quadratic from $\ell\Lambda(-\ell^\dagger)$, the λ-dependence being derived purely from that of Λ. Equating this to our operator $\mathbb{L}$ of (3.1) gives rise to a map between functions v_i and u_i. Different choices of Λ give rise to different maps. Once again we restrict ourselves to the KdV case, referring to [5] for the general discussion. In this case $\alpha_N = 0$ and $v_N = -1$.

The following quantities occur frequently below :

Define

$$V_{kn} = -\alpha_k v_{nx} - \alpha_n v_{kx} - 2v_k v_n , \tag{4.5}$$

and

$$\Lambda_r = \left(\begin{array}{cccc|cccc} 1 & \cdots & \cdots & \lambda^{r-1} & & & & \\ \vdots & & \cdot & & & 0 & & \\ \vdots & \cdot & & 0 & & & & \\ \lambda^{r-1} & & & & & & & \\ \hline & & & & & & & \lambda^r \\ & & & & & 0 & \cdot & \vdots \\ & 0 & & & & \cdot & & \vdots \\ & & & & \lambda^r & \cdots & \cdots & \lambda^N \end{array}\right), \quad r = 0,\ldots,N . \tag{4.6}$$

Making use of the formula $\ell_k\ell_n^\dagger + \ell_n\ell_k^\dagger = -2\alpha_k\alpha_n\partial^2 - \mathcal{V}_{kn}$ one can easily see that the identification :

$$\mathbb{L} = \ell\Lambda_r(-\ell^\dagger) \tag{4.7}$$

gives rise to the equations :

$$\varepsilon_k = \sum_{i=0}^{k} \alpha_{k-i}\,\alpha_i \quad , \quad k = 0,\ldots,r-1 \tag{4.8a}$$

$$\varepsilon_k = \sum_{i=1}^{N-k-1} \alpha_{k+i}\,\alpha_{N-i} \quad , \quad k = r,\ldots,N-1 \tag{4.8b}$$

$$u_k = \frac{1}{2}\sum_{i=0}^{k} \mathcal{V}_{i,k-i} \quad , \quad k = 0,\ldots,r-1 \tag{4.9a}$$

$$u_k = 2v_k + \frac{1}{2}\sum_{i=1}^{N-k-1} \mathcal{V}_{k+i,N-i} \quad , \quad k = r,\ldots,N-1 . \tag{4.9b}$$

Remark

The formulae (4.8) are not a priori consistent. For instance, for $r>1$, there is no choice of α_k which would give $\varepsilon_0 = 0$, $\varepsilon_1 = 1$. Thus, Ito's equation is ruled out of consideration here. Such inconsistencies are, however, exceptional.

When these formulae are consistent, equations (4.9) define a differential mapping from v_i to u_i, sometimes invertible, sometimes not. In fact, (4.9b) always defines an invertible map between $v_r,\cdots,v_{N-1}$ and $u_r,\ldots,u_{N-1}$. The invertibility of M thus rests upon the map (4.9a) between $v_0,\ldots,v_{r-1}$ and $u_0,\ldots,u_{r-1}$, which is invertible if and only if $\varepsilon_0 = 0$.

Subject only to the consistency of (4.8a,b) we have the following important proposition.

Proposition 1

Under the change of variables $\mathbf{u} = \mathbf{M}_r[\mathbf{v}]$ defined by (4.9a,b), the

Hamiltonian operator $\mathbf{B}_r$, given by (3.5b), is the image :

$$\mathbf{B}_r = M_r'\tilde{\mathbf{B}}_r(M_r')^{\dagger}\Big|_{u=M_r[v^{(r)}]} \tag{4.10a}$$

of the constant, first order Hamiltonian operator $\tilde{\mathbf{B}}_r$:

$$\tilde{\mathbf{B}}_r = \frac{1}{4}\left(\begin{array}{ccc|ccc} 0 & & -\partial & & & \\ & \cdot^{\cdot^{\cdot}} & & & 0 & \\ -\partial & & 0 & & & \\ \hline & & & 0 & & \partial \\ & 0 & & & \cdot^{\cdot^{\cdot}} & \\ & & & \partial & & 0 \end{array}\right), \tag{4.10b}$$

where the diagonal blocks are respectively r×r and (N-r)×(N-r).

Here we have used $v^{(r)}$ to denote the modified variables corresponding to the map $\mathbf{u} = M_r[\mathbf{v}^{(r)}]$.

Proof

The Fréchet derivative of the mapping M_r is given by

$$M_r' = \left(\begin{array}{cccc|cccc} m_0 & & & 0 & & & & \\ \vdots & \ddots & & & & & 0 & \\ m_{r-1} & \cdots & \cdots & m_0 & & & & \\ \hline & & & & m_N & \cdots & \cdots & m_{r+1} \\ & & & & & \ddots & & \vdots \\ & 0 & & & & 0 & \ddots & \vdots \\ & & & & & & & m_N \end{array}\right), \tag{4.11a}$$

where

$$m_k = -\alpha_k\partial - 2v_k, \qquad k = 0,\ldots,N \tag{4.11b}$$

(giving $m_N = 2$ when $\alpha_N = 0$, $v_N = -1$). To obtain (4.10a) we use :

$$m_k\partial m_n^{\dagger} + m_n\partial m_k^{\dagger} = -2(\alpha_k\alpha_n\partial^3 + V_{kn}\partial + \partial V_{kn}) \tag{4.11c}$$

The formulae (4.9a,b) then give the result.

Remark

Using (4.11c) one can easily check that the factorisation (4.7) has its counterpart on the level of the third order operator J : $J = m(-\frac{1}{4}\partial)\Lambda m^{\dagger}$, where $m = (m_0,\ldots,m_N)$. The factorisation of J is, in fact, the only one that survives the super extension discussed in section 5.

We now concentrate on the case $\varepsilon_0 \neq 0$, so that the map (4.9a,b) is a genuine Miura map (for r > 0). For clarity, we choose the most interesting case [2] of $\varepsilon_0 = 1$, $\varepsilon_i = 0$ for $i \geq 1$.

Miura maps

Let $\alpha_0 = 1$, $\alpha_i = 0$ for $i \geq 1$, so that $\varepsilon_0 = 1$, $\varepsilon_i = 0$ for $i \geq 1$. The map $\mathbf{u} = \mathbf{M}_0[\mathbf{v}]$, corresponding to Λ_0, is invertible, whilst those corresponding to all other Λ_r $(r > 0)$ are genuine Miura maps. In fact, (4.9b) defines an invertible map whilst (4.9a) is the genuine Miura part. Thus the upper block of Λ_r is the important part when discussing genuine Miura maps. We therefore consider the map M_N corresponding to Λ_N. In this case the Miura map $\mathbf{u} = \mathbf{M}_N[\mathbf{v}]$ is given purely by (4.9a). The Fréchet derivative (4.11a) is then :

$$M_N' = \begin{pmatrix} m_0 & & & 0 \\ \vdots & \ddots & & \\ \vdots & & \ddots & \\ m_{N-1} & \cdots & \cdots & m_0 \end{pmatrix} \tag{4.12a}$$

with $m_0 = -\partial - 2v_0$, $m_i = -2v_i$, $i \geq 1$, and the constant coefficient operator :

$$\tilde{\mathbf{B}}_N = \frac{1}{4}\begin{pmatrix} 0 & & -\partial \\ & \cdot^{\cdot^{\cdot}} & \\ -\partial & & 0 \end{pmatrix} \tag{4.12b}$$

is mapped onto $\mathbf{B}_N$ of (3.5b). It is easy to see that the pre-image of $\mathbf{B}_r$, for $r < N$, is non-local.

The Miura map $\mathbf{u} = M_N[\mathbf{v}^{(N)}]$ can be decomposed into N primitive ones.

Define a sequence of maps $\mathbf{u}^{(k)} = M_N^{k+1}[\mathbf{u}^{(k+1)}]$ by :

$$u_k^{(k)} = \frac{1}{2}\sum_{i=0}^{k} \mathcal{U}_{i,k-i}^{(k+1)}, \qquad u_i^{(k)} = u_i^{(k+1)} \quad \text{for } i \neq k \tag{4.13a}$$

where $\mathcal{U}_{ij}^{(k)}$ is given by (4.5) but with v_i replaced by $u_i^{(k)}$. We can write $\mathbf{u} \equiv \mathbf{u}^{(0)} = M_N[\mathbf{u}^{(N)}] \equiv M_N[\mathbf{v}^{(N)}]$ as the composition of these maps :

$$\mathbf{u}^{(0)} = M_N[\mathbf{u}^{(N)}] = M_N^1 \circ M_N^2 \circ \ldots \circ M_N^N[\mathbf{u}^{(N)}]$$

The Fréchet derivative $\mathbf{M}_N'$ is thus the product of N Fréchet derivatives : $\mathbf{M}_N' = (\mathbf{M}_N^1)' \times \ldots \times (\mathbf{M}_N^N)'$, where :

$$(\mathbf{M}_N^k)' = \begin{pmatrix} 1 & & & & & \\ & \ddots & & & 0 & \\ 0 & & 1 & & & \\ m_{k-1} & \ldots & m_0 & & & \\ & & & 1 & & \\ & 0 & & & \ddots & \\ & & & & & 1 \end{pmatrix} \tag{4.13b}$$

with $m_{k-1}, \ldots, m_0$ on the kth row. It is a very simple calculation to see that the product of these matrices is just (4.12a).

Remark

Each of these maps is non-invertible and injective.

Starting with $\mathbf{B}_N^N = \tilde{\mathbf{B}}_N$ of (4.12b) define $\mathbf{B}_N^k$ inductively by :

$$\mathbf{B}_N^{k-1} = (M_N^k)' \mathbf{B}_N^k ((M_N^k)')^\dagger \Big|_{\mathbf{u}^{(k-1)} = M_N^k[\mathbf{u}^{(k)}]} , \qquad k = N, \ldots, 1. \tag{4.13c}$$

$\mathbf{B}_N^0$ is just our original $\mathbf{B}_N$ of (3.5b).

Direct calculation shows that, as indicated by the notation, each $\mathbf{B}_N^k$ is <u>locally</u> defined in terms of the variables $\mathbf{u}^{(k)}$. Thus, each of the maps M_N^k is a genuine, Hamiltonian Miura map.

Let $M^{(r)} = M_N^1 \circ M_N^2 \circ \ldots \circ M_N^r$. Explicitly, this has the form

$$u_k^{(0)} = \frac{1}{2} \sum_{i=0}^{k} \mathcal{U}_{i,k-i}^{(r)} , \qquad k = 0, \ldots, r-1 , \tag{4.14a}$$

$$u_k^{(0)} = u_k^{(r)} , \qquad k = r, \ldots, N-1 . \tag{4.14b}$$

$M^{(r)}$ and M_r differ by an invertible map [5]. Thus we have :

Proposition 2

There exist local Hamiltonian operators $\mathbf{B}_k^r$ such that $(M^{(r)})' \mathbf{B}_k^r ((M^{(r)})')^\dagger = \mathbf{B}_k^0 \equiv \mathbf{B}_k$ for $k = r, \ldots, N$. These constitute (N-r+1) compatible Hamiltonian structures for the rth modification. The sequence of modified Hamiltonians is defined by $\mathcal{H}_n^r = \mathcal{H}_n \circ M^{(r)}$ and the rth modified hierarchy is written as :

$$\mathbf{u}^{(r)}_{t_n} = \mathbf{B}^r_{N-k}\,\delta\mathcal{H}^r_{n+k}\,, \qquad k = 0,\dots,N-r, \qquad n = 0,1\dots \tag{4.15}$$

Remark

The operator $\mathbf{B}^r_r$ takes the form :

$$B^r_r = \left(\begin{array}{ccc|ccc} 0 & & -\frac{1}{4}\partial & & & \\ & \cdot^{\cdot^{\cdot}} & & & 0 & \\ -\frac{1}{4}\partial & & 0 & & & \\ \hline & & & -J_{r+1} & \cdots & -J_N \\ & 0 & & \vdots & \cdot^{\cdot^{\cdot}} & \\ & & & -J_N & & 0 \end{array}\right) \tag{4.16}$$

and is related to (4.10b) through an invertible transformation.

We can represent these modifications and their Hamiltonian structures schematically as follows :

$$\begin{array}{ccccccccccc}
\mathbf{u} \equiv \mathbf{u}^{(0)} & \xleftarrow{M^1_N} & \mathbf{u}^{(1)} & \xleftarrow{M^2_N} & \cdots\cdots & \xleftarrow{M^r_N} & \mathbf{u}^{(r)} & \xleftarrow{M^{r+1}_N} & \cdots\cdots & \xleftarrow{M^N_N} & \mathbf{u}^{(N)} \equiv \mathbf{v}^{(N)} \\
\mathbf{B}_N \equiv \mathbf{B}^0_N & \longleftarrow & \mathbf{B}^1_N & & \cdots\cdots\cdots\cdots & \longleftarrow & \mathbf{B}^r_N & \longleftarrow & \cdots\cdots & \longleftarrow & \mathbf{B}^N_N \\
\vdots & & & & & & & & & & \\
\mathbf{B}_r \equiv \mathbf{B}^0_r & \longleftarrow & \mathbf{B}^1_r & & \cdots\cdots\cdots\cdots & \longleftarrow & \mathbf{B}^r_r & & & & \\
\vdots & & & & & & & & & & \\
\mathbf{B}_0 \equiv \mathbf{B}^0_0 & & & & & & & & & &
\end{array}$$

Fig 1

Modified Spectral Problem

Generalising the derivation of (4.3b) we can use the factorisation (4.7) to obtain the spectral problem corresponding to each of our modifications (4.14). The first modification leads to :

$$(\partial + u^{(1)}_0)(\partial - u^{(1)}_0)\psi_1 + (\lambda u^{(1)}_1 + \dots + \lambda^{N-1}u^{(1)}_{N-1})\psi_1 = \lambda^N\psi_1 \tag{4.17a}$$

Defining ψ_2 by :

$$(\partial - u^{(1)}_0)\psi_1 = \lambda\psi_2 \tag{4.17b}$$

we find :

$$(\partial + u^{(1)}_0)\psi_2 = (-u^{(1)}_1 - \dots - \lambda^{N-2}u^{(1)}_{N-1} + \lambda^{N-1})\psi_1 \tag{4.17c}$$

Equations (4.17b,c) constitute a 2x2 matrix spectral problem for the first modification. The spectral problems for the remaining modifications are obtained from this one in succession by a series of substitutions and gauge transformations. This is illustrated by the example given below.

Example : Dispersive water waves:

We illustrate the above construction by the simplest nontrivial example, N = 2. The resulting hierarchy is tri-Hamiltonian. It is convenient to introduce, instead of the u_0, u_1 variables used above, a more familiar representation obtained by performing an invertible change of variables:

$$q = u_0 + \frac{1}{4}u_1^2 - \frac{1}{2}u_{1x} \ , \ r = u_1 \ , \tag{4.18}$$

which changes the second order flow into the standard DWW form [16] :

$$q_{t_1} = \frac{1}{2}(-q_x + 2qr)_x \quad , \quad r_{t_1} = \frac{1}{2}(r_x + 2q + r^2)_x \ . \tag{4.19}$$

The 3 Hamiltonian operators then take the form :

$$\bar{\mathbf{B}}_0 = \begin{pmatrix} 0 & \partial \\ \partial & 0 \end{pmatrix} \ , \quad \bar{\mathbf{B}}_1 = \frac{1}{2}\begin{pmatrix} q\partial + \partial q & -\partial^2 + r\partial \\ \partial^2 + \partial r & 2\partial \end{pmatrix} \ , \tag{4.20}$$

$$\bar{\mathbf{B}}_2 = \frac{1}{4}\begin{pmatrix} (r-\partial)(q\partial+\partial q)+(q\partial+\partial q)(r+\partial) & (r-\partial)^2\partial+2(q\partial+\partial q) \\ \partial(r+\partial)^2+2(q\partial+\partial q) & 2(r\partial+\partial r) \end{pmatrix} .$$

The first Miura map, written in these coordinates :

$$q = -w_{0x} - \frac{1}{2}w_{1x} - w_0^2 + \frac{1}{4}w_1^2 \ , \quad r = w_1 \ , \tag{4.21}$$

is easily seen to be equivalent to the Kupershmidt's first modification of DWW hierarchy [16]. The second Miura map :

$$w_0 = v_0 \ , \ w_1 = -v_{1x} - 2v_0v_1 \ , \tag{4.22}$$

however, is not equivalent to either of Kupershmidt's second modifications. The first nontrivial flow is :

$$\begin{aligned} v_{0t_1} &= (\frac{1}{4}v_{1xx} + \frac{1}{2}v_1v_{0x} - v_0^2v_1)_x \ , \\ v_{1t_1} &= (v_0 - v_0v_1^2 - \frac{1}{2}v_1v_{1x})_x \ . \end{aligned} \tag{4.23}$$

The spectral problem for the first modification is given by (4.17), which, in this case, takes the form :

$$\begin{pmatrix} \psi_1 \\ \psi_2 \end{pmatrix}_x = \begin{pmatrix} w_0 & \lambda \\ \lambda - w_1 & -w_0 \end{pmatrix}\begin{pmatrix} \psi_1 \\ \psi_2 \end{pmatrix} . \tag{4.24a}$$

Writing (4.24a) in the variables (v_0, v_1) (using (4.22)) and gauge transforming with $T = \begin{pmatrix} 1 & 0 \\ -v_1 & 1 \end{pmatrix}$ we obtain the spectral problem for the second modification :

$$\begin{pmatrix} \phi_1 \\ \phi_2 \end{pmatrix}_x = \begin{pmatrix} v_0+\lambda v_1 & \lambda \\ \lambda(1-v_1^2) & -v_0-\lambda v_1 \end{pmatrix} \begin{pmatrix} \phi_1 \\ \phi_2 \end{pmatrix} . \tag{4.24b}$$

5 Super Extensions

In this section we generalise the above results to a super extension of the Schrödinger linear problem. To carry this out we just need to include some 'odd' (ie. anti-commuting) variables. We present some useful facts below, referring to [17] for basic super-linear algebra and super matrix theory and to [18] for basic super Hamiltonian theory.

In this paper, we only need the notions of the Fréchet and variational derivatives. Let $\mathbf{U} = F[\mathbf{V}]$ be a change of variables (invertible or otherwise). The Fréchet derivative of F is defined in the usual way (with τ an even parameter):

$$F^0[\mathbf{V}]\mathbf{W} = \frac{d}{d\tau} F[\mathbf{V}+\tau\mathbf{W}]\Big|_{\tau=0} \tag{5.1}$$

where $\mathbf{V}$ and $\mathbf{W}$ are even vectors. The notation F^0 is that used by Kupershmidt [18], who calls this operator the "even" Fréchet derivative. Hamiltonian structures are then transformed through :

$$\tilde{\mathbf{B}} \to \mathbf{B} = F^0\tilde{\mathbf{B}}(F^0)^{s\dagger} \tag{5.2}$$

where $s\dagger$ denotes the superadjoint, defined in the same way as the usual adjoint, but with the transpose of matrices being replaced by supertranspose [17]. With this definition, the skew symmetry of the Poisson bracket (related to $\mathbf{B}$) is <u>not</u> equivalent to the skew-superadjointness of the operator $\mathbf{B}$ (see Leites' discussion of quadratic forms [17]). The conditions implied upon the elements of $\mathbf{B}$, by the skew symmetry of the Poisson bracket, are precisely those used by Kupershmidt to define his superadjoint [18]. The variational derivative is defined in the usual way, but remembering that all derivatives are assumed to act from the left.

We start with Kupershmidt's spectral problem :

$$\begin{aligned} (\partial^2 + u)\psi + \eta\varphi &= \lambda\psi \quad , \\ \partial\phi + \eta\psi &= 0 \quad , \end{aligned} \tag{5.3}$$

for the following bi-superHamiltonian sKdV equation [18]:

$$\begin{aligned} u_t &= \tfrac{1}{4}(u_{xxx} + 6uu_x + 12\eta\eta_{xx}) \quad , \\ \eta_t &= \tfrac{1}{4}(4\eta_{xxx} + 6u\eta_x + 3u_x\eta) \quad , \end{aligned} \tag{5.4}$$

with compatible Hamiltonian structures :

$$\mathbf{B}_0 = \begin{pmatrix} \partial & 0 \\ 0 & 1 \end{pmatrix} , \quad \mathbf{B}_1 = \begin{pmatrix} \partial^3 + 2u\partial + 2\partial u & 2\eta\partial + \partial\eta \\ 2\partial\eta + \eta\partial & \partial^2 + u \end{pmatrix} . \tag{5.5}$$

Kupershmidt's Miura map [18] M, for (5.4) is given by :

$$u = -v_x - v^2 - \vartheta\vartheta_x \quad , \quad \eta = -\vartheta_x - v\vartheta . \tag{5.6}$$

We have the formula :

$$\mathbf{B}_1 = m(-\mathbf{D})m^{s\dagger} \tag{5.7}$$

where $m = M^0$ is the (even) Fréchet derivative of M , $m^{s\dagger}$ is its superadjoint and $\mathbf{D} = \begin{pmatrix} \partial & 0 \\ 0 & 1 \end{pmatrix}$ is the first (and only) Hamiltonian structure of the corresponding sMKdV equation.

The Super Lax Equations and their Hamiltonian Structures

We consider here the isospectral flows of the energy dependent super-Schrödinger equation:

$$(\varepsilon\partial^2 + u)\psi + \eta\varphi = 0 \quad , \quad \varepsilon\partial\phi + \eta\psi = 0 \quad , \tag{5.8a}$$

where, ε, u, ψ are even and η, φ are odd variables. We specify the λ-dependence of ε, u and η to be :

$$\varepsilon = \sum_0^{N-1} \varepsilon_i\lambda^i \quad , \quad u = \sum_0^{N} u_i\lambda^i \quad , \quad \eta = \sum_0^{N} \eta_i\lambda^i \quad , \tag{5.8b}$$

with ε_i being constants and u_i, η_i functions of x and with spectral parameter λ being even. Since we are restricting attention to the sKdV case, we have $u_N = -1$, $\eta_N = -\gamma$ (an odd constant). In [6] it is shown that the linear time evolution of the wave functions ψ, φ takes the form :

$$\psi_t = 2p\psi_x - p_x\psi - \rho\varphi \quad , \quad \varphi_t = \rho\psi_x + \sigma\psi \quad , \tag{5.9}$$

where $\varepsilon\sigma = -2\eta p - \varepsilon\rho_x$, with the variables p (even) and ρ (odd) being functions of the potentials u_i, η_i and of λ. If we time-differentiate (5.8a) (assuming $\lambda_t = 0$) and use (5.9), (5.8) to remove all time and higher order x-derivatives of ψ and φ, we are left with :

$$\begin{aligned} u_t &= \varepsilon p_{xxx} + 4up_x + 2u_x p + 3\eta\rho_x + \eta\rho_x \quad , \\ \eta_t &= 3\eta p_x + 2\eta_x p + \varepsilon\rho_{xx} + u\rho \quad . \end{aligned} \tag{5.10}$$

Introducing the operator $J = \sum_0^N J_k\lambda^k$ with

$$J_k = \begin{pmatrix} \varepsilon_k \partial^3 + 2u_k \partial + 2\partial u_k & 2\eta_k \partial + \partial \eta_k \\ 2\partial \eta_k + \eta_k \partial & \varepsilon_k \partial^2 + u_k \end{pmatrix}, \qquad (5.11)$$

we can rewrite (5.10) as

$$U_t = JP, \qquad (5.12)$$

with $P = (p,\rho)^T$ and $U = (u,\eta)^T$. If we choose P to be a polynomial in λ :

$$P = \sum_{i=0}^{m} P_{m-i}\lambda^i = \left(\sum_0^m p_{m-i}\lambda^i , \sum_0^m \rho_{m-i}\lambda^i\right)^T, \qquad (5.13)$$

with p_i and ρ_i being functions of u_j and η_j only, then (5.12) decomposes into :

$$J_0 P_{k-N} + J_1 P_{k-N+1} + \ldots + J_N P_k = 0, \quad k = 0,1,\ldots,m, \qquad (5.14a)$$

and :

$$\begin{pmatrix} U_0 \\ U_1 \\ \vdots \\ U_{N-1} \end{pmatrix}_{t_m} = \begin{pmatrix} 0 & & J_0 \\ & \iddots & \vdots \\ J_0 & \cdots & J_{N-1} \end{pmatrix} \begin{pmatrix} P_{m-N} \\ \vdots \\ P_m \end{pmatrix}. \qquad (5.14b)$$

Further development is almost the same as in the purely even case. The only difference is that now $U_i \equiv (u_i,\eta_i)^T$ and $P_i \equiv (p_i,\rho_i)^T$ are 2 component vectors rather than scalar functions and J_k are matrix (not scalar) differential operators.

In [6] we prove that :

(i) the recursion relation (5.14a) has a solution of the form (5.13) for any $m \geq 0$.

(ii) $\mathbf{P}^{(m)} = (P_{m-N+1},\ldots,P_m)^T = \delta\mathcal{H}_m$ for some functionals $\mathcal{H}_m$, where δ stands for $(\delta_{u_0},\delta_{\eta_0},\ldots,\delta_{u_{N-1}},\delta_{\eta_{N-1}})^T$.

(iii) there exist (N+1) compatible Hamiltonian operators defined by (3.5b) with J_k given by (5.11), such that the hierarchy of isospectral flows of (5.8) can be written in (N+1)-Hamiltonian form :

$$\mathbf{U}_{t_m} = \mathbf{B}_{N-\ell}\, \delta\mathcal{H}_{m+\ell}, \quad \ell = 0,\ldots,N. \qquad (5.15)$$

Miura Maps

We now generalise our construction of Miura maps associated with the energy dependent Schrödinger spectral problem. The factorisation of that operator does not survive the super extension to (5.8), but we can still factorise our basic operator $J = \sum_0^N J_k \lambda^k$ of (5.12) as :

$$J = (m_0, \ldots, m_N)(-D)\Lambda \begin{pmatrix} m_0^{s\dagger} \\ \vdots \\ m_N^{s\dagger} \end{pmatrix} \tag{5.16}$$

where Λ is a symmetric (λ-dependent) (N+1)x(N+1) matrix, $D=\begin{pmatrix} \partial & 0 \\ 0 & 1 \end{pmatrix}$,

and :

$$m_k = \begin{pmatrix} -\alpha_k \partial - 2v_k & -\vartheta_k \partial + \vartheta_{kx} \\ -\vartheta_k & -\alpha_k \partial - v_k \end{pmatrix}, \quad m_k^{s\dagger} = \begin{pmatrix} \alpha_k \partial - 2v_k & -\vartheta_k \\ -\partial\vartheta_k - \vartheta_{kx} & \alpha_k \partial - v_k \end{pmatrix}, \tag{5.17}$$

are copies of the Fréchet derivative of the elementary Miura map :

$$u = -\alpha v_x - v^2 - \vartheta\vartheta_x \quad , \quad \eta = -\alpha\vartheta_x - v\vartheta . \tag{5.18}$$

Once again, we choose $v_N = -1$, $\vartheta_N = -\gamma$ for the sKdV case, referring to [6] for the general discussion. In fact, for the purposes of these proceedings we present only the formulae for which $\varepsilon_0 = \alpha_0 = 1$, $\varepsilon_k = \alpha_k = 0$, $k \neq 0$.

Choosing Λ to be (4.6) we obtain the following maps $\mathbf{V} \to \mathbf{U}$:

$$u_k = \frac{1}{2} \sum_0^k \mathcal{V}_{i,k-i} \quad , \quad \eta_k = \frac{1}{2} \sum_0^k \Theta_{i,k-i} \quad , \tag{5.19}$$

for $k = 0,\ldots,r-1$, and :

$$u_k = -\alpha_N v_{kx} + 2v_k + \gamma\vartheta_{kx} + \frac{1}{2} \sum_1^{N-k-1} \mathcal{V}_{k+i,N-i} \quad , \tag{5.20a}$$

$$\eta_k = -\alpha_N \vartheta_{kx} + \vartheta_k + v_k \gamma + \frac{1}{2} \sum_1^{N-k-1} \Theta_{k+i,N-i} \quad , \tag{5.20b}$$

for $k = r,\ldots,N-1$, where

$$\begin{aligned} \mathcal{V}_{kn} &= -\alpha_k v_{nx} - \alpha_n v_{kx} - 2v_k v_n - \vartheta_k \vartheta_{nx} - \vartheta_n \vartheta_{kx} \, , \\ \Theta_{kn} &= -\alpha_k \vartheta_{nx} - \alpha_n \vartheta_{kx} - v_k \vartheta_n - v_n \vartheta_k \, . \end{aligned} \tag{5.21}$$

Equations (5.19) and (5.20) result from the formula :

$$m_k(-D)m_n^{s\dagger}+m_n(-D)m_k^{s\dagger} = \begin{pmatrix} 2\alpha_k\alpha_n\partial^3+2V_{kn}\partial+2\partial V_{kn} & 2\Theta_{kn}\partial+\partial\Theta_{kn} \\ 2\partial\Theta_{kn}+\Theta_{kn}\partial & 2\alpha_k\alpha_n\partial+V_{kn}^2 \end{pmatrix} . \quad (5.22)$$

The analogy between the above construction and that for the purely even case is so complete that we need not repeat the details of the calculations here, but concentrate on the results only. The essential difference between the two cases is that J, m_k, D are now matrices rather than scalars.

We denote by $\mathbf{V}^{(r)}$ modified variables corresponding to the Miura map $\mathbf{U} = \mathbf{M}_r[\mathbf{V}^{(r)}]$ defined by (5.19) and (5.20).

Proposition 3

With the factorisation (5.16) with $\Lambda = \Lambda_r$ the operator $\mathbf{B}_r$ is M_r-related to a constant coefficient Hamiltonian operator $\tilde{\mathbf{B}}_r$:

$$\mathbf{B}_r = (M_r^0)\tilde{\mathbf{B}}_r(M_r^0)^{s\dagger}\Big|_{\mathbf{U}=M_r[\mathbf{V}^{(r)}]} \quad (5.23)$$

where $\tilde{\mathbf{B}}_r$ is just (4 times) that of (4.10b), but with ∂ replaced by D.

Remark

The proof of this is the same as that of Proposition 1. The Fréchet derivative of the map M_r is just (4.11a) with m_k replaced by the 2x2 matrices (5.17). Indeed, we can prove exactly the same sequence of results that hold for the purely even case.

Thus Proposition 3 gives us N distinct Hamiltonian Miura maps M_r relating $\mathbf{B}_r$ to the constant coefficient operator $\tilde{\mathbf{B}}_r$, r = 1,...,N. As before, the 'ultimate' Miura map M_N is the most interesting one.

In [6] we prove that :

(i) The Miura map M_N decomposes into N consecutive Miura maps M_N^n, n = 1,...,N,: $M_N = M_N^1\circ\ldots\circ M_N^N$, each M_N^{n+1} being a map $\mathbf{U}^{(n+1)} \to \mathbf{U}^{(n)}$.

(ii) Each Miura map M_N^n is Hamiltonian and the nth modification $\mathbf{U}^{(n)}$ possesses (N+1-n) compatible Hamiltonian structures $\mathbf{B}_k^n$ with the chain of modified systems given in multi-Hamiltonian form :

$$\mathbf{U}^{(n)}_{t_m} = B^{(n)}_{N-\ell}\ \delta\mathcal{H}^{n}_{m+\ell}\ ,\ \ell = 0,\dots,N-n\ ,\ m = 0,1,.., \tag{5.24}$$

where $\mathcal{H}^n_i = \mathcal{H}_i \circ M^1_N \circ \dots \circ M^n_N$ are the modified Hamiltonians.

The situation can be depicted by exactly the same diagram of figure 1.

Example : Super DWW.

These are tri-Hamiltonian with Hamiltonian operators given by :

$$\mathbf{B}_0 = \begin{pmatrix} -J_1 & -J_2 \\ -J_2 & 0 \end{pmatrix}\ ,\ \mathbf{B}_1 = \begin{pmatrix} J_0 & 0 \\ 0 & -J_2 \end{pmatrix}\ ,\ \mathbf{B}_2 = \begin{pmatrix} 0 & J_0 \\ J_0 & J_1 \end{pmatrix}\ , \tag{5.25}$$

where J_2 is now given by $J_2 = \begin{pmatrix} -4\partial & -3\gamma\partial \\ -3\gamma\partial & -1 \end{pmatrix}$. The first few Hamiltonians (with $\gamma = 0$) are :

$$\mathcal{H}_0 = \frac{1}{2}u_1,\ \mathcal{H}_1 = \frac{1}{2}u_0 + \frac{1}{8}u_1^2 + \frac{1}{2}\eta_1\eta_{1x},\ \mathcal{H}_2 = \frac{1}{4}u_0u_1 + \frac{1}{16}u_1^3 + \frac{3}{4}u_1\eta_1\eta_{1x} + \eta_0\eta_{1x}. \tag{5.26}$$

The first nontrivial flow is given by :

$$\begin{aligned}
u_{0t} &= \frac{1}{4}u_{1xxx} + u_0u_{1x} + \frac{1}{2}u_{0x}u_1 + 3\eta_0\eta_{1xx} + \eta_{0x}\eta_{1x}\ ,\\
\eta_{0t} &= \eta_{1xxx} + u_0\eta_{1x} + \frac{1}{2}u_1\eta_{0x} + \frac{3}{4}u_{1x}\eta_0\ ,\\
u_{1t} &= u_{0x} + \frac{3}{2}u_1u_{1x} + 3\eta_1\eta_{1xx}\ ,\\
\eta_{1t} &= \eta_{0x} + \frac{3}{2}u_1\eta_{1x} + \frac{3}{4}u_{1x}\eta_1\ .
\end{aligned} \tag{5.27}$$

An invertible change of variables :

$$\begin{aligned}
q &= \frac{1}{2}u_0 + \frac{1}{8}u_1^2 - \frac{1}{4}u_{1x} + \frac{1}{2}\eta_1\eta_{1x}\ ,\quad r = \frac{1}{2}u_1\ ,\\
\mu &= \eta_0 + \frac{1}{2}u_1\eta_1 - \eta_{1x}\ ,\quad \nu = \eta_1\ ,
\end{aligned} \tag{5.28}$$

transforms (5.27) into the "super long waves" equation of Kupershmidt [19]. The Hamiltonians and first Hamiltonian structure of sDWW hierarchy take a very simple form when written in (q,μ,r,ν) coordinates :

$$\tilde{\mathbf{B}}_0 = \begin{pmatrix} 0 & D \\ D & 0 \end{pmatrix}\ ,\ \tilde{\mathcal{H}}_0 = r,\ \tilde{\mathcal{H}}_1 = q,\ \tilde{\mathcal{H}}_2 = qr + \mu\nu_x\ . \tag{5.29}$$

The sDWW hierarchy possesses two consecutive Hamiltonian modifications with Miura maps :

$$u_0 = -w_{0x} - w_0^2 - \kappa_0\kappa_{0x} \ , \quad u_1 = w_1 \ ,$$
$$\eta_0 = -\kappa_{0x} - w_0\kappa_0 \ , \quad \eta_1 = \kappa_1 \ , \tag{5.30a}$$

and

$$w_0 = v_0 \quad , \quad w_1 = -v_{1x} - 2v_0v_1 - \vartheta_0\vartheta_{1x} - \vartheta_1\vartheta_{0x} \quad ,$$
$$\kappa_0 = \vartheta_0 \quad , \quad \kappa_1 = -\vartheta_{1x} - v_0\vartheta_1 - v_1\vartheta_0 \quad , \tag{5.30b}$$

where we write $(w_0,\kappa_0,w_1,\kappa_1)$ instead of $(u_0^{(1)},\eta_0^{(1)},u_1^{(1)},\eta_1^{(1)})$. The transformation (5.30a) is equivalent to Kupershmidt's single Miura map [19].

In [6] we also present a new super Ito's equation.

6 Conclusions

In this paper we have presented some elementary techniques which can be used to study the Hamiltonian properties of and Miura maps associated with the isospectral flows of a given spectral problem. We have presented these in the context of an energy dependent Schrödinger operator and its super-generalisation, but the approach is more broadly applicable.

In the two cases we present and in others we have calculated, the apparently 'universal' algebraic form (3.5b) of the Hamiltonian operators arises. The only difference from case to case is the form of the operator J_k. In the two cases presented here, these operators were of either scalar or 2x2 matrix type.

Another important aspect of our work is the construction of Miura maps. Our construction is a direct generalisation of the well known factorisation technique used for 'standard' Lax operators [9,10,11,15]. We presented a sequence of Miura maps and 'modified' Hamiltonian operators, which are summarised in figure 1. Another important feature of our Miura maps is that they reduce our Hamiltonian operators (3.5b) to a standard form (4.10b). This is analogous to the reduction of a given finite dimensional Poisson bracket to canonical form by a judicious choice of coordinates (the existence of which is guaranteed by Darboux' theorem in the finite dimension case). This point is discussed more fully in [8].

The spectral problem (1.1) and Hamiltonian structures (3.5b) have recently been given a Lie algebraic interpretation [20,21].

Finally we comment on a result we obtained during this workshop. Recently, Dorfman [22] and Wilson [23] have presented what they respectively call Dirac and quasi-Hamiltonian structures related to the KdV equation through a Miura map. We have generalised

Wilson's quasi-Hamiltonian structures to the case of our energy dependent Schrödinger operators.

Acknowledgements

APF thanks the SERC for their support of his participation in the workshop. MA was partially supported by the Research Problem CPBP 01.03.

References.

[1] M. Antonowicz and A.P. Fordy, A Family of Completely Integrable multi-Hamiltonian Systems, *Phys.Letts.A*, **122**, 95-99 (1987).

[2] M. Antonowicz and A.P. Fordy, Coupled KdV Equations with multi-Hamiltonian Structures, *Physica D*, **28**, 345-58 (1987).

[3] M. Antonowicz and A.P. Fordy, Coupled KdV Equations Associated with a Novel Schrödinger Spectral Problem. Published in : Nonlinear Evolutions , 145-160, ed. J. Leon. World Scientific, Singapore, 1988.

[4] M. Antonowicz and A.P. Fordy, Coupled Harry Dym Equations with multi-Hamiltonian Structures, *J.Phys.A.* **21**, L269-75 (1988).

[5] M.Antonowicz and A.P.Fordy, Factorisation of Energy Dependent Schrödinger Operators : Miura Maps and Modified Systems, *Commun.Math.Phys.* **124,** 465-86 (1989).

[6] M.Antonowicz and A.P.Fordy, Super-extensions of energy dependent Scrödinger operators, *Commun.Math.Phys.* **124,** 487-500 (1989).

[7] P.J.Olver, "Application of Lie Groups to Differential Equations", Springer-Verlag, Berlin, 1986 .

[8] M.Antonowicz and A.P.Fordy, Hamiltonian Structure of Nonlinear Evolution Equations, published in :"Soliton Theory : A Survey of Results", ed. A.P.Fordy, MUP, Manchester 1990.

[9] M.Adler, and J.Moser, On a Class of Polynomials Connected with the KdV Equation, *Commun.Math.Phys.* **61**, 1-30 (1978).

[10] A.P.Fordy and J.Gibbons, Factorization of Operators I. Miura Transformations, *J.Math.Phys.* **21**, 2508-10 (1980).

[11] A.P.Fordy and J.Gibbons, Factorization of operators II, *J.Math.Phys.* **22**, 1170-5 (1981).

[12] F.Magri, A Simple Model of the Integrable Hamiltonian Equation, *J.Math.Phys.* **19**, 1156-1162 (1978).

[13] B.A.Kupershmidt, Is a bi-Hamiltonian System Necessarily Integrable?, *Phys. Letts.A*, **123**, 55-59 (1987).

[14] R.M.Miura, Korteweg-de Vries equation and generalizations I. A remarkable explicit nonlinear transformation, *J.Math.Phys.* **9**, 1202-1204 (1968).

[15] B.A.Kupershmidt and G.Wilson, Modifying Lax Equations and the Second Hamiltonian Structure, *Invent. Math.* **62**, 403-436 (1981).

[16] B.A.Kupershmidt, Mathematics of Dispersive Water Waves, *Commun.Math.Phys.* **99**, 51-73 (1985).

[17] D.A.Leites, Introduction to the Theory of Supermanifolds, *Russ.Math.Surveys* **35**, 1-64 (1980).

[18] B.A.Kupershmidt, "Elements of Superintegrable Systems", D.Reidel, Dordrecht, 1987.
[19] B.A.Kupershmidt, Super Long Waves, *Mech.Res.Commun.* **13**, 47-51 (1986).
[20] A.G. Reyman and M.a.Semenov-Tian-Shansky, Compatible Poisson Structures for Lax Equations : an r-Matrix Approach, *Phys.Letts.A.* **130**, 456-60 (1988).
[21] A.P.Fordy, A.G.Reyman and M.A.Semenov-Tian-Shansky, Classical r-Matrices and Compatible Poisson Brackets for Coupled KdV systems, *Lett.Math.Phys.* **17**, 25-29 (1989).
[22] I.Dorfman, Dirac Structures of Integrable Evolution Equations, *Phys.Letts.A.* **125**, 240-46 (1988).
[23] G.Wilson, On the Quasi-Hamiltonian Formalism of the KdV Equation, *Phys.Letts.A.* **132**, 445-50 (1988).

A.G. Reyman

38. COMPATIBLE POISSON BRACKETS IN TERMS OF R-MATRICES

Abstract

The formalism of classical r-matrix is used to construct families of compatible Poisson brackets for various integrable nonlinear systems associated with Lax equations, zero-curvature equations, and the spectral parameter dependent Schrödinger equation.

Poisson brackets are called compatible if their linear combinations are again Poisson brackets. As is well known, integrable Hamiltonian equations often possess more than one Hamiltonian structure, and these structures turn out to be compatible. Our purpose here is to show how such multi-Hamiltonian equations arise naturally in the r-matrix construction of integrable systems. The main observation is that compatible Poisson brackets result from linear families of r-matrices. The latter are easily constructed if the initial Lie algebra of the r-matrix scheme is a loop algebra. Examples of multi-Hamiltonian equations obtained in this way include generalized AKNS systems (based on the algebra of matrix valued functions) and coupled KdV systems (based on the Virasoro algebra).

This note reports the joint work with A.P. Fordy and M.A. Semenov-Tian-Shansky. Proofs and further details can be found in [3,8]

First recall some basic definitions. Let $\mathfrak{g}$ be a Lie algebra. The Lie-Poisson bracket on the dual space $\mathfrak{g}^*$ is defined by

$$\{\varphi,\psi\}(L) = L([d\varphi(L),d\psi(L)]), \quad L\in\mathfrak{g}^* \quad . \tag{1}$$

For a Hamiltonian H on $\mathfrak{g}^*$, the corresponding Hamilton's equations have the form

$$\frac{d}{dt}L = -\mathrm{ad}^*X\cdot L \tag{2}$$

where $X=dH(L)$ and ad^* denotes the coadjoint action of $\mathfrak{g}$ in $\mathfrak{g}^*$.

A linear operator R on $\mathfrak{g}$ is called a (classical) r-matrix if the skew-symmetric bilinear operation

$$[X,Y]_R = [RX,Y]+[X,RY] \qquad (3)$$

satisfies the Jacobi identity. In this case $[\,,\,]_R$ is a second Lie bracket on $\mathfrak{g}$ and hence defines a second Lie-Poisson bracket $\{\,,\,\}_R$ on $\mathfrak{g}^*$, which will be called the R-bracket. The r-matrix construction of integrable systems is based on the interplay of the two Lie-Poisson brackets.

Theorem 1

1. Coadjoint invariants on $\mathfrak{g}^*$ with respect to the original Lie structure Poisson commute with respect to the R -bracket.

2. The Hamiltonian equations on $\mathfrak{g}^*$ generated by an invariant function H relative to the R -bracket have a generalized Lax form

$$\frac{d}{dt}L = -\mathrm{ad}^*M\cdot L, \qquad M = R(dH(L)), \qquad (4)$$

where ad^* is the coadjoint action for the original Lie structure of $\mathfrak{g}$.

For the proof see, for instance, [9].

If $\mathfrak{g}$ has a nondegenerate invariant inner product which allows to identify $\mathfrak{g}^*$ with $\mathfrak{g}$ and ad^* with ad, then (4) becomes the usual Lax equation.

We shall now define a linear family of R -brackets. A linear operator A on $\mathfrak{g}$ is called intertwining if

$$A\cdot \mathrm{ad}X = \mathrm{ad}X\cdot A, \qquad (5)$$

for any $X\in\mathfrak{g}$, i.e. if $A[X,Y]=[AX,Y]=[X,AY]$.

Theorem 2

If R is an r-matrix and A an intertwining operator then RA is also an r-matrix.

Since the intertwining operators form a linear family, the RA -brackets $\{\,,\,\}_{RA}$ for a given R also form a linear family and so are compatible. All these brackets have a common set of functions in involution which consists of the coadjoint invariants for the original Lie structure on $\mathfrak{g}$. Note that if A is invertible, the linear change of variables $L\to A^*L$ in $\mathfrak{g}^*$ carries the R -bracket into the RA -bracket, which gives a proof of Theorem 2 in this case.

The simplest and most important example of an r-matrix is constructed as follows. Suppose we have a vector space decomposition of $\mathfrak{g}$ as a sum of two Lie subalgebras:

$$\mathfrak{g} = \mathfrak{g}_+ + \mathfrak{g}_- \qquad (6)$$

Let $P_\pm$ be the projection operators onto $\mathfrak{g}_\pm$ parallel to $\mathfrak{g}_\mp$. The r-matrix associated with the decomposition (6) is

$$R = \frac{1}{2}(P_+ - P_-) \tag{7}$$

In this case we have

$$[X,Y]_R = [P_+X, P_+Y] - [P_-X, P_-Y] \tag{8}$$

The construction of Theorem 2 is meaningful only if there are enough intertwining operators on $\mathfrak{g}$. A class of algebras which have a lot of intertwining operators and also have a natural decomposition of the form (6) are the loop algebras defined as follows. Let $\mathfrak{a}$ be a Lie algebra. The loop algebra of $\mathfrak{a}$, denoted by $\mathcal{L}(\mathfrak{a})$, is the Lie algebra of Laurent polynomials $X(\lambda)=\sum x_i\lambda^i$ with coefficients in $\mathfrak{a}$ and with commutator $[x\lambda^i, y\lambda^j]=[x,y]\lambda^{i+j}$.

The natural decomposition of $\mathfrak{g}=\mathcal{L}(\mathfrak{a})$ into subalgebras $\mathfrak{g}_+$ and $\mathfrak{g}_-$ is

$$\mathfrak{g}_+ = \bigoplus_{i\geq 0} \mathfrak{a}\lambda^i, \qquad \mathfrak{g}_- = \bigoplus_{i<0} \mathfrak{a}\lambda^i \tag{9}$$

The dual space $\mathfrak{g}^*=\mathcal{L}(\mathfrak{a})^*$ can be identified with $\mathcal{L}(\mathfrak{a}^*)=\{L(\lambda)=\sum_i u_i\lambda^i, u_i\in\mathfrak{a}^*\}$ via the pairing

$$\langle L, X\rangle = \mathrm{Res}_\lambda \langle L(\lambda), X(\lambda)\rangle = \sum_i u_i(x_{-i-1}). \tag{10}$$

The coadjoint action of $\mathcal{L}(\mathfrak{a})$ on $\mathcal{L}(\mathfrak{a}^*)$ is given by $\mathrm{ad}^*(x\lambda^i)\cdot(u\lambda^j) = (\mathrm{ad}^* x\cdot u)\lambda^{i+j}$.

Clearly, the multiplication operators by scalar Laurent polynomials are intertwining operators in $\mathcal{L}(\mathfrak{a})$. Hence we have a linear family of r-matrices of the form $R\hat{q}$, where R is given by (7) and $\hat{q}$ is the multiplication operator by $q(\lambda)=\sum q_i\lambda^i$. We shall write more explicitly the $R\hat{\lambda^k}$-bracket which we denote by $\{,\}_k$. Choose a basis e^a in $\mathfrak{a}$ and let C^{ab}_c be the corresponding structure constants. Put $u^a_i=u_i(e^a)$, then

$$\left\{u^a_i, u^b_j\right\}_k = \varepsilon_{ij} C^{ab}_c u^c_{i+j+1-k}, \tag{11}$$

where $\varepsilon_{ij}=1$ if $i,j<k$; $\varepsilon_{ij}=-1$ if $i,j\geq k$; otherwise $\varepsilon_{ij}=0$. Such Poisson brackets were studied in [4].

It follows from (11) that Hamiltonian equations for the bracket $\{,\}_k$ have the form

$$\frac{d}{dt} u_i = \sum_j \varepsilon_{ij} J_{i+j+1-k}\left(\frac{\delta H}{\delta u_j}\right), \tag{12}$$

where ε_{ij} is the same as above and the linear operator $J_i:\mathfrak{a}\to\mathfrak{a}^*$ is given by

$$J_i(x) = \mathrm{ad}^* x \cdot u_i \quad . \tag{13}$$

It is easy to determine those invariant (Poisson) subspaces of $\mathcal{L}(\mathfrak{a})^*$ relative to $R\hat{q}$ -brackets which consist of polynomials of bounded degree. For example, the subspace $L_{N,a}=\{L=a\lambda^{N+1}+\sum_0^N u_i\lambda^i\}$ with a fixed is a Poisson subspace for $\{,\}_k$ if $0\le k\le N+1$. Thus there is a (N+2) -dimensional family of compatible brackets on $L_{N,a}$. Hamiltonian equations in $L_{N,a}$ with respect to $\{,\}_k$ have the form $\frac{d}{dt}(u_0,\ldots,u_N)^t = J^{(k)}\left(\frac{\delta H}{\delta u_0},\ldots,\frac{\delta H}{\delta u_N}\right)^t$ where the operator valued $(N+1)\mathrm{x}(N+1)$ matrix $J^{(k)}$ is

$$J^{(k)} = \left(\begin{array}{ccc|cccc} 0 & & J_0 & & & & \\ & \iddots & \vdots & & 0 & & \\ J_0 & \cdots & J_{k-1} & & & & \\ \hline & & & -J_{k+1} & \cdots & -J_N & -J_{N+1} \\ & 0 & & \vdots & & \iddots & \\ & & & -J_N & \iddots & & 0 \\ & & & -J_{N+1} & & & \end{array}\right) \tag{14}$$

with J_i given by (13) and $u_{N+1}=a$.

Let us now discuss the hierarchy of Lax equations associated to $\mathcal{L}(\mathfrak{a})$. The coadjoint invariants of $\mathcal{L}(\mathfrak{a})$ are easily described. Let P be an invariant analytic function on $\mathfrak{a}^*$ and φ a rational function of λ. Then

$$P_\varphi(L) = \mathrm{Res}_\lambda[\varphi(\lambda)P(L(\lambda))] \tag{15}$$

is an invariant function on $\mathcal{L}(\mathfrak{a}^*)$. By (4), Hamilton's equations induced by P_φ with respect to the $R\hat{q}$ -bracket are

$$\frac{d}{dt}L = -\mathrm{ad}^* \; R\,[q\varphi dP(L)]\cdot L =$$

$$= \pm\mathrm{ad}^* \; P_\pm \; [q\varphi dP(L)]\cdot L \quad . \tag{16}$$

Clearly, the Hamiltonians P_φ give rise to the same hierarchy of generalised Lax equations (16) relative to all $R\hat{q}$ - brackets: $P_{\varphi q^{-1}}$ relative to the $R\hat{q}$ - bracket gives the same as P_φ relative to the R - bracket. In particular, the Hamiltonians

$P_K = \mathrm{Res}_\lambda[\lambda^{-K}\varphi(\lambda)P(L(\lambda))]$ relative to $\{,\}_K$ produce the same equation for all k.

All the above constructions extend to twisted loop algebras, and also to the case of rational, trigonometric or elliptic dependence on the spectral parameter λ.

We shall now consider three cases that correspond to different types of the Lie algebra $\mathfrak{a}$.

1. **Lax equations**. If $\mathfrak{a}$ has a nondegenerate invariant inner product $(,)$, then we can also define such a product on $\mathcal{L}(\mathfrak{a})$ by

$$(X,Y) = \mathrm{Res}_\lambda (X(\lambda),Y(\lambda)) = \sum_i (x_i, y_{-i-1}) \tag{17}$$

which identifies $\mathcal{L}(\mathfrak{a}^*)$ with $\mathcal{L}(\mathfrak{a})$ and ad^* with ad, so that equation (16) takes the commutator form $\frac{d}{dt}L=[L,M]$, and the operators J_i in (13) become $J_i=-\mathrm{ad}\, u_i$.

2. **Zero-curvature equations**. Let $\mathfrak{a}$ be as above and let $\tilde{\mathfrak{a}}$ denote the Lie algebra of smooth functions on the circle with values in $\mathfrak{a}$ and pointwise commutator. An invariant inner product on $\tilde{\mathfrak{a}}$ is given by

$$\langle X,Y\rangle = \int (X(x),Y(x))dx \quad . \tag{18}$$

Let $\hat{\mathfrak{a}}$ be the central extension of $\tilde{\mathfrak{a}}$ determined by the 2-cocycle $\omega(X,Y)=\langle X,\partial_x Y\rangle$. Identifying the dual $\hat{\mathfrak{a}}^*$ with $\tilde{\mathfrak{a}}\oplus\mathbb{R}$ we find the coadjoint action of $\tilde{\mathfrak{a}}$ in $\hat{\mathfrak{a}}^*$:

$$\mathrm{ad}^* X\cdot(L,p) = ([X,L]+p\partial_x X, 0) \tag{19}$$

Hence the associated Hamiltonian equations have zero-curvature form. Notice that the variable p lies in the annihilator of the corresponding Lie-Poisson bracket.

Consider now the loop algebra $\mathfrak{g}=\mathcal{L}(\hat{\mathfrak{a}})$. The elements of $\mathfrak{g}$ are represented by pairs $(L(\lambda),p(\lambda))$, where $L(\lambda)=\sum u_i(x)\lambda^i$, and $p(\lambda)=\sum p_i\lambda^i$; the variables p_i annihilate all $R\hat{q}$ - brackets. Now, with the notation of (11), the bracket $\{,\}_k$ takes the form

$$\{u_i^a(x),u_j^b(y)\}_k = \varepsilon_{ij}\Big[C^{ab}_c u^c_{i+j+1-k}(x)\delta(x-y)+ + K^{ab}\sum_s p_s \delta_{i+j+1,k+s}\delta'(x-y)\Big] \quad , \tag{20}$$

where $K^{ab}=(e^a,e^b)$. The operator J_i (13) becomes

$$J_i = p_i\partial_x - \mathrm{ad}\,u_i \quad . \tag{21}$$

Notice that for $p=\lambda^k$, $\{,\}_k$ is ultralocal, i.e. the δ' term vanishes.

The $\hat{Rq}$ - brackets can be written in tensor notation widely used in the inverse scattering method:

$$\{L(\lambda,x)\otimes, L(\mu,y)\}_{\hat{Rq}} = [r_+(\lambda,\mu), L(\lambda,x)\otimes 1 + 1\otimes L(\mu,y)]\delta(x-y)$$

$$+[r_-(\lambda,\mu), L(\lambda,x)\otimes 1-1\otimes L(\mu,y)]\delta(x-y)$$

$$+\frac{q(\lambda)p(\mu)-q(\mu)p(\lambda)}{\lambda-\mu}\delta'(x-y)I, \tag{22}$$

where

$$r_+(\lambda,\mu) = \frac{1}{2}\frac{q(\lambda)+q(\mu)}{\lambda-\mu}I, \quad r_-(\lambda,\mu) = \frac{1}{2}\frac{q(\lambda)-q(\mu)}{\lambda-\mu}$$

$$I = \sum K_{ab}e^a\otimes e^b \quad . \tag{23}$$

The expression (22) in the case $p(\lambda)=1$ was derived in [6] by generalizing the standard case $q(\lambda)=1$.

Finally, the coadjoint invariants on $\hat{\mathfrak{a}}^*$ which give rise to the Hamiltonians of the integrable equations are the invariants of the monodromy matrix of the auxiliary linear problem

$$(p\partial_x-L)\psi = 0 \quad . \tag{24}$$

Example 1. Let $L(\lambda)=s\lambda^{-1}$ and $p(\lambda)=1$. There are two compatible brackets:

$$\{s^a(x), s^b(y)\}_0 = C^{ab}_c s^c(x)\delta(x-y)$$

$$\{s^a(x), s^b(y)\}_1 = K^{ab}\delta'(x-y) \quad . \tag{25}$$

The bracket $\{,\}_0$ defines the usual Poisson structure for systems of the Heisenberg ferromagnet type. Reducing the bracket $\{,\}_1$ to the symplectic leaves of $\{,\}_0$ gives the recursion operator for the Heisenberg ferromagnet, see [5].

Example 2. Let $L(\lambda)=u+v\lambda+a\lambda^2$ with a fixed. There are three compatible structures given by

$$J^{(0)} = -\begin{pmatrix} J_1 & J_2 \\ J_2 & 0 \end{pmatrix}, \quad J^{(1)} = \begin{pmatrix} J_0 & 0 \\ 0 & -J_2 \end{pmatrix}, \quad J^{(2)} = \begin{pmatrix} 0 & J_0 \\ J_0 & J_1 \end{pmatrix}$$

(see (14)). If $p(\lambda)=1$ these become

$$J^{(0)} = \begin{pmatrix} \text{ad } v & \text{ad } a \\ \text{ad } a & 0 \end{pmatrix}, \quad J^{(1)} = \begin{pmatrix} \partial_x - \text{ad } u & 0 \\ 0 & \text{ad } a \end{pmatrix},$$

$$J^{(2)} = \begin{pmatrix} 0 & \partial_x - \text{ad } u \\ \partial_x - \text{ad } u & -\text{ad } v \end{pmatrix}.$$

Observe that if a=0 then v(x) is a Casimir function for the first two structures, and $\{,\}_0$ is the customary bracket for the generalised NLS equation. In this case, reducing $\{,\}_1$ to the symplectic leaves of $\{,\}_0$ gives the well-known recursion operator Λ for NLS, as was first shown in [7]. The reduction of $\{,\}_2$ gives the recursion operator Λ^2.

Example 3. Setting $p(\lambda)=\lambda^2$ in the previous example, so that the auxiliary linear operator is $L=\lambda^2\partial_x-u-v\lambda-a\lambda^2$, we get

$$J^{(0)} = \begin{pmatrix} \text{ad } v & -\partial_x + \text{ad } a \\ -\partial_x + \text{ad } a & 0 \end{pmatrix},$$

$$J^{(1)} = -\begin{pmatrix} \text{ad } u & 0 \\ 0 & \partial_x - \text{ad } a \end{pmatrix}, \quad J^{(2)} = -\begin{pmatrix} 0 & \text{ad } u \\ \text{ad } u & \text{ad } v \end{pmatrix}.$$

3. **Equations based on the Virasoro algebra.** Consider the Lie algebra of vector fields $\varphi(x)\partial_x$ on the circle with commutator $[\varphi\partial_x,\psi\partial_x]=(\varphi\psi'-\psi\varphi')\partial_x$ and take its central extension via the cocycle $\omega(\varphi,\psi)=\int\varphi\psi'''\,dx$. This is the Virasoro algebra $\mathcal{D}$. The dual $\mathcal{D}^*$ is formed by the pairs $(u(x)dx^2,p)$, and the coadjoint action of $\mathcal{D}$ in $\mathcal{D}^*$ is

$$\text{ad}^*(\varphi,c)\cdot(u,p) = \left((p\partial_x^3+u\partial_x+\partial_x u)\varphi,0\right). \tag{26}$$

Applying the general scheme to the loop algebra $\mathcal{L}(\mathcal{D})$ we obtain a sequence of compatible Poisson brackets for the coefficients of $(L(\lambda),p(\lambda))\in\mathcal{L}(\mathcal{D}^*)$ where $L(\lambda)=\sum u_i(x)\lambda^i$ and $p(\lambda)=\sum p_i\lambda^i$:

$$\{u_i(x),u_j(y)\}_k = \varepsilon_{ij}\Big[(u_{i+j+1-k}(x)+u_{i+j+1-k}(y))\delta'(x-y)+$$

$$+\sum_s p_s\delta_{i+j+1,k+s}\delta'''(x-y)\Big] \tag{27}$$

The operator J_i entering in (12),(14) takes the form

$$J_i = p_i\partial_x^3+u_i\partial_x+\partial_x u_i \quad . \tag{28}$$

Coadjoint invariants for the Virasoro algebra are obtained as monodromy invariants of the auxiliary linear problem

$$(p\partial_x^2+u(x))\psi = 0 \quad . \tag{29}$$

In $\mathcal{L}(\mathcal{D}^*)$ this becomes a spectral parameter dependent Schrödinger equation. The ensuing nonlinear equation were studied by Antonowicz and Fordy [1] who found their multi-Hamiltonian structure given by (14) or (27).

As before, the subspace $\mathcal{L}_{N,a}=\{(\mathcal{L}(\lambda)=a\lambda^{N+1}+\sum_0^N u_i\lambda^i\ ,\ p(\lambda)=\sum_0^{N+1} p_i\lambda^i)\}$, where a and $p(\lambda)$ are fixed, is a Poisson subspace in $\mathcal{L}(\mathcal{D})^*$. Setting $p(\lambda)=1$ gives the generalised KdV hierarchy, while setting $p(\lambda)=\lambda^{N+1}$ gives the generalised Harry Dym hierarchy (see [3] for details).

Example. Let $\mathcal{L}(\lambda)=u+v\lambda+a\lambda^2$ with a=const. If $p(\lambda)=1$ we have three compatible Poisson structures for coupled KdV equation [1]:

$$J^{(0)} = -\begin{pmatrix} v\partial_x+\partial_x v & 2a\partial_x \\ 2a\partial_x & 0 \end{pmatrix} \quad , \quad J^{(1)} = \begin{pmatrix} \partial_x^3+u\partial_x+\partial_x u & 0 \\ 0 & -2a\partial_x \end{pmatrix}$$

$$J^{(2)} = \begin{pmatrix} 0 & \partial_x^3+u\partial_x+\partial_x u \\ \partial_x^3+u\partial_x+\partial_x u & v\partial_x+\partial_x v \end{pmatrix} \quad .$$

If a=0 then v(x) is a Casimir function for $J^{(0)}$ and $J^{(1)}$ which in this case (for v=1) are the well-known Poisson structures for KdV. Reducing the bracket given by $J^{(2)}$ to the constraint v=1 gives the squared recursion operator for KdV.

Setting in the above example $p(\lambda)=\lambda^2$, so that the auxiliary linear operator becomes $L=\lambda^2\partial_x^2+u+v\lambda+a\lambda^2$, we get the Poisson brackets for the coupled Harry Dym hierarchy [2].

References

[1] Antonowicz,M. & Fordy,A.P. (1987). Coupled KdV equations with multi-Hamiltonian structures. *Physica D*, **28**,345-358.

[2] Antonowicz,M. & Fordy,A.P. (1988). Coupled Harry Dym equations with multi-Hamiltonian structures. *J.Phys.A.*, **21**,L269-275.

[3] Fordy,A.P., Reyman,A.G. & Semenov-Tian-Shansky M.A. (1989). Classical r-matrices and compatible Poisson brackets for coupled KdV systems. Lett.in *Math.Phys.* **17**,25-29.

[4] Kulish,P.P. & Reyman,A.G. (1983). Hamiltonian structure of polynomial bundles. In: Differential geometry, Lie groups and mechanics V, Zap.Nauchn.Semin. LOMI **123**, 67-76. (English transl.*J.Soviet Math.* **30**,2319-2326 (1985)).

[5] Magri,F., Morosi,C. & Ragnisco,O. (1985). Reduction techniques for infinite dimensional Hamiltonian systems: Some ideas and applications. *Commun.Math.Phys.*, **99**,115-140.

[6] Maillet,J.M. (1986). Hamiltonian structures for integrable classical theories from graded Kac-Moody algebras. *Phys. Lett. B* **167**,401-405.

[7] Reyman,A.G. & Semenov-Tian-Shansky, M.A. (1980). A family of Hamiltonian structures, a hierarchy of Hamiltonians and reduction for first order matrix differential operators. *Funct.Anal.Appl.*, **14**,146-148.

[8] Reyman,A.G. & Semenov-Tian-Shansky, M.A. (1988). Compatible Poisson structures for Lax equations: an r-matrix approach. *Phys.Lett.A* **130**,456-460.

[9] Semenov-Tian-Shansky, M.A. (1983). What is a classical r-matrix? *Funct.Anal.Appl.* **17**,259-272.

I. Ya. Dorfman

39. DIRAC STRUCTURES AND INTEGRABILITY OF NONLINEAR EVOLUTION EQUATIONS

1 Introduction

Zakharov & Faddeev [19] and Gardner [9] were the first to reveal close interrelations between integrability properties of a nonlinear evolution equation and its hamiltonian nature. The Dirac structure presented below was proposed by the author [6] in an attempt to provide the theory of hamiltonian operators, both local and nonlocal ones, with rigorous foundations. The definition of a Dirac structure is given in purely algebraic terms, that make it possible to incorporate the finite-dimensional version of this notion introduced by Courant & Weinstein [2]. Links with integrability are indicated by a generalization of the so-called Lenard scheme [15,16,3,14,11,12].

2 Complexes over Lie algebras

Let $\mathcal{A}$ be a Lie algebra and $(\Omega = \Omega^0\oplus\Omega^1\oplus\ldots,d)$ be a complex of linear spaces, $d\Omega^q\subset\Omega^{q+1}$. If to each element $a\in\mathcal{A}$ there corresponds a linear operator $i_a:\Omega\to\Omega$, $i_a\Omega^q\subset\Omega^{q-1}$, in such a way that $i_ai_b+i_bi_a = 0$ and also $[i_as+di_a,i_b] = i_{[a,b]}$, then (Ω,d) is called a complex over $\mathcal{A}$ [12].

Elements of $\mathcal{A}$ and Ω^q are called vector fields and q-forms, respectively. The result of the pairing between a vector field $a\in\mathcal{A}$ and a 1-form $\xi\in\Omega^1$ given by $i_a\xi\in\Omega^0$ is denoted by (a,ξ) or (ξ,a).

The Lie derivative along a vector field $a\in\mathcal{A}$ is defined by $L_a=i_ad+di_a$. Putting $L_ab=[a,b]$, $b\in\mathcal{A}$ and requiring the validity of the chain rule we can expand L_a to the analogues of tensor fields, namely linear operators acting in the spaces of forms and fields and also between these spaces. If the Lie derivative of some object S along $a\in\mathcal{A}$ vanishes, then we say that a is a symmetry of S, or that S is conserved along a. Symmetries of any object form a

Lie subalgebra in $\mathcal{A}$.

It is sufficient for our purposes to keep in mind two basic examples of complexes over Lie algebras. The first one is the de Rham complex of differential forms on a finite-dimensional manifold X, that is, according to the given definition, a complex over the Lie algebra of vector fields on X. The second example is the so-called complex of formal variational calculus that takes the place of the de Rham complex in the hamiltonian theory of nonlinear evolution equations.

Such a complex is determined by the choice of formal variables $u_\alpha^{(i)}$, α belonging to some set of indices J, i = 0,1,2... and some ring R of functions f, each depending on a finite number of variables. There is a certain reduction procedure that associates with R a special complex $(\Omega=\Omega^0\oplus\Omega^1\oplus\ldots,d)$ over the Lie algebra of derivations of R that commute with d/dx, where

$$\frac{d}{dx} = \sum_{i,\alpha} u_\alpha^{(i+1)}\frac{\partial}{\partial u_\alpha^{(i)}}.$$

In this complex, that we name the complex of formal variational calculus based on R, elements of $\Omega^0=R/(d/dx)R$ represent functionals $\int f dx$. Evolution equations

$$\frac{\partial u_\alpha}{dt} = h_\alpha, \quad h_\alpha \in R, \ \alpha\in J$$

are represented, as usual, by vector fields $h = \{h_\alpha\}$ (see [6] for details).

3 Dirac structure and hamiltonian formalism associated with it

Let (Ω,d) be a complex over some Lie algebra $\mathcal{A}$. Consider $\mathcal{A}\oplus\Omega^1$ equipped with the 'canonical' symmetric bilinear form

$$\langle h_1\oplus\xi_1, h_2\oplus\xi_2\rangle = (h_1,\xi_2) + (h_2,\xi_1).$$

For any linear subspace $\mathcal{L}\subset\mathcal{A}\oplus\Omega^1$ denoted by $\mathcal{L}^\perp$ the subspace $\{h\oplus\xi, \langle h\oplus\xi, h_1\oplus\xi_1\rangle = 0 \ \forall\ h_1\oplus\xi_1\in\mathcal{L}\}$.

Definition. A Dirac structure on $(\mathcal{A},\Omega)$ is a linear subspace $\mathcal{L}\subset\mathcal{A}\oplus\Omega^1$ with the following two properties:

(a) $\mathcal{L}^\perp = \mathcal{L}$;

(b) $(L_{h_1}\xi_2,h_3) + (L_{h_2}\xi_3,h_1) + (L_{h_3}\xi_1,h_2) = 0$ for all $h_1\oplus\xi_1, h_2\oplus\xi_2, h_3\oplus\xi_3\in\mathcal{L}$.

Particular cases of this definition are important, where Dirac structures are graphs. We say that an operator $H:\Omega^1\to\mathcal{A}$ is a hamiltonian one, if its graph $\{H\xi\oplus\xi, \xi\in\Omega^1\}$ is a Dirac structure. We also call an operator $I:\mathcal{A}\to\Omega^1$ symplectic, if its graph $\{h\oplus Ih,\ h\in\mathcal{A}\}$ is a Dirac structure. These definitions come to the ones of Poisson and pre-symplectic structure, respectively, if we assume that (Ω,d) is the de Rham complex of differential forms on a finite-dimensional manifold.

Given a Dirac structure, one can develop a sort of a hamiltonian formalism. We say that $h\in\mathcal{A}$ is a hamiltonian field and $f\in\Omega^0$ is its hamiltonian if $h\oplus df\in\mathcal{L}$. Denote by $\mathcal{H}(\mathcal{L})$ the space of all hamiltonian fields, by π_1 the projection from $\mathcal{A}\oplus\Omega^1$ onto $\mathcal{A}$.

Theorem. Let $\mathcal{L}\subset\mathcal{A}\oplus\Omega^1$ be a Dirac structure. Then

(a) $[h_1,h_2]\oplus[i_{h_1}d\xi_2-i_{h_2}d\xi_1+d(h_1,\xi_2)]\in\mathcal{L}$ for all $h_1\oplus\xi_1, h_2\oplus\xi_2\in\mathcal{L}$;

(b) subspaces $\mathcal{A}_0=\{h\in\mathcal{A}:\ h\oplus 0\in\mathcal{L}\}$, $\mathcal{H}(\mathcal{L})$ and $\mathcal{A}_1=\pi_1\mathcal{L}$ are Lie subalgebras in $\mathcal{A}$, $\mathcal{A}_0\subset\mathcal{H}(\mathcal{L})\subset\mathcal{A}_1$;

(c) the Poisson bracket $\{f_1,f_2\}_{\mathcal{L}}=(h_1,df_2)$ is well-defined on the space of all hamiltonians, providing it with a Lie algebra structure;

(d) $[h_1,h_2]\oplus d\{f_1,f_2\}_{\mathcal{L}}\in\mathcal{L}$.

Note that the last statement is the essence of the conventional finite-dimensional hamiltonian formalism.

A few words about the Lie algebras $\mathcal{A}_0$ and $\mathcal{A}_1$. If the Dirac structure is the graph of a hamiltonian operator $H:\Omega^1\to\mathcal{A}$, then $\mathcal{A}_0=\{0\}$, $\mathcal{A}_1=\mathrm{Im}H$, and we recognize $\mathcal{A}_1$ as the analogue of the Lie algebra that determines the stratification of the Poisson structure into symplectic leaves. If, in the other case, the Dirac structure is the graph of a symplectic operator $I:\ \mathcal{A}\to\Omega^1$, then $\mathcal{A}_0=\mathrm{Ker}\ I$, $\mathcal{A}_1=\mathcal{A}$, so we recognize $\mathcal{A}_0$ as the analogue of the Lie algebra that determines the characteristic foliation of the pre-symplectic structure; symplectic structures are induced on transversal manifolds of complementary dimension.

For general Dirac structures $\mathcal{A}_0$ and $\mathcal{A}_1$ can be both nontrivial.

4 Pairs of Dirac structures and Lenard scheme

The fact that an evolution equation possesses a Dirac structure does not, of course, guarantee its integrability. However the exisence of two distinct but compatible Dirac structures becomes a source of integrability.

We say that two Dirac structures $\mathcal{L},\mathcal{M}\subset\mathcal{A}\oplus\Omega^1$ form a pair (are compatible) if the set

$$A_{\mathcal{L}\mathcal{M}}=\{a_1\oplus a_2:\ \exists\xi\in\Omega^1,\ a_1\oplus\xi\in\mathcal{M},\ a_2\oplus\xi\in\mathcal{L}\}\subset\mathcal{A}\oplus\Omega^1$$

is a Nijenhuis relation.

The notion of a Nijenhuis relation was introduced by Gelfand & Dorfman [12] as a natural generalization of the one of a Nijenhuis operator, i.e. an operator $A:\ \mathcal{A}\to\mathcal{A}$ that satisfies the condition

$$[Aa,Ab] - A[a,Ab] - A[Aa,b] + A^2[a,b] = 0$$

The special role of Nijenhuis operators in integrability theory is determined by their close relations with deformations of Lie algebras [4].

Important particular cases of the general definition are the notions of a hamiltonian pair of operators and a symplectic pair of operators, i.e. such ones that their graphs form Dirac pairs. It is possible to prove that $H,K:\Omega^1\to\mathcal{A}$ form a hamiltonian pair if all the linear combinations $H+\mu K$ are hamiltonian operators (this property Gelfand & Dorfman [11] used as the definition). The notion of a symplectic pair is somewhat more complicated.

The following theorem is the generalization of the scheme known as the Lenard scheme of integrability.

Theorem. Let $\mathcal{L},\mathcal{M}$ be a pair of Dirac structures and $h_0,h_1,\ldots\in\mathcal{A}$, $\xi_{-1},\xi_0,\ldots\in\Omega^1$ such that $h_i\oplus\xi_{i-1},\in\mathcal{L},h_i\oplus\xi_i\in\mathcal{M}$. Suppose $d\xi_{-1} = d\xi_0 = 0$. Let the following condition be satisfied: from $\xi\in\Omega^1$, $d\xi(a,b)=0$ $\forall a,b\varepsilon\pi_1\mathcal{A}_{\mathcal{L}\mathcal{M}}$ it follows that $d\xi = 0$. Then

(a) all ξ_i are closed, $d\xi_i = 0$;

(b) if $\xi_i = df_i$, $f_i\in\Omega^0$ (which is guaranteed if the cohomology group $H^1(\Omega)$ is trivial), then all f_i are conserved along all h_j;

(c) all f_i are in involution with respect to both structures:

$$\{f_i,f_j\}_{\mathcal{L}} = \{f_i,f_j\}_{\mathcal{M}} = 0;$$

(d) all $[h_i,h_j]$ lie in the intersection of the $\mathcal{A}_0$-type Lie algebras (see above) generated by $\mathcal{L}$ and M, respectively.

5 Some examples, comments and open problems

The theory of local hamiltonian operators and hamiltonian pairs seems rich and developed in comparison with the theory of local symplectic operators and symplectic pairs. The first examples in the symplectic framework were given by the author in 1987, namely, the local symplectic pair

$$I = u_x^{-2}\frac{d}{dx} - u_x^{-3}u_{xx}$$

$$J = u_x^{-2}\left(\frac{d}{dx}\right)^3 - 3u_x^{-3}u_{xx}\left(\frac{d}{dx}\right)^2 + (3u_x^{-4}u_{xx}^2 - u_x^{-3}u_{xxx})\frac{d}{dx}$$

that determines the Lenard scheme of the Krichever-Novikov equation

$$u_t = u_{xxx} - \frac{3}{2}u_x^{-1}u_{xx}^2$$

and the local symplectic pair

$$I = \frac{d}{dx}$$

$$J = \left(\frac{d}{dx}\right)^3 + 4u_x \frac{d}{dx} + 2u_{xx}$$

determining the Lenard scheme of the potentiated KdV equation

$$u_t = 3u_x^2 + u_{xxx}.$$

There is now a wide field of action in searching such pairs, not only in the case of one unknown function but also for operators represented by matrices.

Some special classes of local hamiltonian operators are well-investigated due to their importance in integrability theory, such as infinite-dimensional Kirillov-Kostant structures [13], hydrodynamic structures [7], the so-called second hamiltonian structure of Lax equations [10] and others. Similar work on symplectic operators and also general Dirac structures may bring new interesting examples.

Another problem is still open: what is the correct analogue of the Darboux theorem for local hamiltonian operators? All the results yet achieved are enumerated in Mokhov's [17] paper and refer to one unknown function and differential operators of the first or the third order. The same problem arises now also for symplectic operators and general Dirac structures.

Two recent results obtained by the author may somehow be useful for the case of one unknown function and higher order differential operators. Consider a differential operator

$$P = \sum_{k=0}^{N} p_k(u, u_x, \ldots, u^{(n_k)}) \left(\frac{d}{dx}\right)^k$$

of an odd order N with coefficients p_k that belong to some ring R of functions of formal variables $u, u_x, \ldots,$ as above. Put

$$\operatorname{ord} p_k \overset{\text{def}}{=} \min \{\tau : \frac{\partial p_k}{\partial u^{(\tau+1)}} = \frac{\partial p_k}{\partial u^{(\tau+2)}} \cdots = 0\}.$$

Theorem. If P is a symplectic operator then $\max_k \{\operatorname{ord} p_k + k - 1\}$ is either 3N or ≤3N-2.

Theorem. If P is a hamiltonian operator then $\max_k \{\operatorname{ord} p_k + k - 1\}$ is either 2N or ≤2N-2.

Another point worth indicating is the so-called τ-scheme of integrability [1,8,18,5], that substitutes the Lenard scheme in some cases when the latter fails to work. Interrelations between the τ-scheme, the notion of a hamiltonian pair of operators and the corresponding Lenard scheme are more or less clear. It occurs that usually one of the operators of the hamiltonian pair is the

Lie derivative of the other along some element τ; that very τ determines the τ-scheme (see [5]). It seems of interest to build up the corresponding theory of Lie derivatives of Dirac structures and to trace similar interrelations between Dirac pairs and the τ-scheme.

References

[1] Chen, H.H. & Lin, Y.C. & Lin, J.E. (1982). On a new hierarchy of symmetries for the integrable nonlinear evolution equations. Preprint PL 83-002, University of Maryland.

[2] Courant, T. & Weinstein, A. (1986). Beyond Poisson structures. Preprint, University of California, Berkeley.

[3] Dorfman, I. Ya. (1978). On formal variational calculus in the algebra of smooth cylindric functions. *Funct. Anal. & Appl.* **12(2)**, 32.

[4] Dorfman, I. Ya. (1984). Deformations of hamiltonian structures and integrable systems. In *Nonlinear Phenomena and Turbulence. Proceedings of Kiev 1983 Workshop*, p. 1313. Zakharov, V. (ed.). Gordon & Breach: New York.

[5] Dorfman, I. Ya. (1986). τ-scheme in the theory of integrable equations. In *Group-theoretical Methods in Physics*, vol. 1, p. 435. Markov, M. (ed.). VNU Science Press: Utrecht.

[6] Dorfman, I. Ya. (1987). Dirac structures of integrable evolution equations. *Phys. Lett. A* **125**, 240.

[7] Dubrovin, B.A. & Novikov, S.P. (1984). On Poisson brackets of a hydrodynammic type. *Doklady Academii Nauk SSSR* **279**, 294.

[8] Fokas, A.S. & Fuchssteiner, B. (1981). The hierarchy of the Benjamin-Ono equation. *Phys. Lett. A* **86**, 341.

[9] Gardner, C.S. (1971). Korteweg-de Vries equation and generalizations, IV. *J. Math. Phys.* **12**, 1548.

[10] Gelfand, I.M. & Dikii, L.A. (1978). A family of hamiltonian structures associated with nonlinear integrable equations. Preprint of the Institute of Applied Mathematics, No. 136, Moscow.

[11] Gelfand, I.M. & Dorfman, I. Ya. (1979). Hamiltonian operators and algebraic structures associated with them. *Funct. Anal. & its Appl.* **13**(4), 13.

[12] Gelfand, I.M. & Dorfman, I. Ya. (1980). The Schouten bracket and hamiltonian operators. *Funct. Anal. & its Appl.* **14**(3), 71.

[13] Gelfand, I.M. & Dorfman, I. Ya. (1981). Hamiltonian operators and infinite-dimensional Lie algebras. *Funct. Anal. & its Appl.* **15**(3), 23.

[14] Kulish, P.P. & Reyman, A.G. (1978). The hierarchy of symplectic forms for the Schrödinger and Dirac equation on the axis. *Trans. Leningrad Branch Math. Inst.* **77**, 134.

[15] Lax, P.D (1976). Almost periodic solutions of the KdV equation. *SIAM Rev.* **18**, 351.

[16] Magri, F. (1978). A simple model of the integrable hamiltonian equations. *J. Math. Phys.* **19**, 1156.

[17] Mokhov, O.I. (1987). Hamiltonian differential operators and contact geometry. *Funct. Anal. & its Appl.* **21**(3), 53.

[18] Oevel, W. & Fuchssteiner, B. (1982). Explicit formulas for symmetries and conservation laws of the Kadomtsev-Petviashvili equation. *Phys. Lett. A* **88**, 323.

[19] Zakharov, V.E. & Faddeev, L.D. (1971). The KdV equation as a completely integrable system. *Funct. Anal. & its Appl.* **5**(4), 18.

C. Athorne

40. REDUCTION OF NON-LOCAL HAMILTONIAN STRUCTURES

We discuss the behaviour of Hamiltonian structures for a generalised AKNS hierarchy under componentwise reductions to equations of KdV and mKdV type, identifying those cases where a new local structure arises.

The isospectral deformations of the AKNS p,q- system [1,2].

$$\left\{ \partial_x + \frac{1}{2}\begin{pmatrix} a\zeta & 0 \\ 0 & -a\zeta \end{pmatrix} + \begin{pmatrix} 0 & q \\ p & 0 \end{pmatrix} \right\} \begin{pmatrix} \psi_1 \\ \psi_2 \end{pmatrix} = 0 \tag{1}$$

are the members of a hierarchy of partial differential evolution equations, the first two members of which are written:

$$\begin{aligned} aq,_{t_2} &= q,_{xx} - 2q^2p \\ -ap,_{t_2} &= p,_{xx} - 2p^2q \end{aligned} \tag{2}$$

$$\begin{aligned} a^2q,_{t_3} &= q,_{xxx} - 6pqq,_x \\ a^2p,_{t_3} &= p,_{xxx} - 6qpp,_x \end{aligned} \tag{3}$$

Each member of the hierarchy possesses a Hamiltonian structure,

$$\begin{pmatrix} q \\ p \end{pmatrix},_{t_n} = \begin{pmatrix} 0 & 1 \\ -1 & 0 \end{pmatrix} \begin{pmatrix} \delta/\delta q \\ \delta/\delta p \end{pmatrix} \mathbb{H}_n \equiv \omega_0(\mathbb{H}_n) \tag{4}$$

where,

$$\mathbb{H}_2 = \frac{1}{a}\int dx \; \{ -\frac{1}{2}\, p,_x q,_x - q^2p^2 \}$$

$$\mathbb{H}_3 = \frac{1}{a^2}\int dx \; \{ \frac{1}{2}\, p,_{xx}q,_x - \frac{1}{2}\, p,_x q,_{xx}$$

$$+ \frac{3}{2} q^2 pp,_x - \frac{3}{2} p^2 qq,_x \}$$

and so on. This structure is called (ultra-) local as it involves only the functional derivatives with respect to p and q.

The pde's (3), and indeed all the higher odd order flows, have as certain admissible reductions algebraic constraints on p and q which are preserved under the time evolutions. These are: (i) q = -1 and (ii) q = p . They give rise to the hierarchies corresponding to the KdV and mKdV equations :

$$u,_{t_3} = u,_{xxx} + 6uu,_x \tag{5i}$$

$$v,_{t_3} = v,_{xxx} + 6v^2 v,_x \tag{5ii}$$

Their Hamiltonian structures are local, that is, polynomial in

∂_x:

$$u,_{t_3} = \partial_x \, {\delta}/{\delta u} \int dx \, \{ -\frac{1}{2} u,_x^2 + u^3 \} \tag{6i}$$

$$= (\partial_x + 2u\partial_x + 2\partial_x u) \, {\delta}/{\delta u} \int dx \, \{ -\frac{1}{2} u^2 \} \tag{6i'}$$

$$v,_{t_3} = \partial_x \, {\delta}/{\delta v} \int dx \, \{ -\frac{1}{2} v,_x^2 + \frac{1}{2} v^4 \} \tag{6ii}$$

A natural question to ask is: How do these local (bi-) Hamiltonian structures arise from the AKNS structure under reduction?

The simplest generalisation of the AKNS hierarchy is as follows [3,4,5]. We take the linear problem,

$$\{ \partial_x + \zeta A + Q \} \Psi = 0 \tag{7}$$

where A and Q are (n+m)×(n+m) matrices of the forms:

$$A = \left(\begin{array}{c|c} \frac{an}{n+m} I_m & \\ \hline & \frac{-am}{n+m} I_n \end{array} \right) , \quad Q = \left(\begin{array}{c|c} & \mathbf{q} \\ \hline \mathbf{p} & \end{array} \right) \tag{8}$$

Here **q** and **p** are respectively m×n and n×m matrix potentials and the adjoint action of A is regular (invertible) on the space of matrices of the form Q. We then write $Q = Q^+ + Q^-$ where $(\mathrm{ad}A)Q^\pm = \pm aQ^\pm$. Such linear problems are naturally associated with hermitian symmetric spaces. Again the first two members of the hierarchy are written:

$$\pm aQ^\pm,_{t_2} = Q^\pm,_{xx} - 2[Q^\pm,[Q^\mp,Q^\pm]] \tag{9}$$

$$a^2 Q^\pm,_{t_3} = Q^\pm,_{xxx} - 3[Q^\pm,_x,[Q^\mp,Q^\pm]] \tag{10}$$

Their Hamiltonian structure is of the form :

$$Q^{\pm}_{,t_n} = \mathrm{ad}A\, \delta/_{\delta Q^{\mp}}\, \mathbb{H}_n \tag{11}$$

The odd numbered flows admit at least the reductions:

(i) Q^{-} = constant, to an mn component generalised KdV equation, and

(ii) $Q = Q^{T}$, to an mn component generalised mKdV equation. However it does not appear that these equations have, in general, local Hamiltonian structures [5,6] and so the question arises: When do generalised (m)KdV equations in this class have local (bi-) Hamiltonian structures ?

To answer these questions we recall the recursion operator for the p,q- system, writing ∂ for ∂_x:

$$\begin{pmatrix} q \\ p \end{pmatrix}_{,t_{n+1}} = \frac{1}{a} \begin{pmatrix} \partial - 2q\partial^{-1}p & -2q\partial^{-1}q \\ 2p\partial^{-1}p & -\partial + 2p\partial^{-1}q \end{pmatrix} \begin{pmatrix} q \\ p \end{pmatrix}_{,t_n} \tag{12}$$

$$\equiv \mathfrak{R} \begin{pmatrix} q \\ p \end{pmatrix}_{,t_n} \tag{13}$$

We may use this to construct from the single local structure (4) a lattice of non- local Hamiltonian structures connecting each time flow to each Hamiltonian:

$$\begin{pmatrix} q \\ p \end{pmatrix}_{,t_n} = \mathfrak{R}^{n-m}\omega_0(\mathbb{H}_m) \equiv \omega_{n-m}(\mathbb{H}_m) \tag{14}$$

This hierarchy of Hamiltonian structures is discussed in [7].

To effect the mKdV reduction put $q = v+\eta$ and $p = v-\eta$ where η is small. Then, as will be shown shortly and in more generality , $\mathbb{H}/_{2n+1} = O(\eta)$ and $\mathbb{H}_{2n} = \mathbb{H}_{2n}[v,v] + O(\eta^2)$. Consequently we lose the local AKNS Hamiltonian structure as $\eta \to 0$ but the ω_1 structure has O(1) term,

$$\begin{pmatrix} v \\ v \end{pmatrix}_{,t_{2n+1}} = \frac{1}{2a} \begin{pmatrix} \partial - 2v\partial^{-1}v & -2v\partial^{-1}v \\ 2v\partial^{-1}v & -\partial + 2v\partial^{-1}v \end{pmatrix} \begin{pmatrix} 0 & 1 \\ -1 & 0 \end{pmatrix} \begin{pmatrix} \delta/_{\delta v} \\ \delta/_{\delta v} \end{pmatrix} \mathbb{H}_{2n}[v,v]$$

in which the non- local terms cancel to leave the usual local structure:

$$v_{,t_{2n+1}} = \frac{1}{2a}\, \partial\, \delta/_{\delta v}\, \mathbb{H}_{2n}[v,v] \tag{15}$$

For the KdV reduction we put $p = -1 + \eta$ and note that again $\mathbb{H}_{2n+1} = O(\eta)$. Again we lose the AKNS local structure. If the $O(\eta)$ term in $\mathbb{H}_{2n}$ is $\mathbb{H}'_{2n}$ then $\delta/_{\delta\eta}\mathbb{H}'_{2n}$ is O(1) and the O(1) terms from the ω_1 structure give:

Reduction

$$\begin{pmatrix} q \\ 0 \end{pmatrix}_{,t_{2n+1}} = \frac{1}{a}\begin{pmatrix} \partial + 2q\partial^{-1} & -2q\partial^{-1}q \\ 2\partial^{-1} & -\partial - 2\partial^{-1}q \end{pmatrix}\begin{pmatrix} 0 & 1 \\ -1 & 0 \end{pmatrix}\begin{pmatrix} \frac{\delta \mathbb{H}}{\delta q}_{2n} \\ \frac{\delta \mathbb{H}'}{\delta \eta}_{2n} \end{pmatrix}$$

We may now solve for the η-derivative and obtain finally the usual third order structure:

$$q_{,t_{2n+1}} = -\frac{1}{2a}\left(\partial^3 + 2q\partial + 2\partial q \right) \frac{\delta \mathbb{H}}{\delta q}_{2n} \tag{16}$$

The other local stucture arises from the ω_{-1} structure which we write to O(1) as,

$$\frac{1}{a}\begin{pmatrix} \partial + 2q\partial^{-1} & -2q\partial^{-1}q \\ 2\partial^{-1} & -\partial - 2\partial^{-1}q \end{pmatrix}\begin{pmatrix} q \\ 0 \end{pmatrix}_{,t_{2n+1}} = \begin{pmatrix} \frac{\delta \mathbb{H}'}{\delta \eta}_{2n+2} \\ -\frac{\delta \mathbb{H}}{\delta q}_{2n+2} \end{pmatrix}$$

from which, without using the η-derivative at all:

$$q_{,t_{2n+1}} = -\frac{a}{2}\,\partial\, \delta/_{\delta q}\, \mathbb{H}_{2n+2} \tag{17}$$

Summarising these results: under reductions the local AKNS Hamiltonian structure ω_0 disappears and (i) for KdV reductions the non-local ω_1 and ω_{-1} structures collapse to local structures whereas (ii) for mKdV reductions ω only becomes local.

We now consider reductions of the generalised hierarchy, (11). The same picture of the hierarchy of non-local Hamiltonian structures arises where now the recursion operator is:

$$\mathfrak{R} = \left(\partial - (\mathrm{ad}Q)\partial^{-1}(\mathrm{ad}Q) \right) (\mathrm{ad}A)^{-1} \tag{18}$$

The linear problem (7) is rewritten as a matrix Riccati equation,

$$\gamma_{,x} + \zeta[A,\gamma] + Q - \gamma Q \gamma = 0 \tag{19}$$

where,

$$\gamma = \gamma^{+} + \gamma^{-} = \sum_{n=1}^{\infty} (a\zeta)^{-n}\gamma_n$$

and a recursion for γ_n yeilds for Hamiltonian densities the coefficients of the series:

$$\mathfrak{H} = \mathrm{Tr}(AQ\gamma) \tag{20}$$

In particular:

$$\mathbb{H}_n = \mathrm{const}_n \int_{-\infty}^{\infty} dx\, \mathfrak{H}_{n+1} \tag{21}$$

In addition it is useful to define the object $\mathfrak{K} = \mathrm{Tr}\,(Q\gamma)$ and easy to show, from (19), that it is a total x-derivative.

We wish to show that the reductions are admissible and Hamiltonian and to construct the local Hamiltonian structures. The first part of the following argument is a generalisation of an argument in [8] which suggests that we decompose $\mathfrak{H}$ into even and odd parts, $\mathfrak{H}^{(+)}$ and $\mathfrak{H}^{(-)}$, corresponding respectively to odd and even numbered Hamiltonian densities.

From consideration of the linear problem with Q^T replacing Q and the associated Riccati equation, one deduces the identities,

$$\gamma^{(\pm)}(Q^T) = \pm\, \gamma^{(\pm)T}(Q) \tag{22}$$

which lead to relations between the Hamiltonian densities:

$$\mathfrak{H}^{(\pm)}(Q^T) \pm \mathfrak{H}^{(\pm)}(Q) = \pm a_{\frac{n-m}{n+m}}\, \mathfrak{K} \cong 0$$

Here $\cong$ means equal up to total x-derivatives. Now put $Q = v + \eta$ where $v = v^T$ and $\eta = -\eta^T$ in which case $\mathfrak{H}^{(+)}(v) \cong 0$, implying that $\mathbb{H}_{2n+1} = 0$, and $\mathfrak{H}^{(-)}(v+\eta) \cong \mathfrak{H}^{(+)}(v-\eta)$, implying $\mathbb{H}_{2n}(v+\eta) = \mathbb{H}_{2n}(v) + O(\eta^2)$. This is just the behaviour indicated earlier for the mKdV type reduction.

The reduced ω_1 Hamiltonian structure is,

$$\frac{1}{2}\Big\{\delta_{il}\delta_{jk}\partial - v_{ir}\partial^{-1}v_{lr}\delta_{jk} - v_{sj}\partial^{-1}v_{sk}\delta_{il} + v_{ik}\partial^{-1}v_{lj} + v_{lj}\partial^{-1}v_{ik}\Big\}\frac{\delta}{\delta v_{kl}}$$

where k,j and r run over 1,..n and l,i and s run over 1,..m. The non-local part of this structure will vanish only if n = m = 1 as can be seen by performing contractions over j,k and i,l. Hence we conclude that, within this class, only the single component mKdV has a local Hamiltonian structure.

The KdV type reduction does not give rise to a relation of the form (22) but it is not difficult to show that again the odd numbered Hamiltonians vanish [9]. The even numbered Hamiltonians have a part of $O(\eta)$ under the reduction $\mathbf{p} = \mathbf{r} + \eta$, $\mathbf{r}$ constant, called $\mathbb{H}'_{2n}$. Inorder to solve for the η-derivative of $\mathbb{H}'_{2n}$ in the ω_1 structure as before , $\mathbf{r}$ has to be invertible and we may choose it to be the identity matrix by a gauge transformation. The Hamiltonian structure arising has a third order local part and a non-local part,

$$\frac{1}{2}\,[\mathbf{q},\partial^{-1}[\mathbf{q},\frac{\delta\mathbb{H}_{2n}}{\delta\mathbf{q}}]]$$

which will vanish only for n =m = 1. Thus we conclude that only the single component KdV has a third order local structure and hence that, within this class of generalisations, only the single component KdV is locally bihamiltonian.

Finally the KdV type reduction of the ω_{-1} Hamiltonian structure gives rise to the equation,

$$\mathbf{r}\partial^{-1}(\mathbf{q},_{t_{2n+1}})\mathbf{r} = -\frac{a}{2}\frac{\delta \mathbb{H}_{2n+2}}{\delta \mathbf{q}}$$

from which it is seen that $\mathbf{r}$ must be invertible (w.l.o.g the identity) and the usual first order structure persists. So only the generalised KdV with square matrix potentials have local (first order) structures. These are the equations,

$$\mathbf{q},_t = \mathbf{q},_{xxx} + 3\mathbf{q}\mathbf{q},_x + 3\mathbf{q},_x\mathbf{q}$$

studied in [10,11].

Note that one may make further reductions of this last case which preserve the local nature of the Hamiltonian structure. Admissible reductions are e.g. $\mathbf{q} = \pm\mathbf{q}^T$ and $\mathbf{q} = \mathbf{r}\mathbf{q}^T\mathbf{r}^{-1}$. Thus, for example,

$$\mathbf{q} = \begin{pmatrix} q_1 & q_2 \\ q_2 & q_3 \end{pmatrix} = \mathbf{q}^T$$

gives,

$$q,_{t_3} = \begin{pmatrix} 1 & 0 & 0 \\ 0 & 1/2 & 0 \\ 0 & 0 & 1 \end{pmatrix} \partial \, \nabla \mathbb{H} \tag{23}$$

where q now represents the vector $(q_1 \; q_2 \; q_3)^T$, ∇ the functional derivative with respect thereto, and $\mathbb{H}$ is given by :

$$\mathbb{H} = \int dx \; \{ -\tfrac{1}{2}q_1,_x^2 - q_2,_x^2 - \tfrac{1}{2}q_3,_x^2 - (q_1+q_3)^2 + 3(q_1+q_3)(q_1q_3-q_2^2) \}$$

An example of the second kind of reduction is,

$$\mathbf{r} = \begin{pmatrix} 0 & 1 \\ 1 & 0 \end{pmatrix} , \qquad \mathbf{q} = \begin{pmatrix} q_2 & q_3 \\ q_1 & q_2 \end{pmatrix}$$

for which, in the same notation as (23),

$$q,_{t_3} = \begin{pmatrix} 0 & 0 & 1 \\ 0 & 1/2 & 0 \\ 1 & 0 & 0 \end{pmatrix} \partial \, \nabla \mathbb{H} \tag{24}$$

and $\mathbb{H}$ is given by :

$$\mathbb{H} = \int dx \; \{ -q_2,_x^2 - q_1,_x q_3,_x - 2q_2^3 - 6q_1q_2q_3 \}$$

We close with one or two remarks.

In summary, within this class of generalisations, only the single component KdV is locally bihamiltonian and only the single component mKdV and square many component KdV are locally Hamiltonian.

The above does not constitute a proper proof of these things because the Hamiltonian structure of a reduced equation may not come from a Hamiltonian structure of the unreduced equation. One might afterall imagine a non-Hamiltonian system which has Hamiltonian reductions.

Finally, these results might be thought of as somewhat negative. Of more interest would be an AKNS-like system having the usual hierarchy of non-local Hamiltonian structures which reduces to locally multi-Hamiltonian systems of the variety discussed in [12].

I would like to thank V.S.Gerdjikov and Allan Fordy for discussions and the SERC for financial support.

References

[1] Zakharov, V.E. & Shabat, A.B. (1972) Exact theory of two-dimensional self focussing and one- dimensional self-modulation of waves in nonlinear media. *Soviet Physics JETP*, **34**, 62-69.

[2] Ablowitz, M.J., Kaup, D.J., Newell, A.C. & Segur, H. (1974) The inverse scattering transform - Fourier analysis for nonlinear problems. *Studies in Applied Mathematics*, **53**, 249-315.

[3] Fordy, A.P. & Kulish P.P. (1983) Nonlinear Shrödinger equations and simple Lie algebras. *Communications in Mathematical Physics*, **89**, 427-443.

[4] Gerdjikov, V.S. (1986) Generalised Fourier transforms for soliton equations. *Inverse Problems*, **2**, 51-74.

[5] Athorne, C. & Fordy, A.P. (1987) Generalised KdV and mKdV equations associated with symmetric spaces. *Journal of Physics A*, **20**, 1377-1386.

[6] Athorne, C. (1988) Some integrable equations associated with symmetric spaces. In: *Nonlinear Evolutions*, ed. Jérôme J-P Leon, pp 191-198. World Scientific, Singapore.

[7] Kulish, P.P. & Reyman, A.G. (1983) Hierarchy of symplectic forms for the Schrödinger and the Dirac equations on a line. *Journal of Soviet Mathematics*, **22**, 1627-1637.

[8] Novikov, S., Manakov, S.V., Pitaevskii, L.P. & Zakharov, V.E. (1984) *Theory of Solitons*, Plenum, New York.

[9] Athorne, C. (1988) Local Hamiltonian Structure of multi-component KdV equations. *Journal of Physics*, A,**21**, 4549-4556.

[10] Wadati, M. & Kamijo, T. (1974) On the extension of the inverse scattering method. *Progress in Theoretical Physics*, **52**, 397-414.

[11] Calogero, F. & Degasperis, A. (1977) Nonlinear evolution equations solvable by the spectral transform II. *Nuovo Cimento B*, **39**, 1-53.

[12] Antonowicz, M. & Fordy, A.P. (1987a) A family of completely integrable multi-Hamiltonian systems. *Physics Letters A*, **122**, 95-99; (1987b) Coupled KdV equations with multi-Hamiltonian structures. *Physica D*, **28**, 345-358.

I. Marshall

41. CANONICAL COORDINATES AND AFFINE LIE ALGEBRAS

We consider some concrete examples of integrable systems in affine Lie algebras and give canonical descriptions for them. One of these examples provides an interesting case of an integrable constrained system, another one has the form given by a natural Hamiltonian with magnetic terms.

1 **Introduction**

By applying a well known theorem of Adler, Kostant and Symes about the splitting of Lie algebras to loop algebras many known - and some unknown -results can be recovered. Perhaps the most notable of these is the recent construction of a Lax representation for the Kovalevsky top [1]. We concentrate here on the family of systems obtainable as the stationary flows of the N-wave equations in [2]. A smaller family - associated with hermitian symmetric spaces - was found in [3]; a description of the same results in terms of orbits, and by use of the AKS theorem, was given in [4] and in [5].

The possibilities for the form taken by the equations of motion are quite complicated. The whole procedure would be amply justified if any system were found to have a physical interpretation; to do this it may help to have a canonical coordinate description of the equations. I want to show here how to do this by discussing a few examples. It is also to be hoped that seeing the explicit interplay between a coordinate description (geometrically less clear, but physically more relevant), and the more sophisticated Lie algebraic description of the same system, may prove useful to people who feel uncomfortable with some of the more mathematics-bound parts of the literature. In the course of doing this we discover an example which seems to call for an extension of the usual notion of integrability according to Liouville's theorem.

2 **The general scheme - briefly**

We take as our starting point the Garnier system, this is the following set of o.d.e.s:

$$q_i = -2\sum_{k=1}^{n} q_k^2 q_i - \varepsilon_i q_i \tag{1}$$

where $\varepsilon_1,\dots,\varepsilon_n$ are n parameters.
These correspond to the Hamiltonian

$$H(q,p) = \frac{1}{2}\sum_{k=1}^{n} p_k^2 + \frac{1}{2}\Big(\sum_{k=1}^{n} q_k^2\Big)^2 + \frac{1}{2}\sum_{k=1}^{n} \varepsilon_k q_k^2 \tag{2}$$

with the canonical equations of motion for q and p,

$$\dot{q} = \frac{\partial H}{\partial p}, \quad \dot{p} = -\frac{\partial H}{\partial q} \tag{3}$$

In the standard way this is expressed by saying that (1) is equivalent to the Hamiltonian system (M,H,ω) with $M=\mathbb{R}^{2n}$, $\omega=dq\wedge dp$ and H given by (2). Then (1) or equivalently (3) is the flow given by the vector field χ_H where χ_H satisfies

$$\chi_H \lrcorner\, \omega = dH \tag{4}$$

Equations (1), (3) and (4) are all the same.

We can write this system as a Lax system (L,B); i.e. we can find matrices L and B such that (3) is the same as $\dot{L}=[B,L]$. This is a very convenient way of proving that the Garnier system is integrable [6].

In [3] the Garnier system is shown to be one of a larger class of integrable systems, all having a Lax representation and all corresponding to stationary flows of generalised NLS equations. In [5] we see how to represent a yet larger class of systems corresponding now to stationary flows of N-wave equations.The systems under study benefit in [5], from being endowed with a more elegant geometrical formulation; it is found that they are just an example of a now rather well known set-up due separately to Adler, Kostant, Symes.

In effect what we have is a large class of systems of o.d.e.s of the form of (1) and an equivalent class of Hamiltonian systems of the form of (4) - each of which has a Lax representation (this is a direct consequence of the fact that our manifolds are coadjoint orbits of Lie groups). The involutivity of the invariants of the Lax matrix is guaranteed in each case by the AKS theorem. In the perhaps rather vain hope of finding concrete physical applications for some of this (by now quite extensive) class of integrable systems, it is interesting to find the equivalent of the canonical variables (q,p) in (2). This corresponds to putting coordinates on the manifold in which ω takes the diagonal form $\omega=dx\wedge dy$.

This is quite easy to do. We consider some examples, all generated by the procedure given in [5].

Examples

1) $M = G\circ\alpha$ with G the Lie group generated by the matrices

$$\exp\Big(\sum_{i=1}^{r}\lambda^{-i}\xi_i\Big) \text{ with } \xi_i\in\mathfrak{su}(n+1),\ r\in\mathbb{N}_+$$

and

$$\alpha = i\lambda^2\mathrm{diag}(n;-1,\ldots,-1) + i\,\mathrm{diag}(a;\alpha_1,\ldots,\alpha_n) \qquad (a=-\sum_{k=1}^{n}\alpha_k)$$

and

$$g\circ\alpha = (g\alpha g^{-1})_+, \tag{5}$$

where + means discard the terms in negative powers of λ.

Consider the Lie algebra

$\mathfrak{G} = \{\sum_{i=r}^{s}\lambda^i\xi_i \,|\, \xi_i\in\mathfrak{su}(n+1),$ r,s integers$\}$ having the bracket

$$[\sum_{i=q}^{r}\lambda^i\ \xi_i, \sum_{i=s}^{t}\lambda^i\ \eta_i] = \sum_{i=q}^{r}\ \sum_{j=s}^{t}\lambda^{i+j}\ [\xi_i\eta_j]\ .$$

Define $<\ ,\ >:\mathfrak{G}\times\mathfrak{G}\longrightarrow\mathbb{C}$ by

$$<\sum_{i=q}^{r}\lambda^i\xi_i, \sum_{i=s}^{t}\lambda^i\eta_i> = \sum_{i+j=-1}\mathrm{tr}(\xi_i\eta_j) = \mathrm{tr}\Big(\sum_{i=q}^{r}\lambda^i\xi_i, \sum_{i=s}^{t}\lambda^i\eta_i\Big)\Big|_{-1},$$

where $|_{-1}$ means take the coefficient of λ^{-1}.

Let $\mathfrak{G}_\pm$ be the subalgebras,

$$\mathfrak{G}_+ = \{\sum_{i=r}^{s}\lambda^i\xi_i \,|\, \xi_i=0 \text{ if } i<0\}, \qquad \mathfrak{G}_- = \{\sum_{i=r}^{s}\lambda^i\xi_i \,|\, \xi_i=0 \text{ if } i\geq 0\}.$$

We can identify $\mathfrak{G}_-^*$ with $\mathfrak{G}_+$ using the non-degenerate inner product $<\ ,\ >$ defined above; we then interpret M as the coadjoint orbit of G in the dual $\mathfrak{G}_-^*$ of its Lie algebra $\mathfrak{G}_-$.

For $f\in C^\infty(\mathfrak{G})$ we define $\nabla_f(x)$, $x\in\mathfrak{G}$, by $\frac{d}{dt}\Big|_{t=0} f(x+ty) = <y,\nabla_f(x)>$ for any $y\in\mathfrak{G}$.

(eg. if $f(x)=\mathrm{tr}(x^2\lambda^k)|_{-1}$ then $\frac{d}{dt}\Big|_{t=0} f(x+ty) = \mathrm{tr}(2xy\lambda^k)|_{-1}$ and $\nabla_f(x)=2x\lambda^k$.)

Then ω is defined by

$$\omega(\sigma(X),\sigma(Y))(\beta) = <\beta,[X,Y]> \quad \beta\in M,\ X,Y\in\mathfrak{G}_- \tag{6}$$

σ is a mapping from $\mathfrak{G}_-$ to TM given by

$$(\sigma(x)f)(\beta) = \frac{d}{dt}\Big|_{t=0} f(e^{xt}\circ\beta)$$

where $\circ$ means the coadjoint action as in (5). Although σ is not injective (6) is well defined on M, see eg.[8].

The Poisson bracket corresponding to (6) is given by

$$\{f,g\}(\beta) = \langle\beta,[\nabla_f(\beta),\nabla_g(\beta)]\rangle \tag{7}$$

(This is a consequence of the fact that $\sigma(\nabla_f)=\chi_f$ where $\chi_f \lrcorner\, \omega = df$.)

Let $I(\mathfrak{G})$ be the set of Ad-invariant functions on $\mathfrak{G}$, ie. $f\in I(\mathfrak{G}) \Leftrightarrow f(g\circ X)=f(X)\ \forall X\in\mathfrak{G}, \forall g\in G$, where now $\circ$ is the Ad action given by $g\circ X=gXg^{-1}$.

The equation of motion for the Hamiltonian function $H\in I(\mathfrak{G})$ is given by

$$\dot\beta = -[\nabla_H(\beta)_+,\beta] \tag{8}$$

A typical point in M is

$$\beta = i\lambda^2\mathrm{diag}(n;-1,\ldots,-1) + \lambda\begin{pmatrix}0 & \mathbf{q}^\dagger\\ -\mathbf{q} & 0\end{pmatrix} + \begin{pmatrix}0 & \mathbf{p}^\dagger\\ -\mathbf{p} & 0\end{pmatrix} - \frac{i}{n+1}\begin{pmatrix}|\mathbf{q}|^2 & 0\\ 0 & -\mathbf{q}\mathbf{q}^\dagger\end{pmatrix} +$$

$$+\ i\mathrm{diag}(a;\alpha_1,\ldots,\alpha_n)$$

$$\mathbf{q},\mathbf{p}\in\mathbb{C}^n. \tag{9}$$

Clearly then we can expand ω in terms of the one-forms dq_i, dq_i^*, dp_i, dp_i^*. The easiest way to do this is to calculate the Poisson brackets. For this we need ∇q_i, ∇q_i^*, ∇p_i, ∇p_i^*:

It is easy to see that $q_i(\beta)=\mathrm{tr}\left(-\lambda^{-2}\begin{pmatrix}0 & \mathbf{i}^T\\ 0 & 0\end{pmatrix}\beta\right)\Big|_{-1}$, etc., with obvious notation, so that

$$\nabla q_i = -\lambda^{-2}\begin{pmatrix}0 & \mathbf{i}^T\\ 0 & 0\end{pmatrix},\quad \nabla p_i = -\lambda^{-1}\begin{pmatrix}0 & \mathbf{i}^T\\ 0 & 0\end{pmatrix},\quad \nabla q_i^* = -\nabla q_i^T,\quad \nabla p_i^* = -\nabla p_i^T.$$

Then for example

$$\{q_i,q_j^*\}(\beta) = \langle\beta,-\lambda^{-4}\begin{pmatrix}\delta_{ij} & 0\\ 0 & -\mathbf{j}\mathbf{i}^T\end{pmatrix}\rangle = -\mathrm{tr}\left(\beta\begin{pmatrix}\delta_{ij} & 0\\ 0 & -\mathbf{j}\mathbf{i}^T\end{pmatrix}\lambda^{-4}\right)\Big|_{-1} = 0$$

We find that $\{q_i,p_j^*\} = i(n+1)\delta_{ij}$, $\{q_i^*,p_j\} = -i(n+1)\delta_{ij}$, and all other Poisson brackets are zero. In this simple case then it follows that

$$\omega = -\frac{i}{n+1}(d\mathbf{q}\wedge d\mathbf{p}^* + d\mathbf{p}\wedge d\mathbf{q}^*) \tag{10}$$

In fact the flow corresponding to the Hamiltonian $\frac{1}{2}\mathrm{tr}(\beta^2\lambda^{-1})|_{-1}$ just gives the Garnier system on $\mathbb{C}^n$,

$$q_i = -|q|^2 q_i + \tfrac{1}{4}(n+1)(a-\alpha_i)q_i. \tag{11}$$

2) $M = G\circ\alpha$ as before:

G is the Lie group generated by $\left\{\exp\left(\sum_{i=1}^{r} \lambda^{-i}\xi_i\right),\ \xi_i \in su(n+1),\ r\in\mathbb{N}_+\right\}$,

$$\alpha = i\lambda^3 \mathrm{diag}(n;-1,\dots,-1) + \frac{i}{(n+1)^2}\begin{pmatrix}\Lambda_0 & 0\\ 0 & \bar{\Lambda}\end{pmatrix}$$

with $\Lambda_0 \in \mathbb{R}$, $\bar{\Lambda}^{\dagger}(=\bar{\Lambda}^{*T})=\bar{\Lambda} \in iu(n)$, and $\Lambda_0+\mathrm{tr}\bar{\Lambda}=0$.
In other words the only difference compared with 1) is a slightly different choice of α.
This time we find that a generic point of M is

$$\beta = i\lambda^3\mathrm{diag}(n;-1,\dots,-1) + \lambda^2\begin{pmatrix}0 & \mathbf{q}^{\dagger}\\ -\mathbf{q} & 0\end{pmatrix} + \lambda\begin{pmatrix}0 & \pi^{\dagger}\\ -\pi & 0\end{pmatrix} - \frac{i}{n+1}\lambda\begin{pmatrix}|\mathbf{q}|^2 & 0\\ 0 & -\mathbf{q}\mathbf{q}^{\dagger}\end{pmatrix} +$$

$$+ \begin{pmatrix}0 & \tau^{\dagger}\\ -\tau & 0\end{pmatrix} + \frac{i}{n+1}\begin{pmatrix}-\pi^{\dagger}q & 0\\ 0 & q\pi^{\dagger}\end{pmatrix} + \frac{i}{n+1}\begin{pmatrix}-q^{\dagger}\pi & 0\\ 0 & \pi\mathbf{q}^{\dagger}\end{pmatrix} + \frac{i}{(n+1)^2}\begin{pmatrix}\Lambda_0 & 0\\ 0 & \bar{\Lambda}\end{pmatrix} \qquad (12)$$

We have $\nabla q_i=-\lambda^{-3}N_i$, $\nabla q_i^{*}=\lambda^{-3}N_i^{T}$, $\nabla\pi_i=-\lambda^{-2}N_i$, $\nabla\pi_i^{*}=\lambda^{-2}N_i^{T}$,

$\nabla\tau_i=-\lambda^{-1}N_i$, $\nabla\tau_i^{*}=\lambda^{-1}N_i^{T}$, with $N_i = \begin{pmatrix}0 & \mathbf{i}^{T}\\ 0 & \mathbf{0}\end{pmatrix}$

and we get the Poisson brackets:

$$\{q_i,\tau_j^{*}\}=i\,\mathrm{tr}\left(\mathrm{diag}(n;-1,\dots,-1)\begin{pmatrix}-\delta_{ij} & 0\\ 0 & \mathbf{ij}^{T}\end{pmatrix}\right) =-i(n+1)\delta_{ij} = \{\tau_i,q_j^{*}\}.$$

$$\{\pi_i,\pi_j^{*}\} = -i(n+1)\delta_{ij}.$$

$$\{\pi_i,\tau_j^{*}\} = -\mathrm{tr}\left(\begin{pmatrix}0 & \mathbf{q}^{\dagger}\\ -\mathbf{q} & 0\end{pmatrix}\begin{pmatrix}-\delta_{ij} & 0\\ 0 & \mathbf{ij}^{T}\end{pmatrix}\right) = 0.$$

$$\{\tau_i,\tau_j^{*}\} = -\frac{i}{n+1}\mathrm{tr}\left(\begin{pmatrix}|\mathbf{q}|^2 & 0\\ 0 & -\mathbf{q}\mathbf{q}^{\dagger}\end{pmatrix}\begin{pmatrix}-\delta_{ij} & 0\\ 0 & \mathbf{ij}^{T}\end{pmatrix}\right) = \frac{i}{n+1}|q|^2\delta_{ij} + \frac{i}{n+1}q_iq_j^{*}$$

All others zero.

In these coordinates it is a difficult task to invert the above relations to write ω. However if we make the transformation

$$t=\tau+\frac{1}{2}\frac{1}{(n+1)^2}|q|^2q$$

we find the Poisson brackets:

$\{q_i,t_j^{*}\}=-i(n+1)\delta_{ij}=\{t_i,q_j^{*}\}$, $\{\pi_i,\pi_j^{*}\}=-i(n+1)\delta_{ij}$, all others zero,

and these are easy to invert, giving

$$\omega = \frac{i}{n+1}(dq_\wedge dt^\dagger + d\pi_\wedge d\pi^\dagger + dt_\wedge dq^\dagger). \tag{13}$$

In this case $H(\beta)=-\frac{1}{2}(\beta^2\lambda^{-2})|_{-1}$ gives the equation of motion;

$$\ddot{q} = -3|q|^2\dot{q} - 3(q^\dagger\dot{q})q - i\Lambda_0 q + i\bar{\Lambda}q \tag{14}$$

If we write $(\mathbb{R}^{6n},H,\omega)$ in coordinates it is convenient to replace (H,ω) by $\hat{H}=(n+1)^3H$ and $\hat{\omega}=(n+1)^3\omega$. Thus $\hat{H}=(\sigma^\dagger\rho+\rho^\dagger\sigma) + \frac{3}{2}|q|^2(\sigma^\dagger q+q^\dagger\sigma) - \Lambda_0|q|^2 + q^\dagger\bar{\Lambda}q$, and $\hat{\omega}= i(dq\wedge d\rho^\dagger+d\sigma\wedge d\sigma^\dagger+d\rho\wedge dq^\dagger)$.

Equation (14) raises I think, an interesting question (see [5]): Consider the equation

$$\ddot{\eta}_i = -\frac{3}{4}\sum_{k=1}^{n}\eta_k^4\eta_i - \gamma_i\eta_i^{-3} \qquad (\eta_i\in\mathbb{R}) \tag{15}$$

with $\gamma_1,\ldots,\gamma_n$ some constants. Let us also suppose that this system has the initial condition

$$\{\eta(0)=\alpha,\ \dot{\eta}(0)=\beta\}. \tag{16}$$

Set $x_i = \eta_i^2$, then we have

$$\ddot{x}_i = \frac{1}{2}x_i^{-1}\dot{x}_i^2 - \frac{3}{2}\sum_{k=1}^{n}x_k^2x_i - 2\gamma_i x_i^{-1}$$

$$\Rightarrow \qquad \ddot{x}_i = -3\sum_{k=1}^{n}x_k\dot{x}_k x_i \quad -3\sum x_k^2\dot{x}_i \tag{17}$$

Now we can say that (17), which is in turn equivalent to (15) (together with the appropriately corresponding initial condition), is equivalent to

$$\ddot{x}_i = -3\sum_{k=1}^{n}(x_k^2+y_k^2)\dot{x}_i - 3\sum_{k=1}^{n}(x_k\dot{x}_k+y_k\dot{y}_k)x_i + 3\sum_{k=1}^{n}(x_k\dot{y}_k-y_k\dot{x}_k)y_i$$

$$\ddot{y}_i = -3\sum_{k=1}^{n}(x_k^2+y_k^2)\dot{y}_i - 3\sum_{k=1}^{n}(x_k\dot{x}_k+y_k\dot{y}_k)y_i - 3\sum_{k=1}^{n}(x_k\dot{y}_k-y_k\dot{x}_k)x_i \tag{18}$$

with the initial condition

$$\{x_i(0)=\alpha_i,\quad \dot{x}_i(0)=2\alpha_i\beta_i,\quad x_i(0)=2\alpha_i\beta_i^2-\ \tfrac{3}{2}\sum\alpha_k^2\alpha_i-\ 2\gamma_i\alpha_i^{-1},\quad y_i(0)=0,$$
$$\dot{y}_i(0)=0,\ y_i(0)=0\}. \tag{19}$$

It can easily be seen by setting $\mathbf{q}=\mathbf{x}+i\mathbf{y}$ that (18) is the same as (14) with $\Lambda_o=0$, $\bar{\Lambda}=\mathbf{0}$. Hence (18) is an integrable system, i.e. it can be solved by quadratures for an arbitrary initial condition;

in particular for the one given above, leading to a solution by quadratures of (15) for an arbitrary initial condition (α,β). It is not however at all clear how to write (15) as a Hamiltonian system intrinsically and it certainly is not obtained by restricting $H=(\sigma^\dagger\rho+\rho^\dagger\sigma)+\frac{3}{2}|q|^2(\sigma^\dagger q+q^\dagger\sigma), \omega=i(dq\wedge d\rho^\dagger+d\sigma\wedge d\sigma^\dagger+d\rho\wedge dq^\dagger)$ to the submanifold

$$\{\mathrm{Im}q_i=\mathrm{Re}\sigma_i=\mathrm{Im}\rho_i=0,\ F_i=q_i\rho_i-\mathrm{Im}(\sigma_i)^2-2\gamma_i=0\}.$$

This is just

$$\{\mathrm{Im}q=\mathrm{Im}\dot{q}=\mathrm{Im}\ddot{q}=0,$$

$$\mathrm{Re}\ddot{q}_i=\frac{1}{2}(\mathrm{Re}q_i)^{-1}(\mathrm{Re}\dot{q}_i)^2-\frac{3}{2}\sum_{k=1}^{n}((\mathrm{Re}q_k)^2\mathrm{Re}q_i-2\gamma_i(\mathrm{Re}q_i)^{-1}\}$$

I suggest that the above procedure justifies calling (15) <u>integrable</u>, even though it may not itself be a Hamiltonian system; ie.such that $\exists H(Q,P)$ so that

$$\left[\dot{Q}=\frac{\partial H}{\partial P},\ \dot{P}=-\frac{\partial H}{\partial Q}\right]\Leftrightarrow\left[\ddot{\eta}_i=-\frac{3}{4}\sum_{k=1}^{n}\eta_k^4\eta_i-\gamma_i\eta_i^{-3}\right]\text{for some mapping }(Q,P)\longrightarrow\eta.$$

The geometrical picture is as follows:

Suppose we have a manifold N, dimN=2n, and consider the dynamical system on N given by

$$\dot{x}=v(x)\qquad\text{for some vector field } v \text{ on N.}$$

Suppose there is an embedding $\Psi:N\longrightarrow M$ of N into a symplectic manifold (M,ω), ie. $\exists\ \phi_1,\ldots,\phi_r\in C^\infty(M)$, where $r=2(m-n)$ and $2m=\dim M$, such that the submanifold given by $\{\phi_i=0, i=1,\ldots,2r\}$ is the same as $\mathrm{Im}\Psi$.

Suppose that there exists a function $H\in C^\infty(M)$ satisfying the conditions $X_H|_{\Psi(N)}\phi_i=0$ $i=1,\ldots,2r$ where $X_H\lrcorner\,\omega=dH$, and $d\Psi\cdot v=X_H|_{\Psi(N)}$.

(To compare this with the example, v is to be compared with (15) and the ϕ's are $\mathrm{Im}q_i,\mathrm{Re}\sigma_i,\mathrm{Im}\rho_i,F_i$.)

Call the system $\dot{x}=v(x)$ <u>integrable</u> if (M,ω,H) is integrable - ie. $\exists K_1,\ldots,K_n\in C^\infty(M)$ with $X_HK_i=0$ and $X_{K_i}K_j=0$ $\forall i,j$.

All this means that $\dot{x}=v(x)$ is a kind of constrained system on M. The above is not the same as the procedure described by Moser [9]; we do not require that $\det\{\phi_i,\phi_j\}=\det(X_{\phi_i}\phi_j)\neq0$ -making N a symplectic submanifold- and $\phi_1,\ldots\phi_r$ satisfying $\{\phi_i,\phi_j\}=0$ and $\{H,\phi_i\}=0$ $i=1,\ldots,r$; we do require however that $\{H,\phi_j\}=0$ on N for $i=1,\ldots,2r$.

The reader may find the arguments used in the example -making use of initial conditions- more accessible. The point is that the system $\dot{p}=X_H(p)$ may have its flow contained in a submanifold $N\subset M$ as long as it starts off in N. Meanwhile the flow in N considered *intrinsically* without reference to its inclusion in M may not be Hamiltonian at all - let alone integrable. Indeed N may not even be a symplectic submanifold of (M,ω) (although it may have an independent -and irrelevant- symplectic structure of its own).

Note that in the case of Hamiltonian flow in the dual of a Lie algebra, as in the example, what is being suggested above is just what arises for certain (non generic) choices of initial condition $x(0)=x_0$ if the symplectic leaf containing x_0 (=the coadjoint orbit through x_0) is not identified. In such a situation the foregoing argument is equivalent to the naïve statement that the equation of motion corresponding to the flow through x_0 is integrable if the generic case is integrable. It appears, for example therefore, that we are justified in calling the vector KdV and MKdV equations discussed in [7] integrable, as they arise in just the way described.

3) $M = G\circ\alpha$:

G is generated by $\exp\left(\sum_{i=1}^{r}\lambda^{-i}\xi_i\right)$, $\xi_i\in su(n+2)$,

$$\alpha = \frac{i}{n+2}\lambda^2(\varepsilon_1+\varepsilon_2)\mathrm{diag}\left(\frac{(n+1)\varepsilon_1-\varepsilon_2}{\varepsilon_1+\varepsilon_2}, \frac{-\varepsilon_1+(n+1)\varepsilon_2}{\varepsilon_1+\varepsilon_2}, -1,\dots,-1\right)$$

$$+\, i\begin{pmatrix} \Lambda_{-1} & 0 & 0 \\ 0 & \Lambda_0 & 0 \\ 0 & 0 & \bar{\Lambda} \end{pmatrix}$$

$\varepsilon_1,\varepsilon_2 \in \mathbb{R}$, $\Lambda_{-1},\Lambda_0 \in \mathbb{R}$, $\bar{\Lambda}^{\dagger}=\bar{\Lambda} \in iu(n)$, with $\Lambda_{-1}+\Lambda_0+\mathrm{tr}\bar{\Lambda}=0$.

A generic point of M is

$$X = \alpha + \lambda\begin{pmatrix} 0 & a^* & x^{\dagger} \\ -a & 0 & y^{\dagger} \\ -x & -y & \mathbf{0} \end{pmatrix} + \begin{pmatrix} 0 & \chi^* & \xi^{\dagger} \\ -\chi & 0 & \eta^{\dagger} \\ -\xi & -y & \mathbf{0} \end{pmatrix} + \frac{i}{\varepsilon_1-\varepsilon_2}\begin{pmatrix} -|a|^2 & 0 & 0 \\ 0 & |a|^2 & 0 \\ 0 & 0 & \mathbf{0} \end{pmatrix}$$

$$+\frac{i}{\varepsilon_1}\begin{pmatrix} -x^{\dagger}x & 0 & 0 \\ 0 & 0 & 0 \\ 0 & 0 & xx^{\dagger} \end{pmatrix} + \frac{i}{\varepsilon_2}\begin{pmatrix} 0 & 0 & 0 \\ 0 & -y^{\dagger}y & 0 \\ 0 & 0 & yy^{\dagger} \end{pmatrix}$$

Let $M = \begin{pmatrix} 0 & 1 & 0^T \\ 0 & 0 & 0^T \\ 0 & 0 & \mathbf{0} \end{pmatrix}$, $N_i = \begin{pmatrix} 0 & 0 & \mathbf{i}^T \\ 0 & 0 & 0^T \\ 0 & 0 & \mathbf{0} \end{pmatrix}$, $L_i = \begin{pmatrix} 0 & 0 & 0^T \\ 0 & 0 & \mathbf{i}^T \\ 0 & 0 & \mathbf{0} \end{pmatrix}$.

We get $\nabla a=-\lambda^{-2}M$, $\nabla a^*=\lambda^{-2}M^T$, $\nabla x_i=-\lambda^{-2}N_i$, $\nabla x_i^*=\lambda^{-2}N_i^T$, $\nabla y_i=-\lambda^{-2}L_i$, $\nabla y_i^*=\lambda^{-2}L_i^T$, $\nabla\chi=-\lambda^{-1}M$, $\nabla\chi^*=\lambda^{-1}M^T$, $\nabla\xi_i=-\lambda^{-1}N_i$, $\nabla\xi_i^*=\lambda^{-1}N_i^T$, $\nabla\eta_i=-\lambda^{-1}L_i$, $\nabla\eta_i^*=\lambda^{-1}L_i^T$,

and thence the Poisson bracket relations:

$$\{a,\chi^*\}=-i(\varepsilon_1-\varepsilon_2),\quad \{x_i,\xi_j^*\}=-i\varepsilon_1\delta_{ij},\quad \{y_i,\eta_j^*\}=-i\varepsilon_2\delta_{ij},\quad \{\chi,\xi_i^*\}=y_i,$$
$$\{\chi,\eta_i\}=-x_i,\quad \{\xi_i,\eta_j^*\}=a\delta_{ij},$$

together with their complex conjugates. All others are zero.

This time we need to define

$$\hat{\xi}_i = \xi_i + \frac{i}{\varepsilon_1-\varepsilon_2}ay_i,\qquad \hat{\eta}_i = \eta_i + \frac{i}{\varepsilon_1-\varepsilon_2}a^*x_i,$$

to diagonalise the above.

It is convenient to introduce the following change of coordinates:

$$a=-i(\varepsilon_1-\varepsilon_2)b,\quad x=-i\varepsilon_1u,\quad y=-i\varepsilon_2v$$

Then $\omega = (db\wedge d\chi^*-d\chi\wedge db^*) + (du\wedge d\hat{\xi}^\dagger-d\hat{\xi}\wedge du^\dagger) + (dv\wedge d\hat{\eta}^\dagger-d\hat{\eta}\wedge dv^\dagger)$

The Hamiltonian $H=-\frac{1}{2}\mathrm{tr}(\lambda^{-1}X^2)|_{-1}$ is

$$H = |\chi|^2+ (\hat{\xi}^\dagger\hat{\xi}) + (\hat{\eta}^\dagger\hat{\eta}) -i\varepsilon_2b^*(v^\dagger\hat{\xi}) +i\varepsilon_2b(\hat{\xi}^\dagger v) +\varepsilon_2^2|b|^2(v^\dagger v)$$
$$+i\varepsilon_1b(u^\dagger\hat{\eta}) -i\varepsilon_1b^*(\hat{\eta}^\dagger u) +\varepsilon_1^2|b|^2(u^\dagger u) +(\varepsilon_1-\varepsilon_2)^2|b|^4$$
$$+\varepsilon_1^2(u^\dagger u)^2 +\varepsilon_2^2(v^\dagger v)^2 +(\varepsilon_1-\varepsilon_2)\varepsilon_1|b|^2(u^\dagger u) -(\varepsilon_1-\varepsilon_2)\varepsilon_2|b|^2(v^\dagger v)$$
$$+\varepsilon_1\varepsilon_2(u^\dagger v)(v^\dagger u) -(\varepsilon_1-\varepsilon_2)(\Lambda_{-1}-\Lambda_0)|b|^2 -\varepsilon_1\Lambda_{-1}(u^\dagger u) +\varepsilon_1u^\dagger\bar{\Lambda}u$$
$$-\varepsilon_2\Lambda_0(v^\dagger v) +\varepsilon_2v^\dagger\bar{\Lambda}v$$

giving the equations of motion:

$$\ddot{b} = i\varepsilon_2v^\dagger\dot{u} +i\varepsilon_1\dot{v}^\dagger u -2(\varepsilon_1-\varepsilon_2)^2|b|^2b -(\varepsilon_1-\varepsilon_2)\varepsilon_1b(u^\dagger u)$$
$$+(\varepsilon_1-\varepsilon_2)\varepsilon_2b(v^\dagger v) +(\varepsilon_1-\varepsilon_2)(\Lambda_{-1}-\Lambda_0)b$$

$$\ddot{u} = i\varepsilon_2\dot{b}v - i(\varepsilon_1-\varepsilon_2)b\dot{v} - 2\varepsilon_1^2(u^\dagger u)u - \varepsilon_1\varepsilon_2(v^\dagger u)v - (\varepsilon_1-\varepsilon_2)\varepsilon_1|b|^2u$$

$$+\varepsilon_1\Lambda_{-1}u \qquad\qquad -i\varepsilon_1\bar{\Lambda}u$$

$$\ddot{v} = -i\varepsilon_1\dot{b}^*u - i(\varepsilon_1-\varepsilon_2)b\dot{v} - 2\varepsilon_2{}^2(v^\dagger v)v + (\varepsilon_1-\varepsilon_2)\varepsilon_2|b|^2v$$

$$+\varepsilon_1\varepsilon_2(u^\dagger v)u + \varepsilon_2\Lambda_0 v - \varepsilon_2\bar{\Lambda}v$$

A similar calculation starting off with $M=G\circ\alpha$, G generated by $\exp(\sum_{i=1}^{r}\lambda^{-i}\xi_i)$, $\xi_i\in su(n+2)$, and $\alpha = \frac{i}{n+2}\lambda(\varepsilon_1+\varepsilon_2)\mathrm{diag}\left(\frac{(n+1)\varepsilon_1-\varepsilon_2}{\varepsilon_1+\varepsilon_2}, \frac{-\varepsilon_1+(n+1)\varepsilon_2}{\varepsilon_1+\varepsilon_2}; -1,\ldots -1\right)$ with $H(\beta)=-\frac{1}{3}\mathrm{tr}(\lambda^{-1}\beta^3)|_{-1}$, gives the well-known 3-wave interaction system, see [10].

As this example shows, it is possible to produce quite complicated systems with many free parameters. In this case the Hamiltonian function has terms which are linear in momentum as required for motion in an electromagnetic field. (This also comes about by starting off with $M=G\circ\alpha$ and $\alpha=\lambda^2A+\lambda B+C$, instead of $\alpha=\lambda^2A+C$ as in all the examples above.)

References

[1] A.G.Reyman & M.Semenov-Tian-Shansky, *Lett.Math.Phys.* **14** (1987) 55-61.

[2] A.P.Fordy & P.Kulish, *Comm.Math.Phys.* **89** (1983) 427-443.

[3] A.P.Fordy, S.Wojciechowski & I.Marshall, *Phys.Lett.A* **113** (1986) 395.

[4] A.G.Reyman, Proceedings of the Steklov Institute (1986) **155**.

[5] I.Marshall, *Phys.Lett.A* **127** (1988) 19-26.

[6] S.Wojciechowski, *Physica Scripta* **31** (1985) 433-438.

[7] C.Athorne & A.P.Fordy, *J.Phys.A* **20** (1987) 1377-1386.

[8] V.Guillemin & S.Sternberg, *Symplectic Techniques in Physics,* Cambridge University Press (1984).

[9] J.Moser, *Integrable Hamiltonian Systems and Spectral Theory*, Lezioni Fermiane 1981. Published by the Scuola Normale Superiore Pisa

[10] R.K.Bullough, Z.Jiang & S.Wojiechowski, *Phys.Lett.A* **117** (1986) 399-404.

A. Kundu and B. Basu Mallick

42. ON SOME PROBLEMS OF THE KORTEWEG-DE-VRIES EQUATION

"It isn't that they can't see the solution. It is that they can't see the problem."

G.K. Chesterton. The Scandal of Father Brown
"The Point of a Pin"

Some recent problems on the Poisson bracket (PB) structure and the integrability of the KdV equation are analysed and their possible solutions are found. In particular, the required modifications of the PB proposed before are obtained in a systematic way through an alternative r-s matrix approach exploring also the topological reasons and the source behind such modifications. Some paradoxes regarding the action-angle variables of the KdV system are resolved by extending the earlier results. An analogous Levinson type theorem is proved for this system and the corresponding modified PB's are shown as the constrained Dirac bracket pointing out the errors in the existing literature.

1 Introduction

The Korteweg-de-Vries equation, the first equation recognized as an integrable system more than twenty years ago [1,2], is undoubtedly the most famous in the integrable family of non-linear systems. A vast amount of research work has been devoted to this single model and the leading sections of almost all the monographs and review articles on soliton physics, whose number is already significant are represented by this particular equation. The KdV equation is known (i) to be an integrable system with soliton solutions; (ii) to possess infinite set of conserved quantities in involution; (iii) to be a Hamiltonian system with a definite Poisson bracket structure allowing action-angle variables, etc., etc.

So where are the **problems** of the KdV system? Before answering this question let us first take a brief look into the known **solutions** [3].

1.1 *Solutions to the KdV system*

It is well known that the KdV equation

$$u_t + u_{xxx} - 6uu_x = 0 \qquad (1)$$

for rapidly decreasing fields at infinities is an integrable system with an infinite set of conserved charges $\{C_n\}$, $n = 1,2,3,\ldots\ldots\ldots\ldots\ldots$ given by

$$C_1 = \int u dx,\ C_2 = \int u^2 dx,\ C_3 = \int (\tfrac{1}{2} u_x^2 + u^3) dx \qquad (2)$$

etc. One may associate a linear Schrödinger operator

$$(-\partial_{xx}^2 + u)\Psi = \lambda^2 \Psi \qquad (3)$$

to the KdV equation and solve it by inverse scattering transform. It is shown that the field solutions have a one to one correspondence with the scattering data and canonical action angle variables are present, with discrete modes associated with solitonic sectors and the continuous modes with the radiations. The scattering element $a(\lambda)$, being a constant in time, yields the infinite set of conserved quantities $\{C_n\}$ through $\arg a(\lambda) = \sum_{n=1}^{\infty} C_n/(2i\lambda)^{2n-1}$. Using the dispersion relation

$$\arg a(\lambda) = \frac{1}{i} \left[\sum_{l=1}^{N} \ln \frac{\lambda - i\text{æ}_l}{\lambda + i\text{æ}_l} \right] - \frac{1}{\pi} \int_{-\infty}^{\infty} \frac{\ln|a(\lambda')|}{\lambda' - \lambda} d\lambda' \qquad (4)$$

the conserved quantities can be expressed through the action variables $p_l = 2\text{æ}_l^2$ and $p(\lambda) = \frac{4\lambda}{\pi} \ln|a(\lambda)|$ only, as for example (4) yields

$$C_1 = \sum_{1}^{N} (2p_l)^{1/2} - \int_{0}^{\infty} (2\lambda)^{-1} p(\lambda) d\lambda. \qquad (5)$$

Conventionally the Poisson bracket of the KdV system is given by the Gardner bracket

$$\{f,g\}_G = \frac{1}{2} \int_{-\infty}^{\infty} dx \left[\frac{\delta f}{\delta u(x)} \left(\frac{\delta g}{\delta u(x)} \right)_x - \left(\frac{\delta f}{\delta u(x)} \right)_x \frac{\delta g}{\delta u(x)} \right] \qquad (6)$$

leading to

$$\{u(x), u(y)\}_G = \frac{1}{2} (\delta_x(x-y) - \delta_y(x-y)) \qquad (7)$$

for the KdV fields. The corresponding Hamiltonian may be given by $H = C_3$.

Moreover, the Gardner bracket (6) was shown [4,5] to be a Dirac bracket with constrained dynamics.

All these things can be found in great detail in the existing literature. The above investigations on the KdV systems were believed to be exhaustive and all solutions to this system completely known, until recently, when Faddeev and Takhtajan [6] discovered the problems with the Poisson structure of the KdV equation.

2 The problems

It was shown by Faddeev and Takhtajan [6] that the Gardner bracket (6) for the KdV system violates the Jacobi identity (JI) and hence, it cannot be a proper bracket of the KdV system. As a remedy they suggested a modified bracket satisfying the JI in the form

$$\{f,g\}_{FT} = \{f,g\}_G + \frac{\epsilon}{2}(f_+g_- - f_-g_+); \quad \epsilon = +1, \tag{8}$$

where $f_\pm \equiv \lim_{L\to\pm\infty} \frac{\delta f}{\delta u(L)}$. But this in turn created other problems. Firstly C_1 turns out to be an annihilator for this bracket. Secondly, the action-angles exhibit noncanonical nature along with a PB coupling between discrete and continuous modes in the form $\{p(\lambda), q_n\} = \frac{4}{\sqrt{p_n}}\lambda\delta(\lambda)$. But the latter relation leads to some paradoxical conclusions, since the situation $p \equiv 0$, $q_n \neq 0$ turns out to be an impossible one, which means apparently the nonexistence of pure soliton sectors in the KdV system. Subsequently, however Arkadiev et al. [7] guessed yet another PB: $\{\ ,\ \}_{APP}$, which may be given by (8) but with $\epsilon = -1$ and is free from the above drawbacks.

To clarify this confusing situation we put forward the following natural questions.

(i) Can the required modifications of the PB be found in a systematic way without involving any guess work?

(ii) Can one find any deeper reason behind the existence of three different PB's?

(iii) Do any other objects along with the PB also get modified and can the paradoxical conclusion drawn from Faddeev and Takhtajan [6] be resolved?

(iv) Does a Levinson type theorem exist for the KdV system due to its association with the Schrödinger operator?

(v) Is the Gardner bracket, as shown in the existing literature, possible to express as a Dirac bracket in the KdV system, since the former one violates the JI while the latter one by construction must satisfy it? Also whether the FT and APP brackets are representable as the constrained Dirac brackets.

The aim of our investigation is to analyse these problems of the KdV system and answer the questions posed above.

3 Exploring the answers to the peculiarities of the KdV system

In this section we present our results seeking answers to the above questions concerning the peculiarities of the KdV system.

3.1 *Systematic derivation of the modified PB's and symmetry criteria behind their existence*

We adopt here a somewhat non-conventional but highly effective r-s matrix approach [8,9,10] for deriving the required PB's in a systematic way. Our key observation is that canonical structure of the KdV system is non-ultralocal (i.e. contains derivative of δ-functions) and at finite interval: $x \in [-L,L]$ the JI is satisfied even for the Gardner bracket. Therefore, we start with the evaluation of the Poisson brackets of the associated monodromy matrices $T^x_y(\lambda)$ satisfying the linear equation

$$\partial_x T^x_y(\lambda) = \upsilon(x,\lambda)\, T^x_y(\lambda), \quad T^x_x(\lambda) = 1 \tag{9a}$$

with

$$\upsilon(x,\lambda) = i(\lambda\sigma^3 - u(x)\sigma^- + \sigma^+) \tag{9b}$$

and subsequently take the limit $L \to \infty$, after demanding some symmetry criteria. Finally, we arrive at the PB's between scattering elements yielding the canonical action angle variables. Briefly the steps involved are as follows (for details see [11])

$$\{T^x_y(\lambda) \otimes_, T^{x'}_{y'}(\mu)\} = \int_y^x dz \int_{y'}^{x'} dz'\, T^x_z(\lambda) \otimes T^{x'}_{z'}(\mu) \{\upsilon(z,\lambda) \otimes \upsilon(z',\mu)\} T^z_y(\lambda) \otimes T^{z'}_{y'}(\mu), \tag{10}$$

which with the use of (7) and (9b) and after extracting the integrals reduces for $T^L_{-L} \equiv T_L$ to

$$\{T_L \otimes_, \tilde{T}_L\}_{N\text{-per.}} = r_L T_L \otimes \tilde{T}_L - T_L \otimes \tilde{T}_L r_{-L} \tag{11a}$$

for a **non-periodic** model and

$$\{T_L \otimes_, \tilde{T}_L\}_{\text{per.(anti-per)}} = r_L T_L \otimes \tilde{T}_L - T_L \otimes \tilde{T}_L\, r_{-L} + \eta[(T_L \otimes \mathbf{1}) s_L (\mathbf{1} \otimes \tilde{T}_L) - (\mathbf{1} \otimes \tilde{T}_L) s_L (T_L \otimes \mathbf{1})], \tag{11b,c}$$

with $\eta = +1$ for **periodic**: $u(x+L) = u(x)$ and $\eta = -1$ for

anti-periodic: $u(x+L) = -u(x)$ models, where

$$r_L = -\frac{1}{2(\lambda+\mu)}\,[\sigma^3\otimes\sigma^- - \sigma^-\otimes\sigma^3 - (\lambda-\mu)\sigma^-\otimes\sigma^- + \mathbb{P}/(\lambda-\mu)] \tag{12a}$$

and

$$s_L = \frac{1}{2}\,\sigma^- \otimes \sigma^-\,, \qquad \mathbb{P} = \frac{1}{2}(\mathbf{1} \otimes \mathbf{1} + \sigma^a \otimes \sigma^a). \tag{12b}$$

Now, passing on to the limit $L \to \infty$, for the scattering matrices $T(\lambda) = \begin{pmatrix} a & \bar{b} \\ b & \bar{a} \end{pmatrix} = E_+^{-1}(x,\lambda)T_y^x(\lambda)E_-(y,\lambda)$, where $E_\pm$ are the asymptotic values of the Jost functions, we get three different PB's as the limits of (11a-c) with different symmetry criteria. It is curious to note that at the infinite limit (11a) yields $\{\ ,\ \}_G$ while (11b) gives exactly $\{\ ,\ \}_{FT}$ and (11c) reduces to $\{\ ,\ \}_{APP}$. This systematic derivation of three different PB's seems to be possible only because the models somehow "remember" their topological properties present at the finite interval, even in the infinite limit, which are difficult to guess otherwise. The consistent way of getting the modified PB's as explained above allows us also to analyse the Poisson structure in detail revealing the fact that the results of Faddeev & Takhtajan [6] are valid only when the scattering data are singular at $\lambda = 0$, i.e. when $m \neq 0$ (see (13)). But these sectors always contain continuous modes [12] and consequently, the situation of $p \equiv 0$, $q_n \neq 0$, i.e. existence of pure solitons remains outside the domain of applicability of the earlier results. On the other hand we have found that in the case $m=0$ the earlier results get modified and the action-angles become canonical and since the pure solitons "live" only in this sector, their separate existence again does not lead to any paradox.

3.2 *Source of peculiarities of the KdV Poisson brackets*

It has been found that the source of peculiarities of the KdV system is the singularity behaviour of the scattering elements at $\lambda=0$ given by [13]

$$a(\lambda) = -\frac{m}{2i\lambda} + 1 - n(\lambda), \quad b(\lambda) = \frac{m}{2i\lambda} + \rho(\lambda). \tag{13}$$

As a result of this singularity at $\lambda = 0$ a term like $-\frac{\mathrm{Re}}{\pi}\int_c \frac{\ln a(\lambda)d\lambda}{\lambda'-\lambda}$, where c is a small half circle around $\lambda = 0$ may be added in general in the dispersion relation (4) for $a(\lambda)$. Consequently, the conserved quantity C_1 also gets modified. Note that in the strong limit these modifications vanish irrespective of the singularity of $a(\lambda)$. However, when arg $a(\lambda)$ or C_1 are involved in a PB operation the limits shoud be taken in the final stage and interestingly, here this leads to nontrivial contributions. In handling these types of singular theories one should always be careful about extracting the limits and whenever there is a possibility of discrepancies the limiting values should be taken only at the final step after carrying out all the operations. We find that the singularity behaviour (13) in FT model yields the action angle relation in the form

$$\{p(\lambda),\ q(\mu)\} = \delta(\lambda-\mu) - \theta(\lambda)\frac{\lambda}{2\mu}\delta(\lambda)$$

and

$$\{p(\lambda).q_n\} = \theta(\lambda)\frac{\lambda\delta(\lambda)}{\sqrt{p_n}} .$$

Here also the limit should be taken after evaluating the integrals with singular functionals, if finally such an operation yields a nontrivial result. It is interesting to note that for $m\neq0$ $\lim_{\lambda\to0}\theta(\lambda) = -1$, while for $m=0$, $\lim_{\lambda\to0}\theta(\lambda) = 0$ even after the required integration with allowed singular functionals [11]. Therefore for the trivial m sector, which includes also pure solitons, the action angles become canonical with no mixing between continuous and discrete modes.

3.3 *Levinson-type therem for the KdV system*

Another interesting observation is that the singularity of $a(\lambda)$ not only changes the PB and modifies the dispersion relation, but also allows a Levinson-type theorem in the KdV system related to its soliton number and the singularity index m [12].

The radial part of the well-known 3-d Schrödinger equation for a spherically symmetric potential is also given by the same Schrödinger operator (3) as in the KdV case, but in the half line. It is, therefore, surprising to observe that while the singularity structure of the scattering elements in the standard half-line Schrödinger operator problem is well studied and as a consequence, the Levinson theorem is well established [14], there was no attempt, as far as we know, to bring this knowledge to the KdV system. Our aim is to fill up this gap and explore the significance of such a theorem in the context of the KdV equation. Hence, repeating the same line of argument as in Newton [14], we start with $\frac{1}{2\pi i}\int_\Gamma d(\ln a(\lambda)) = N$, where N is now the soliton number of the KdV equation corresponding to the number of "bound states" and Γ is the contour enclosing the zeroes of $a(\lambda)$. The pole-type singularity of $a(\lambda)$ at $\lambda=0$ leads to the value $\oint_c d(\ln a(\lambda)) = i\pi q$, where $q = 0$ for $m = 0$ and $q = 1$ for $m \neq 0$. Therefore we get finally the Levinson theorem for the KdV system in the form

$$\delta \equiv -\arg a(0^+) = \begin{cases} \pi N & , \text{ if } m = 0 \\ \pi(N-1/2) & , \text{ if } m \neq 0 \end{cases} , \tag{14}$$

from which one may conclude, that arg a(0) carries important information abuot the presence of soliton and radiation like solutions. In particular, if it is integer times π, the KdV solution is purely solitonic, while for $\frac{1}{2}$ integer x π, along with solitons the "radiations" are also present in the field solution. The singularity index may be given in various forms like

$$m = \lim_{x \to +\infty} \partial_x \Psi_-(x,\lambda)\Big|_{\lambda=0} = -\int_{-\infty}^{\infty} u(x)\,\Psi_-(x,o)dx \tag{15a}$$

$$= \frac{1}{2\pi i}\int_{-\infty}^{\infty} \frac{r(\lambda)}{\lambda}\,\partial_x(\Psi_- e^{i\lambda x})\Big|_{x\to+\infty} d\lambda \,, \tag{15b}$$

which along with (13) induce the following interesting properties on m (for details see [12]):

(i) m is a real valued conserved quantity and determined only by the initial field configurations;

(ii) $m = 0$ for $r(o) = 0$. Therefore, for reflectionless potentials or pure solitons m is always trivial;

(iii) In the absence of solitons we always have $m > 0$. Solitons may only change the sign of m keeping its absolute value unaltered. If soliton number N is odd, the sign of m is flipped, while even N keeps the sign unchanged.

From the above results it is clear that the singularity structure of $a(\lambda)$ playing a significant role in various aspects of the KdV system also, in particular, subdivides the solution manifold into different sectors labelled by the values of the singularity index m. Pure solitons constitute $m = 0$ sector, "radiations" along with even N solitons belong to $m > 0$, while with odd N's to $m < 0$ sectors. Different non-trivial values of m are responsible for different types of radiation. Moreover, m being a conserved quantity, this picture remains unaltered in time.

3.4 *Different PB's of the KdV system as the Dirac bracket*

The fact that the KdV system is a constrained system becomes clear from the corresponding Lagrangian

$$L = \frac{1}{2}\int(\varphi_t\varphi_x + \frac{1}{3}\varphi_x^{\;3} - \varphi_{xx}^{\;2})dx, \quad \varphi_x = u(x), \tag{16}$$

yielding the canonical momentum $p(x) = \frac{\delta L}{\delta\dot{\phi}} = \varphi_x$, which puts a field constraint $S(x) = p(x) - \varphi_x$ on the system. Consequently, the related Poisson bracket may be given by the Dirac bracket, generalized to include field variables in the form

$$\{f,g\}^* = \{f,g\} - \int\int_{-\infty}^{\infty} dxdy\ \{f,S(x)\}\ k(x,y)\{S(y),g\} \qquad (17)$$

with $k(x,y) = e^{-1}(x,y)$, where $e(x,y) = \{S(x),S(y) = \frac{1}{2}[\delta_y(x-y)-\delta_x(x-y)]$. In earlier investigations [4,5] it was concluded that the Dirac bracket (17) yields the standard Gardner bracket $\{\ ,\ \}_G$. However, we find [12] that the earlier solution $k(x,y) = -\frac{1}{2}\epsilon(x-y)$ is actually not correct and consequently, the result found before concerning the KdV model is not exact, as suspected by us in Section 2. We observe further that it is difficult as such to find a proper solution fo $k(x,y)$, unless we demand some extra topological constraints like (i) $S_1 = \varphi(L^-)$ and $S_2 = \varphi(-L^+)$ or (ii) $\tilde{S}_1 = \varphi(L^-) + \varphi(-L^+)$ along with the dynamical constraint $S(x)$. This allows exact solutions to the extended K with $(k(x,y), k(x,\alpha), k_{\alpha\beta})$ and modifies the Dirac bracket (17) to add extra terms like

$$\sum_\alpha \int dx(\{f,S(x)\}\{S_\alpha,g\} - \{f,S_\alpha\}\{S(x),g\})k(x,y) + \sum_{\alpha\beta} \{f,S_\alpha\}\{S_\beta,g\}k_{\alpha\beta}.$$

One should also note, that the above topological constraints, in fact follow systematically while minimizing the action to get the KdV equation and they are related to the periodic and anti-periodic properties of the KdV field. It is fascinating to find finally that if one confines to the KdV systems only, then in the first case the Dirac bracket reduced exactly to the FT-bracket and in the second case to the APP-bracket, which are suggested recently and satisfy the JI. The conventional Gardner bracket is possible to obtain as a Dirac bracket only approximately by neglecting some crucial boundary terms (as done before), which, however, is responsible for the violation of the JI in the KdV case.

4 Conclusions

The KdV system, as understood currently, still contains many subtleties. We explore here different aspects of this model exhibiting such peculiarities. In particular, we derive in a systematic way different PB structures suggested recently exploring also some deeper topological reasons behind the existence of three different Poisson brackets. The source of such unusual features is the singularity of the scattering elements at $\lambda = 0$. As a consequence of this a Levinson type theorem exists for the KdV system related to its soliton number and the singularity index m, which is found to be a conserved quantity subdividing the solution manifold into different sectors. Pure solitons fall into the trivial m sector. Earlier attempts to show the Gardner bracket for the KdV system as the constrained Dirac bracket are found to be inaccurate. The proper treatment of the Dirac bracket should include some additional topological constraints yielding actually their investigation includes only nontrivial m sectors where there are no pure solitons. We find that in the trivial m sector the action angles become canonical with no mixing between discrete and continuous modes and hence no problem occurs with the existence of pure solitons in the KdV model.

References

[1] Zabusky, N.I. & Kruskal, M.D. (1965). Interaction of solitons in a collisionless plasma and the recurrence of initial states. *Phys. Rev. Lett.* **15**, 240.

[2] Gardner, C.S., Greene, J.M., Kruskal, M.D. & Miura, R.M. (1967). Method for solving the Korteweg-de-Vries equation. *Phys. Rev. Lett.* **19**, 1095.

[3] Novikov, S.P. (ed.) (1984). *Theory of Solitons*. Plenum: New York.

[4] Kentwell, G.W. (1988). The Hamiltonian structure of the modified and fifth-order Kortweg-de-Vries equation. *J. Math. Phys.* **20**, 46.

[5] Lund, F. (1986). Constrained Hamiltonian dynamics and the Poisson structure of some integrable systems. *Physica* **18D**, 420.

[6] Faddeev, L.D. & Takhtajan, L.A. (1985). Poisson structure for the KdV equation. *Lett. Math. Phys.* **10**, 183.

[7] Arkadiev, V.A., Podgrebkov, A.K. & Polivanov, M.C. (1988). On the Posson structure for the Korteweg-de Vries equation. *Sov. Phys. Dokl.* **32**,

[8] Faddeev, L.D. (1984). Integrable models in (1+1)-dimensional quantum field theory. In *Recent Advances in Field Theory and Statistical Mechanics*, Zuber, J.B. & Stora, R. (eds.). North Holland.

[9] Maillet, J.M. (1985). Kac-Moody algebra and extended Yang-Baxter relations in the O(N) non-linear σ-model. *Phys. Lett.* **162B**, 137.

[10] Tsyplyaev, S.A. (1981). Commutation relations of the transition matrix in the classical and quantum inverse scattering methods (local case). *Teor. Mat. Fiz.* **48**, 24.

[11] Kundu, A. & Basu Mallick, B. (1988). Investigation of the Hamiltonian structure of the KdV system through r-s matrix formalism revealing some new aspects. *Preprint* SINP/TNP/28-88.

[12] Basu Mallick, B. & Kundu, A. (1989). Levinson type theorem for the KdV system and its consequences, *Phys. Lett.A* **135**, 113; (1988). Faddeev-Takhtajan and Arkadiev et al. brackets for the KdV system through Dirac's constrained Hamiltonian formalism. *Preprint SINP/TNP/40-88.*

[13] Taflin, E. (1984). Dynamical symmetries for KdV equation. *Rep. Math. Phys.* **20**, 171.

[14] Newton, R. (1966). *Scattering Theory of Waves and Particles.* McGraw-Hill: New York.

[13] Taflin, E. (1984). Dynamical symmetries for KdV equation. *Rep. Math. Phys.* **20**, 171.
[14] Newton, R. (1966). *Scattering Theory of Waves and Particles.* McGraw-Hill: New York.

Part 5

NON-INTEGRABLE SYSTEMS

C.R. Doering, J.D. Gibbon, D.D. Holm

43. FINITE DIMENSIONALITY IN THE COMPLEX GINZBURG-LANDAU EQUATION

1 Introduction

Those who have worked with integrable systems for a number of years have become familiar with the analytical methods, such as the many variations of the IST, needed to solve Hamiltonian systems with an infinite number of degrees of freedom. Less attention has been given to the fact that once one steps out of the domain of complete integrability, it is much more difficult to prove anything about infinite dimensional problems analytically. Despite the great advances in understanding the qualitative nature of the various bifurcations which nonlinear systems can undergo, the main results have generally been confined to finite dimensional problems. The infinite dimensional nature of p.d.e's has caused obvious difficulties. The question of whether *ostensibly* infinite dimensional systems can, in fact, behave in a *finite* dimensional way without introducing severe approximations is one which has taxed both applied and pure mathematicians for two decades. The truncation necessary to integrate a given equation on a machine could easily introduce behaviour which is purely an artefact of that truncation. While a proof of finite dimensionality in even two dimensional Navier Stokes equations is still beyond us there has been some considerable progress in simpler problems. Indeed, it has been this problem of finite or 'low' dimensionality which has started to engage the minds of analysts in the last two or three years. The two closely interlinked sets of work on the establishment of a *finite* attractor dimension for p.d.e's [1-8] and the work on so-called *inertial manifolds* for the Kuramoto-Sivashinsky (KS) equation [2, 6,7,8] has stimulated activity in this area. The KS equation

$$u_t + uu_x + u_{xx} + u_{xxxx} = 0 \tag{1.1}$$

periodic on $[0\ L]$, has a linear growth rate which is

$$\lambda = k^2 - k^4 \tag{1.2}$$

Discrete modes for which $k^2 > 1$ are stable since $\lambda < 0$. The idea of the so-called 'cone condition' which appears in [6] is simple in physical terms. The system is not in equilibrium since energy is being pumped in through the unstable modes $k^2 < 1$. The stable modes lose energy by diffusion with the energy loss increasing as k increases, as equation (1.2) shows. The nonlinear term pumps energy into the higher modes and so the two processes compete. There is, however, no *a priori* reason to suppose that a *finite* number of modes can overcome the nonlinear terms, but in fact it turns out that this can be achieved. The aim is to estimate a value of N, derived from a cone condition, which shows when the modes k_n for $n < N$ control the modes for which $n \geq N+1$. This control, which is achieved through a cone condition, slaves the high modes to the low modes by means of a global slaving function Φ. The inertial manifold is essentially the graph of this function Φ: it is a smooth invariant exponentially attracting manifold which contains the universal attractor. On this manifold the p.d.e. is a finite dimensional dynamical system. This idea of slaving an infinite number of degrees of freedom to a finite number of degrees of freedom establishes a direct connection between infinite dimensional problems and the elegant and important work of recent years on finite dimensional dynamical systems. This paper will concentrate on showing how to construct a cone condition [10], thereby establishing the existence of an inertial manifold for the complex Ginzburg-Landau (CGL) equation on periodic boundary conditions on [0,1]:

$$A_t = RA + \beta A_{xx} + \gamma A|A|^2 \qquad \gamma,\ \beta \in \mathbb{C} \tag{1.3}$$

However, we do not go into the details of how to construct this manifold, which can be found in [11]. Details of inertial manifolds in general can be found in [4].

This is a simple example of an equation which has a good physical pedigree which can be used as an archetype for understanding how p.d.e's which possess rotating wave solutions behave. It too has a bandwidth of unstable modes since the growth rate of rotating waves solutions is

$$\lambda = R - k^2 \tag{1.4}$$

Hence all modes $k^2 < R$ are unstable and $k^2 > R$ are stable. There is a gap in the spectrum

$$\lambda_n - \lambda_{n+1} = (2n+1)k_1^2 \tag{1.5}$$

which increases linearly with n. Clearly therefore it is an ideal candidate for an inertial manifold calculation. The CGL equation occurs as an amplitude equation on long space and time scales in a variety of problems and was first derived by Newell and Whitehead [14] using the finite bandwidth concept. One set of circumstances is in Taylor vortex flow and convection [12,13,15] where the coefficients β and γ are real. For Poiseuille flow [16-18,20] these coefficients are complex. In the latter case $\gamma_r > 0$ and the equation blows up in finite time [17-19]. There are, however, two cases when $\gamma_r < 0$ [20]. The CGL equation also occurs in studies in chemical turbulence [21-23]. The stability of rotating wave solutions to sidebands has been studied widely [24-27] and various papers have investigated numerically the possibility of chaos and pattern formation [28-33]. Since the intention of this paper is to consider various aspects of finite dimensional behaviour, we shall choose $\gamma_r < 0$ and take the equation in the form

$$A_t = RA + (1+i\nu)A_{xx} - (1+i\mu)A|A|^2 \tag{1.6}$$

2 The cone condition for the CGL equation

The idea of a cone condition, introduced in [6,4,5], enables us to estimate the number of Fourier modes needed to span the attractor. This was achieved for the CGL equation in [9]. We will firstly try to explain the general ideas behind this and then go on to prove that there is indeed an absorbing set in the Hilbert space $\mathcal{H} \equiv L^2[0,1]$. When we have obtained bounds on the necessary norms we will then establish the estimate of the number of Fourier modes needed to span the attractor. These Fourier modes will be eigenfunctions of the Laplacian so let us suppose that we require N of them. We will define P_N to be a projection operator which projects solutions on the universal attractor onto the first N Fourier modes $|k_n| \le k_N$. $Q_N = I-P_N$ is the infinite dimensional projection onto all other modes $|k_n| \ge k_{N+1}$. The idea of a "cone property" is a means of controlling the high modes by the low modes. Figures 1 and 2 give the idea. Let A and A' be two solutions on the universal attractor. Defining $a = A'-A$ with $p = P_N a$ and $q = Q_N a$, we can construct a cone by means of a quality L which is defined by

$$L = \|q\|_2^2 - \|p\|_2^2 \quad . \tag{2.1}$$

Inside the cone we define $L > 0$, on the cone $L = 0$ and outside $L < 0$. $\|\cdot\|$ denotes the norm in $\mathcal{H} \equiv L^2[0,1]$. We now note that outside the cone, when $\|q\|_2^2 < \|p\|_2^2$, the high modes are controlled by the low modes since low wave number modes can vanish only when high modes vanish. We can now define a function (see [6,4.5]):

$$\Phi\colon P_N(attractor) \subset P_N\mathcal{H} \to Q_N\mathcal{H} \tag{2.2}$$

which maps a subset of the finite dimensional space $P_N\mathcal{H}$, i.e. $P_N(attractor)$, into the remainder of the configuration space

$$Q_N A = \Phi(P_N A) \tag{2.3}$$

and the nature of the finite dimensionality can be seen by a slaving of the high modes by the low modes

$$A = \underbrace{P_N A}_{\text{low modes}} + \underbrace{Q_N A}_{\text{high modes}} = P_N + \Phi(P_N A) \tag{2.4}$$

We can at least give a descriptive account of how the cone idea allows a projection which is well defined i.e. (2.2) yields a unique value for Φ. In simplified terms, let us think of two solutions A and A' crossing in configuration space such that they may not touch but there may be at least one point or one which lies vertically above or below a point on the other. At this point $P_N a = 0$ and hence $L > 0$. This can only occur inside the cone by definition and there is no unique projection at these points. However, outside the cone we can project the attractor one to one onto its image. The task is therefore, to find N such that we can make $L < 0$ as $t \to \infty$. If we can achieve this, then asymptotically we can achieve a projection onto N Fourier modes. An inertial manifold is a smooth (Lipschitz) exponentially

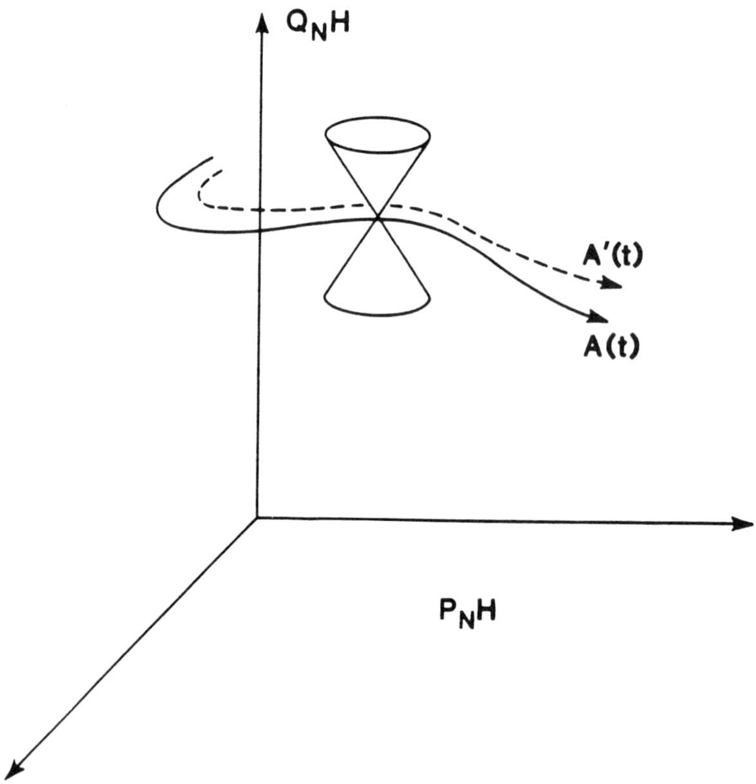

Figure 1. Illustration of a cone in $\mathcal{H}$. At each point an orbit $A(t)$ a cone can be drawn and all other orbits [e.g. $A'(t)$] are excluded from the interior ($L > 0$) of these cones.

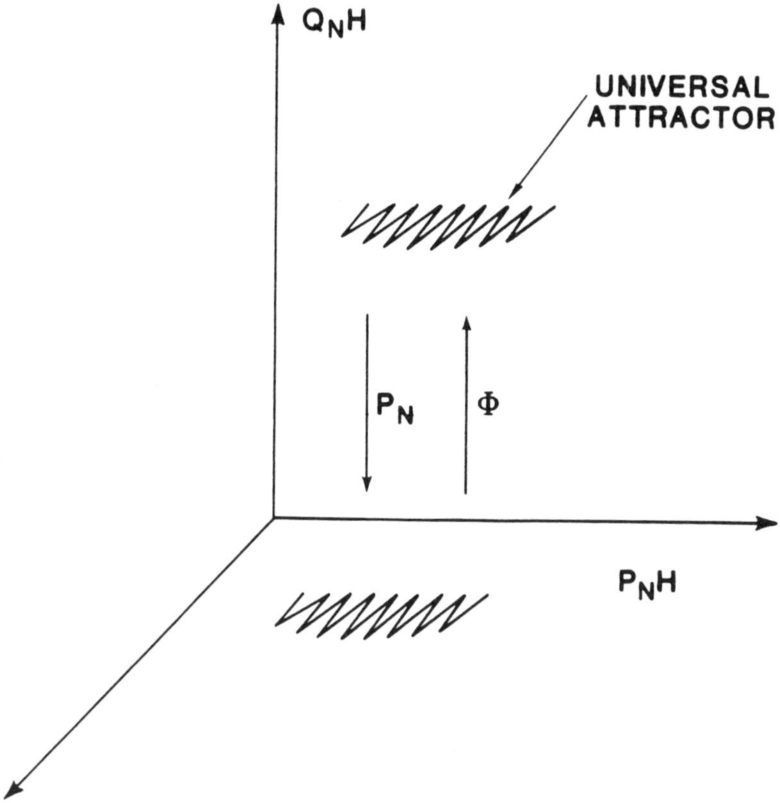

Figure 2. The cone conditon on the attractor implies that the projection P_N of the attractor onto the finite dimensional space $P_N\mathcal{H}$ is one-to-one onto its image and hence invertible there ($\Phi = P_N^{-1}$).

attracting invariant manifold which contains the universal (global) attractor. Restricted to this manifold the CGL equation is equivalent to a finite dimensional dynamical system. This projection now allows the universal attractor to be spanned by *2N+1* Fourier modes. This provides an upper bound on the attractor dimension although this estimate is not necessarily good. With a cone condition, it is possible to use standard invariant manifold techniques, such as centre manifold methods [4,5,11] to perform the required extension to an inertial manifold. This suggests the identification of an inertial manifold, which is the graph of the extension over the new domain, as a global centre manifold. The geometric idea of an inertial manifold is given in Figure 3. It is now necessary to prove that there exists a compact absorbing set, which is a necessary condition for the existence of a universal attractor. Let us define a mass $M(t)$ and an energy $E(t)$ (which would be conserved

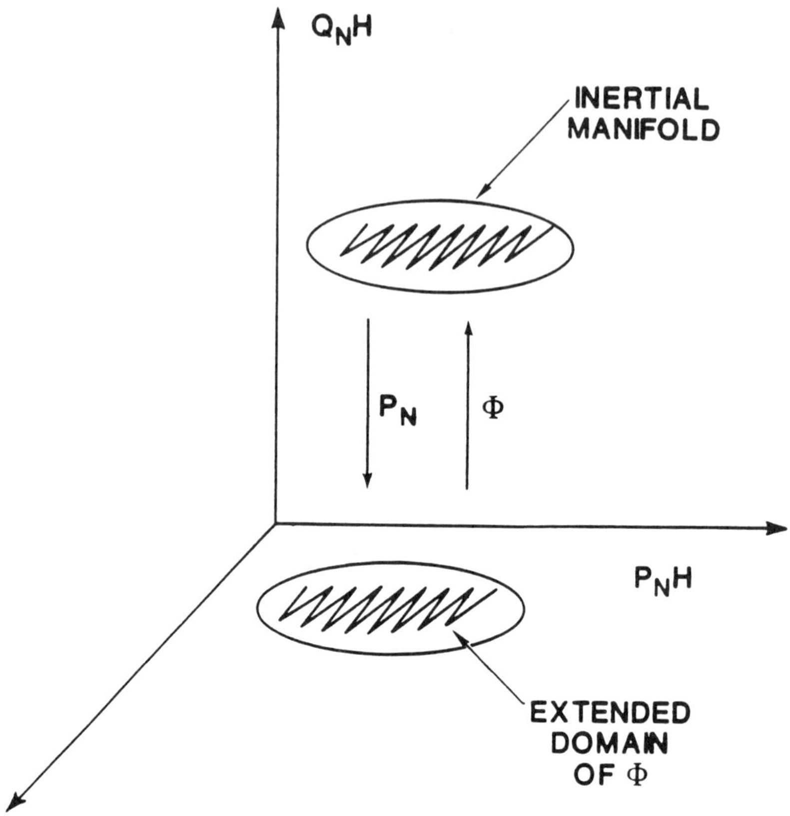

Figure 3. an inertial manifold is the extension of the domain of definition of Φ off of the projection of the attractor but preserving the reduced dynamics. the graph of Φ is an invariant, exponentially attracting manifold in the configuration space that contains the attractor.

quantities for the NLS equation)

$$M = \int_0^1 |A|^2 dx \qquad\qquad E = \int_0^1 |A_x|^2 dx \ . \tag{2.5}$$

Differentiation of M w.r.t. t gives

$$\tfrac{1}{2}\dot{M}_t = RM - E - \int |A|^4 dx \tag{2.6}$$

whence the use of the Cauchy-Schwarz inequality gives

$$\tfrac{1}{2}M_t \le RM - E - M^2 \le RM - M^2 \tag{2.7}$$

so

$$M(t) \leq R[1-\exp(-2Rt)]^{-1} \tag{2.8}$$

Hence in our Hilbert space $\mathcal{H} \equiv L^2[0,1]$, the flow contracts into a ball of radius $R(\sigma) = [(1+\sigma)R]^{\frac{1}{2}}$ in a finite time $t(\sigma) = -\{\ln[\sigma/1+\sigma]\}/2R$. In other words, $M \leq R$ as $t \to \infty$, for all ν and μ. When we turn to E, a calculation whose details we will not give, we have to make a distinction between $|\mu| \gtrless \sqrt{3}$. We find that we have an absorbing ball such that as $t \to \infty$

$$E \leq (1+\sigma)^2 R^2 \qquad |\mu| \leq \sqrt{3} \tag{2.9a}$$

$$E \leq \delta^2 R^2 \{1+[1+\delta^{-2}(1+\delta)/R]^{1/2}\}^2 \qquad \mu > \sqrt{3} \tag{2.9b}$$

where $\delta = \max\{0,-2+|1+i\mu|\}$. It is now necessary to find a uniform L^∞ norm for $A(x,t)$. These turn out to be, for ν arbitrary (see [9]).

$$\|A\|_\infty^2 = \lim_{t\to\infty} \sup_x |A(x,t)|^2$$

$$\|A\|_\infty^2 = \leq \begin{cases} R + 2R^{3/2} & |\mu| \leq \sqrt{3} \quad (2.10a) \\ R + 2\delta R^2\{1+[1+\delta^{-2}(1+\delta)/R]^{1/2}\} & |\mu| > \sqrt{3} \quad (2.10b) \end{cases}$$

It is also possible to find bounds when $\nu = 0$ and μ arbitrary [9]. Let us now return again to the cone condition. In terms of $a=A'-A$, the difference between two solutions, we have

$$a_t = Ra + (1+i\nu)a_{xx} - (1+i\mu)[A'|A'|^2 - A|A|^2] \tag{2.11}$$

Operating on this equation with P_N and Q_N to form p and q, we have

$$p_t = Rp + (1+i\nu)p_{xx} - (1+i\mu)P_N[A'|A'|^2 - A|A|^2] \tag{2.12a}$$

$$q_t = Rq + (1+i\nu)q_{xx} - (1+i\mu)Q_N[A'|A'|^2 - A|A|^2] \tag{2.12b}$$

Multiplying (2.12a) by p^*, integrating over $[0,1]$ and taking the real part and repeating the process for (2.12b), we obtain

$$\frac{1}{2}\frac{\partial}{\partial t}\|p\|_2^2 = R\|p\|_2^2 + \mathrm{Re}\{(1+i\nu)\int_0^1 p^* p_{xx}\,dx\} \tag{2.13a}$$

$$- Re\{(1+i\mu)\int_0^1 p^* P_N[A'|A'|^2 - A|A|^2]dx\}$$

$$\frac{1}{2}\frac{\partial}{\partial t}\|q\|_2^2 = R\|q\|_2^2 + Re\{(1+i\nu)\int_0^1 q^* q_{xx}dx\} \qquad (2.13b)$$

$$- Re\{(1+i\mu)\int_0^1 q^* Q_N[A'|A'|^2 - A|A|^2]dx\}$$

In order to construct a differential inequality for $L = \|q\|_2^2 - \|p\|_2^2$, we need to estimate not only the nonlinear terms but the terms:

$$Re\{(1+i\nu)\int_0^1 [q^* q_{xx} - p^* p_{xx}]dx\} = \int_0^1 (|p_x| - |q_x|^2)dx \quad . \qquad (2.14)$$

However, we know that on periodic boundary conditions

$$\int_0^1 |p_x|^2 dx \le k_N^{\,2}\,\|p\|_2^2; \qquad \int_0^1 |q_x|^2 dx \ge k_{N+1}^{\,2}\|q\|_2^2 \qquad (2.15)$$

The right hand side of (2.14) now obeys the inequality

$$\int_0^1 (|p_x|^2 - |q_x|^2)dx \le k_N^{\,2}\,\|p\|_2^2 - k_{N+1}^{\,2}\|q\|_2^2 \qquad (2.16)$$

Subtracting (2.13a) from (2.13b) and using (2.16) we have

$$\frac{1}{2}\frac{\partial}{\partial t}(\|q\|_2^2 - \|p\|_2^2) \le (R-k_N^{\,2})(\|q\|_2^2 - \|p\|_2^2) - (k_{N+1}^{\,2} - k_N^2)\|q\|_2^2$$

$$+ \text{ nonlinear term} \qquad (2.17)$$

Now we can observe how the gap in the spectrum of the Laplacian competes with the nonlinear terms

$$\frac{1}{2}L_t \le (R-k_N^{\,2})L - (2N+1)k_1^{\,2}\|q\|_2^2 \quad + \text{ nonlinear term} \qquad (2.18)$$

There are two points here. Firstly, as we explained in §1, *stable* modes occur when $k^{\,2} > R$. Hence the sign on the nonlinear term L is negative. If we can bound the nonlinear terms in terms of $\|q\|_2^2$ then we can overcome these terms by choosing N large enough. Bounds on the nonlinear terms can be found in the following way:

$$\text{Nonlinear term} \;=\; - Re\{(1+i\mu)\int_0^1 [q^* Q_{\mathbb{N}} - p^* P_{\mathbb{N}}][A'|A'|^2 - A|A|^2]dx\}$$

$$(2.19)$$

Using the fact that the projections are self-adjoint, we find

$$\text{Nonlinear term} = -\,Re\{(1+i\mu)\int_0^1 (q^*-p^*)(A'|A'|^2-A|A|^2)dx\} \tag{2.20}$$

Part of the integrand of (2.20) can be written as

$$A'|A'|^2-A|A|^2 = a(|A'|^2+|A|^2) + a^*AA' = (p+q)(|A'|^2+|A|^2) + a^*AA' \tag{2.21}$$

and we find that

$$\begin{aligned}\text{Nonlinear term} &= -\,Re\Big\{(1+i\mu)\int_0^1 [-|q|^2+|p|^2-2iIm(q^*p)] \\ &\quad\times(|A'|^2+|A|^2) + (1+i\mu)\int_0^1 (p^{*2}-q^{*2})AA'dx\Big\} \\ &\le 2\|A\|_\infty^2\|p\|_2^2 + 4\mu\|A\|_\infty^2\int |q||p|dx \\ &\quad + |1+i\mu|(\|p\|_2^2+\|q\|_2^2)\|A\|_\infty^2 \\ &\le \|A\|_\infty^2\,[2\|p\|_2^2 + 4\mu\|q\|_2\|p\|_2 + \sqrt{1+\mu^2}\,(\|p\|_2^2+\|q\|_2^2)]\end{aligned} \tag{2.22}$$

Consequently we have

$$\begin{aligned}\tfrac{1}{2}L_t &\le (R-k_N^2)L - (2N+1)k_1^2\|q\|_2^2 \\ &\quad + \|A\|_\infty^2\{2\|p\|_2^2 + 4\mu\|p\|_2\|q\|_2 + \sqrt{1+\mu^2}\,(\|p\|_2^2+\|q\|_2^2)\}\end{aligned} \tag{2.23}$$

Inside the cone, when $\|q\|_2^2 > \|p\|_2^2$ $(L > 0)$ we can choose N large enough such that

$$\tfrac{1}{2}L_t < (R-k_N^2)L \tag{2.24}$$

and so L is decreasing as $t \to \infty$ for those modes (stable modes) for which $k_N^2 > R$. When L reaches zero on the cone, we find that $L_t < 0$ and to L decreases further to become negative. Once negative, it must remain negative. In summary therefore, if we choose

$$2\pi\, N_{max} > max\{R^{1/2},\ \|A\|_\infty^2\,[1+\sqrt{1+\mu^2}+2|\mu|]/k_1 - \tfrac{1}{2}k_1\} \tag{2.25}$$

where k_1 is the fundamental wave number, then trajectories which are inside the cone are expelled from the cone in an exponential way as $t \to \infty$ and once outside remain outside. We note that we have the estimates for $\|A\|_\infty^2$ in equation (2.10). This estimate is an *upper bound* on the attractor dimension uniform in ν. Equation (2.10) shows that this upper bound goes like $R^{3/2}$ for $|\mu| \le \sqrt{3}$ and like R^2 for $|\mu| > \sqrt{3}$. In fact in [9], we find that N_{max} goes like R when $\nu = 0$ for all μ. These bounds are obviously dependent upon the estimates for $\|A\|_\infty^2$.

3 The modulational instability: lower bounds on the Fourier spanning and attractor dimensions

The CGL equation is well known to possess so-called rotating wave solutions. These must be contained in the universal attractor - indeed they form a lower bound on both the Fourier spanning dimension *(D)* and the universal attractor dimension *(d)*. These solutions take the form

$$A_n = a_n \exp[i(k_n x - \omega_n t)] \qquad n = 0,1,2,... \tag{3.1a}$$

$$\omega_n = R\nu + (\mu+\nu)k_n^2 \tag{3.1b}$$

$$|a|^2 = R - k_n^2 \tag{3.1c}$$

The first set of bifurcation points are $R^{1/2} = k_n^2$ ($k_n = \pm 2\pi n$) for each n, including $n = 0$, which grow off the zero solution. Although these solutions appear trivial, they are expressed in terms of Fourier modes and so they give a lower bound on the Fourier spanning dimension

$$D \ge D_{rot} \equiv 1+2[R^{1/2}/2\pi] \tag{3.2}$$

Each Fourier mode k_n may be stable or unstable to a neighbour k_n. To test for this we use the method described in Stuart & DiPrima [26] for discussing the Eckhaus instabilty [24-25]. This requires us to look at the stability of A_n. Lack of space precludes us from doing this calculation here although it can be found in [9]. We find that the value of R at which the spatially homogeneous rotating wave (k = 0) becomes unstable to side-band perturbations of wavenumber k_m is

$$R = -k_n^2(1+\nu^2)/2\epsilon \tag{3.4}$$

Using this, we now proceed with the task of this section, to develop lower bounds on the attractor. First, for the attractor dimension, the linearized stability analysis (including neutral modes) gives the dimension of the unstable manifold of each of the rotating wave solutions. For each specific set of the parameters μ and ν, this set of dimensions for the rotating wave solutions can be determined as a function of R simply by counting, using the plot of neutral stability curves. Although in any particular situation the counting procedure described above can be carried out for any specific rotating wave and set of parameters, there is no convenient explicit expression for the dimension of the unstable manifold of an arbitrary rotating wave solution. However, the spatially homogeneous rotating wave *is* amenable to computation, and the maximum of its and the trivial solution's unstable manifold dimensions provides a lower bound on the universal attractor dimension. In the modulationally unstable case ($\varepsilon = 1 + \mu\nu < 0$) the number of unstable directions for the spatially homogeneous rotating wave is computed from (3.4).

$$m = [\{-2\varepsilon/(1+\nu^2)\}^{1/2}(R^{1/2}/2\pi)] \quad . \tag{3.5}$$

Since the rotating wave itself is one-dimensional, a lower bound on the universal attractor dimension is found to be

$$d \geq max\{1+[(-2\varepsilon/(1+\nu^2))^{1/2}R^{1/2}/2\pi)],\ D_{rot}(R)\}. \tag{3.6}$$

We may bound this lower bound from above, uniformly in ν, by recalling that the closest point of approach of the intersections of the neutral curves and the R axis to the origin occurs for $\nu = -\{1+(1+\mu^2)^{1/2}\}/\mu$. Hence, a lower bound on the universal attractor dimension is given by

$$\sup_{\nu} d \geq max\{1+[\{\mu^2/(1+(1+\mu^2)^{1/2})\}^{1/2}(R^{1/2}/2\pi)],\ D_{rot}(R)\} \ . \tag{3.7}$$

For $|\mu| \leq \surd 3$, the second term in (3.7) is always greater than or equal to the first. When $|\mu| > \surd 3$, the first term can dominate the second for some values of R, since $\surd 3$ is the critical value of μ for which there exists a ν such that the homogeneous rotating wave is unstable to a side-band *before* the trivial solution goes unstable to the second ($k = 2\pi$) rotating wave as R is increased from 0. When $|\mu|$ is larger than $\surd 24$ the first term in (3.7) is always larger than the second. We summarise these lower bounds on the universal attractor dimension, uniform in ν:

$$\sup_{\nu} d \geq 1+2[R^{1/2}/2\pi] \ , \qquad |\mu| \leq \surd 3 \ , \tag{3.8a}$$

$$\sup_{\nu} d \geq max\{1+[\{\mu^2/(1+(1+\mu^2)^{1/2})\}(R^{1/2}/2\pi)],\ D_{rot}(R)\} \ , \qquad ()$$

$$\surd 3 < |\mu| \leq \surd 24 \ , \tag{3.8b}$$

$$\sup_{\nu} d \geq 1+[\{\mu^2/(1+(1+\mu^2)^{1/2})\}^{1/2}(R^{1/2}/2\pi)], \qquad \surd 24 < |\mu| \ . \tag{3.8c}$$

We remark that the border-line situation $|\mu| = \surd 3$ has appeared in the previous section in the analysis of the cone condition. We now make some final remarks about the attractor dimension. While an inertial manifold exists for this problem, the dimension of which gives an upper bound on the attractor dimension, it is still possible to calculate an upper bound on the fractal and Hausorff dimensions of the universal attractor. As in [1] it is possible to show that the Lyapunov exponents on the attractor of a dynamical system describe the exponential rate of divergence of trajectories that start at nearby points. A positive Lyapunov exponent indicates sensitive dependence on initial conditions and hence chaos, for in this case the prediction of a trajectory after a long time requires an exponentially accurate knowledge of its initial conditions. Computation of the global Lyapunov exponents give an upper bound on the fractal and Hausdorff dimensions. It is possible to show that [36, 34] this dimension can be reduced to const x $R^{1/2}$ for all values of μ-ν. It can also be shown that using Lyapunov functionals associated with the conservation laws of the NLS equation, the dimension of the inertial manifold can be reduced to const x $R^{4/3}$ in certain areas of the μ-ν plane (even when $\varepsilon < 0$) [37] and even down to const x R close to the ν-axis.

References

[1] P. Constantin & C. Foias, *Comm. Pure Appl. Math.*, **38**, (1985), 1.

[2] P. Constantin, C. Foias, B. Nicolaenko & R. Temam; 'Integral & inertial manifolds in dissipative pde's', Springer series in Applied Math. Vol **70**, 1988.

[3] C. Foias, B. Nicolaenko, G.R. Sell & R. Temam, C.R. Acad. Sci. Paris, Serie 1, **301**, (1985), 285.

[4] 'Integral and inertial manifolds in dissipative pde's'. P.Constantin, C.Foias, B.Nicolaenko & R.Temam; Springer Applied Math. Series **69**, 1988.

[5] R. Temam, 'Infinite & finite dimensional attractors in mathematics & mechanics' Springer series in Appl Math Vol **68** 1988.

[6] C. Foias, B. Nicolaenko, G.R. Sell & R. Temam, J. Math. Pure & Appl.to appear 1987.

[7] B. Nicolaenko, B. Scheurer & R. Temam, Physica **16D**, (1985), 155.

[8] J.M. Hyman & B. Nicolaenko, Physica **18D**, (1986), 113.

[9] J. Mallet-Paret & G.R. Sell, Brown University Preprint (1987).

[10] C. Doering, J.D. Gibbon, D.D. Holm & B. Nicolaenko, Nonlinearity, **1**, 279-309, 1988.

[11] P.Constantin. "Construction of Inertial Manifolds", Proc. of Boulder Conference 1987; to appear AMS, Providence, Rhode Island.

[12] A.C. Newell & J.A. Whitehead, J. Fluid Mech., **38**, (1969), 279.

[13] R.C. DiPrima, V. Eckhaus & L.A. Segel, J. Fluid Mech, **49**, (1971), 705.

[14] A.C. Newell & J.A. Whitehead, Proc. IUTAM Symposium Herrenhalb 1969, Springer, Berlin, 1971.

[15] S. Kogelman & R.C. DiPrima, Phys. Fluids, **13**, (1970), 1.

[16] K. Stewartson, & J.T. Stuart, J. Fluid Mech, **48**, (1971),

529.
[17] L. Hocking, K. Stewartson & J.T. Stuart, J. Fluid Mech., **51**, (1972),705.
[18] L. Hocking & K. Stewartson, Proc. Roy. Soc. Lond. **A326**, (1972), 289.
[19] J. Ball, Quart. J. Math. Oxford, **28**, (1977), 473.
[20] P. Blennerhassett, Phil. Trans. Roy. Soc., **298**, (1980), 451., and H. Brandt, P. Lomdahl & A.C. Newell; Physica **23D**, (1986), 345.
[21] Y. Kuramoto & T. Yamada, Prog. Theor. Phys. Suppl., **56**, (1976), 679.
[22] Y. Kuramoto, Prog. Theor. Phys. Suppl., **64**, (1978), 346.
[23] Y. Kuramoto & S. Koga, Prog. Theor. Phys. Suppl., **66**, (1981), 1081.
[24] W. Eckhaus, Studies in Nonlinear Stability Theory, Springer, Berlin, 1965.
[25] C. Lange & A.C. Newell, SIAM J. Appl. Math., **27**, (1974), 441.
[26] J.T. Stuart & R.C. DiPrima, Proc. Roy. Soc. Lond. **A362**, (1978), 27.
[27] Catherine Holmes, Proc. Roy. Soc. Lond. **A402**, (1985), 229.
[28] H.T. Moon, P. Heurre & L. Redekopp, Phys. Rev. Lett., **49**, (1982),458.
[29] H.T. Moon, P. Heurre & L. Redekopp, Physica **7D**, (1983), 135.
[30] L. Keefe, Stud. Appl. Math., **73**, (1985), 91.
[31] K. Nozaki & N. Bekki, Phys. Rev. Lett., **51**, (1983), 2171.
[32] P. Holmes, Physica **23D**, (1986), 84.
[33] J. Kaplan & J. Yorke, in Functional Differential Equations and Approximations of Fixed Points, edited by H.O. Peitgen & H.O. Walther,Springer, Berlin, 1979.
[34] C. Doering, J.D. Gibbon, D.D. Holm & B. Nicolaenko; Exact Lyapunov dimension of the universal attractor for the CGL equation: Phys. Rev.Lett., **59** (1987), 2911-14.
[35] C. Doering, J.N. Elgin, J.D. Gibbon & D.D. Holm; Low dimensionality in the laser equations in the good cavity limit; Phys Letts A **129**, (1988) 310.
[36] J.M. Ghidaglia & B. Heron, Physica **28D**, (1987), 282.
[37] M. Bartuccelli, C. Doering and J. D. Gibbon; 'An improvement on the dimension of the inertial manifold in the CGL equation' Preprint 1989.

S.V. Manakov and V.G. Kamensky

44. ON THE CREATION OF VORTICES UNDER PHASE TRANSITIONS

In this paper we shall be concerned with the problem of solving the so-called Ginzburg-Landau equation.

$$\psi_t = \Delta\psi - \alpha\psi - |\psi|^2\psi \tag{1}$$

which is known to be a reasonable model for the dynamics of the phase transitions. More precisely, we shall look for the solutions of the inital value problem for this equation which express some non-trivial topological properties, while the initial data do not.

The time-behaviour (and especially the <u>long-time</u> behaviour) of the solutions of (1) is drastically dependent on the sign of the parameter α. If α is positive, then any solution which decreases at infinity in space, tends to zero fast enough with time and $\psi=0$ is the only solution which is stable and attractive. This can be easily proved if one mentions that (1) can be written in the form

$$\psi_t = -\frac{\delta\mathcal{F}}{\delta\bar{\psi}}\ , \tag{2}$$

where functional $\mathcal{F}$ is given by

$$\mathcal{F} = \int(|\nabla\psi|^2+\alpha|\psi|^2+\frac{|\psi|^4}{2})dx^3 \quad (\geq 0 \text{ if } \alpha\geq 0) \tag{3}$$

(which is called "the free energy"), and that the time-derivative of $\mathcal{F}$ is non-positive :

$$\frac{d\mathcal{F}}{dt} = -2\int\left|\frac{\delta\mathcal{F}}{\delta\bar{\psi}}\right|^2 dx^3 = -2\int|\psi_t|^2 dx^3\ . \tag{4}$$

So only stationary solutions can play the role of an attractor. But, there is no such solution with zero boundary condition at infinity. Indeed, for time-independent solutions we have

$$\Delta\psi = \alpha\psi + |\psi|^2\psi$$

and by multiplying this equality by $\bar{\psi}$ and integrating over the whole space we get the equality

$$-\int|\nabla\psi|^2dx^3 = \int(\alpha|\psi|^2+|\psi|^4)dx^3 ,$$

which indicates a contradiction ("-" = "+") unless ψ is equal to zero for all **x**. At this point our analysis stops and we fail to add anything else for the positive α case. But for the qualitative analysis it is quite enough.

The situation with $\alpha<0$, which corresponding to the case of real physical significance, is much more complicated and demonstrates a rather interesting behaviour. First of all, $\psi=0$ is no more a stable solution. There is a variety of stable solutions of (1) which are bounded for all **x**, but do not decrease at infinity. The most stable (i.e. locally most attractive) among them are the form

$$\psi(\theta) = (\alpha)^{1/2}\exp(i\theta) \tag{5}$$

where θ is an arbitrary x-independent parameter. These solutions correspond to the absolute minimum of the free energy functional (3). So, if we start with initial condition decreasing at infinity, we can expect that the solution for time large enough will be of the form given by expression (5) almost everywhere with the parameter θ, probably, being a slowly varying function of **x** (and, consequently, of t). This guess is absolutely natural and, as we shall show later, is correct.

Apart from the solutions (5) there exists another family of stable stationary solutions of (1) which are of real physical interest. The following example is typical of such solutions. Let ψ be independent of z and have the form

$$\psi = \mathcal{A}(r)e^{\pm i\varphi} , \tag{6}$$

where r and φ are polar co-ordinates on the x-y plane and $\mathcal{A}$ is real. Then from (1) one gets

$$\frac{1}{r}\frac{\partial}{\partial r}\left(r\frac{\partial\mathcal{A}}{\partial r}\right) - \frac{\mathcal{A}}{r^2} + \alpha\mathcal{A} - \mathcal{A}^3 = 0 . \tag{7}$$

As has been shown by L.P. Pitaevskii (many years ago), (7) has a unique solution bounded for all r from zero to infinity. It behaves like cr^t for small r and tends to $|\alpha|^{1/2}$ when $r\to\infty$. The essential feature of this solution is that if one moves along any simple closed curve around the z-axis, the phase of $\psi(x,y,z)$ changes by $\pm2\pi$, but if the curve does not go around the z-axis no change takes place. Thus one gets a homotopic singularity (which, in some ways, is responsible for the stability of the solution).

Solutions of the type (6) and (7) are called vortex-lines and play an important role in the theories of superconductivity and superfluidity. In the first case they correspond to Abrikosov's vortices; in the theory of superfluidity they can be associated with Feynman vortices. We should point out that there exists no theory which provides a mechanism of vortex creation. Also we must say that all the above facts are more or less well known. Now we turn to the topic of this paper.

We consider the initial-value-problem for equation (1)

$$\psi_t=\Delta\psi+\psi-|\psi|^2\psi \ . \tag{8}$$

For convenience we scale away the parameter α, putting $\alpha=-1$. We take the initial value of ψ small enough, $|\psi(\mathbf{x},0)|\ll 1$ and decreasing at infinity faster than any exponential. Then, if time is not very large, we can replace (8) by its linear part:

$$\psi_t = \Delta\psi+\psi \ . \tag{9}$$

The solution of (9) is given by

$$\psi(x,t) = \int c(k)\exp[(1-k^2)t+i\mathbf{k}\cdot\mathbf{x}]dk^3 \tag{10}$$

where

$$c(k) = \frac{1}{(2\pi)^3}\int\psi(x,0)\exp(-i\mathbf{k}\cdot\mathbf{x})dx^3. \tag{11}$$

Remark

Since $\psi(\mathbf{x},0)$ vanishes rapidly at infinity $c(\mathbf{k})$ is an entire function of all three complex variables k_1,k_2,k_3.

Following our previous paper [1], we shall look at $\psi(x,t)$, given by (10), for large t and $|\mathbf{x}|$, putting $\mathbf{x}=\mathbf{v}t+\mathbf{y}$. Then, for t large, but not very large (so that $|\psi|$ is still small), we can compute the integral (10) via the stationary-point approximation:

$$\psi(\mathbf{x},t) \simeq \left(\frac{\pi}{t}\right)^{3/2} c\left(\frac{i}{2}\mathbf{v}\right)\exp\{(1-\frac{v^2}{4})t-\frac{\mathbf{v}\cdot\mathbf{y}}{2}\} \ . \tag{12}$$

From (5) we see that if $|\mathbf{v}|>2$ then $\psi(t)\to 0$, while for $|\mathbf{v}|<2$ it increases. This definitely indicates the so-called KPP (Kolmogorov, Petrovsky, Piskunov [2]) behaviour of the solution; that is, one should expect the radius of the wave-front solution to increase with speed equal to two. More precisely, let us put $\mathbf{v}=2\mathbf{n}$, $|\mathbf{n}|=1$. Then from (12),

$$\psi(x,t) \simeq \left(\frac{\pi}{t}\right)^{3/2} c(i\mathbf{n})\exp(-\mathbf{n}\cdot\mathbf{y}) \ . \tag{13}$$

i.e. the solution of the linearised equation (9) with $\mathbf{x}=2\mathbf{n}t+\mathbf{y}$ is practically time-independent and, moreover, depends only on one spatial variable $z=\mathbf{n}\cdot\mathbf{y}$. $\psi(x,t)$ of (3) decreases exponentially as $z\to+\infty$, that is ahead of what we expect to be the wave front. (In fact, one can estimate the position of the front of the wave from (13), setting $|\psi|$ to be of order one, which gives

$$z_w \sim -\frac{3}{2}\ell g t+\ell g|c(i\mathbf{n})| \ .$$

Behind the front the solution ψ of the full nonlinear equation (8) should tend to some stable solution of this equation.

All these words are nothing but a motivation for an attempt to match a linear tail of the solution (13) with a special solution of (8) which depends only on $z=\mathbf{n}\cdot(\mathbf{x}-2\mathbf{n}t)=\mathbf{n}\cdot\mathbf{x}-2t$. Looking for $\psi(z)=\psi(\mathbf{n}\cdot\mathbf{x}-2t)$ as a solution of (8) one finds that $\psi(z)$ satisfies

$$\psi_{zz}+2\psi_z+\psi-|\psi|^2\psi = 0 . \tag{14}$$

Let us, for a while, study this equation. Putting $\psi=u(z)\exp(i\theta(z))$, with $u(z)$ and $\theta(z)$ real, one obtains from (14)

$$\begin{aligned} &u_{zz}-u\theta_z^2+2u_z+u-u^3 = 0 , \\ &2u_z\theta_z+u\theta_{zz}+2u\theta_z = 0 . \end{aligned} \tag{15}$$

From the second of these equations it follows that $\theta_z=\frac{c}{u^2}\exp(-2z)$, where c is some constant of integration. On substitution of the last expression into the first of the equations (15) we get

$$u_{zz}+2u_z+u-u^3-\frac{c^2}{u^3}e^{-4z} = 0 .$$

This equation has no bounded solutions unless $c=0$, which means that the phase of ψ is z-independent. This fact is very significant for us.

So, the only bounded solutions of (14) are the form

$$\psi = u(z)e^{i\theta} ,$$

with $u(z)$ solving

$$u_{zz}+2u_z+u-u^3 = 0 . \tag{16}$$

We fix the solution of (16) by the condition $\bar{u}(0)=\frac{1}{2}$. This <u>bounded</u> solution tends to one when $z\to-\infty$, and decreases monotonically as z increases, having asymptotic form

$$\bar{u}(z) = Ae^{-z}+Bze^{-z} , \quad z\to\infty . \tag{17}$$

The positive constants A and B can be obtained only numerically. All the other desirable solutions of (16) are of the form

$$u(z,z_0) = \pm\bar{u}(z+z_0) .$$

Now finally, we can match the linear tail (13), which is

$$\psi(z) = \left(\frac{\pi}{t}\right)^{3/2}c(i\mathbf{n})\exp(-z) , \tag{18}$$

with the solution of (14) given by

$$\psi(z) = e^{i\theta}\bar{u}(z+z_0) . \tag{19}$$

Comparing expressions (17) and (18) one can see that z_0 in (19) must be large. The leading term in z_0 in the expansion (17) is

$$\bar{u}(z+z_0) = Bz_0 e^{-z_0} e^{-z} , \quad z_0 \gg |z| . \tag{20}$$

Then, comparing (20) and (18) we have

$$Bz_0 e^{-z_0} e^{i\theta} = \left(\frac{\pi}{t}\right)^{3/2} c(i\mathbf{n}) . \tag{21}$$

From this equation we obtain

$$\theta = \theta(\mathbf{n}) = \arg c(i\mathbf{n}) . \tag{22}$$

Also we get $z_0(\mathbf{n},t)$ from the equation: $z_0 e^{-z_0} = B^{-1}(\pi/t)^{3/2} c(i\mathbf{n})$, so that

$$z_0(\mathbf{n},t) \sim -\lg\left[\frac{c(i\mathbf{n})}{Bt^{3/2}}\right] . \tag{23}$$

$\theta(\mathbf{n})$ defines the phase of the solution behind the wave front and z_0 gives us the position of the front :

$$z_w = -z_0 . \tag{24}$$

Taking into account that $\mathbf{n}$ is an arbitrary unit vector, we can write down the asymptotic $t\to\infty$ solution of our initial value problem:

$$\psi(\mathbf{x},t) = \bar{u}(|\mathbf{x}|-2t+z_0(\mathbf{n},t))\exp(i\theta(\mathbf{n})) , \tag{25}$$

where $\mathbf{n}=\frac{\mathbf{x}}{|\mathbf{x}|}$. This asymptotic solution is valid for $|\mathbf{x}|/t=O(1)$, $t\to\infty$ and for all directions $\mathbf{n}$ if function $c(i\mathbf{n})$ has no zeros. Let us recall $c(i\mathbf{n})$ from (11):

$$c(i\mathbf{n}) = \frac{1}{(2\pi)^3}\int\psi(\mathbf{x},0)e^{\mathbf{n}\cdot\mathbf{x}}dx^3 . \tag{26}$$

$c(i\mathbf{n})$ is a complex valued function on S^2 and, in the generic case has an even number of zeros. ($c=c_R+ic_{Im}$, equations $c_R(i\mathbf{n})=0$, $c_{Im}(i\mathbf{n})=0$ define two closed curves on S^2, and crossings of these curves give zeros of $c(i\mathbf{n})$; then generically, two closed curves cross each other an even number of times). If $c(i\mathbf{n}_0)=0$, expansion (25) breaks down because $z_0=\infty$. Nevertheless, (25) is still valid for any $\mathbf{n}$ in a neighbourhood of $\mathbf{n}_0$ (non-uniformly in time, of course).

Now, let us look at the $\theta(\mathbf{n})$ when $|\mathbf{n}-\mathbf{n}_0|\ll 1$. Putting $\mathbf{n}=\mathbf{n}_0+\mathbf{x}$, $|\mathbf{x}|\ll 1$, $\mathbf{n}\cdot\mathbf{x}=0$, we can write for $c(i\mathbf{n})$ ($c(i\mathbf{n}_0)=0$)

$$c(i\mathbf{n}) = (\mathbf{a}\cdot\mathbf{x})+O(x^2) , \tag{27}$$

where $\mathbf{a}$ is a complex-valued vector. By introducing $\xi=x_1+ix_2$ we can rewrite (27) in the form

$$c(i\mathbf{n}) = a_1\xi + a_2\bar{\xi} \ . \tag{28}$$

From this expression it is clear, that if one moves along any closed curve around zero in the complex ξ-plane, the arguement of c(i**n**), (28) changes by $\pm 2\pi$ if $|a_1| \neq |a_2|$, but this case is rather exceptional. Now, from (25) we see that our solution provides homotopic singularities of the vortex type described above. So we can say that the solution (25) contains, in general, an even numbers of vortices. This statement leads to the conclusion that vortices are created, generically, through a process of blowing-up of the unstable "vacuum" $\psi=0$; that is, through a transition from initially an unstable solution to a stable one which is a typical process in the dynamics of the phase transitions.

Acknowledgements

The authors are extremely grateful to A.P. Fordy and P.M.Santini for their great help in the translation and printing of this paper.

References

[1] V.G. Kamensky, S.V. Manakov JETP Lett. **45**, 638 (1987).

[2] A.N. Kolmogorov, I.G. Petrovskii, N.S. Piskunov, Bull. Moscow State University, *Math. and Mechanics*, **1**, 1 (1937).

C.J. Camacho and F. Lund

45 GENERATION OF SOLITONS IN NON-INTEGRABLE SYSTEMS

The generation of solitons in non-integrable systems is studied. Specifically, we have solved the Boussinesq equation on a half-line with box-type boundary conditions and we have considered the scattering of Toda solitons by an interface, consisting of an infinite Toda lattice in which the spring constant in one half differs from the spring constant in the other half. In both cases we find a very efficient soliton generating mechanism with most of the initial energy generating solitons with a very small amount of radiation.

1 Introduction

One of the most interesting challenges of soliton physics is to understand how these nonlinear objects are coupled to an external source or inhomogeneity: what is their excitation mechanism?

From a mathematical point of view, one is interested in solving a differential equation such as KdV, sine-Gordon, Boussinesq, Toda and the like, with additional terms representing dissipation or external forcing, or in which the coefficients are space dependents [6,7,8,9]. From a physical point of view, one has localized lumps carrying energy and momentum, and the question arises, how can they be excited and in which way will they evolve? [1,2,5,11]

One possible line of attack is to start with the exact solution of the ideal case, obtained from the inverse scattering method, and suitably perturb it to obtain information useful to the nonideal case. Another method that has proven useful is numerical experiment; particularly useful when deviations from integrability are strong. In this work we take the second approach, studying two different problems. The first one is the solution of the Boussinesq equation in the half line with a time dependent boundary condition. In the second one, we studied what happens to a Toda soliton as it hits an interface.

2 Boussinesq equation

Chu, Xiang & Baransky [4] solved numerically the KdV equation with box-type boundary conditions. The interesting result found by them was that a number of identical solutions peel off this boundary condition, their amplitude being twice the height of the boundary perturbation, and their number depending on the duration of the perturbation.

In order to understand the generality of these results we have solved numerically the Boussinesq equation

$$U_{tt} = \beta^2 U_{xx} + \gamma(U^2)_{xx} + \varepsilon U_{xxxx} \tag{1}$$

on the half-line with boundary conditions similar to those considered by Chu et al. [4], which are also relevant to the generation of Love wave solutions [10]. The boundary condition $U(0, t)$ we take is shown in Fig. 1. Initial conditions are $U(x, 0) = U_t(x, 0) = 0$.

The numerical scheme we use away from $x = 0$ is:

$$\begin{aligned}(\Delta t)^{-2}(U_j^{n+1} - 2U_j^n + U_j^{n-1}) &= \beta^2(U_{j+1}^n - 2U_j^n + U_{j+1}^n)(\Delta x)^{-2} \\ &+ \gamma[(U_{j+1}^n)^2 - 2(U_j^n)^2 + (U_{j-1}^n)^2](\Delta x)^{-2} \\ &+ \varepsilon(U_{j+2}^n - 4U_{j+1}^n + 6U_j^n - 4U_{j-1}^n + U_{j-2}^n)(\Delta x)^{-4}\end{aligned} \tag{2}$$

where we have defined $U_j^n \equiv U(j\Delta x, n\Delta t)$. However we should modify

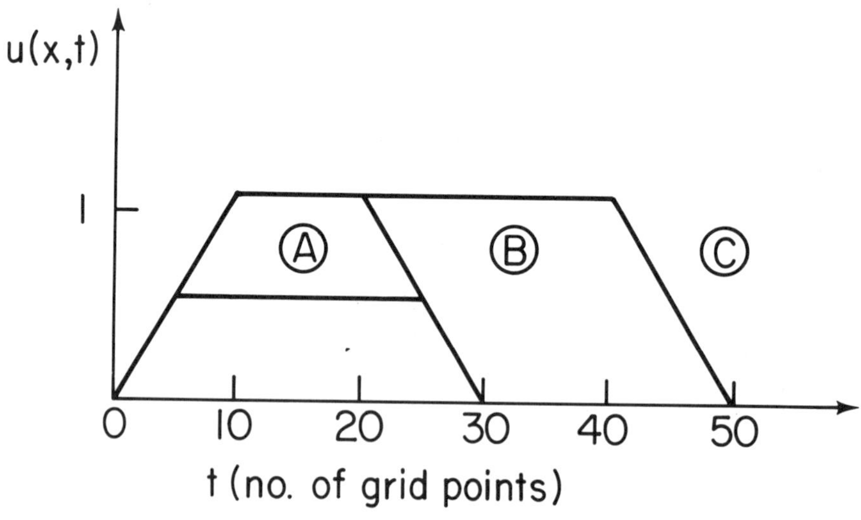

Figure 1. The three different boundary conditions used to solve eq. (1) on the half-line.

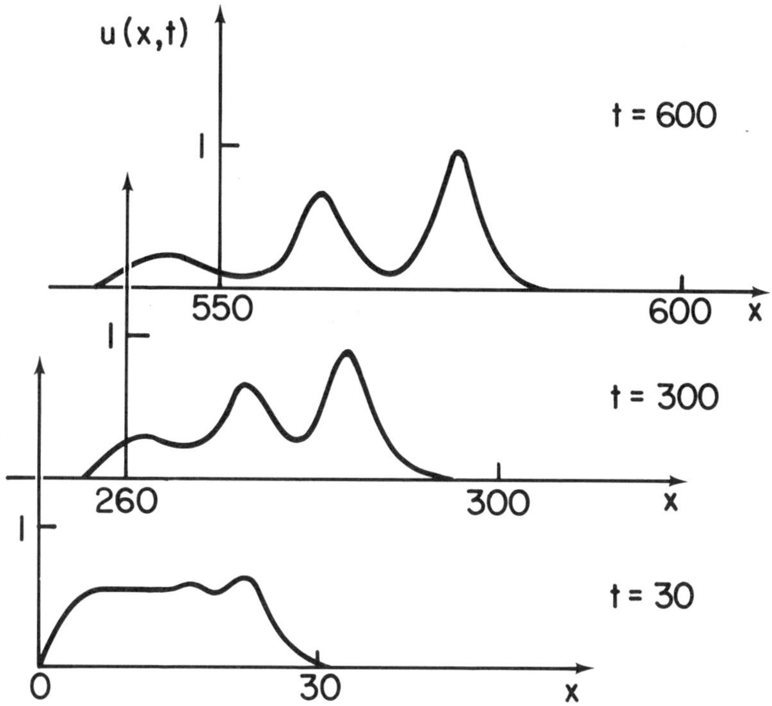

Figure 2. The time evolution of Boundary condition A. Three solitons are formed. The amplitude of the leading one is about twice the height of the boundary perturbation. Time and space are given in number of grid points.

the discretization for j = 1, where we use a scheme similar to the one of Chu et al. [4], approximating values at the point $x = 3\Delta x/2$, $t = (N + 1/2)\Delta t$ by

$$U_{xx} \simeq (2\Delta x)^{-2}(\bar{U}_0 - \bar{U}_1 - \bar{U}_2 + \bar{U}_3),$$

$$(U^2)_{xx} \simeq (2\Delta x)^{-2}((\bar{U}_0^2) - (\bar{U}_1^2) - (\bar{U}_2^2) + (\bar{U}_3^2)),$$

$$U_{xxxx} \simeq (\Delta x)^{-4}(\bar{U}_0 - 4\bar{U}_1 + 6\bar{U}_2 - 4\bar{U}_3 + \bar{U}_4)$$

$$U_{tt} \simeq (2\Delta t)^{-2}(1/2U_0^{n+2} - U_0^{n+1} + 1/2U_0^n + 3/2U_2^{n+2}$$

$$- 2U_2^{n+1} - 1/2U_2^n + U_1^{n+1} - 2U_1^n + U_1^{n-1} + U_2^{n-1}), \qquad (3)$$

where $\bar{U}_j \equiv 1/2(U_j^{n+1} + U_j^n)$.

We have chosen $\beta = 5.5$, $\gamma = 4$ and $\varepsilon = 100$; $\Delta t = 0.68$ and $\Delta x = 4$. For numerical stability of the wave equation one must have $\Delta x >$

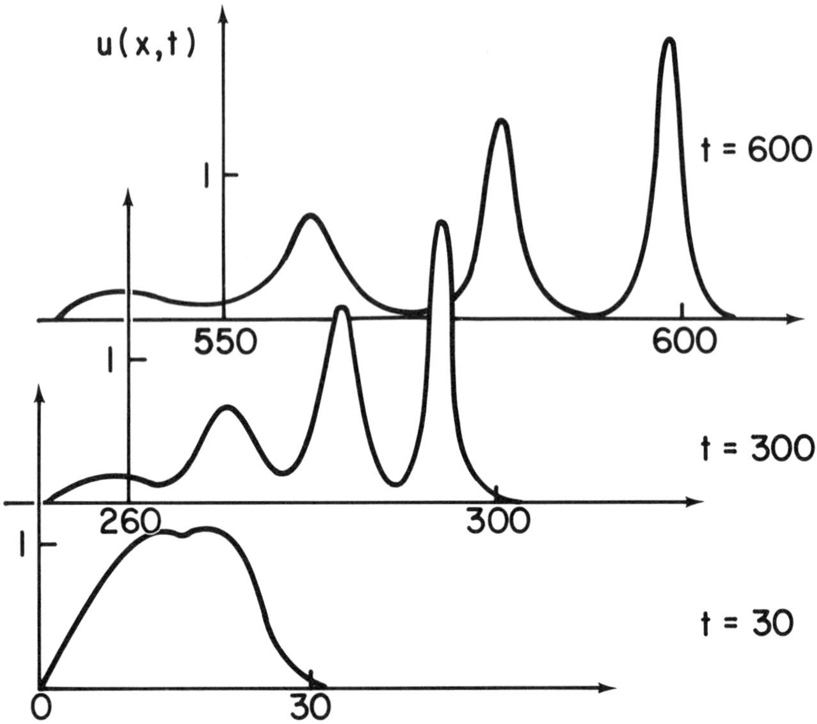

Figure 3. The time evolution of Boundary condition B. Four solitons are formed. The amplitude of the leading one is about twice the height of the boundary perturbation. Time and space are given in number of grid points.

$\beta\Delta t$. Moreover, short wavelengths have to be cut off and we found that for β, γ and ε as given above, good results were obtained.

The evolution for the three different boundary conditions of Fig. 1 is given in Figs. 2-4. Clearly, for small times the $U_{tt} = \beta^2 U_{xx}$ portion of the equation dominates as it should. Afterwards the nonlinear and dispersive terms become dominant and the original perturbation breaks up to form a number of solitons. As for KdV [4] their number increases with height and width of the boundary condition. As opposed to what happens in the KdV case, however, they are not all of the same height; rather, their amplitudes increase roughly linearly. The amplitude of the leading soliton is quite sensitive to the height of the boundary perturbation but not to its width, and is roughly twice the amplitude of the boundary perturbation, as for the KdV solitons.

As a check on the numerics we verified that

$$\int_0^\infty \frac{\partial U}{\partial t}\, dx = 0$$

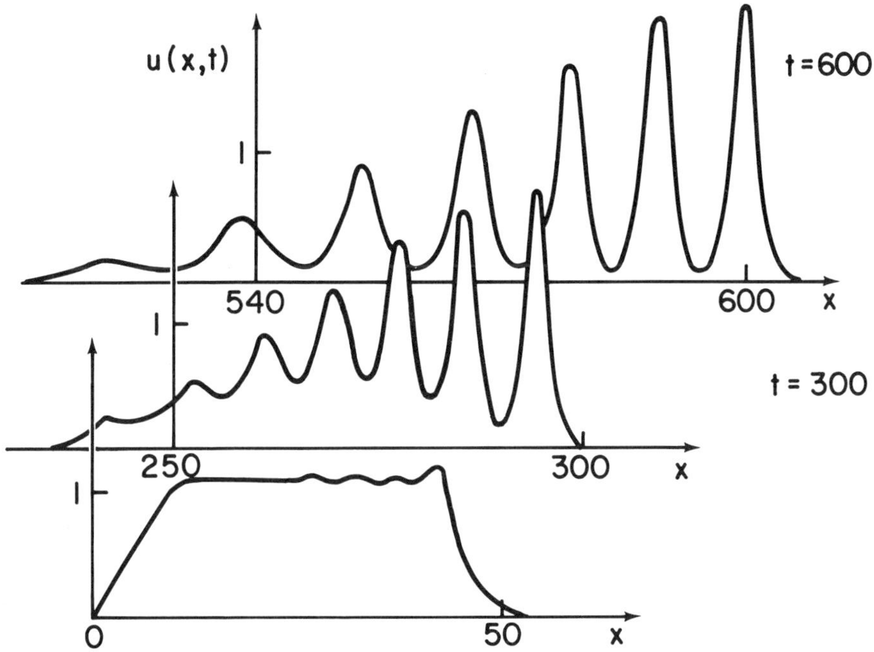

Figure 4. The time evolution of Boundary condition C. Seven solitons are formed. Here as well the leading soliton grows to twice the amplitude of the boundary perturbation. Time and space are given in number of grid points.

after the perturbation had died out. Also the bumps in Figs. 2-4 were fitted with the analytical solution and an excellent agreement was found, well within the numerical accuracy.

3 Scattering of Toda solitons by an interface

The Toda lattice is a one dimensional system of masses connected by nonlinear springs. We have considered as a model system for an interface, a surface where the properties of the medium have a discontinuous jump, an infinite Toda lattice with two different spring constants. The reason for this choice is that it is already discrete (in space), as appropriate for numerical work, and the case $b = b'$ is integrable and has well known soliton solutions [12]. Its Hamiltonian is

$$
\begin{aligned}
H = {} & 1/(2m) \sum_{n=-\infty}^{\infty} p_n^2 \\
& + a \sum_{n=-\infty}^{-1} \{ b^{-1} (e^{-b(q_{n+1} - q_n)} - 1) + q_{n+1} - q_n \} \\
& + a \sum^{\infty} \{ b'^{-1} (e^{-b'(q_{n+1} - q_n)} - 1) + q_{n+1} - q_n \}
\end{aligned} \qquad (4)
$$

the energy and momentum of this system are conserved but, to the best of our knowledge, there are no higher conserved quantities.

Toda solitons centred at the origin at time $t = 0$ have only one free parameter, say K, which determines their energy, amplitude and speed. Their explicit form is [12]

$$e^{-br_n} - 1 = [m\beta^2/(ab)] \operatorname{sech}^2(Kn \pm \beta t),$$

where $r_n = q_n - q_{n-1}$ and $\beta = (ab/m)^{1/2} \operatorname{sh} K$.

They have an amplitude

$$q_{-\infty} - q_{+\infty} = 2K/b,$$

which is the property we use to measure K for the scattered solitons. Their energy E and momentum P are

$$E = (2a/b)(\operatorname{sh} K \operatorname{ch} K - K),$$

$$P = \pm\, 2(ma/b)^{1/2} \operatorname{sh} K.$$

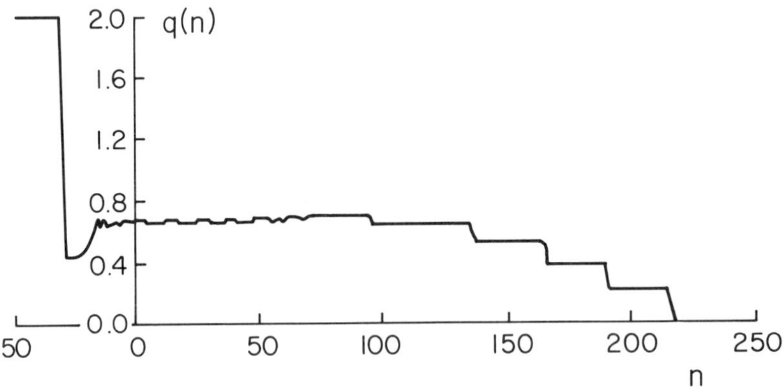

Figure 5. Solutions of the equations of motion for m = a = 1, (b,b') = (2,30), time step = 0.008, after 3000 iterations. The interface is at n = 0. There is one large amplitude reflected soliton travelling to the left, five smaller amplitude transmitted solitons travelling to the right and the slower radiation is in between.

In our case, $b \neq b'$, solutions localized far to the left or to the right behave as if the lattice were infinite, and it makes sense to speak of solitons in these regions. A soliton incident from the left will evolve, for large times, into a solution consisting of radiation and new solitons travelling to the left ("reflected") or to the right ("transmitted"). We have thus an inelastic scattering of a Toda soliton by an interface and we shall investigate the final states appearing for a variety of initial conditions and interface discontinuities.

The dynamical equations for the Hamiltonian (4) were solved numerically by a straightforward replacement of time derivatives by finite differences, with initial conditions corresponding to a soliton incident from the far left, and its evolution was followed. A typical situation is shown in Fig. 5. In general, new solitons, reflected and transmitted, were generated by the interface, and the evolution was followed until it appeared that soliton generation was complete. As a check on the numerics we verified that the energy (4) was preserved in time to within 0.2%.

A first type of experiment consisted in changing the energy of the incident soliton with fixed b, b′ and a second, changing b′ for fixed b and incident energy. The notation K_i, K_r, K_t will refer to the incident, reflected, and transmitted solitons, respectively.

3.1 *Incident solitons of variable energy; spring constants fixed*

Figure 6 shows results for a soliton incident on a softer lattice, $b = 2$, $b' = 0.5$, for several K_i. There is one reflected and one

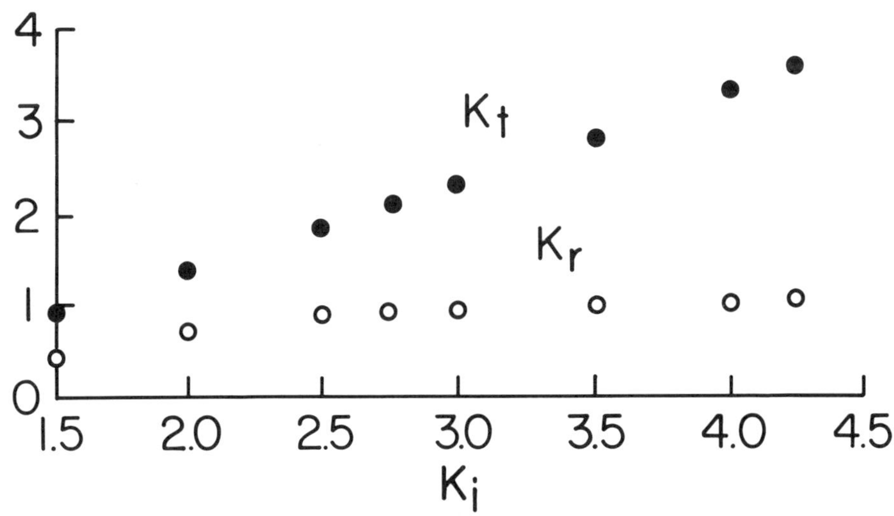

Figure 6. K_r and K_t are functions of K_i for a soliton incident on a softer lattice: $b = 2$, $b' = 0.5$. There is one reflected and one transmitted soliton in all cases.

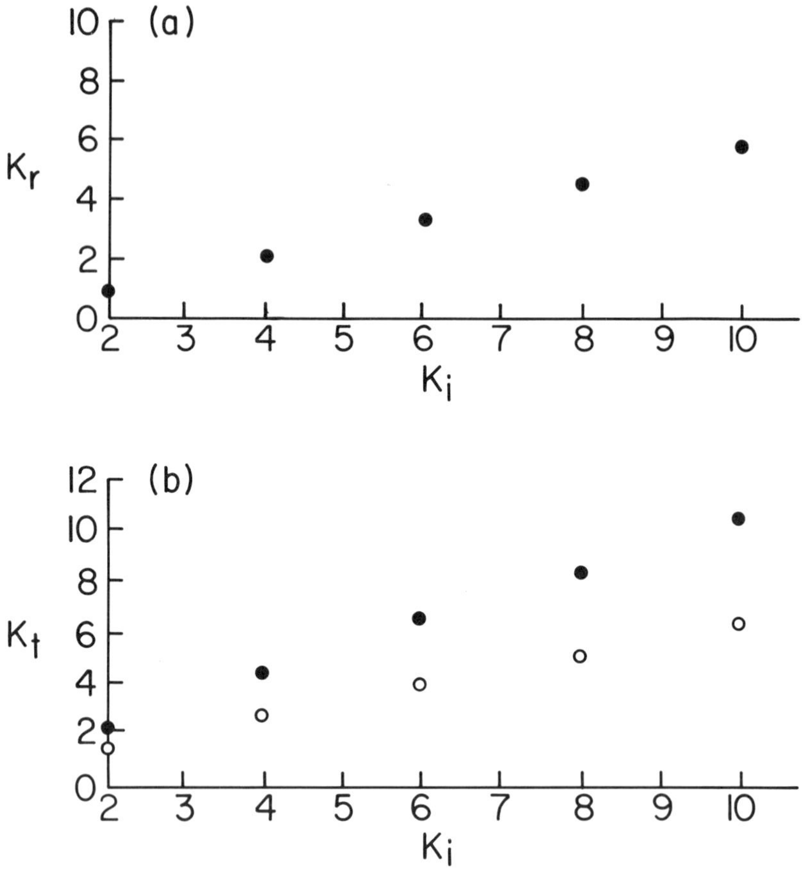

Figure 7. The case $b = 2$, $b' = 5$ at varying K_i. There is one reflected (a) and two transmitted (b) solitons.

transmitted soliton and their K's are plotted as functions of K_i. Results for a soliton incident on a harder lattice are shown in Figs. 7-8 in order of increasing spring constant contrast.

3.2 *Incident soliton of fixed energy; b fixed, b' variable*

Results for $K_i = 2$, $b = 2$ are shown on Fig. 9, in which K_r and K_t are plotted in terms of b'. There is one reflected soliton and the number of transmitted solitons increases with b' from one to five. Figure 10 has similar information for $K_i = 6$. In this case we did not find a reflected soliton. The features of the transmitted solitons are similar to the case $K_i = 2$.

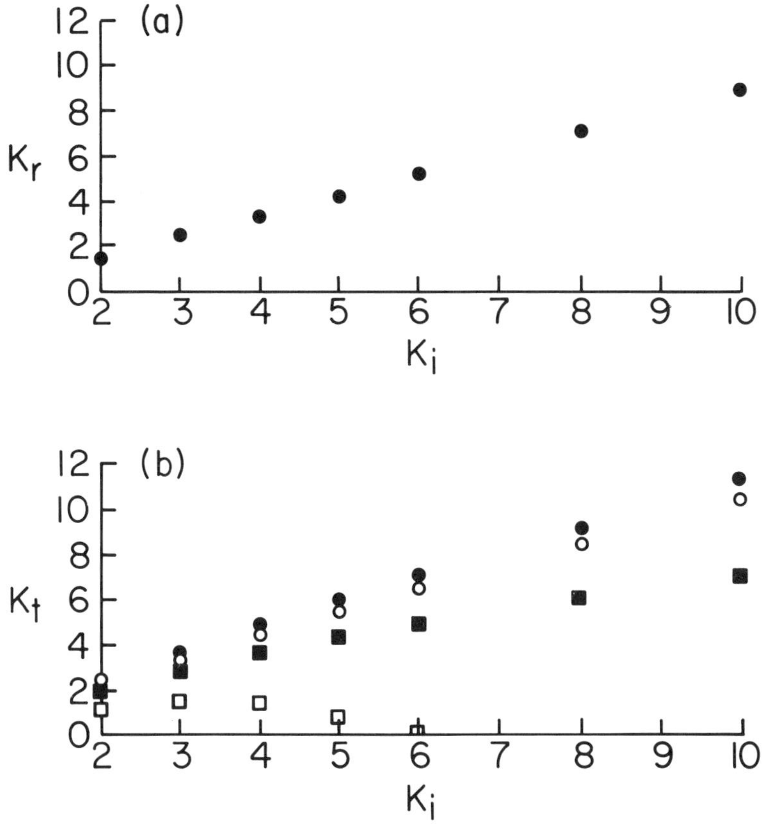

Figure 8. The case b = 2, b′ = 20 at varying K_i. There is one reflected soliton (a) and four (resp. three) transmitted ones (b) when $K_i \lesssim 6$ (resp. $K_i \gtrsim 6$).

Discussion

For the parameter range considered in this work, soliton behaviour at an interface is very different according to whether it hits a harder (b′ > b) or softer (b′ < b) lattice. For a soliton incident on a softer lattice, it is seen from Fig. 6 that there is only one reflected and one transmitted soliton, whose energies increase with the energy of the incident soliton. K_r increases more slowly than K_t in accord with the intuitive notion that a more energetic soliton should be less disturbed by the interface. For incidence on a harder lattice, the number of transmitted solitons increases, being higher the harder the lattice. However, for b′ = 20 we found that the number of transmitted solitons decreased as the incident

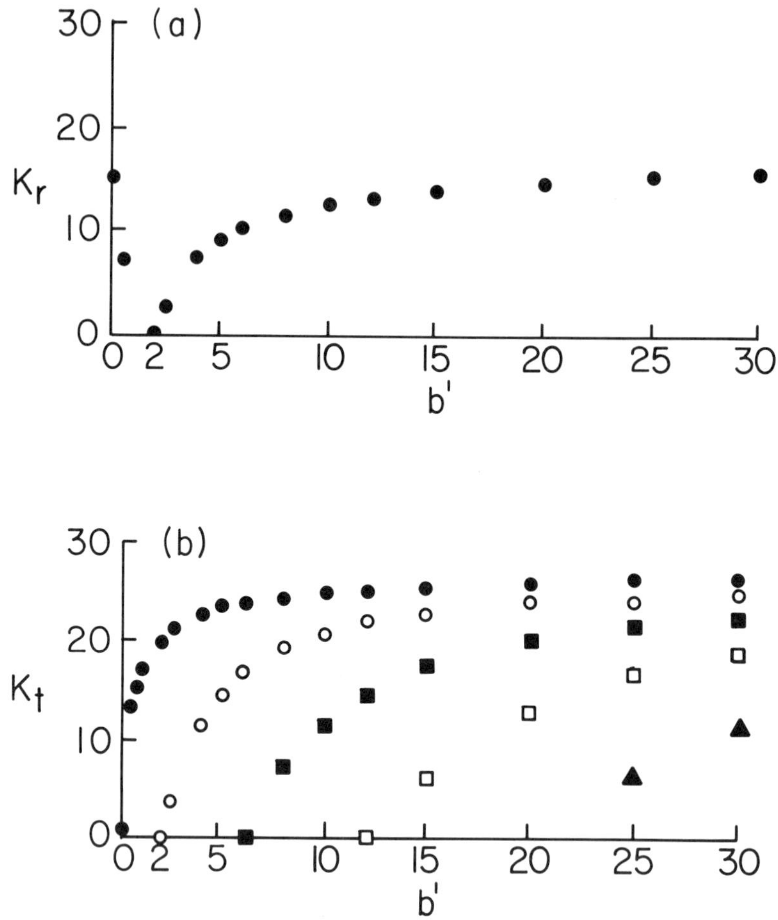

Figure 9. One incident soliton of K_i = 2, b = 2 with b' variable. (a): K_r in terms of b', there is ony one reflected soliton; (b): there are form one to five transmitted solitons whose K_t are plotted in terms of b'.

energy increased. This, and the fact that the energy of trailing transmitted solitons increases with K_i more slowly than the energy of the leading transmitted soliton (cf. Figs. 7-8) confirms the notion already mentioned that solitons of high enough energy should hardly be affected by an interface.

Another point worthy of note and in which soliton behaviour for incidence on a softer or harder lattice is qualitatively different, is the amount of energy that is scattered in the form of radiation. For b' < b and low K_i a significant fraction of the incident energy is radiated away, the amount of which rapidly

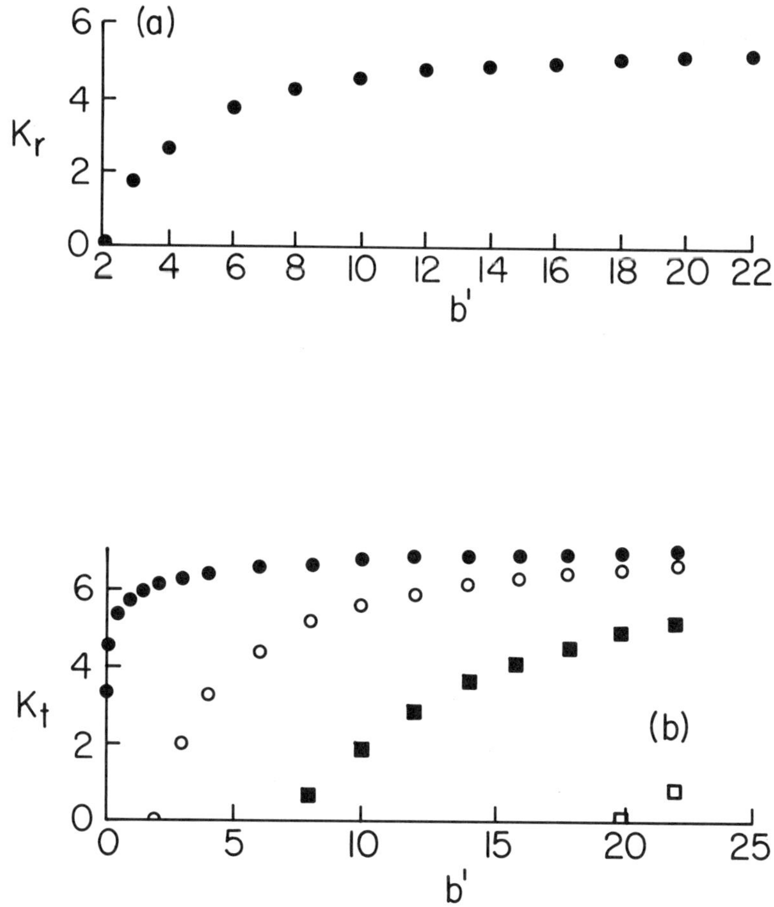

Figure 10. For an incident soliton of K_i = 6, b = 2, variable b' there is one reflected soliton (a) if b' > 2 and from one to four transmitted ones (b).

decreases as K_i is increased; again, the interface is only a small perturbation for a high-energy soliton. When b' > b the energy that goes into radiation is negligible [3].

At fixed incident energy, Figs. 9-10 show that the number of transmitted solitons steadily increases with b' for b' > b. That is, a steep interface is an efficient soliton generator. For K_i = 2, b' < 2 there is only one reflected soliton whose energy increases as b' decreases, that is as the interface gets sharper, but there is also a lot of radiation [3]. For K_i = 6, b' < 2 we did not find a reflected soliton, in all likelihood because it would take more time than we have available for it to appear.

Again, for b′ > b the incident soliton is scattered mostly into new solitons, the energy going into radiation being less than 2%.

4 Concluding remarks

We have studied the generation of solitons solving the Boussinesq equation on a half-line with box-type boundary conditions and the inelastic scattering of Toda solitons by an interface, consisting of two semi-infinite Toda lattices of different spring constants. Apart from the quantitative behaviour described above, the main qualitative observation is that most of the initial energy is scattered as solitons with a very small amount of radiation and in both cases we found a very efficient soliton generation mechanism.

In solving the Boussinesq equation we found some similarities with the results for the KdV equation (dependence of the number and amplitude of solitons with the boundary condition) and some discrepancies (height distribution of solitons) believed to come from the wave equation part of the Boussinesq equation which requires a transient time before solitons can appear.

The scattering in the Toda lattice gives us an insight of how solitons are generated by an interface. Particularly interesting is the role that the parameter b′ plays in this process, where K_t as a function of b′ produces bifurcation type diagrams (cf. Figs. 9-10).

References

[1] Bishop, A.R., Campbell, D.K., Lomdahl, P.S., Horowitz, B. & Phillpot, S.R. (1984). Non-linear dynamics, breathers and photoinduced absorption in polyacetylene. *Synthetic Metal II* **9**, 223.

[2] Camacho, C. & Lund, F. (1985). Generation of Boussinesq solitons via boundary conditions. *Phys. Lett. A* **108**, 1.

[3] Camacho, C. & Lund, F. (1987). Inelastic scattering of Toda solitons by an interface. *Physica D* **26**, 379.

[4] Chu, C.K., Xiang, L.W. & Baransky, Y. (1983). Solitary waves induced by boundary motion. *Comm. Pure & Appl. Math.* **36**, 495.

[5] Druhl, K.J., Carlsten, J.L. & Wentzel, R.G. (1985). Aspects of soliton propagation in stimulated Raman scattering. *J. Stat. Phys.* **39**, 615.

[6] Ferguson, Jr., W.E., Flashka, H. & McLaughlin, D.W. (1982). Nonlinear normal modes for the Toda chain. *J. Comput. Phys.* **45**, 157.

[7] Fogel, M.B., Trullinger, S.E., Bishop, A.R. & Krumhansl, J.A. (1977). Classical particlelike behaviour of sine-Gordon solitons in scattering potentials and applied fields. *Phys. Rev. B* **15**, 1578.

[8] Kaup, D.J. (1984). The forced Toda lattice: an example of an almost integrable system. *J. Math. Phys.* **25**, 277.

[9] Knickerbocker, C.J. & Newell, A.C. (1985). Propagation of solitary waves in channels of decreasing depths. *J. Stat. Phys.* **39**, 653.

[10] Lund, F. (1983). Interpretation of the precursor to the 1960 great Chilean earthquake as a seismic solitary wave. *Pure Appl. Geophys.* **121**, 18.
[11] Sethna, J.P. & Kivelson, S. (1982). Photoinduced soliton pair production in polyacetylene: an instanton approach. *Phys. Rev. B* **26**, 3513.
[12] Toda, M. (1981). *Theory of Nonlinear Lattices.* Springer: New York.

Kimiaki Konno

46. COMPLEX DYNAMICAL SYSTEM FOR TODA LATTICE

1 Introduction

In a series of papers nonlinear interactions between solitons for the Korteweg-de Vries (KdV) equations, the Boussinesq equation [5,7] and the Hirota-Ito equation [6] have been investigated by extending one of the independent variables to complex. We have obtained much information by this extension. Indeed, soliton solutions for these equations are expressed as a sum of poles in the complex plane. Observing behaviour of the poles, we can clearly understand interactions among solitons and anti-solitons. As an example, some poles giving contributions to one soliton change into poles belonging to another soliton after collision. This is one of the essential mechanisms to explain the nonlinear interaction between solitons.

In this paper we will first discuss nonlinear interactions between solitons for the Toda lattice in the same way as the previous papers by extending an independent variable into complex. We introduce an auxiliary function in order to transform poles of soliton solution to zeros of the function. The zeros are numerically solved by using the Newton's method. We can identify the method with a dynamical system in the complex plane where these zeros correspond to attractive fixed points. Our problem is to determine regions of the plane such that an initial point converges ultimately to one of the attractive fixed points. This is, so called, the Cayley's problem [1,2,3]. The set of all points whose orbits converge to the fixed points is called the basin of attraction. We shall refer to it as the Fatou set. The main purpose of this paper is to present numerical results of the Fatou set on the system in order to investigate dynamics of the zeros of the function, which means that each zero does not exist independently from other zeros of the same soliton and also from those of other solitons. The Julia set can be obtained as a complement of the Fatou set. Since our dynamical system is constructed on the basis of the Toda lattice equation, soliton solutions contain the exponential functions which have essential singularity at infinity. Then, the Fatou set shows different behaviour from that of the polynomial and the rational functions.

In fact it will be shown that the immediate basin of attraction, which is a component of the Fatou set that contains one of the fixed points, stretches to infinity. Another important point is that, according to motion of solitons, the Fatou set and the Julia set change drastically.

Nonlinear interactions between solitons for the Toda lattice are discussed on the basis of behaviour of zeros of an auxiliary function in section 2. In section 3 a complex dynamical system is introduced and the Fatou set and the Julia set of the system are studied and the last section is devoted to the concluding remarks.

2 Nonlinear interations between solitons

The Toda lattice equation is given by

$$\frac{d^2}{dt^2} \log (1 + V_n) = V_{n+1} + V_{n-1} - 2 V_n. \tag{1}$$

We introduce an auxiliary function ϕ_n [4,8]:

$$V_n = \frac{d^2}{dt^2} \log \phi_n(t). \tag{2}$$

One soliton solution is derived from

$$\phi_n = 1 + A \exp(\eta), \tag{3}$$

with $\eta = kn-\beta t$ and $\beta^2 = 4 \sinh^2(k/2)$. Extending the variable t to complex, we study behaviour of simple zeros of the function. The zeros are placed at $t_m(n,k) = t_{Rm}(n,k) + i\ t_{Im}(n,k)$ ($m = 0, \pm 1, \pm 2, \ldots..$) with

$$\begin{aligned} t_{Rm} &= (kn + \log A)/\beta, \\ t_{Im} &= (2m+1)\pi/\beta. \end{aligned} \tag{4}$$

The zeros are spaced at constant intervals $2\pi/\beta$ along a line parallel to the imaginary axis and move along the real axis by depending upon n. The problem is to investigate how zeros of two parallel lines for two solitons behave when they collide.

The auxiliary function for a two soliton solution is given by

$$\phi_n = 1 + A_1\exp(\eta_1) + A_2\exp(\eta_2) + A_3\exp(\eta_1+\eta_2), \tag{5}$$

with $\eta_i = k_i n - \beta_i t$, $\beta_i^{\ 2} = 4 \sinh^2(k_i/2)(i = 1,2)$ and

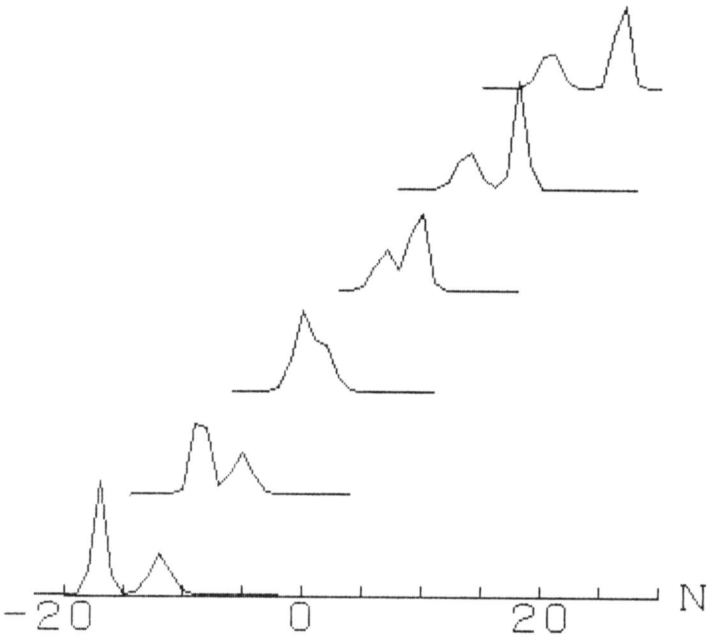

Figure 1. (a) Profiles of collision for two solitons running in the same direction from $t = -12$ to 18 with $k_1 = 3$, $k_2 = 1.8$, $A_1 = 1$ and $A_2 = 1$, (b) trajectories of zeros with real t axis vs. n and (c) with imaginary t axis vs. n. (For (b) and (c) see the following two pages.)

$$A_3 = - \frac{(\beta_1 - \beta_2)^2 - 4 \sinh^2 ((k_1 - k_2)/2)}{(\beta_1 + \beta_2)^2 - 4 \sinh^2 ((k_1 + k_2)/2)} A_1 A_2.$$

A_i and K_i are positive constants.

In Fig. 1a we illustrate profiles of collision of solitons with $k_1 = 3$ and $k_2 = 1.8$ running in the same direction where amplitudes among neighbouring positions are connected by lines. In Fig. 1b we show trajectories of zeros by projecting information of the imaginary axis in Fig. 1c to the real axis. We number trajectories of the zeros in Fig. 1c from one and upward from the bottom. The trajectories 3, 6 and 9 of the zeros for the large soliton at $n = -20$ change into the trajectories 2, 5, and 8 of zeros for a new small soliton at $n = 20$. (In the following we denote briefly a zero contribution to a soliton by a zero for the soliton.) The remainder of the zeros for the large soliton are rearranged and form an intermediate state with a wave number k ~

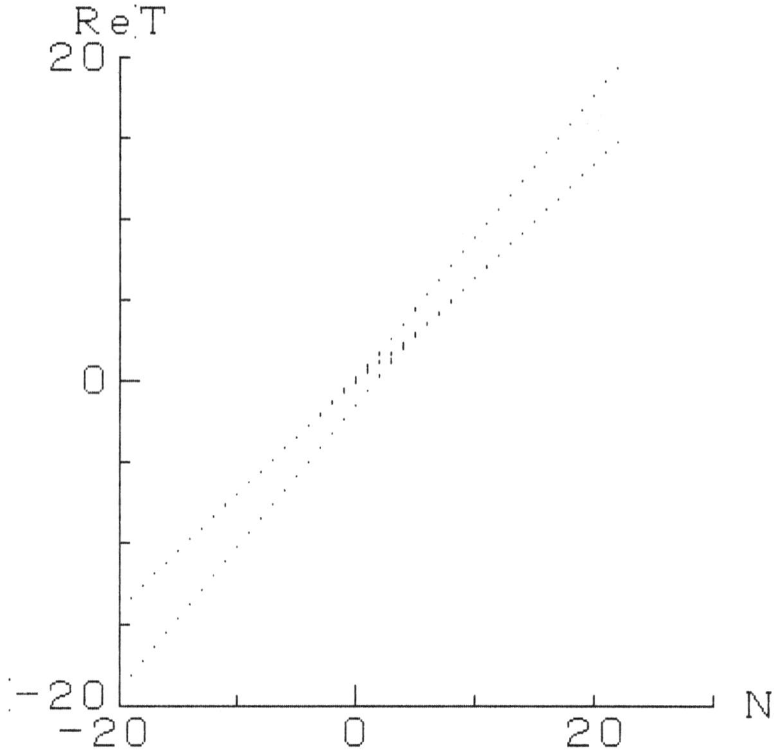

Figure 1(b).

$k_1 - k_2$ and an energy $\beta \sim \beta_1 - \beta_2$. It absorbs the zeros for the original small soliton of the trajectories 2, 5 and 8 at negative n and recovers zeros for the large soliton at the final stage.

We can also discuss head-on collision of solitons where we see a large peak of amplitude just at collision. We can understand this fact in such a way that the zeros for two solitons form an intermediate state with a wave number $k \sim k_1 + k_2$ and an energy $\beta \sim \beta_1 + \beta_2$ at $n = 0$ so that the imaginary part t_I of the first zero's trajectory decreases and affects the amplitude directly.

3 Complex dynamical system

The zeros of the auxiliary function are numerically calculated by the following Newton's method as

$$t_{i+1} = t_i - \frac{\phi_n(t_i)}{\frac{d\phi_n(t_i)}{dt_i}} \equiv f(t_i). \qquad (6)$$

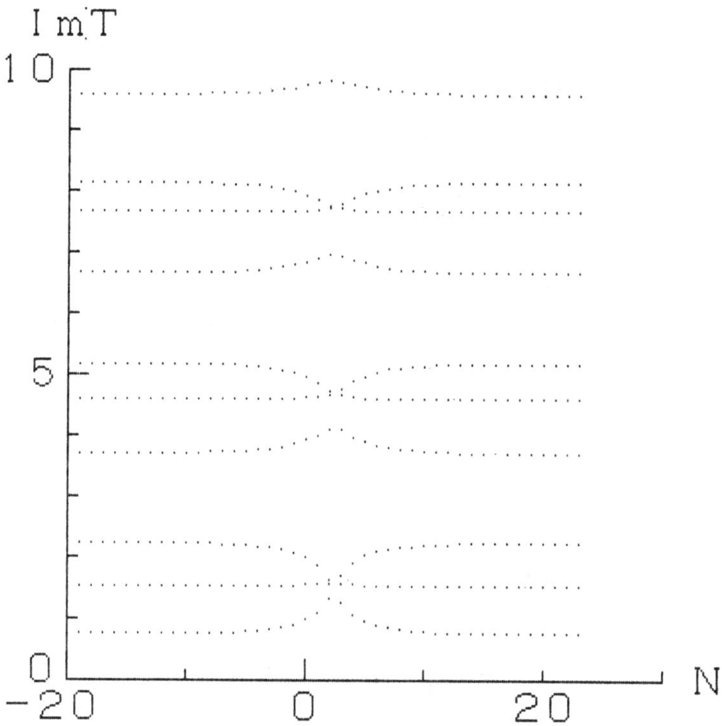

Figure 1(c).

We can identify this formula with a complex dynamical system. Fixed points of the map correspond to the zeros of ϕ_n. They are super-attracting since a derivative of f(t) with respect to t at t = $t_m(n,k)$ is vanishing identically. In order to obtain much information on dynamics of the zeros, we are interested in

searching for domain in the complex plane such that an initial point converges ultimately to one of the fixed points. This is the Cayley's problem based on the Toda lattice. Mostly the Julia set and the Fatou set of the polynomial and the rational functions are discussed [1,2,3]. Since ϕ_n has the exponential function, which has an essential singularity at infinity, these sets arouse interest.

We show the Fatou set of one soliton solution with n = 0 and k = 3 for initial conditions within $-10 \le t_R \le 5$ and $-6 \le t_I \le 6$. We find large disconnected immediate basins of attraction stretching to negative real infinity. In each of the basins one of t_m is included. On surrounding the basins, we find an infinite number of small basins of attraction, which are components of the immediate attracting basins. They show self-similar structure. Since we cannot use colour print in this book, we cannot distinguish fixed

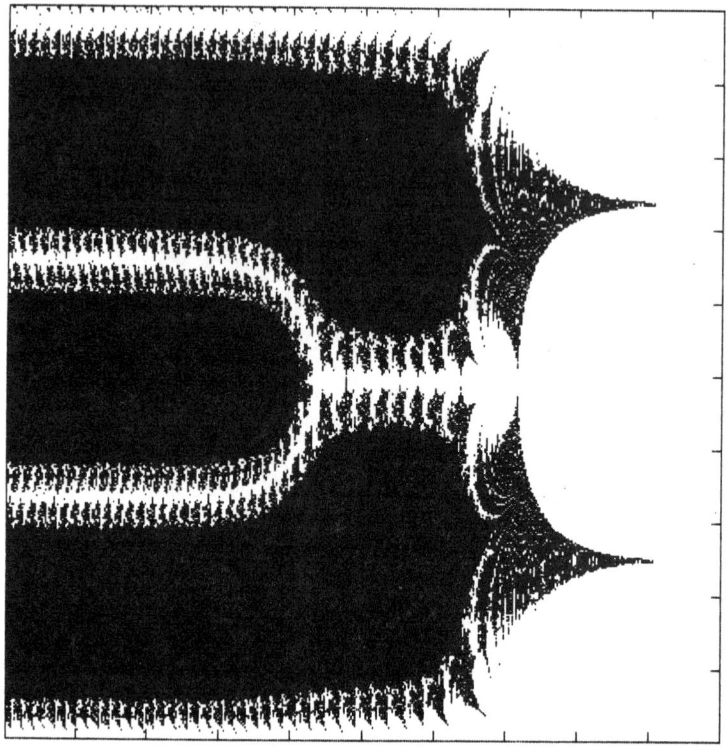

Figure 2. The Fatou set for one soliton of $k = 3$ and $A = 1$ with real t axis vs. imaginary t axis for $n = 0$ within $-10 \le t_R \le 5$ and $-6 \le t_I \le 6$.

points where the initial point converges.

The Fatou set for two soliton solution at $n = -4$ with $\beta_1/\beta_2 = 2$ of $k_1 = 3$ and $k_2 = 1.8527$ is shown in Fig. 3. An immediate basin of attraction for the small soliton is found at the middle of those for the large soliton. On surrounding those immediate basins, an infinite number of basins of attraction are found as in the case of one soliton. For $n = 3$ the immediate basin of attraction for new small soliton covers those for the large soliton. We find that the Fatou set has a self-similar structure.

For the case of head-on collision entire positive real infinity is not the Julia set. The structure becomes more complicated. For negative n the immediate basin of attraction for the small (large) soliton can be found in the right (left) half plane, respectively, and infinitely many components of the basins of attraction appear in the opposite side of the corresponding immediate basins. For positive n the immediate basin of attraction for the small soliton is in a limited region while those for the large soliton cover it

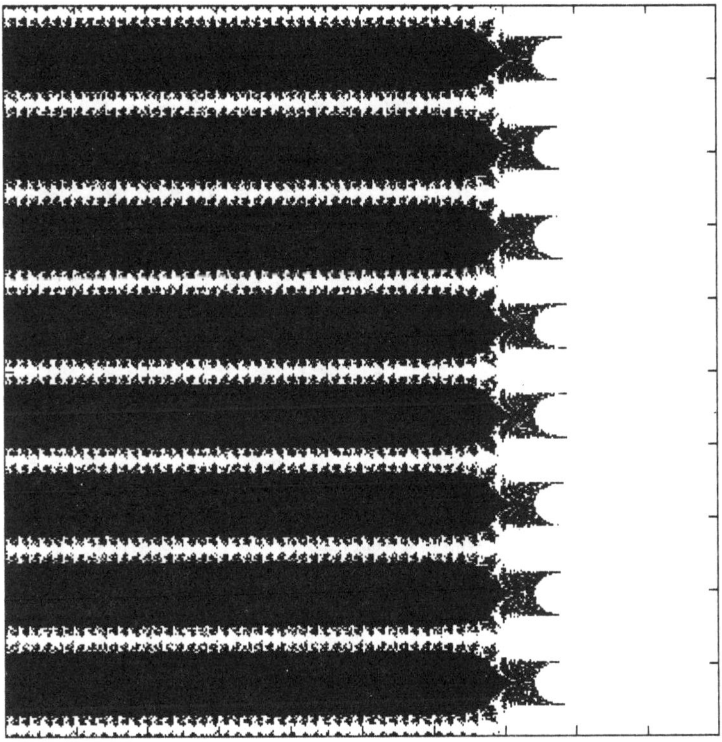

Figure 3. Fatou set for two solitons of $k_1 = 3$ and $k_2 = 1.8527$ ($\beta_1/\beta_2 = 2$), $A_1 = 1$ and $A_2 = 1$ with real t axis vs. imaginary t axis for $n = -4$ within $-6 \le t_R \le 4$ and $0 \le t_I \le 3$.

and still extend to both real infinite directions.

4 Concluding remarks

We have studied nonlinear interactions between solitons for the Toda lattice by extending an indpendent variable to complex with the help of the auxiliary function. For the case of collision of solitons running in the same direction, some of the zeros for the large soliton produce zeros for a new small soliton. The remainder of the zeros form an intermediate state with an energy $\beta_1 - \beta_2$ and absorb the zeros for the original small soliton. For the case of head-on collision we have found no such mechanism of exchanging zeros, but have seen the total formation of an intermediate state with an energy $\beta_1 + \beta_2$. These results are similar to those for the Boussinesq equations since $0 < A_3/A_1A_2 < 1$.

The Newton's method to calculate numerically zeros for the

auxiliary function has been identified with a complex dynamical system. We have investigated the Fatou set and the Julia set as its complement for one and two soliton solutions. Since the function has the exponential function, we have observed that the behaviour at infinity is very different from those for the polynomial and the rational functions. For one soliton solution, we have found large immediate basins of attraction which stretch to negative real infinity. Surrounding each of the basins, we have seen infinitely many distinct components of the other basins of attraction. They show the self-similar structure. For two soliton solution, the Fatou set changes as motion of solitons. Especially in the head-on collision, the Julia set changes drastically. For negative n it has interior, but for positive n it seems not to have.

The self-similarity of the Fatou set denotes that any neighbourhood of the basins of attraction is a miniature of the total system. Then if we search for the neighbourhood, we can obtain all information on the system. This means that each of the zeros has been influenced mutually by the zeros for all other solitons as well as by other zeros for the same soliton. This is our physical interpretation of the structure of the Fatou set.

In summary, we have totally discussed the nonlinear interactions between solitons by investigating behaviour of the zeros of the auxiliary function and the structure of the Fatou set and also the Julia set of the function.

References

[1] Blanchard, P. (1984). Complex analytic dynamics on the Riemann sphere. *Bull. Am. Math. Soc.* **11**, 85.

[2] Brolin, H. (1965). Invariant sets under iteration of rational functions. *Arkiv für Matematik* **6**, 103.

[3] Devaney, R.L. (1986). *An introduction to chaotic dynamical systems.* The Benjamin/Cummings Publishing Co. Inc.

[4] Hirota, R. (1973). Exact n-soliton solution of a nonlinear lumped network equation. *J. Phys. Soc. Jpn.* **35**, 286.

[5] Konno, K. (1987). Nonlinear interactions between solitons in complex t-plane. II. *J. Phys. Soc. Jpn.* **56**, 1334.

[6] Konno, K. (1989). Nonlinear interactions among solitons and anti-solitons in complex t plane for Hirota-Ito equation. In *Nonlinear wave motion,* Jeffrey, A. (ed.), p. 118. Pitman Monographs and Surveys in Pure and Applied Mathematics. Longman Scientific & Technical.

[7] Konno, K. & ito, H. (1987). Nonlinear interactions between solitons in complex t-plane. I. *J. Phys. Soc. Jpn.* **56**, 897.

[8] Toda, M. & Wadati, M. (1973). A soliton and two solitons in an exponential lattice and related equations. *J. Phys. Soc. Jpn.* **34**, 18.

H. Hasegawa

47. COMPLETE INTEGRABILITY AND STOCHASTIZATION OF YUKAWA'S EQUATION FOR QUANTUM CHAOS

1 Introduction

By the "Yukawa's equation for quantum chaos" we mean a set of multi-dimensional equations of motion which T. Yukawa [9] proposed to analyse in order to give a dynamical framework to the random-matrix theory familiar to statistical description of nuclear levels, in particular to generalize its basic notion of Gaussian ensembles [1]. It has turned out, by the author's own note [10], that the set of equations describes a complex dynamical system which is equivalent to the Euler-Calogero-Moser (ECM) system after the naming by Wojciechowski [8]. This dynamical system is of particular interest because of its complete integrability, a claim by itself of value to study as a rare example of complex Hamiltonian systems. Hence we have a look of curious coexistence of **randomness** and **complete integrability**. The purpose of this note is to introduce a logic to reconcile these opposite properties to discover a useful development of physics.

2 The ECM system and a quantum mechanical perturbation theory

This system can be defined by the direct product of a Hamiltonian system of N-degree of freedom and another system with the Euler's structure that its dynamical variables comprise the components of angular momentum which generate an N-dimensional rotation group O(N), i.e. $(q_n, p_n)_{n=1}^{N} \otimes (\ell_{nm}\ (=-\ell_{mn}))_{n,m=1}^{N}$. The equations of motion for this system are given by

$$\dot{q}_n = p_n, \quad \dot{p}_n = \sum_{m(\neq n)} \frac{2\ell_{nm}^2}{(q_n-q_m)^3} \tag{1}$$

$$\dot{\ell}_{nm} = \sum_{r(\neq n,m)} \left\{ -\frac{\ell_{nr}\ell_{rm}}{(q_n-q_r)^2} + \frac{\ell_{nr}\ell_{rm}}{(q_m-q_r)^2} \right\} \quad (n\neq m). \tag{2}$$

The first set of equations, (1), represents clearly a canonical system with Hamiltonian

$$\mathcal{H} = \frac{1}{2}\sum_{n=1}^{N} p_n^2 + \frac{1}{2}\sum_{n,m(n\neq m)} \frac{\ell_{nm}^2}{(q_n-q_m)^2} \tag{3}$$

whose potential part consists of a sum of inverse-squared distances between each pair of one-dimensional N-particles. We note that this potential is non-uniform in general as regards the particles, depending on the second set of variables quadratically, so that it is even time-dependent.

Wojciechowski [8] has given a satisfactory construction of the above dynamics by exploiting two kinds of Poisson brackets adapted to each subdynamics:

I. $$\{f,g\}_C = \sum_{n=1}^{N} \left(\frac{\partial f}{\partial q_n}\frac{\partial g}{\partial p_n} - \frac{\partial f}{\partial p_n}\frac{\partial g}{\partial q_n}\right) \tag{4}$$

in particular $\{q_n,q_m\}_C=\{p_n,p_m\}_C=0$, $\{q_n,p_m\}_C=\delta_{nm}$ for the canonical system $\mathscr{C}=(:q_n,p_n)$,

II. $$\{\ell_{nm},\ell_{rs}\}_L=\delta_{ns}\ell_{rm}+\delta_{nr}\ell_{ms}+\delta_{ms}\ell_{nr}+\delta_{mr}\ell_{sn} \tag{5}$$

as the structure relations of the Lie algebra L of the group O(N) and for the extended algeba $\mathscr{A}$ of L in which $\{\ ,\ \}_L$ is defined by the Leibnitz product formula.

Three main features of his result can be summarized as follows:

(a) A natural definition of Poisson bracket is introduced to $\mathscr{C} \otimes \mathscr{A}$ by $\{F,G\} \equiv \{F,G\}_C + \{F,G\}_L$ in which $\{F,G\}=0$ for any such F's and G's; $F\varepsilon\mathscr{C}$ and $G\varepsilon\mathscr{A}$. Then, the full equations of motion (1) and (2) can be put into the Hamiltonian form $\dot{F}=\{F,H\}$ inasmuch as

$$\dot{q}_n=\{q_n,H\},\ \dot{p}_n=\{p_n,H\},\ \dot{\ell}_{nm}=\{\ell_{nm},H\}. \tag{1'}$$

(b) Introduce three NxN real matrices $X=(X_{nm})$, $L=(L_{nm})$ and $\ell=(\ell_{nm})$, the first two symmetric, defined by

$$X_{nm} = \delta_{nm}q_n,\quad L_{nm} = \delta_{nm}p_n - (1-\delta_{nm})\frac{\ell_{nm}}{q_n-q_m}, \tag{5}$$

and the last antisymmetric by definition. There exists a Lax representation of the equations of motion for the second and third matrices in terms of another antisymmetric matrix A such that

$$\dot{L} = [A,L] \text{ and } \dot{\ell} = [A,\ell] \tag{6}$$

and that these conform, in part, to the starting equations of motion (1) and (2) or (1′). The antisymmetric matrix $A=(A_{nm})$ is given as

$$A_{nm} = -(1-\delta_{nm}) \frac{\ell_{nm}}{(q_n-q_m)^2} . \tag{7}$$

It implies the existence of a one-parameter family of N-dimensional rotations $R_t \varepsilon O(N)$, depending on time t, such that for an initial value problem of (6)

$$L_t = R_t L_0 R_t^T \text{ and } \ell_t = R_t \ell_0 R_t^T , \tag{8}$$

and that R_t satisfies (R^T denotes the transpose of R)

$$\dot{R}_t = A_t R_t, \quad R_t R_t^T = 1; \quad R_{t=0} = 1. \tag{9}$$

This does not give the full answer to the construction of dynamics. However,
(c) X_t satisfies an inhomogeneous Lax form, so to speak, i.e.

$$\dot{X}_t = [A_t, X_t] + L_t; \; X_{t=0}(\equiv X_0) = \text{independent of } L_0 \text{ and } \ell_0 \tag{10}$$

which yields an explicit solution (again in terms of R_t) as

$$X_t = R_t(X_0 + tL_0)R_t^T. \tag{11}$$

The two Lax representations in (6) and the inhomogeneous one (10) yield a set of matrix form of the equations of motion now equivalent to (1) and (2) or (1′). One may note that (11) represents a diagonalization process of an arbitrary real symmetric matrix X_0+tL_0 by means of an N-dimensional rotation $R_t \varepsilon O(N)$.

In the above summary, a deduction of (a) and (b) requires merely some matrix manipulations. To establish (c), Wojciechowski employed several useful identities which include $\{L,H\}=[A,L]$, $\{\frac{1}{2}\Sigma q_n^2,L\}=X$, $\{\frac{1}{2}\Sigma q_n^2,H\}=\Sigma p_n q_n$, $\{\Sigma p_n q_n,L\}=L$ and the Jacobi identity to deduce the inhomogeneous Lax form (10). Here, we give a supplementary argument about the most important result, namely the **complete integrability** of the ECM system characterized by (1) and (2).

The summary in (a)-(c) actually does not convince us that the system is in fact integrable, because it has not been prescribed yet how to integrate R_t from the initial data. The prescription should be contained in (11), which is discussed by Nakamura & Lakshmanan [5].

(d) Consider the eigenvalue problem associated with (11)

$$q_n(t)\, u_n(t) = (X_0 + tL_0)u_n(t). \tag{12}$$

For simplicity we assume the non-degeneracy of the eigenvalues. There exists a set of equations of motion for every raw vector of R_t; a linear differential equation for the vector with coefficients fully prescribed by the eigenvalues and eigenvectors, i.e. $\{q_n(t)\}$ and the representation in $\{u_n(t)\}$ of the initial matrix L_0 or ℓ_0:

$$(\dot{R}_t)_{in} = \sum_{m(\neq n)} (R_t)_{im} \frac{L_{0;mn}(t)}{q_m(t)-q_n(t)} \tag{13a}$$

$$= \sum_{m(\neq n)} (R_t)_{im} \frac{-\ell_{0;mn}(t)}{(q_m(t)-q_n(t))^2} \tag{13b}$$

Remark. From (5), one has $-\ell_t=[X_t,L_t]$ and, by substituting (8) and (11) into this, also $-\ell_0=[X_0+tL_0,L_0]=[X_0,L_0]$. Thus, all the initial data are contained in X_0 and L_0. It also assures the equivalence of the two representations (13a) and (13b), in which the latter is obtained from $\dot{R}_t=R_tA_0(t)$ with $A_0(t)=R_t^T\ A_tR_t\ A_t$ given in (7)); the equation of motion (9) for R_t **intertwined** into this form.

We can now observe that the dynamics of the ECM system is equivalent to the transformation to diagonalize a real symmetric matrix which depends on time linearly. This allows an interpretation that such a matrix reresents the Hamiltonian of a finite quantum system perturbed, the time t just representing the perturbation strength. The complete integrability of the ECM system, therefore, can be seen from that of such matrices (NxN real, symmetric) whose degree of freedom is equal to $\frac{1}{2}N(N+1)$ and each matrix element proportional to t of which gives a constant of motion [10].

3 Yukawa's reformulation of GOE and a generalization

The Gaussian orthogonal ensemble (GOE), a central notion of the random matrix theory [7], can be defined as a class of those NxN real symmetric matrices H (Hamiltonians with time reversal, without spin) to satisfy

(i) all the matrix elements of H are random variables independent of each other,

(ii) the distribution of these elements is invariant by arbitrary orthogonal transformations made to H.

Let the (density of the) joint distribution of such GOE matrices be denoted by P(H). The above two requirements fix its form as

$$P(H) = C_N\, e^{-\frac{1}{4a^2}\mathrm{Tr}(H-\bar{E}1)^2} \tag{14}$$

where $\bar{E}$ is the mean (energy), $\bar{E} = \frac{1}{N}\langle \mathrm{Tr} H\rangle$, (14a)

and a^2 is related to the variances

$$a^2 = \langle H_{ij}^2\rangle \;(i\neq j), \; = 2^{-1}\langle H_{ii}^2\rangle. \tag{15}$$

In these averaging manipulations, the measure dH is defined by

$$dH = \prod_{i=1}^{N} dH_{ii} \prod_{i,j(i>j)}^{N} dH_{ij} \tag{16}$$

so that

$$\langle\cdot\rangle = \int dH\, P(H)\cdot \; . \tag{16a}$$

which fixes the normalization constant C_N to be $(4\pi a^2)^{-\frac{N(N+1)}{2}}$.

It is then possible to find the joint distributin of the eigenvalue energies of H merely by getting the Jacobian J of the measure, when going over to the diagonal representation H_D of H, i.e. $H_D = OHO^T$, from dH to dH_D:

$$J_N\, dH_D \equiv J_N\, dE_1 \ldots dE_N\, \phi(\alpha) \prod_{i>j} d\alpha_{ij} = dH. \tag{17}$$

We note that the degrees of freedom $\frac{1}{2}N(N+1)$ of NxN real symmetric matrices are partitioned, in the H_D representation, to the N eigenvalues and the rest $\frac{1}{2}N(N-1)$ angle variables α_{ij}'s which only enter the orthogonal matrix O to diagonalize H, not the distribution function $P(H_D)$. Thus it is always possible to deal with the distribution for En's by integrated-out density with these angle variables such that

$$\begin{aligned} P(E_1, \ldots E_N) dE_1 \ldots dE_N &= P(H_D) J_N dE_1 \ldots dE_N \\ &= C_N\, e^{-\frac{1}{4a^2}\Sigma_{n=1}^{N}(E_n-\bar{E})^2} J_N dE_1 \ldots dE_N. \end{aligned} \tag{18}$$

Porter [7] gave a simple argument that J_N must be of the form

$$J_N \propto \prod_{i,j(i>j)} |E_i - E_j| \tag{19}$$

so that the joint distribution (density) of N eigenvalues is given by

$$P(E_1,\ldots E_N) = C'_N \prod_{i,j(i>j)} |E_i - E_j| e^{-\frac{1}{4a^2}\Sigma_{n=1}^{N}(E_n-\bar{E})^2} . \tag{20}$$

This purely static formulation is now put into a dynamical one following Yukawa [9,10].

His basic idea is to consider that any GOE Hamiltonian can be regarded as being developed from a fixed one by perturbation with an increase of its strength: $H = H(t) = H_0 + tV$. Thus it immediately associates itself with the dynamics of the ECM system discussed in Section 2, by changing the notations

$$X_0 \to H_0,\ L_0 \to V,\ l_0 \to [H_0, -V] \text{ and } q_n(t) \to E_n(t).$$

Also,

$$\mathcal{H} = \frac{1}{2}\sum_{n=1}^{N} p_n^2 + \frac{1}{2}\sum_{n,m(n\neq m)} \frac{\ell_{nm}^2}{(E_n-E_m)^2}$$

$$= \frac{1}{2}\sum_{n=1}^{N} V_{nn}^2 + \frac{1}{2}\sum_{n,m(n\neq m)} V_{nm}^2 = \frac{1}{2}\,\mathrm{Tr}V^2. \tag{21}$$

Thus the dynamics means here a Liouville dynamics in the phase space whose dimensionality is doubled, i.e. equal to $N(N+1)$. It starts from the original phase space comprising $(q_i^0, p_{ij}^0) = (E_i^0, V_{ij})$ and then proceeds to $(:E_n(t),\ V_{nm}(t))$, to which $\frac{1}{2}N(:N-1)$ angle variables (α_{ij}) are added, the coordinates conjugate to $p_{ij}^0 = V_{ij}$, $i>j$. One then expects the starting distribution ρ to tend, by an addition of extra ingredients if any, to the GOE distribution. But what is the advantage of such rewriting of the already established physical result? This will be answered at the end of this short communication.

Our concern here is that by rewriting the statistical problem in such a dynamical frame, at least, one can gain a transparent reasoning of why Porter's Jacobian (19) arises, as first noticed by Pechukas [6] and then by Yukawa [9]. The reasoning can be achieved from the equations of motion written in two ways:

$$\frac{dp_n}{dt} = \sum_{m(\neq n)} \frac{2V_{nm}^2}{E_n-E_m},\quad \frac{dV_{nm}}{dt} = \sum_{r(\neq n,m)} V_{nr}V_{rm}\left(\frac{1}{E_n-E_r} + \frac{1}{E_m-E_r}\right)$$

$$- V_{nm}(p_n-p_m)/E_n-E_m, \tag{22a}$$

and

$$\frac{dp_n}{dt} = \sum_{m(\neq n)} \frac{2\ell_{nm}^2}{(E_n E_m)^3}, \quad \frac{d\ell_{nm}}{dt} = \sum_{r(\neq n,m)} \ell_{nr}\ell_{rm} \left(\frac{-1}{(E_r-E_n)^2} + \frac{1}{(E_r-E_m)^2} \right) \tag{22b}$$

besides $\frac{dE_n}{dt} = p_n$.

The second equation is just a recall of (1) and (2), which is deduced from the first by a change of the variables from V_{nm} to ℓ_{nm} with

$$\ell_{nm} = (E_m - E_n)V_{nm}. \tag{23}$$

Let a set of equations of motion be expressed in the form $\frac{dy_i}{dt} = \dot{y}_i(y)$ with right hand side $\dot{y}$ being called a flow velocity. Then, to go over from (22a) to (22b) can be characterized by

$$\text{div } \dot{y} = - \sum_{n>m} \frac{p_n - p_m}{E_n - E_m} = \frac{-d}{dt} \log \prod_{n>m} |E_n - E_m| \rightarrow \quad \text{div } \dot{y} = 0$$

which makes the flow incompressible (actually, a Hamiltonian flow). In the first set of coordinates (E_n, V_{nm}), therefore, $\frac{1}{J}\frac{dJ}{dt} = -\text{div } \dot{y}$ if J denotes $\prod_{n>m} |E_n - E_m|$, and any stationary density $\rho(y)$ can be expressed as $\rho(y)=J(E)\text{x}[\text{any function of constants of motion}]$. Thus in the space ($E_n$, V_{nm}), a proper measure can be introduced to write

$$\begin{aligned} d\mu(E,V) &= \prod_{n=1}^{N} dE_n \, dV_{nn} \prod_{n>m} |E_n - E_m| dV_{nm} \\ &= \sum_{n=1}^{N} dE_n \, dp_n \prod_{m>m} d\ell_{nm} \equiv d\mu(E,p,\ell), \end{aligned} \tag{24}$$

the latter being nothing but the Lebesgue measure defined in the second space (still, integrated out with respect to the rest angles α_{ij}'s).

Yukawa introduced his most important hypothesis, presenting a result: among $\frac{1}{2}N(N+1)$ constants of motion in eqs. of motion (22b), take up two; $\mathcal{H}$ given by (21) and $Q \equiv -\frac{1}{2}\text{Tr}\ell^2$, and consider the canonical distribution with two respective temperatures β^{-1} and γ^{-1}, i.e. $e^{-\beta\mathcal{H}-\gamma Q}$.

$$\int_{p,\ell} d\mu(E,p,\ell)e^{-\beta\mathcal{H}-\gamma Q} = \left(\prod_{m>m}\left[1+\frac{\gamma}{\beta}(E_n-E_m)^2\right]^{-1/2}\right)J_N dE_1 \ldots dE_N \tag{25}$$

where the integration is made over the $\frac{1}{2}N(N-1)$ momenta p and ℓ. This right hand side expression tends, for the limit $N\to\infty$ and $\gamma\to 0$ but $N\gamma$=fixed, to the GOE distribution (20) with $a^2 = \frac{\beta}{2N\gamma}$.

Our comments on Yukawa's result above are as follows:

(a) The GOE distribution has been derived on a kind of **central limit theorem** (N>>1). This looks much more physical than the formal derivation (i) and (ii).

(b) The complete integrability of the ECM dynamics plays a role at least by one constant of motion other than $\mathcal{H}$, i.e. Q whose temperature parameter γ^{-1} determines the GOE exponent. Can one find a positive reason that all the other constants are unnecessary?

(c) Consider the opposite limit $\beta\to 0$ in the derived distribution (25). Formally, this exhibits an effect of cancelling J_N, a level-clustering effect of the GOE. This seems to promise a describability of regular spectra by the distribution (25), which in fact has been discussed by Yukawa [9] in an attempt to associate it with the Poisson distribution. It is highly desirable to construct such an interpolation scheme.

4 A stochastization of Yukawa's equation

Complete integrability is valuable to explore, but it is by itself a property irrelevant to statistical mechanics. Thus if one wants to assure a smooth approach to Yukawa's hypothesized equilibrium in (25), one needs to add some ergodic elements to the complete-integrable dynamics by the Hamiltonian (21). A practical method of doing this is to add external noise forces which destroy the integrability. The resulting dynamics becomes stochastic, which we call a **stochastization**. The basic idea is not new: it is sufficient to recall Dyson's Brownian motion model [2]. Indeed, the recent works by Pechukas [6], Yukawa [9] and Nakamura & Lakshmanan [5] were motivated by this model, in which a formation of the chaotic spectra is regarded as a dynamics with the fictitious time t (Dyson's fictitious time). We have made a contribution [4] along the line of our comments (a)-(c) above, giving an answer to the question about Yukawa's two temperatures and a scheme of interpolation between Poisson and Wigner distributions (for a detailed account, see [3]).

References

[1] Bohigas, O. & Giannoni, M.-J. (1984). Chaotic motion and random matrix theories. *Lecture Notes in Physics* **209**, 1. Springer Verlag: Berlin, Heidelberg, New York, London, Paris, Tokyo.

[2] Dyson, F.J. (1962). A brownian-motion model for the eigenvalues of a random matrix. *J. Math. Phys.* **3**, 1191.

[3] Hasegawa, H. (1987). Stochastic formulation of energy level statistics: application to the hydrogen atom in a magnetic field. In *Physics of Atoms and Molecules*, Taylor, K. (ed.). Plenum: New York, to be published.

[4] Hasegawa, H., Mikeska, H.J. & Frahm, H. (1988). Stochastic formulation of energy-level statistics. *Phys. Rev. A* **38**, 395.

[5] Nakamura, K. & Lakshmanan, M. (1986). Complete integrability in a quantum description of chaotic systems. *Phys. Rev. Lett.* **57**, 1661.

[6] Pechukas, P. (1983). Distribution of energy eigenvalues in the irregular spectrum. *Phys. Rev. Lett.* **51**, 943.

[7] Porter, C.E. (1965). Fluctuations of quantal spectra. In *Statistical Theories of Spectra: Fluctuations*, p. 2. Academic Press Inc.: London.

[8] Wojciechowski, S. (1985). An integrable marriage of the Euler equations with the Calogero-Moser system. *Phys. Lett.* **111A**, 101.

[9] Yukawa, T. (195). New approach to statistical properties of energy levels. *Phys. Rev. Lett.* **54**, 1883.

[10] Yukawa, T. (1986). Lax form of the quantum mechanical eigenvalue problem. *Phys. Lett. A* **116**, 227.

K. Nakamura

48. INTEGRABILITY AND NONINTEGRABILITY ASPECTS OF QUANTUM CHAOS

Quantum mechanics of nonintegrable and chaotic systems is investigated from a viewpoint of integrable dynamical systems. Singularities at degenerate points in the case characterized by several nonintegrability parameters are found to be incorporated into a high-dimensional "field theory".

1 **Introduction**

The subject of quantum chaos has attracted a growing interest [11]. The terminology of quantum chaos is currently defined for the quantum mechanics of systems which exhibit the dynamical chaos in the classical limit. While the linear nature of the Schrödinger equation suppresses a chaotic diffusion and eventually leads to the vanishing K-S entropy, many fundamental problems remain unsolved, e.g. those of irregular energy spectra (e.g. [9]) and of complicated quantum diffusion even in the recurrent time regime [5,8]. Theoretical tools borrowed from a field of random systems are not always effective here. The Wigner-Dyson distribution in random matrix theory is sometimes inconsistent with level-spacing distribution of quantum chaos [4]. The concept of fractals is not always practical in quantifying "chaotic" wave functions [5,8], because of the finiteness of the Planck constant $\hbar$. Recent challengers [10] have derived coupled dynamical equations for energy eigenvalues and matrix elements, following a philosophy due to Dyson [2]. Most of their efforts, however, have been directed towards confirming the effectiveness of the random matrix theory in the field of quantum chaos. But, there is no persuasive reason why the random matrix theory should be applicable to the analysis of quantum chaos.

In Section 2, we shall give a brief summary of our recent work [6,7], in which a set of "dynamical" equations for eigenfunctions (rather than matrix elements) and eigenvalues will be shown to be equivalent to the completely-integrable Calogero-Moser-Sutherland's system with internal dual vector space. In Section 3, we shall derive a high-dimensional "field-theoretical" equation from the eigenvalue problem which includes several nonintegrability parameters. The role of singularities at degenerate points will be discussed there. The final section will be devoted to summary and discussions.

2 Dynamical system behind the quantum chaos

We first confine ourselves to autonomous and nonintegrable systems. Let $H = H_0 + tV$ be the Hamiltonian for a quantum bound system which is classically nonintegrable. H_0 and V are a classically integrable part and a nonintegrable perturbation, respectively. Both H and H_0 are Hermitian, i.e. $H^+ = H$ and $H_0^+ = H_0$. Hence $V^+ = V$. A single parameter t denotes the strength of nonintegrability. A manifold with a definite symmetry is considered. Then we can assume discrete eigenvalues $\{x_n(t)\}$ to be nondegenerate, suppressing a negligible possibiity of accidental degeneracies. (When the Hamiltonian includes several nonintegrability parameters, this assumption is not valid which will be described in Section 3.) Eigenfunctions $\{|n(t)\rangle\}$ are complex orthonormal and form a complete set. If we take t as "time", equations of motion for $x_n(t)$, $|n(t)\rangle$ and $\langle n(t)|$ can be obtained from the time-independent Schrödinger equation $H(t)|n(t)\rangle = x_n|n(t)\rangle$ as

$$dx_n/dt = p_n \; [\equiv V_{nn} \equiv \langle n(t)|V|n(t)\rangle], \tag{1a}$$

$$dp_n/dt = 2 \sum_{m(\neq n)} V_{nm}V_{mn}(x_n - x_m)^{-1}, \tag{1b}$$

$$d|n\rangle/dt = \sum_{m(\neq n)} |m\rangle V_{mn}(x_n - x_m)^{-1}, \tag{1c}$$

$$d\langle n|/dt = \sum_{m(\neq n)} \langle m|V_{nm}(x_n - x_m)^{-1}. \tag{1d}$$

Eqs. (1c & d) have been derived as follows: taking the t derivative of the Schrödinger equation, we have $V|n\rangle + Hd|n\rangle/dt = p_n|n\rangle + x_n d|n\rangle/dt$. Its left-hand side equals

$$\sum_m |m\rangle V_{mn} + \sum_m |m\rangle x_m \langle m|(d|n\rangle/dt) = p_n|n\rangle + \sum_{m(\neq n)} |m\rangle V_{mn} + \sum_{m(\neq n)} |m\rangle x_m V_{mn}(x_n - x_m)^{-1},$$

where $\langle m|(d|n\rangle/dt) = V_{mn}(x_n - x_n)^{-1}$ for $m \neq n$ and $\langle n|(d|n\rangle/dt) = (d\langle n|/dt)|n\rangle = 0$ are used. Noting $x_n(t) \neq$ const ($\equiv 0$), we arrive at eqs. (1c & d). Here, the equality $\langle n|(d|n\rangle/dt) = 0$ and its Hermitian conjugate prove to be valid so long as a single parameter is changed (see Section 3).

For the study below, it is convenient to rewrite eqs. (1) in a perfectly canonical formalism. Let us define Λ_{nm} as

$$\Lambda_{nm} = i^{-1}V_{nm}(x_n - x_m), \tag{2}$$

which reduces to $\Lambda_{nm} = \langle n|i^{-1}[H_0,V]|m\rangle \equiv \langle n|\Lambda|m\rangle$. The operator Λ is t-independent and proves to be Hermitian. Let λ_0 be the lowest eigenvaluve of Λ; then the operator $\bar{\Lambda} \equiv \Lambda - \lambda_0 I$ becomes a t-independent and nonnegative Hermitian. $\bar{\Lambda}$ can thereby be decomposed as $\bar{\Lambda} \equiv L^+L$. Here L is an appropriate t-independent operator which has its unique inverse L^{-1}. Λ_{nm} thus becomes

$$\Lambda_{nm} = \langle n|L^+L|m\rangle + \lambda_0\delta_{nm}. \tag{3}$$

By using eqs. (2) and (3), eqs. (1b-d) become

$$dp_n/dt = 2 \sum_{m(\neq n)} \Lambda_{nm}\Lambda_{mn}(x_n - x_m)^{-3}. \tag{1b'}$$

$$d(L|n\rangle)/dt = -i \sum_{m(\neq n)} L|m\rangle\Lambda_{mn}(x_n - x_m)^{-2}, \tag{1c'}$$

$$d(\langle n|L^+)/dt = i \sum_{m(\neq n)} \langle m|L^+\Lambda_{mn}(x_n - x_m)^{-2} \tag{1d'}$$

Eqs. (1c' & d'), though substantially the same as eqs. (1c & d), take advantageous forms for our study on the complete integrability below. $L|n\rangle$ and $\langle n|L^+$ are now read as complex dual vectors. In contrast with previous results [10], eqs. (1a) and (1b'-d') describe the dynamics for both eigenfunctions and eigenvalues and suggest a perfectly canonical formalism.

We now show that these equations can also be attained from an effective "classical" Hamiltonian. Let us introduce a classical N-particle Hamiltonian with internal complex dual space in N dimensions for each particle:

$$\tilde{H} = \sum_{n=1}^{N} \frac{1}{2}p_n^2 + \frac{1}{2}\sum_{n=1}^{N} \sum_{\substack{m=1 \\ (m\neq n)}}^{N} \Lambda_{nm}\Lambda_{mn}(x_n - x_m)^{-2} \tag{4}$$

with Λ_{nm} $(n\neq m)$ defined as a scalar product of complex vectors $\langle n|L^+$ and $L|m\rangle$, consistent with eqs. (2) and (3). Applying Poisson brackets given by

$$\{A,B\} \equiv \sum_n\left(\frac{\partial A}{\partial p_n}\frac{\partial B}{\partial x_n} - \frac{\partial A}{\partial x_n}\frac{\partial B}{\partial p_n}\right) + \sum_n\left(\frac{\partial A}{\partial \langle n|L^+}\frac{\partial B}{\partial iL|n\rangle} - \frac{\partial A}{\partial iL|n\rangle}\frac{\partial B}{\partial \langle n|L^+}\right), \tag{5}$$

one gets a set of canonical equations from eq. (4) as

$$dx_n/dt = \{\tilde{H}, x_n\} = \delta\tilde{H}/\partial p_n, \tag{6a}$$

$$dp_n/dt = \{\tilde{H}, p_n\} = -\partial\tilde{H}/\partial x_n, \tag{6b}$$

$$d(iL|n>)/dt = \{\tilde{H}, iL|n>\} = \partial\tilde{H}/\partial<n|L^+, \tag{6c}$$

$$d(<n|L^+)/dt = \{\tilde{H}, <n|L^+\} = -\partial\tilde{H}/\partial iL|n>. \tag{6d}$$

Eqs. (6) together with eq. (4) are exactly the same with eqs. (1a) and (1b′-d′), if the number N in the latter is finite. Clearly, this equivalence holds even in the limit $N = \infty$. Eqs. (6) with (4) prove to be a Calogero-Moser N-particle system in 1+1 dimensions with internal complex vector space, which happened to be proposed by Gibbons & Hermsen [3] and later by Wojciechowski [14] in a completely different context of nonlinear dynamics, though in the present case Λ_{mn}, $L|n>$ and $<n|L^+$ are defined in a much more intricate way. The number of degrees of freedom of the system, eqs. (6), is $N(N + 1)$, consisting of N for positions and momenta of N particles (levels) and N^2 for internal freedom (complex eigenstates) possessed by particles (levels).

The complete integrability of eqs. (6) will be elucidated, merely by rewriting the novel findings of Gibbons & Hermsen [3] in our notations. Let us define N x N matrices:

$$P_{nm} = \delta_{nm}p_n + (1 - \delta_{nm})i\Lambda_{nm}(x_n - x_m)^{-1},$$

$$X_{nm} = \delta_{nm}x_n,$$

$$\Gamma_{nm} = (1 - \delta_{nm})i\Lambda_{nm}(x_n - x_m)^{-2},$$

$$L = (L|1>, L|2>, \ldots, L|n>, \ldots, L|N>),$$

where $L|n>$ are N-component column vectors. Then, Lax representation of eq. (6) can be obtained. Here we show its complete algebraic solution:

$$\begin{aligned} X(t) &= U(t)[X(0) + P(0)t]U^{-1}(t), \\ P(t) &= U(t)P(0)U^{-1}(t), \\ L(t) &= L(0)U^{-1}(t), \\ L^+(t) &= U(t)L^+(0). \end{aligned} \tag{7}$$

U(t) is a unitary matrix defined in terms of the time-ordering operator T, i.e.

$$U(t) \equiv T \exp [\int_0^t \Gamma(\tau) d\tau]. \tag{8}$$

Thus, if we would have a knowledge of eigenfunctions and eigenvalues at t = +0 (i.e. in the integrable or weak-coupling limit), those at t > 0 (i.e. in the nonintegrable or strong-coupling regime) will be provided by the solution in eqs. (7).

While we have so far been confining ourselves to the autonomous system, the same argument as above holds even for nonintegrable driven systems in the case when the driving field consists of periodic pulse trains [7]. The problem of the time-dependent Schrödinger equation is now reduced to an eigenvalue problem for the one-period propagator:

$$\mathrm{ex}(-itV)\cdot\exp(-2\pi i H_0)|n(t)\rangle = \exp(-ix_n(t))|n(t)\rangle, \tag{9}$$

where t is a nonintegrability parameter, representing the pulse strength. By taking the t derivative of Eq. (9), we obtain:

$$dx_n/dt = p_n, \tag{10a}$$

$$dp_n/dt = \frac{1}{4} \sum_{m\neq n} \langle n|L^+L|m\rangle\langle m|L^+L|n\rangle \cos[(x_n-x_m)/2]\sin^{-3}[(x_n-x_m)/2], \tag{10b}$$

$$d(L|n\rangle)/dt = (-i/4) \sum_{m\neq n} L|m\rangle\langle m|L^+L|n\rangle \sin^{-2}[(x_n - x_m)/2], \tag{10c}$$

$$d(\langle n|L^+)/dt = (i/4) \sum_{m\neq n} \langle n|L^+L|m\rangle\langle m L^+ \sin^{-2}[(x_n - x_m)/2]. \tag{10d}$$

L and L^+ are t-independent operators, though their definition here is a bit different from eq. (3). We find that eqs. (10) constitute a generalized Sutherland system in 1+1 dimensions whose complete integrability was proved by Nakamura & Mikeska [7].

In this way, we conclude: there exists a universal "classical" dynamical system which underlies the quantum chaos. It is a completely-integrable many-particle system with internal complex vector space, whose effective classical Hamiltonian is given by

$$\tilde{H} = \sum_{n=1} \frac{1}{2} p_n^2 + \frac{1}{2} \sum_{n=1} \sum_{\substack{m=1 \\ (m\neq n)}} \langle n|L^+\cdot L|m\rangle\langle m|L^+\cdot L|n\rangle\, \mathcal{P}(x_n-x_m|\omega,\omega'). \tag{11}$$

In eq. (11), $\mathcal{P}(z|\omega,\omega')$ is the doubly-periodic Weierstrass function. In fact, $\mathcal{P}(z|\infty,\infty)=z^{-2}$ and $\mathcal{P}(z|0,i\pi/2)=\frac{1}{4}\sin^{-2}(z/2)$ correspond to the nonintegrable autonomous and driven (pulsed) nonautonomous systems, respectively.

3 Extension to "field theory"

So long as a single parameter is to be changed in the Hamiltonian, the desymmetrized spectra exhibit no degeneracy except for accidental ones. We can see only avoided crossings. The presence of a multitude of avoided crossings is a quantum-mechanical manifestation of the classical chaos. In Section 2, this situation was examined from a viewpoint of nonlinear dynamics. when several parameters are to be changed, however, we see degeneracies in general [13]. Let us consider the eigenvalue problem:

$$H(\vec{t})|n(\vec{t})\rangle \equiv (H_0 + \sum_{\mu=1}^{d} t_\mu V_\mu)|n(\vec{t})\rangle = x_n(\vec{t})|n(\vec{t})\rangle, \qquad (12)$$

where $\vec{t} = (t_1, t_2, ..., t_d)$ denotes a set of nonintegrability parameters. Just as in Section 2, the Hamiltonian $H(\vec{t})$ in eq. (12) is assumed to be Hermitian and to have bound spectra. Let us consider the desymmetrized spectra of eq. (12). The spectra are now constructed in the high-dimensional parameter space $(\vec{t})$, in which each eigenvalue constitutes a hyper energy surface. In general, we can see the contacts of adjacent energy surfaces at (diabolical) degenerate points. These points are the source of topological singularities: let us consider a closed circuit around a degenerate point in $\vec{t}$ space. Then, the eigenfunction $|n(\vec{t})\rangle$ joining the degeneracy acquires the topological phase [1,12] given by

$$i\oint\langle n|\nabla_{\vec{t}}|n\rangle\cdot d\vec{t}. \qquad (13)$$

This phase is a manifestation of the fact that eigenfunctions are multi-valued in a high-dimensional parameter space, though of course single-valued in the configuration space. We may expect that the presence of many (diabolical) degenerate points in energy spectra would be an indicator of the quantum chaos whose Hamiltonian is characterized by several nonintegrability parameters.

We take $\{t_\mu\}$, $\{x_n(\vec{t})\}$, and $\{|n(\vec{t})\rangle\}$ as d-dimensional "space-time coordinates", components of a vector field and elements of a matrix field, respectively. By making the t_μ derivative of eq. (12), we obtain

$$\partial^2_{t_\mu} x_n = 2\sum_{m(\neq n)} \langle n|L^+_\mu\cdot L_\mu|m\rangle\langle m|L^+_\mu\cdot L_\mu|n\rangle(x_n - x_m)^{-3}, \qquad (14a)$$

$$\partial_{t_\mu}(L_\mu|n\rangle) = -i\sum_{m(\neq n)} L_\mu|m\rangle\langle m|L^+_\mu\cdot L_\mu|n\rangle(x_n - x_m)^{-2} - iL_\mu|n\rangle A_{n,\mu}, \qquad (14b)$$

Hermitian conjugate of eq. (14b), (14c)

where $A_{n,\mu} \equiv i\,(n|\partial_{t_\mu}|n>$, describing the topologically singular structures around degenerate points. This is a kind of "gauge potential", which satisfies

$$\partial_{t_\mu} A_{n,\nu} - \partial_{t_\nu} A_{n,\mu} = \sum_{m(\neq n)} \{<n|L_\mu^+ \cdot L_\mu|m><m|L_\nu^+ \cdot L_\nu|n>(x_n - x_m)^{-4}$$

$$- [\mu \longleftrightarrow \nu]\}. \qquad (14d)$$

In eqs. (14a-d), we do not take the Einstein's contraction for suffix μ, ν; L_μ with $\mu = 1,2,\ldots,d$ is a t_μ-independent operator satisfying $i^{-1}[H,V_\mu] \equiv L_\mu^+ L_\mu$.

Eqs. (14a-d) represent a universal dynamical system behind the quantum-mechanical counterpart of chaos characterized by several nonintegrability parameters. In contrast with the issue in Section 2, eqs. (14) describe a d-dimensional field theory coupled with a "gauge field". We can construct a corresponding "field-theoretical" Lagrangean and also extend the present theory to a driven system.

4 Summary and discussions

Despite the accumulation of studies on quantum chaos, most of the activity has rather concentrated on phenomenological descriptions, e.g. on level-spacing distributions and localizations. Even a challenge by Pechukas [10] was eventually directed towards confirming the effectiveness of the random matrix theory. In this paper, we have derived a set of dynamical equations for eigenvalues and eigenfunctions, by taking a nonintegrability parameter as quasi-time. These equations possess the complete information about eigenfunctions which was partly lost in the treatment due to Pechukas [10]. We have shown their complete integrability. In the case when a quantum Hamiltonian includes several nonintegrability parameters, we have derived an effective field theory for field variables (i.e. eigenfunctions and eigenvalues), by taking those parameters as "space-time coordinates". Singularities at degenerate points are found to be described by a "gauge field" coupled with "matter field" above. A detailed analysis of this field theory will be made elsewhere.

Acknowledgements

Regarding the material in Section 2, I would like to thank collaborators, in particular M. Lakshmanan and H.J. Mikeska.

References

[1] Berry, M.V. (1984). *Proc. Roy. Soc. London* **A392**, 45.
[2] Dyson, F.J. (1962). *J. Math. Phys.* **3**, 1191.
[3] Gibbons, J. & Hermsen, T. (1984). *Physica* **D11**, 337.
[4] Izrailev, F.M. (1986). *Phys. Rev. Lett.* **56**, 541.

[5] Nakamura, K., Bishop, A.R. & Shudo, A. (1989). *Phys.* Rev. **B39**, 12422.
[6] Nakamura, K. & Lakshmanan, M. (1986). *Phys. Rev. Lett.* **57**, 1661.
[7] Nakamura, K. & Mikeska, H.J. (1987). *Phys. Rev.* **A35**, 5294.
[8] Nakamura, K., Okazaki, Y. & Bishop, A.R. (1986). *Phys. Rev. Lett.* **57**, 5.
[9] Nakamura, K. & Thomas, H. (1988). *Phys. Rev. Lett.* **61**, 247.
[10] Pechukas, P. (1983). *Phys. Rev. Lett.* **51**, 943.
[11] Seligman, T.H. & Nishioka, H. (eds.) (1986). *Quantum Chaos and Statistical Nuclear Physics.* Springer: Berlin.
[12] Simon, B. (1983). *Phys. Rev. Lett.* **51**, 2167.
[13] von Neumann, J. & Wigner, E.P. (1929). *Phys. Z.* **30**, 467.
[14] Wojciechowski, S. (1986). *Physica Scr.* **34**, 304.

Part 6

PHYSICAL APPLICATIONS

M. Wadati, T. Deguchi and Y. Akutsu

49. EXACTLY SOLVABLE MODELS AND NEW LINK POLYNOMIALS

1 The Yang-Baxter relation and exactly solvable models

Recently, we have developed a general theory to construct link polynomials from exactly solvable models in statistical mechanics. In this paper, we explain how the soliton theory has given a solution to one of the most fundamental problems in mathematics.

We start from a brief introduction to the exactly solvable models in statistical mechanics. We can define two kinds of models, vertex and IRF models, on a two-dimensional square lattice. A state in vertex model is specified by the configuration of site variables (often, we call them "spins") on four bonds around a vertex (Fig. 1). We denote its Boltzmann weight by $w(i,j,k,\ell;u)$ where u is called spectral parameter. It is known that the Boltzmann weight $w(i,j,k,\ell;u)$ is interpreted as the S-matrix element $S^{ik}_{j\ell}(u)$ of an integrable system with u being the rapidity difference. When $w(i,j,k,\ell;u)$ is non-zero only for the case $i+j = k+\ell$, we say that the model has "charge conservation" property. In IRF model, a state is specified by the configuration of spins on four corners around a face (Fig. 2). Its Boltzmann weight is denoted by $w(a,b,c,d;u)$ where u is the spectral parameter. The Yang-Baxter relation is a sufficient condition for the solvability of the model. It reads for the vertex model (the factorized S-matrix)

$$\sum_{abc} S^{bq}_{cr}(v)S^{ap}_{kc}(u+v)S^{ia}_{jb}(u) = \sum_{abc} S^{ap}_{bq}(u)S^{ia}_{cr}(u+v)S^{jb}_{kc}(v), \qquad (1a)$$

and for the IRF model

$$\sum_{c} w(b,d,c,a;u)w(a,c,f,g;u+v)w(c,d,e,f;v)$$

$$= \sum_{c} w(a,b,c,g;v)w(b,d,e,c;u+v)w(c,e,f,g;u). \qquad (1b)$$

The model whose Boltzmann weights satisfy the Yang-Baxter relation

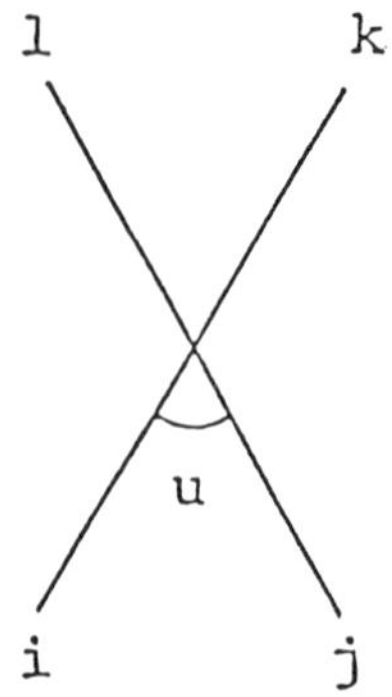

Figure 1. Boltzmann weight $w(i,j,k,\ell;u)$ for vertex model.

is referred to as an exactly solvable model.

We assume that the Boltzmann weights satisfy the following relations:

(i) Standard initial condition

$$S^{ik}_{j\ell}(u=0) = \delta_{i\ell}\delta_{jk}, \quad w(a,b,c,d;u=0) = \delta_{ac}. \tag{2a,b}$$

(ii) Inversion relation (unitarity condition)

$$\sum_{m,p} S^{mk}_{p\ell}(u)S^{ip}_{jm}(-u) = \rho(u)\rho(-u)\delta_{i\ell}\delta_{jk}, \tag{3a}$$

$$\sum_{e} w(e,c,d,a;u)w(b,c,e,a;-u) = \rho(u)\rho(-u)\delta_{bd}, \tag{3b}$$

where $\rho(u)$ is a model-dependent function.

(iii) Second inversion relation (second unitarity condition)

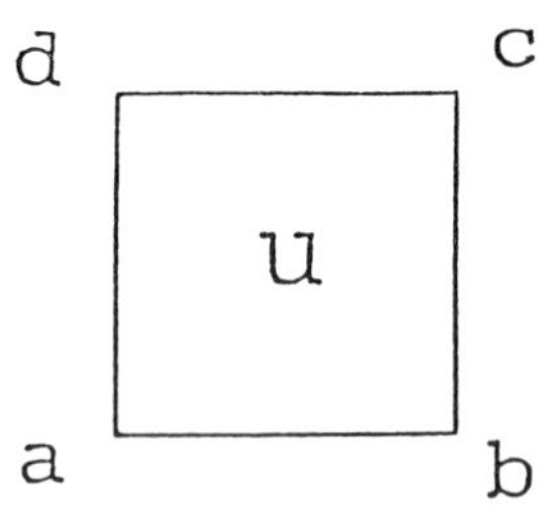

Figure 2. Boltzmann weight $w(a,b,c,d;u)$ for IRF model.

$$\sum_{m,p} S^{im}_{p\ell}(\lambda-u)S^{kp}_{mj}(\lambda+u)\cdot r(p)r(m)/r(k)r(j) = \rho(u)\rho(-u)\delta_{ij}\delta_{k\ell}, \tag{4a}$$

$$\sum_{e} w(c,e,a,b;\lambda-u)w(a,e,c,d;\lambda+u)\cdot\psi(e)\psi(b)/\psi(a)\psi(c) = \rho(u)\rho(-u)\delta_{bd}. \tag{4b}$$

We call the parameter λ crossing point and the quantities $r(i)$ and $\psi(a)$ crossing multipliers.

(iv) Crossing symmetry

$$S^{ik}_{j\ell}(u) = S^{j\ \ell}_{\bar{k}\ \bar{i}}(\lambda-u)\cdot\{r(i)r(\ell)/r(j)r(k)\}^{1/2}, \tag{5a}$$

$$w(a,b,c,d;u) = w(d,a,b,c;\lambda-u)\cdot(\psi(a)\psi(c)/\psi(b)\psi(d)\}^{1/2}. \tag{5b}$$

Here we have used the notation $\bar{j} = -j$ etc.

(v) Reflection symmetry

$$S^{ik}_{j\ell}(u) = S^{\ell j}_{ki}(u), \quad w(a,b,c,d;u) = w(c,b,a,d;u). \tag{6a,b}$$

We remark that the above relations come from the physical requirements for the model [9].

2 The Yang-Baxter operators

The Yang-Baxter relation can be cast into an operator formalism. We define Yang-Baxter operator for vertex model [1,2,7] by

$$X_i(u) = \sum_{k,\ell,m,p} S^{km}_{\ell p}(u) I^{(1)} \otimes\cdots\otimes E^{(i)}_{pk} \otimes E^{(i+1)}_{m\ell} \otimes I^{(i+2)} \cdots\otimes I^{(n)}, \tag{7a}$$

and for IRF model [1,8] by

$$[X_i(u)]^{p_0\cdots p_n}_{\ell_0\cdots\ell_n} = \delta^{p_0}_{\ell_0} \cdots \delta^{p_{i-1}}_{\ell_{i-1}} w(\ell_i,\ \ell_{i+1},\ p_i,\ \ell_{i-1};u)$$
$$\times\ \delta^{p_{i+1}}_{\ell_{i+1}} \cdots \delta^{p_n}_{\ell_n} \tag{7b}$$

The Yang-Baxter operators $\{X_i(u)\}$ satisfy the following relation (Yang-Baxter algebra)

$$X_i(u)X_{i+1}(u+v)X_i(v) = X_{i+1}(v)X_i(u+v)X_{i+1}(u),$$

$$X_i(u)X_j(v) = X_j(v)X_i(u), \quad |i-j| \geq 2. \tag{8}$$

The first relation in (8) is nothing but the Yang-Baxter relation (1) in operator form.

To discuss the theory as generally as possible, we consider the Wu-Kadanoff-Wegner transformation which relates an IRF model to a vertex model (Fig. 3). In the unrestricted IRF model, the Boltzmann weights contain an additional parameter $\vec{\omega}_0 = \omega_0\vec{n}_0$, $\vec{n}_0$ being a unit vector, since the origin of the spin space is arbitrary. For instance, the Boltzmann weights for the unrestricted 8VSOS model are

$$w(\ell,\ell\pm1,\ell,\ell\pm1;u) = h(\lambda-u)/h(\lambda),$$

$$w(\ell\pm1,\ell,\ell\mp1,\ell;u) = h(u)/h(\lambda)\cdot\{h(\omega_{\ell-1})h(\omega_{\ell+1})\}^{1/2}/h(\omega_\ell),$$

$$w(\ell\pm1,\ell,\ell\pm1,\ell;u) = h(\omega_\ell\pm u)/h(\omega_\ell), \qquad (9)$$

where

$$\omega_\ell = \ell\lambda + \omega_0, \qquad \ell = 0, \pm1, \pm2, \ldots, \qquad (10)$$

$$h(u) = \sin u\cdot\prod_{n=1}^{\infty}(1-2p^{2n}\cos 2u + p^{4n})(1-p^{2n}). \qquad (11)$$

In this case, ω_0 is an arbitrary parameter. It has been known [1] that, at the criticality (for instance, p=0 for the unrestricted 8VSOS model), an IRF model becomes equivalent to the corresponding vertex model in the limit $\omega_0\to\pm i\infty$. There, the Boltzmann weights, the crossing multipliers and the spins are related by

$$w(a,b,c,d;u) \to S^{ik}_{j\ell}(u),\ \psi(b)/\psi(a) \to r^2(j),$$

$$i = a-d,\quad j = b-a,\quad k = b-c,\quad \ell = c-d. \qquad (12)$$

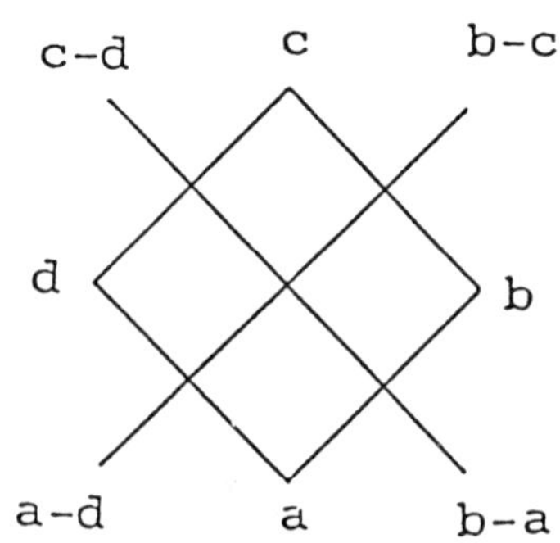

Figure 3. Wu-Kadanoff-Wegner transformation.

3 Braid group and solvable models

We generally call a one-dimensional object a string. A knot is a closed string which does not cross with itself. As a more general object, an assembly of knots with mutual entanglements is called a link. A problem we consider is a topological classification of knots and links. We solve the problem by constructing explicitly a link polynomial, which is an isotopy invariant, from the Yang-Baxter operators.

We define braid and braid group. Braids are formed when n points on a horizontal line are connected by n strings to n points on another horizontal line directly below the first n points. A general n-braid is constructed from a trivial n-braid by successive applications of the operations $b_1,\ldots,b_{n-1}$ and their inverses (Fig. 4). The braid group B_n is defined by a set of generators, $b_1,\ldots,b_{n-1}$ which satisfy

$$b_i b_{i+1} b_i = b_{i+1} b_i b_{i+1}, \quad b_i b_j = b_j b_i, \quad |i-j| \geq 2. \tag{13}$$

Each topologically equivalent class of the n-braids is identified with an element in B_n.

Given a braid, one may form a link by tying opposite points. Conversely, any link is represented as a closed braid. However, the expression of a link as a closed braid is not unique. The equivalent braids expressing the same link are mutually transformed by applications of two types of operations (Markov moves I and II):

$$\text{I.} \quad AB \rightarrow BA, \qquad A,\ B \in B_n,$$

$$\text{II.} \quad A \rightarrow Ab_n, \quad A \rightarrow A\, b_n^{-1}, \quad A \in B_n, \quad b_n \in B_{n+1}. \tag{14}$$

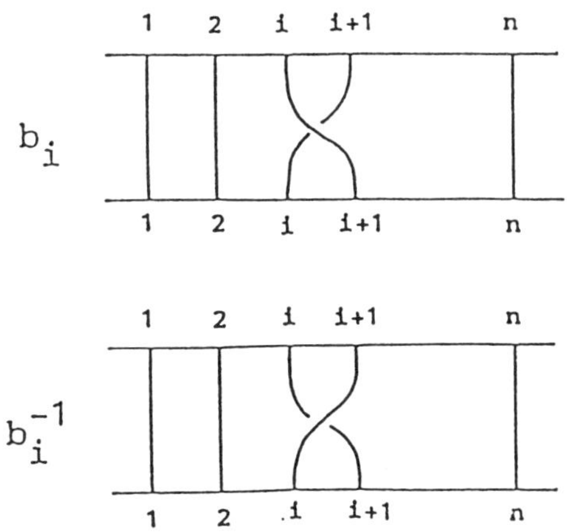

Figure 4. Operations b_i and b_i^{-1}.

Hence, to obtain a link polynomial which is a topological invariant, we first construct a representation of the braid group and then find a Markov move invariant quantity defined on the representation.

The link polynomial $\alpha(\cdot)$ must satisfy the conditions (we use the same notations for generators and their representations):

$$\text{I.}\quad \alpha(AB) = \alpha(BA), \qquad A,\ B \in B_n,$$

$$\text{II.}\quad \alpha(A) = \alpha(A\ b_n) = \alpha(A\ b_n^{-1}), \quad A \in B_n,\ b_n \in B_{n+1}. \tag{15}$$

This quantity is readily obtained if we can find a linear functional $\phi(\cdot)$, called the Markov trace, which has the following properties (the Markov properties):

$$\text{I.}\quad \phi(AB) = \phi(BA, \qquad A,\ B \in B_n,$$

$$\text{II.}\quad \phi(Ab_n)= \tau\phi(A),\ \phi(Ab_n^{-1}) = \bar{\tau}\phi(A), \quad A\in B_n,\ b_n \in B_{n+1},$$

$$\tau = \phi(b_i),\quad \bar{\tau} = \phi(b_i^{-1}). \tag{16}$$

With the Markov trace $\phi(\cdot)$, the link polynomial $\alpha(\cdot)$ is given by

$$\alpha(A) = (\tau\bar{\tau})^{-(n-1)/2}(\bar{\tau}/\tau)^{e(A)/2}\phi(A), \qquad A \in B_n. \tag{17}$$

Here $e(A)$ is the exponent sum of b_i's in the braid A. For instance, if $A = b_1^4 b_2^{-2} b_3$, then $e(A) = 4-2+1 = 3$.

Comparing (8) and (13), we notice that if $u = u+v = v$ in (8) we obtain the representation (G_i) of the braid group from the Yang-Baxter operator $X_i(u)$. Thus, we find [1,2] that, after suitable normalization if necessary,

$$X_i(0) = I\ ,\quad G_i = \lim_{u\to\infty} X_i(u). \tag{18}$$

The limit $u\to\infty$ implies that the model should be critical whose Boltzmann weights are parametrized by hyperbolic or trigonometric functions.

4 Link polynomials

We explicitly construct the Markov trace. We first consider the vertex model. We assume that

$$\phi(A) = \mathrm{Tr}(HA)/\mathrm{Tr}(H), \qquad A \in B_n, \tag{19a}$$

$$H = h^{(1)}\otimes h^{(2)}\otimes \cdots \otimes h^{(n)},\ h_{pq} = r^2(p)\delta_{pq}. \tag{20a}$$

Here $r(p)$ is the crossing multiplier. In order that $\phi(\cdot)$ satisfies the Markov properties I and II, we need the following conditions. For the Markov property I, it is sufficient to have $[H,b_i] = 0$, which follows from that the model should have the "charge conservation" [2]. For the Markov property II, it is sufficient that

$$\lim_{u\to\infty} \sum_{\ell} S_{\ell k}^{k\ell}(u) r^2(\ell) = \chi(\lambda) \quad \text{(independent of k)}, \tag{21a}$$

$$\sum_{\ell} r^2(\ell) = \xi(\lambda). \tag{22a}$$

The condition (22a) is obvious. Thus, for any solvable vertex model with (21a) and charge conservation property the Markov trace is given by (19a) and (20a).

We next consider the IRF model. We assume that

$$\phi(A) = \mathrm{Tr}(HA)/\mathrm{Tr}(H), \quad A \in B_n, \tag{19b}$$

$$[H]_{\ell_0\cdots\ell_n}^{p_0\cdots p_n} = \delta_{\ell_0}^{p_0}\psi(\ell_1)/\psi(\ell_0)\cdot\delta_{\ell_1}^{p_1}\psi(\ell_2)/\psi(\ell_1)\;\cdots$$

$$\cdots\;\psi(\ell_n)/\psi(\ell_{n-1})\delta_{\ell_n}^{p_n}. \tag{20b}$$

Here $\psi(\ell)$ is the crossing multiplier. $\mathrm{Tr}(\cdot)$ is the constrained trace defined by

$$\mathrm{Tr}(A) = \sum_{\substack{\ell_1\cdots\ell_n \\ \ell_0:\text{fixed}}}{}' A_{\ell_0\cdots\ell_n}^{\ell_0\cdots\ell_n}, \tag{23}$$

where the symbol Σ' represents the summation under the constraint imposed on the model. Then, Tr(HA) has the form [1]

$$\mathrm{Tr}(HA) = \sum_{\substack{\ell_1\cdots\ell_n \\ \ell_0:\text{fixed}}}{}' A_{\ell_0\cdots\ell_n}^{\ell_0\cdots\ell_n}\cdot\psi(\ell_n)/\psi(\ell_0). \tag{24}$$

A similar argument as for the vertex model leads to the conditions

$$\lim_{u\to\infty} \sum_{b\sim a} w(a,b,a,c;u)\psi(b)/\psi(a) = \chi(\lambda) \quad \text{(independent of a and c)}, \tag{21b}$$

$$\sum_{b\sim a} \psi(b)/\psi(a) = \xi(\lambda) \quad \text{(independent of a)}, \tag{22b}$$

where the summations are over all values of b admissible to a. Thus, for any solvable IRF model with (21b) and (22b) the Markov trace is given by (19b) and (20b). The Markov trace (19b) for the IRF model reduces to the one in (19a) for the corresponding vertex model in the limit $\omega_0 \to \pm i\infty$ [1].

Since the representation of the braid group is obtained by the formula (18) and the link polynomial is related to the Markov trace as (17), we arrive at a conclusion: we have a general method to construct link polynomials from solvable vertex models and from solvable IRF models.

The conditions (21) and (22) are combined into a single relation with a finite spectral parameter u:

$$\sum_{\ell} S^{k\ell}_{\ell k}(u) r^2(\ell) = H(u;\lambda) \text{ (independent of k)}, \tag{25a}$$

$$\sum_{b\sim a} w(a,b,a,c;u)\psi(b)/\psi(a) = H(u;\lambda) \text{ (independent of k)}. \tag{25b}$$

We call (25) extended Markov invariance and $H(u;\lambda)$ characteristic function. When a vertex model and an IRF model are related by the Wu-Kadanoff-Wegner transformation, both characteristic functions have the same form with the replacement $\lambda \to i\lambda$. For the model which has the crossing symmetry, (25) is equivalent to the projection property [7]

$$X_i(u)X_i(\lambda) = \beta(u)X_i(\lambda), \quad \beta(u) = H(\lambda-u;\lambda)\rho(\lambda-u). \tag{26}$$

Explicit forms of characteristic functions $H(u;\lambda)$ are the following:

(i) N-state vertex model or composite SU(2) model [1,2,3]

$$H(u;\lambda) = \prod_{n=0}^{N-2} \mathrm{sh}((n+2)\lambda-u)/\mathrm{sh}\lambda. \tag{27}$$

The N = 2 case corresponds to the well-known Jones polynomial.

(ii) A_{m-1} IRF model [8]

$$H(u;\lambda) = \sin(m\omega-u)/\sin(\omega-u). \tag{28}$$

(iii) B_m, C_m, D_m IRF models [8]

$$H(u;\lambda) = \sigma\sin(2\lambda-u)\sin(\sigma\omega+\lambda-u)/\sin(\lambda-u)\sin(\omega-u). \tag{29}$$

Here $\sigma = 1$ and $\lambda = (2m-1)/2$ for the B_m model, $\sigma = -1$ and $\lambda = (m+1)\omega$ for the C_m model, and $\sigma = 1$ and $\lambda = (m-1)\omega$ for the D_m model. With the replacement $\lambda \to i\lambda$, we obtain $H(u;\lambda)$ for the corresponding vertex model.

Two-variable extension of the link polynomial constructed from the N-state vertex model has been reported [2,6,9]. The N = 2 case is the HOMFLY polynomial. Composition (or fusion) of the A type model leads to the same two-variable link polynomials.

5 Algebras associated with the solvable models

In Section 3, we have shown that we can construct the representation $\{G_i\}$ of the braid group through the formula (18). In addition to the defining relation (8) of the braid group, the representation $\{G_i\}$ satisfies the following reduction relation.

(i) N-state vertex model [2,3,6]

$$(G_i-C_1I)(G_i-C_2I)\cdots(G_i-C_NI) = 0,$$

$$C_r = (-1)^{r+1}t^{N(N-1)/2-(N-r)(N-r+1)/2}, \quad t = \exp(2\lambda). \tag{30}$$

(ii) A type IRF model [8]

$$(G_i-I)(G_i+tI) = 0, \quad t = \exp(-2i\omega). \tag{31}$$

(iii) B, C, D type IRF models [8]

$$(G_i-\beta I)(G_i-I)(G_i+tI)= 0,$$

$$t = \exp(-2i\omega), \quad \beta = \sigma\exp(-i(2\lambda+\omega(1+\sigma))). \tag{32}$$

For the vertex model, λ and ω should be replaced by $i\lambda$ and $i\omega$. We notice that the 2-state vertex model and the A type model give the Hecke algebra.

Furthermore, with monoid operator E_i

$$E_i = X_i(u=\lambda), \tag{33}$$

a set of operators $\{G_i,E_i\}$forms a closed algebra. For instance, we have the following:

(i) N-state vertex model

$$E_i = \{k_0/(C_N-C_1)\cdots(C_N-C_{N-1})\}\cdot(G_i-C_1I)\cdots(G_i-C_{N-1}I),$$

$$E_i^2 = k_0E_i, \; G_iE_i = E_iG_i = C_NE_i,$$

$$G_iG_{i+1}G_i = G_{i+1}G_iG_{i+1}, \; E_iE_{i\pm1}E_i = E_i,$$

$$E_iG_{i\pm1}G_i = G_{i\pm1}G_iE_{i\pm1} = k_1E_iE_{i\pm1},$$

$$G_{i\pm1}E_iG_{i\pm1} = k_1^2G_i^{-1}E_{i\pm1}G_i^{-1}, \; G_{i\pm1}E_iE_{i\pm1} = k_1G_i^{-1}E_{i\pm1},$$

$$E_{i\pm1}E_iG_{i\pm1} = k_1E_{i\pm1}G_i^{-1}, \; E_iG_{i\pm1}E_i = k_1/C_N\cdot E_i,$$

$$G_iG_j = G_jG_i, \quad E_iE_j = E_jE_i, \qquad |i-j| \geq 2, \tag{34}$$

where

$$k_0 = t^{(N-1)/2} + t^{(N-3)/2} + \cdots + t^{-(N-1)/2}, \tag{35}$$

(ii) B, C, D type IRF models

$$E_i = 1/\beta(t-1)\cdot(G_i-I)(G_i+tI),$$

$$E_i^2 = (\beta-1)(\beta+t)/\beta(t-1)\cdot E_i, \quad G_iE_i = E_iG_i = \beta E_i,$$

$$G_iG_{i+1}G_i = G_{i+1}G_iG_{i+1}, \quad E_iE_{i\pm1}E_i = E_i,$$

$$E_iG_{i\pm1}G_i = G_{i\pm1}G_iE_{i\pm1} = t\, E_iE_{i\pm1}$$

$$G_{i\pm1}E_iG_{i\pm1} = t^2G_i^{-1}E_{i\pm1}G_i^{-1}, \quad G_{i\pm1}E_iE_{i\pm1} = tG_i^{-1}E_{i\pm1},$$

$$E_{i\pm1}E_iG_{i\pm1} = tE_{i\pm1}G_i^{-1}, \quad E_iG_{i\pm1}E_i = t/\beta\cdot E_i,$$

$$G_iG_j = G_jG_i, \quad E_iE_j = E_jE_i, \quad |i-j| \geq 2. \tag{36}$$

From the reduction relation, we obtain skein relation which relates the link polynomials with different numbers of twist.

(i) N-state vertex model

$$\alpha(L_+) = (1-t)t^{1/2}\alpha(L_0) + t^2\alpha(L_-), \quad (N=2)$$

$$\alpha(L_{2+}) = t(1-t^2+t^3)\alpha(L_+) + (t^4-t^5+t^7)\alpha(L_0)-t^8\alpha(L_-), \quad (N=3)$$

$$\begin{aligned}\alpha(L_{3+}) &= t^{3/2}(1-t^3+t^5-t^6)\alpha(L_{2+}) \\ &\quad + t^6(1-t^2+t^3+t^5-t^6+t^8)\alpha(L_+) \\ &\quad + t^{25/2}(-1+t-t^3+t^6)\alpha(L_0) - t^{20}\alpha(L_-), \quad (N=4)\end{aligned} \tag{37}$$

(ii) A_{m-1} type IRF model

$$\alpha(L_+) = (1-t)t^{(m-1)/2}\alpha(L_0) + t^m\alpha(L_-), \quad t = \exp(-2i\omega). \tag{38}$$

(iii) B_m, C_m, D_m type IRF models

$$\alpha(L_{2+}) = (1-t+\beta)e^{-i(2\lambda+\omega(\sigma-1))}\cdot\alpha(L_+)$$
$$+ (t+\beta t-\beta)e^{-2i(2\lambda+\omega(\sigma-1))}\cdot\alpha(L_0)$$
$$- t\beta e^{-3i(2\lambda+\omega(\sigma-1))}\cdot\alpha(L_-),$$

$$t = \exp(-2i\omega),\ \beta = \sigma\exp(-i(2\lambda+\omega(1+\sigma))). \qquad (39)$$

In the above, by L_{3+}, L_{2+}, L_+, L_0 and L_-, we have denoted links which have, at a particular intersection, the configuration represented by b_i^3, b_i^2, b_i^1, b_i^0 and b_i^{-1}, respectively. For the corresponding vertex model, λ and ω in (31), (32), (38) and (39) are replaced by $i\lambda$ and $i\omega$.

References

[1] Akutsu, Y. & Wadati, M. (1987a). Knot invariants and the critical statistical systems. *J. Physical Soc. Jap.* **56**, 839.
[2] Akutsu, Y. & Wadati, M. (1987b). Exactly solvable models and new link polynomials. I. N-state vertex models. *J. Physical Soc. Jap.* **56**, 3039.
[3] Akutsu, Y., Deguchi, T. & Wadati, M. (1987c). Exactly solvable models and new link polynomials. II. Link polynomials for closed 3-braids. *J. Physical Soc. Jap.* **56**, 3464.
[4] Akutsu, Y., Deguchi, T. & Wadati, M. (1988a). Exactly solvable models and new link polynomials. IV. IRF models. *J. Physical Soc. Jap.* **57**, 1173.
[5] Akutsu, Y. & Wadati, M. (1988b). Knots, links, braids and exactly solvable models in statistical mechanics. *Comm. Math. Phys.* **117**, 243.
[6] Deguchi, T., Akutsu, Y. & Wadati, M. (1988a). Exactly solvable models and new link polynomials. III. Two-variable topological invariants. *J. Physical Soc. Jap.* **57**, 757.
[7] Deguchi, T., Wadati, M. & Akutsu, Y. (1988b). Exactly solvable models and new link polynomials. V. Yang-Baxter operator and braid-monoid algebra. *J. Physical Soc. Jap.* **57**, 1905.
[8] Deguchi, T., Wadati, M. & Akutsu, Y. (1988c). Link polynomials constructed from solvable models in statistical mechanics. *J. Physical Soc. Jap.* **57**, 2921.
[9] Wadati, M. & Akutsu, Y. (1988). From solitons to knots and links. *Prog. Theor. Phys. (Suppl.)* **94**, 1.
Note added in proof: we have recently published the following papers
[10] Akutsu, Y., Deguchi, T. & Wadati, M. (1989). The Yang-Baxter relation: a new tool for knot theory. In *Braid Group, Knot Theory and Statistical Mechanics*, Yang, C.N. & Ge, M.L. (eds.), p. 151. World Scientific.

[11] Wadati, M., Deguchi, T. & Akutsu, Y. (1989). Exactly solvable models and knot theory. *Physics Reports* **180**, 247.
[12] Deguchi, T., Wadati, M. & Akutsu, Y. (1989). Knot theory based on solvable models at criticality. *Adv. Stud. in Pure Math.* **19**, 155.

J.-M. Maillet and F. Nijhoff

50. MULTIDIMENSIONAL LATTICE INTEGRABILITY AND THE SIMPLEX EQUATIONS

The simplex equations are multidimensional generalizations of the Yang-Baxter (triangle) equations, specific examples being the Zamolodchikov (or tetrahedron) equations and the Bazhanov-Stroganov (or 4-simplex) equations. We discuss the structure of the hierarchy of simplex equations and we show how this leads to a notion of classical integrability for discrete systems defined on multidimensional lattices.

1 Lattice equations in 1 + 1 dimensions

Among the many types of integrable systems that we are acquainted with nowadays there is a very interesting class, namely the class of Lattice equations by which we mean integrable nonlinear partial differential equations, where both variables ("space" and "time") take on discrete values, i.e. they form a square two-dimensional lattice, cf. e.g. [1],[2] and references therein. By taking proper continuum limits these equations go over into some of the well-known integrable partial differential equations (PDE's [1]), but one may argue that in some respects the lattice equations have an even richer structure than the corresponding PDE's, because they contain information not only of a single PDE but of an entire hierarchy of PDE's, cf. [3]. A linear system for the lattice system may be formulated in terms of a matrix Φ_k depending on a (spectral) parameter $k\in\mathbf{C}$, by giving Φ_k values on the vertices of the lattice and by connecting the different vertices by a transport matrix L_k living on the links of the lattices (cf. Fig. 1).

$$\Phi_k(n + 1) = L_k(n + 1,n)\cdot\Phi_k(n), \qquad n\in\mathbf{Z}. \tag{1}$$

On a two dimensional lattice $(n,m), n,m\in\mathbf{Z}$, we are supposed to have a Lax pair by specifying the transport matrices L_k in the two directions on the lattice, which we write as follows,

$$\Phi_k\begin{bmatrix} n+1 \\ m \end{bmatrix} = L_k\begin{bmatrix} n+1, & n \\ m, & m \end{bmatrix} \cdot \Phi_k\begin{bmatrix} n \\ m \end{bmatrix},$$

$$\Phi_k\begin{bmatrix} n \\ m+1 \end{bmatrix} = L_k\begin{bmatrix} n, & n \\ m+1, & m \end{bmatrix} \cdot \Phi_k\begin{bmatrix} n \\ m \end{bmatrix} \tag{2}$$

The compatibility condition for the overdetermined system (2) reads

$$L_k\begin{bmatrix} n+1, & n \\ m+1, & m+1 \end{bmatrix} \cdot L_k\begin{bmatrix} n, & n \\ m+1, & m \end{bmatrix} = L_k\begin{bmatrix} n+1, & n+1 \\ m+1, & m \end{bmatrix} \cdot L_k\begin{bmatrix} n+1, & n \\ m, & m \end{bmatrix}. \tag{3}$$

Equation (3) leads to integrable equations if we specify the dependence of L_k in the two directions as a function of k. The most simple example is obtained by choosing the L_k to depend linearly on k, i.e.

$$L_k\begin{bmatrix} n+1, & n \\ m, & m \end{bmatrix} = kI + Q(n,m), L_k\begin{bmatrix} n, & m \\ m+1, & m \end{bmatrix} = kJ + P(n,m), \tag{4}$$

in which I and J are constant (i.e. independent of n and m) matrices with $[I,J] = 0$. In that case (3) reduces to the pair of equations (for simplicity we restrict ourselves below to the case that $I=J=1$)

$$Q(n,m+1)\cdot P(n,m) = P(n+1,m)\cdot Q(n,m), \tag{5a}$$

$$Q(n,m+1) + P(n,m) = P(n+1,m) + Q(n,m). \tag{5b}$$

Solving on the one hand (5a) by putting $Q(n,m) = g(n+1,m)\cdot g(n,m)^{-1}, P(n,m) = g(n,m+1)\cdot g(n,m)^{-1}$ we find from (5b) the following lattice equation in terms of the matrix g(n,m).

$$g(n+1,m+1)\cdot g(n,m+1)^{-1} + g(n,m+1)\cdot g(n,m)^{-1}$$

$$= g(n+1,m+1)\cdot g(n+1,m)^{-1} = g(n+1,m)\cdot g(n,m)^{-1}. \tag{6}$$

On the other hand if we solve (5b) by taking $Q(n,m) = H(n+1,m) - H(n,m), P(n,m) = H(n,m+1) - H(n,m)$ we obtain from (5a) the lattice equation

$$[H(n+1,m+1) - H(n,m+1)]\cdot[H(n,m+1) - H(n,m)] =$$

$$[H(n+1,m+1) - H(n+1,m)]\cdot[H(n+1,m) - H(n,m)]. \tag{7}$$

Equations (6) and (7) can be considered to be a lattice generalization of equations in the AKNS-hierarchy and its gauge equivalent counterparts [1].

The purpose of this note is to present some ideas on how to generalize the type of systems presented above into dimensions d >

2. One such way to do this is along lines similar to the ones that lead from PDE's in 1 + 1 dimensions to PDE's in 2 + 1 dimensions (typical examples being the KP equation or the DS equation). Such a program has been proposed, e.g. in [4]. It is, however, not clear whether systems of KP type are genuinely higher-dimensional equations in the sense that the underlying algebraic structures to these equations really have a higher-dimensional character [5]. On the contrary, there are many indications that this is probably not the case [6].

It is well known that the basic algebraic structure underlying most integrable systems in 1 + 1 dimensions is given by the (classical or quantum) Yang-Baxter equations [7]. It is natural therefore, if one is interested in extracting the algebraic structure of integrable nonlinear systems, to consider the possibility of generalizing the Yang-Baxter equations. This leads us to the study of the so-called d-simplex equations which we are about to discuss next.

2 The simplex equations

The notion of the d-simplex equations was introduced by Bazhanov & Stroganov [8] in their study of the tetrahedron equation. One may think of them as being generalizations of the condition for commutativity of matrices, when the objects of interest no longer depend on a single pair of matrix-indices but on a collection of pairs of matrix-indices [9]. Thus we have tensor-like objects $\left(S^{j_1 j_2,\dots,j_d}_{i_1 i_2,\dots,i_d}\right)$, $d = 1,2,\dots,$ and the product of d + 1 of such objects is given by

$$\left(S^{(1)}\right)^{k_1 k_2,\dots,k_d}_{i_1 i_2,\dots,i_d} \left(S^{(2)}\right)^{j_1 k_{d+1},\dots,k_{2d-1}}_{k_1 i_{d+1},\dots,i_{2d-1}} \cdots \left(S^{(d+1)}\right)^{j_d j_{2d-1}\cdots j_D}_{k_d k_{2d-1}\cdots k_D}, \tag{8}$$

($D \equiv \frac{1}{2}d(d+1)$), in which summation over repeated indices is understood. The essential feature of this generalized matrix product is that every object $S^{(p)}$ is coupled once and only once to each of the other d objects via a matrix multiplication. Alternatively, we can regard the $S^{(p)}$ as being matrices acting on the tensor product $V_{\alpha_1} \otimes \cdots \otimes V_{\alpha_d}$ of d vector spaces V_{α_1}, $1 \le \alpha_1 < \alpha_2 < \cdots < \alpha_d \le D$. Introducing the tensor product $V^{(d)} \equiv V_1 \otimes V_2 \otimes V_D$, and letting any $S_{\alpha_1\dots\alpha_d}$ denote a matrix setting on $V^{(d)}$ which acts nontrivially only on the vector spaces $V_{\alpha_1},\dots,V_{\alpha_d}$ in the tensor product $V^{(d)}$, then we can rewrite (8) as the matrix product of matrices setting on $V^{(d)}$:

$$S^{(1)} * S^{(2)} *\dots* S^{(d+1)} \equiv S^{(1)}_{1,d+1,\dots,2d-1}\cdot\dots\cdot S^{(d+1)}_{d,2d-1,\dots,D}. \tag{9}$$

Now we take all the $S^{(p)}$ to be one and the same function S but each $S^{(p)}$ depending on a different subset of variables $\{\theta_{ij}\}_{1\le i<j\le d}$. In fact, the θ_{ij} are the angles in a hyperspherical (d-1)-simplex on S^{d-1} (i.e. the (d-1)-dimensional hypersphere). For more details cf. [9]. The d-simplex equation is now simply the condition of commutativity for a (d + 1)-fold product as in (9) of such a parameter-family of matrices S, i.e.

$$S^{(1)} * S^{(2)} *\dots* S^{(d+1)} * \dots * S^{(2)} * S^{(1)}. \tag{10}$$

As examples let us write down explicitly the first four d-simplex equations, with $a = S^{(1)}, B = S^{(2)}, C = S^{(3)}$, etc.

d = 1: $A^{k}_{i}B^{j}_{k} = B^{k}_{i}A^{j}_{k}$ (matrix commutativity) (11a)

d = 2: $A^{k_1k_2}_{i_1i_2}\,B^{j_1k_3}_{k_1i_3}\,C^{j_2j_3}_{k_2k_3} = C^{k_2k_3}_{i_2i_3}\,B^{k_1j_3}_{i_1k_3}\,A^{j_1j_2}_{k_1k_2}$

(Yang-Baxter equation) (11b)

d = 3: $A^{k_1k_2k_3}_{i_1i_2i_3}\,B^{j_1k_4k_5}_{k_1i_4i_5}\,C^{j_2j_4k_5}_{k_2k_4i_6}\,D^{j_3j_5j_6}_{k_3k_5k_6}$

$= D^{k_3k_5k_6}_{i_3i_5i_6}\,C^{k_2k_4j_6}_{i_2i_4k_6}\,B^{k_1j_4j_5}_{i_1k_4k_5}\,A^{j_1j_2j_3}_{k_1k_2k_3}$

(Zamolodchikov equation) (11c)

d = 4: $A^{k_1k_2k_3k_4}_{i_1i_2i_3i_4}\,B^{j_1k_5k_6k_7}_{k_1i_5i_6i_7}\,C^{j_2j_5k_8k_9}_{k_2k_5i_8i_9}\,D^{j_3j_6j_8k_{10}}_{k_3k_6k_8i_{10}}\,E^{j_4j_7j_9j_{10}}_{k_4k_7k_9k_{10}} =$

$= E^{k_4k_7k_9k_{10}}_{i_4i_7i_9i_{10}}\,D^{k_3k_6k_8j_{10}}_{i_3i_6i_8k_{10}}\,C^{k_2k_5j_8j_9}_{i_2i_5k_8k_9}\,B^{k_1j_5j_6j_7}_{i_1k_5k_6k_7}\,A^{j_1j_2j_3j_4}_{k_1k_2k_3k_4}.$

(Bazhanov-Stroganov equation) (11d)

For d = 2 we have the Yang-Baxter (YB) or triangle equation [10], which arises either as the condition for the factorizability of the scattering-matrix of three particles on a line [11], or as the generalized star-triangle equation for the Boltzmann weights in two-dimensional statistical mechanical spin- or vertex models [12]. These equations are widely studied and it is impossible to give here an account of all its aspects. Let us only mention that many of its solutions are known and that these so-called quantum R-matrices form the basic ingredients in the quantum inverse scattering method (cf. [14] for a review), which has been devised to solve integrable nonlinear quantum systems in d = 2. A beautiful reformulation of this method has led to the interesting theory of quantum groups [14,15].

For d = 3 we have the tetrahedron equation, which was first introduced by Zamolodchikov [16]. It describes the factorizability conditions for the scattering of four straight strings into three-string amplitudes in a plane, or the condition for commutation of layer-to-layer transfer matrices in a spin mode on a three dimensional cubic lattice [17,19]. In contrast to the YB equations there is to date only one known nontrivial solution [18], which is equivalent to a free-fermion model on a 3d cubic lattice [19], and for which the corresponding free energy has been calculated by Baxter [20]. So far, none of the interesting relations with (classical) integrable systems that one has for the YB equations has been established for the d-simplex equations for $d \geq 3$. In particular, there is no knowledge of a relation between the tetrahedron equations and integrable PDE's in three dimensions, there is no relation with hamiltonian structures and there is no quantum inverse scattering scheme for $d \geq 3$. These questions form part of a program undertaken by the authors, cf. [21].

We will not go into great detail here concerning the d-simplex equations. For more details cf. [9]. We will only focus on one aspect, namely the way in which the hierarchy of d-simplex equations is built up in successive steps by regarding the d-simplex equation as the consistency condition for an object that describes the violation of the (d - 1)-simplex equation in an appropriate way. In fact, suppose that we start out from the (d - 1)-simplex equations for objects $T^{(p)}$ similar to the objects $S^{(p)}$ in equation (10). Now there is a very simple way to introduce a violation or breaking of the simplex equation for the matrices $T^{(p)}$ by assuming that every $T^{(p)}$ depends additionally on an extra pair of indices, i.e. $T^{(p)}{}^{j_0}_{i_0} \equiv \left(T^{(p)}\right)^{j_0 j_1 \ldots j_d}_{i_0 i_1 \ldots i_d}$ such that we can couple these extra indices carried by the d matrices $T^{(p)}, p = 1, \ldots, d$, that appear in the (d - 1)-simplex equation, to a new matrix $S = S^{j_0 j'_0 j''_0 \ldots}_{i_0 i'_0 i''_0 \ldots}$. Thus writing

$$S^{k_0 k'_0 k''_0 \ldots}_{i_0 i'_0 i''_0 \ldots} \left(T^{(1)}{}^{j_0}_{k_0} * T^{(2)}{}^{j'_0}_{k'_0} * \ldots * T^{(d)}{}^{j_0^{(d-1)}}_{k_0^{(d-1)}} \right) =$$

$$= \left(T^{(d)}{}^{k_0^{(d-1)}}_{i_0^{(d-1)}} * \ldots * T^{(2)}{}^{k'_0}_{i'_0} * T^{(1)}{}^{k_0}_{i_0} \right) S^{j_0 j'_0 j''_0 \ldots}_{k_0 k'_0 k''_0 \ldots}\,, \qquad (12)$$

where the product between the $T^{(p)}$ is the same as in (9), cf. (8), for (d - 1 instead of d), we have introduced a breaking of the (d - 1)-simplex equation, and the matrix S is a measure of the breaking. Now the matrix S′ depends on parameters [9], such as the $T^{(p)}$ do, and as such the equation (12) gives rise to a consistency condition coming from the fact that in a product of proper combinations of $D = \frac{1}{2}d(d + 1)$ different $T^{(p)}$ the order in which the $T^{(p)}$ appear can be inverted making repeated use of equation (12). This consistency condition gives an equation for S which is exactly the d-simplex equation. The simplest example of this procedure is provided if we break the 1-simplex equation for $T^{(1)}, T^{(2)}$, in which case (12) becomes

$$S^{k_0 k'_0}_{i_0 i'_0}\, T^{(1)}{}^{j_0 k_1}_{k_0 i_1}\, T^{(2)}{}^{j'_0 j_1}_{k'_0 k_1} = T^{(2)}{}^{k'_0 k_1}_{i'_0 i_1}\, T^{(1)}{}^{k_0 j_1}_{i_0 k_1}\, S^{j_0 j'_0}_{k_0 k'_0}. \tag{13}$$

The 2-simplex equations for S follow by identifying the two different ways of inverting the order in the product

$$T^{(1)}{}^{j_0 k_1}_{i_0 i_1}\, T^{(2)}{}^{j'_0 l_1}_{i'_0 k_1}\, T^{(3)}{}^{j''_0 j_1}_{i''_0 l_1}, \tag{14}$$

by applying equation (13) repeatedly. We will not discuss here how the compatibility of equation (12) is implemented for the higher-order d-simplex equations, for details we refer to [22]. We finish this section by the remark that in order to introduce a violation of the (d - 1)-simplex equation of the form (12) we have to extend the objects $T^{(p)}$ to include extra indices, meaning that they become objects that act on a tensor product with additional vector spaces. This idea of increasing the number of matrix-indices is something we shall use also in the next section.

3 Generalization of Lax equations

We now return to the problem of arriving at a notion of **classical** integrability on lattices of dimension $d > 2$. It is a well-known fact that the presence of a Lax pair, like equation (2), ensures one that in principle one may be able to integrate the equations of motion coming from (3) by solving the linear spectral problem given by either one of the members of (2), under suitably chosen boundary conditions. (In practice, however, formidable problems of a technical nature may arise, which we shall not dwell upon here.) In particular, from the Lax pair an infinite series of constants of the motion can be found. We are interested in generalizations of equations (2) and (3) that will preserve in some way these basic features.

Our first aim is to extend the notion of a Lax pair by trying to generalize the basic algebraic structure underlying (3). Our suggestion is that we have to use the structure of the hierarchy of the d-simplex equations as a guideline. Considering the list (11) of simplex equations it is suggestive to compare equation (3)

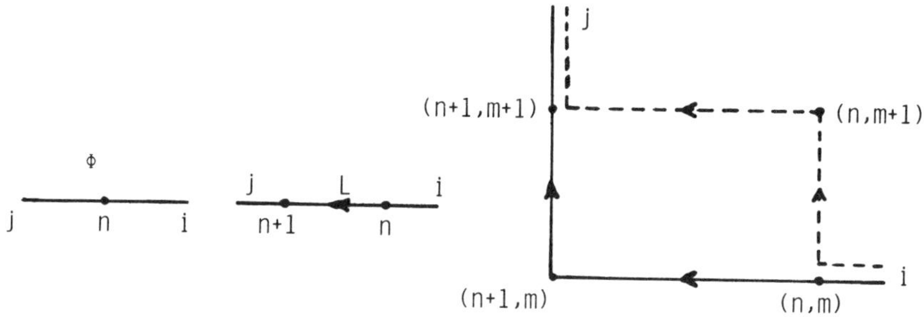

Figure 1. The vertex $\Phi_k(n)^j_i$, *and the transport-matrix* $L_k(n+1,n)^j_i$ *on a 1d-lattice, resp. the commutativity condition along a plaquette on a 2d-lattice.*

with the 1-simplex equation, the only difference being that the objects in (11a) are numerical matrices, whereas in equation (3) we have functions of the sites on a lattice, and, furthermore, there are lattice translations involved in the equation. Nevertheless, this gives a strong suggestion that in one dimension higher we have to consider the 2-simplex equation to get an idea of the form of the Lax equations in d = 3. More generally we expect that the form of the Lax equations in d dimensions should resemble the one of the (d - 1)-simplex equation. In order to write down an explicit proposal let us give some arguments that will lead to this end. We shall restrict our discussion here to the case d = 3. The argument, however, is quite general and extensions to any dimension are straightforward [22].

One may argue that the main feature that provokes the occurrence of a compatibility condition as in equation (3), is the fact that the number of ways available to couple matrices to each other (namely by matrix products) is smaller than the number of ways of going about on a two-dimensional lattice. Pictorially this is indicated in Fig. 1c. To say it in a more fanciful way: the number of "matrix directions" is inferior to the number of space-time directions. If, now, we want to increase the number of lattice directions in such a way that we arrive at a genuinely higher-dimensional equation, we have to avoid the possibility of getting closed equations already on a lower-dimensional level. This means that we have to violate the Lax equations that describe the compatibility conditions of (2) on a plane lattice. The only natural way of doing this is by introducing, as in the previous section, extra indices that allow us to couple the matrices in different ways, increasing the number of "matrix directions". This idea suggests that we might try to initiate the construction presented in Section 2 where we were led to the d-simplex equation by breaking the (d - 1)-simplex equation.

Introducing additional matrix directions would mean that the matrix Φ_k in (2) would become an object $\Phi^{j_1 j_2}_{i_1 i_2}$ (we will suppress the dependence on spectal parameters in the following), see Fig. 2c. Strictly speaking, however, if (3) is no longer satisfied Φ will become a path-dependent object: the solutions of (2) will depend on the way we go about on the lattice, cf. the discussion in [21]. In that case Φ will no longer be an object that lives on the vertices of the lattice, and there is no sense in taking Φ to be the fundamental object in the theory. Its role would be replaced now by the link objects L in both directions which are now of the form

$$L\begin{bmatrix} n+1, & n \\ m, & m \end{bmatrix}^{J\,j}_{I\,i} \text{ and } L\begin{bmatrix} n, & n \\ m+1, & m \end{bmatrix}^{J\,j}_{I\,i}. \tag{15}$$

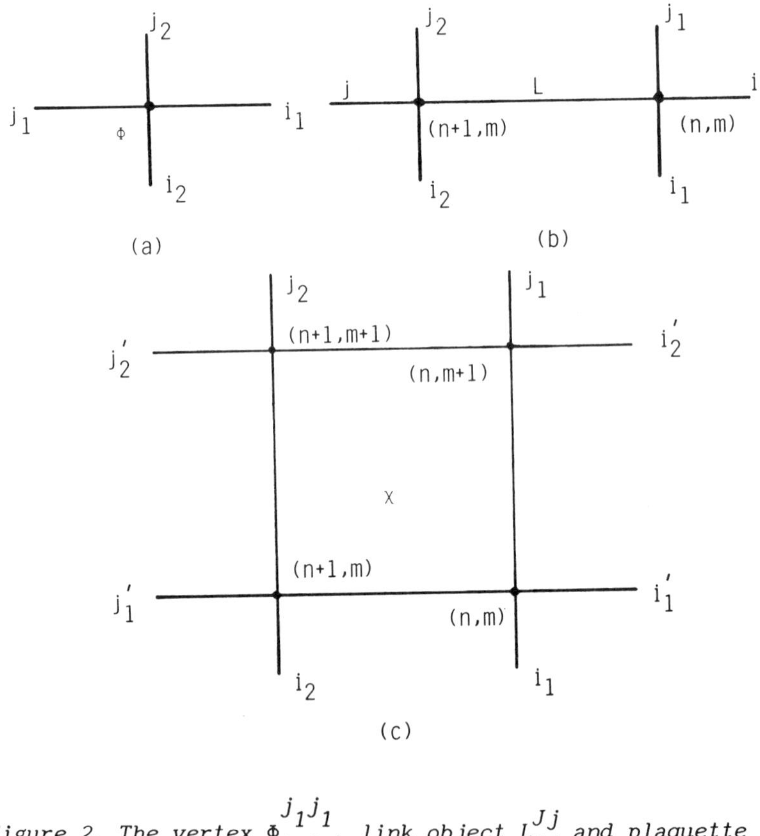

Figure 2. The vertex $\Phi^{j_1 j_1}_{i_1 i_2}$, *link object* L^{Jj}_{Ii} *and plaquette object* $\chi^{JJ'}_{II'}$ *as in equations (15) and (16),* $I = (i_1, i_2), J = (j_1, j_2)$, *etc.*

As is clear from Fig. 2b the elementary link objects, which are obtained by glueing together two vertices, should depend on two indices i,j along the link and two pairs of indices $I = (i_1,i_2)$, $J = (j_1,j_2)$ perpendicular to the link. Now as the two link objects (15) become the fundamental objects in the dynamics we would like to write down an analogue of a linear system in terms of these quantities. In order to describe the dynamics of link variables on a plane we introduce objects χ and $\tilde{\chi}$ that live on plaquettes of the lattice, e.g.

$$\chi = \chi \begin{bmatrix} n+1, & n \\ m+1, & m \end{bmatrix}^{JJ'}_{II'} ,$$

where I,J are as above and $I' = (i'_1,i'_2), J = (j'_1,j'_2)$, see Fig. 2c. With these objects we are able to describe how a couple of link-objects (15) are transported along a plaquette to give a new pair of translated link variables. This is achieved by the following relation

$$\begin{aligned} &\chi\begin{bmatrix} n+1, & n \\ m+1, & m \end{bmatrix}^{KK'}_{II'} L\begin{bmatrix} n+1, & n \\ m+1, & m \end{bmatrix}^{Jk}_{Ki} L\begin{bmatrix} n, & n \\ m+1, & m \end{bmatrix}^{J'j}_{K'k} = \\ &= L\begin{bmatrix} n+1, & n+1 \\ m+1, & m \end{bmatrix}^{K'k}_{I'i} L\begin{bmatrix} n+1, & n \\ m, & m \end{bmatrix}^{Kj}_{Ik} \tilde{\chi}\begin{bmatrix} n+1, & n \\ m+1, & m \end{bmatrix}^{JJ'}_{KK'} . \end{aligned} \tag{16}$$

The summations in equation (16) are performed over a sigle matrix index k between the L and over pairs of indices $K = (k_1,k_2), K' = (k'_1,k'_2)$ which couple the χ and $\tilde{\chi}$ to the L. Note that equation (16) is precisely a relation that describes the breaking of equation (3), in much the same way as in which equation (13) is a breaking of the 1-simplex equation. We expect therefore that equation (16) also leads to a compatibility condition by a similar connection as in (14). In order to have an overdetermined system, however, we have now to introduce another lattice direction, because (16) will never lead to closed equations on a lattice in d = 2. We consider therefore a 3d-lattice, where the lattice points are given by $(n,m,h), n,m,h \in \mathbf{Z}$, and we impose an equation like (16) in every one of the three directions. Thus we have three objects

$$\chi\begin{bmatrix} n+1, & n \\ m+1, & m \\ h, & h \end{bmatrix}, \ \chi\begin{bmatrix} n+1, & n \\ m, & m \\ h+1, & h \end{bmatrix} \ \text{and} \ \chi\begin{bmatrix} n, & n \\ m+1, & m \\ h+1, & h \end{bmatrix} , \tag{17}$$

each of which is associated with one of the three different types of plaquette on a 3d-lattice, and similarly for $\tilde{\chi}$. Following (14) we consider now the following combination of link objects:

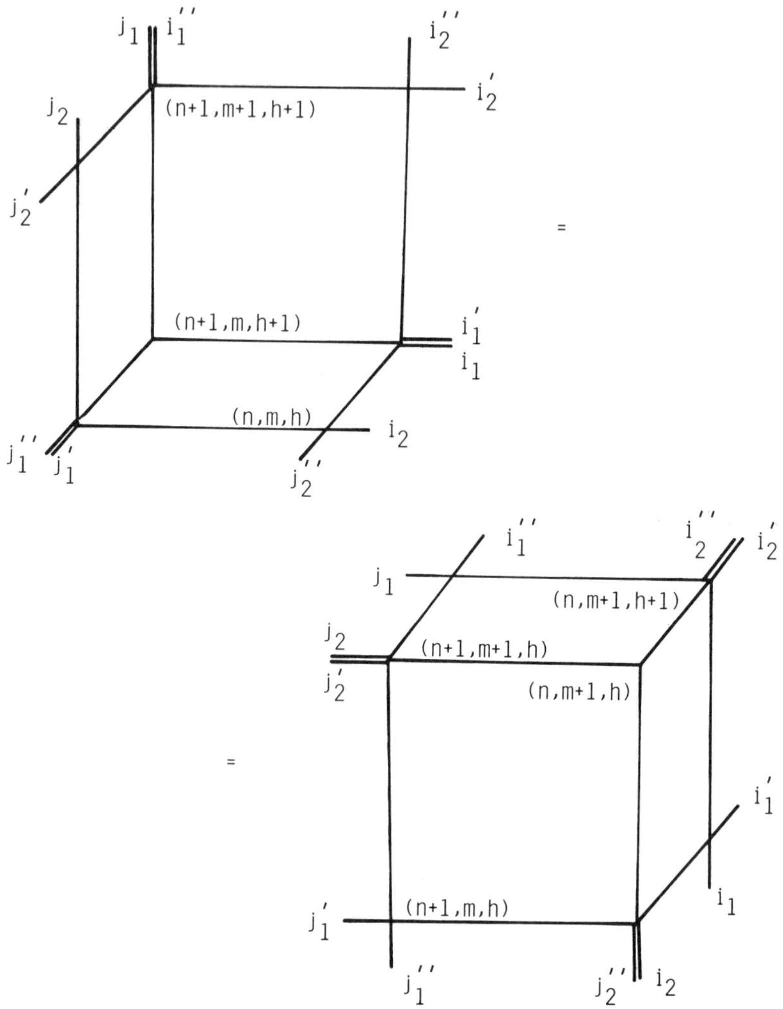

Figure 3. Graphical representation of the compatibility equation (20) for plaquettes along a cube on 3d-lattice.

$$L\begin{bmatrix} n+1, & n+1 \\ m+1, & m+1 \\ h+1, & h \end{bmatrix}_{Ii}^{Jk} \quad L\begin{bmatrix} n+1, & n+1 \\ m+1, & m \\ h, & h \end{bmatrix}_{I'k}^{J'l} \quad L\begin{bmatrix} n+1, & n \\ m, & m \\ h, & h \end{bmatrix}_{I''l}^{J''j} . \qquad (18)$$

Using now equation (16) for the three different plaquette objects (17) we can reverse the order in (18) leading to

$$L\begin{bmatrix} n+1, & n \\ m+1, & m+1 \\ h+1, & h+1 \end{bmatrix}^{J''k}_{I''i} \quad L\begin{bmatrix} n, & n \\ m+1, & m \\ h+1, & h+1 \end{bmatrix}^{J'l}_{I'k} \quad L\begin{bmatrix} n, & n \\ m, & m \\ h+1, & h \end{bmatrix}^{Jj}_{Il} , \qquad (19)$$

and this reversion can be obtained in two different ways, using a different combination of $\chi, \tilde{\chi}$. If we identify these two different ways we obtain the following equation for the χ

$$\chi\begin{bmatrix} n+1, & n \\ m+1, & m \\ h+1, & h+1 \end{bmatrix}^{K'K''}_{I'I''} \quad \chi\begin{bmatrix} n+1, & n \\ m, & m \\ h+1, & h \end{bmatrix}^{KJ''}_{IK''} \quad \chi\begin{bmatrix} n+1, & n+1 \\ m+1, & m \\ h+1, & h \end{bmatrix}^{JJ'}_{KK'} =$$

$$= \chi\begin{bmatrix} n, & n \\ m+1, & m \\ h+1, & h \end{bmatrix}^{KK'}_{II'} \quad \chi\begin{bmatrix} n+1, & n \\ m+1, & m+1 \\ h+1, & h \end{bmatrix}^{JK''}_{KI''} \quad \chi\begin{bmatrix} n+1, & n \\ m+1, & m \\ h, & h \end{bmatrix}^{J'J''}_{K'K''} , \qquad (20)$$

and a similar equation for $\tilde{\chi}$. Equation (20) describes the consistency for plaquette-objects along a cube on a 3d-lattice, cf. Fig. 3. It is our proposal for a generalization of equations (3) on a lattice in d = 3. By construction it is clear that (20) resembles in form the 2-simplex equation. The only difference is now that the objects are no longer numerical quantities but functions of lattice points (n,m,h), and, furthermore, there are relative shifts on the lattice involved. If these relative shifts were not present equation (20) would be exactly the YB equation. Because of the shifts, however, we have a highly nontrivial generalization of the YB equation. In fact, we believe that due to the shifts equation (20) is **as a classical lattice equation** related to the tetrahedron equation (11d) rather than to the triangle equation terms of one or more "spectral" parameters. In order to extract nonlinear equations from (20) we have to make an appropriate ansatz on such a parameter dependence in $\chi, \tilde{\chi}$. Due to the trilinear structure of equation (20), however, this is a nontrivial problem. It is obviously related to the one of finding solutions to the YB equations, but now in terms of functions on the lattice instead of numerical matrices. In fact, we conjecture that the proper parametrization of equation (20) is related to the one for the tetrahedron equations, i.e. in terms of spherical angles on S^2. These questions are under current investigation by the authors.

References

[1] Nijhoff, F.W., Quispel, G.R.W., Capel, H.W. & Wiersma, G.L. (1984). *Phys. Lett.* **103A**, 293; Quispel, G.R.W., Nijhoff, F.W., Capel, H.W. & van der Linden, J. (1984). *Physica* **125A**, 344.

[2] Nijhoff, F.W., Capel, H.W. & Wiersma, G.L. (1985). *Springer Lect. Notes Phys.* **239**, 263.

[3] Wiersma, G.L. & Capel, H.W. (1987). *Physica* **142A**, 199; Wiersma, G.L. (1987). *Thesis*, University of Amsterdam.
[4] Nijhoff, F.W. (1985). *Lett. Mat. Phys.* **9**, 237; (1986). *Physica* **18D**, 380.
[5] Schultz, C.L., Ablowitz, M.J. & Bar Yaacov, D. (1987). *Phys. Rev. Lett.* **59**, 2825.
[6] Santini, P.M., in this volume.
[7] Faddeev, L.D. (1984). In *Recent Advances in Field Theory and Statistical Mechanics*, p. 561. Stora, R. & Zuber, J.B. (eds.). North-Holland: Amsterdam.
[8] Bazhanov, V.V. & Stroganov, Yu. G. (1982). *Theor. Math. Phys.* **52**, 685.
[9] Maillet, J.M. & Nijhoff, F.W. *The Tetrahedron Equation and the Four-Simplex Equation,* Fermi-lab Preprint 88/71-T.
[10] Yang, C.N. (1967). *Phys. Rev. Lett.* **19**, 1312; Baxter, R.J. (1972). *Ann. Phys.* **70**, 193.
[11] Zamolodchikov, A.B. (1980). *Sov. Sci. Revs.* **A2**, 2.
[12] Baxter, R.J. (1982). *Exactly Solved Models in Statistical Mechanics.* Academic Press: London.
[13] Kulish, P.P. & Sklyanin, E.K. (1982). *Springer Lect. Notes Phys.* **151**, 61.
[14] Drinfel'd, V.G. (1985). *Sov. Math. Dokl.* **32**, 254; (1986). *Proc. Int. Congr. Math, Berkeley.*
[15] Jimbo, M. (1985). *Lett. Math. Phys.* **10**, 63; (1986). **11**, 247; (1986). *Commun. Math. Phys.* **102**, 537.
[16] Zamolodchikov, A.B. (1980). *Sov. JETP* **52**, 325.
[17] Jaekel, M.I. & Maillard, J.M. (1982). *J. Phys.* **A15**, 1309.
[18] Zamolodchikov, A.B. (1983). *Commun. Math. Phys.* **79**, 489; Baxter, R.J. (1983). *Commun. Math. Phys.* **88**, 185.
[19] Baxhanov, V.V. & Stroganov, Yu. G. (1984). *Nucl. Phys.* **B230** [FS10], 435.
[20] Baxter, R.J. (1984). *Phys. Rev. Lett.* **53**, 1795; (1986). *Physica* **18D**, 321.
[21] Nijhoff, F.W. & Maillet, J.M. (1988). In *Proc. VIth Int. Workshop on Nonlinear Evolution Equations and Dynamical Systems,* p. 281. Léon, J. (ed.). World Sci. Publ. Co.: Singapore.
[22] Maillet, J.M. & Nijhoff, F.W., to be published.

M. Lakshmanan, K. Porsezian and M. Daniel

51. ON THE NONLINEAR DYNAMICS OF THE ONE DIMENSIONAL CLASSICAL HEISENBERG FERROMAGNETIC SPIN CHAIN

The Heisenberg spin chain is a fascinating nonlinear dynamical system which exhibits coherent [6,15] as well as chaotic [12] structures depending on the nature of the interactions. Here, we will be mainly concerned with the coherent structures and the associated integrability properties of ferromagnetic chains.

1 Description of classical spin chains

1.1 *Models*

The total energy of a typical ferromagnetic spin system [10] stems mainly from (i) interaction between magnetic ions in the form of bilinear, biquadratic or even higher order exchange, (ii) interaction due to crystal field anisotropy, (iii) interaction with internal or external fields, and (iv) interaction between magnetic and elastic degrees of freedom. A standard model Hamiltonian which incorporates most of these interactions is of the form

$$H = -\sum_{ij} J[\vec{S}_i.\vec{S}_j + \alpha(\vec{S}_i.\vec{S}_j)^2] + A\sum_i (S_i^z)^2 - \mu\sum_i \vec{S}_i.\vec{B}. \qquad (1)$$

Here J is the exchange integral and α represents the biquadratic exchange parameter. "A" corresponds to the single site anisotropy coefficient with easy axis along the z-direction, μ is the gyromagnetic ratio and $\vec{B}$ stands for the external magnetic field. It is of particular importance to treat the spin angular momentum operators $\vec{S}_i$'s as classical 3-component vectors of unit (constant) length which corresponds, for example, to the case of very large values of spins and the limit $\hbar \longrightarrow 0$.

1.2 *Equations of motion*

1.2.1. *Discrete spin chains.* The evolution equation for the spin systems can be obtained in the following way. Starting from the Heisenberg equation of motion for the quantum spin operators $\vec{S}_n$,

$$i\hbar\frac{d\vec{S}_n}{dt} = [\vec{S}_n, H], \quad n = 1,2,3,\ldots,N, \tag{2}$$

where the square bracket stands for the usual commutators, the classical limit can be obtained by taking

$$(\vec{S}_n)_{\text{quantum}} \underset{\substack{\hbar \to 0 \\ S \to \infty}}{\longrightarrow} (\vec{S}_n)_{\text{classical}}. \tag{3}$$

Then the quantum commutator becomes the classical Poisson bracket

$$(1/i\hbar)\ [\vec{S}_n, H] \underset{\hbar \to 0}{\longrightarrow} \{S_n, H\}_{PB} \tag{4}$$

and the dynamical equation for the classical ferromagnetic spin chain becomes [3]

$$\frac{d\vec{S}_n}{dt} = \{\vec{S}_n,\ H\}_{PB.},\quad \vec{S}_n = (S_n^x,\ S_n^y,\ S_n^z),\quad \vec{S}_n^2 = 1. \tag{5}$$

The dynamics of spins can also be considered from an altogether classical point of view by developing a Hamiltonian formalism from the outset. Generalizing the classical Poisson bracket by including spin as a canonical variable, suitable canonical equations can be obtained in analogy with the spinless nonrelativistic particles [7,18]. Thus for given functions $f(\vec{S}_n)$ and $g(\vec{S}_n)$ of the spin variable, the Poisson bracket may be defined in terms of the spin components as

$$\{f,g\}_{PB} = \sum_{n=1}^{N} \sum_{\alpha\beta\gamma=1}^{3} \varepsilon_{\alpha\beta\gamma} \frac{\partial f}{\partial S_n^\alpha} \frac{\partial g}{\partial S_n^\beta} S_n^\gamma. \tag{6}$$

Here $\varepsilon_{\alpha\beta\gamma}$ represents the complete antisymmetric Levi-Civita tensor. It can be easily checked that the spin canonical Poisson bracket (6) satisfies the same algebraic relations as the usual canonical Poisson bracket,

$$\{f,g\}_{PB} = -\ \{g,f\}_{PB} \tag{7}$$

and if k is another arbitrary function of spins, then

$$\{f,g+k\}_{PB} = \{f,g\}_{PB} + \{f,k\}_{PB}, \qquad (8)$$

$$\{f,gk\}_{PB} = g\{f,k\}_{PB} + \{f,g\}_{PB}k \qquad (9)$$

and that the Jacobi's identity

$$\{f,\{g,k\}_{PB}\}_{PB} + \{g,\{k,f\}_{PB}\}_{PB} + \{k,\{f,g\}_{PB}\}_{PB} = 0 \qquad (10)$$

holds. As a result, the canonical equation of motion can again be written in the form (5).

Now, using (5), the equation of motion for the Heisenberg ferromagnetic spin system (1) can be written as

$$\frac{d\vec{S}_n}{dt} = \vec{S}_n \times [J(\vec{S}_{n+1} + \vec{S}_{n-1}) + \alpha J\{\vec{S}_{n+1}(\vec{S}_n.\vec{S}_{n+1}) + \vec{S}_{n-1}(\vec{S}_n.\vec{S}_{n-1})\} - 2AS^z_n\hat{m} + \mu\vec{B}], \qquad (11)$$

$$\hat{m} = (0,0,1),\ \vec{S}^2_n = 1,\ n = 1,2,3,\ldots,N.$$

While deriving eq. (11), we have taken the exchange interactions only between nearest neighbour spins.

1.2.2 *Continuum approximation.* Since the nonlinear character of eq. (11) renders a general discussion difficult, one may consider the continuum limit of the system which is considered to be a satisfactory description in the long wavelength, low temperature limit. We allow the lattice parameter 'a' between nearest neighbour sites to approach zero and assume a slow variation of $\vec{S}_n$ over a lattice distance. Introducing a one dimensional spin field $\vec{S}(x,t)$ at position x and time t so that when x = na, $\vec{S}(na,t) = \vec{S}_n(t)$ and expanding in a Taylor series, we have

$$\vec{S}_{n\pm1}(t) = \vec{S} \pm a\frac{\partial\vec{S}}{\partial x} + \frac{a^2}{2!}\frac{\partial^2\vec{S}}{\partial x^2} \pm \frac{a^3}{3!}\frac{\partial^3\vec{S}}{\partial x^3} + \frac{a^4}{4!}\frac{\partial^4\vec{S}}{\partial x^4} + O(a^5). \qquad (12)$$

Using the above expansion, the continuum version of eq. (11) to $O(a^4)$ can be written after suitable rescaling of t and redesignation of the parameters A and μ by a factor $(Ja^2)^{-1}$ as

$$\vec{S}_t = \vec{S} \times \{\vec{S}_{xx} + \nu\vec{S}_{xxxx} - \beta[(\vec{S}_x)^2\vec{S}_{xx} + 2(\vec{S}_x.\vec{S}_{xx})\vec{S}_x] - 2AS^z\hat{m} + \mu\vec{B}\}, \qquad (13)$$

where

$$\nu = a^2/12,\quad \beta = [\alpha a^2/(1+2\alpha)],\quad \vec{S}^2 = 1.$$

Here suffixes correspond to partial derivatives.

2 Dynamics of the discrete spin chains

In studying the dynamics of the discrete spin chains, eq. (11) is in general not known to be solvable except under certain special situations. For example the Heisenberg model of the isotropic chain governed by the equation of motion

$$\frac{d\vec{S}_n}{dt} = \vec{S}_n \times (\vec{S}_{n+1} + \vec{S}_{n-1}),\ \vec{S}_n^2 = 1 \tag{14}$$

is not known to be completely integrable. However, a slightly modified form

$$\frac{d\vec{S}_n}{dt} = \vec{S}_n \times \left\{ \frac{\vec{S}_{n+1}}{(1+\vec{S}_n.\vec{S}_{n+1})} + \frac{\vec{S}_{n-1}}{(1+\vec{S}_n.\vec{S}_{n-1})} \right\} \tag{15}$$

which is derivable from the Hamiltonian

$$H = -2 \sum_i \log\,(1+\vec{S}_i.\vec{S}_{i+1}) \tag{16}$$

is completely integrable, admits the Lax pair

$$L_n = \frac{\lambda + \lambda^{-1}}{2} I + \frac{\lambda - \lambda^{-1}}{2} S_n,$$

$$B_n = i\left(1 - \frac{\lambda^2 + \lambda^{-2}}{2}\right) \frac{S_n + S_{n-1}}{(1+\vec{S}_n.\vec{S}_{n-1})} - i\,\frac{\lambda^2 - \lambda^{-2}}{2}\, \frac{I + S_{n-1}S_n}{(1+\vec{S}_n.\vec{S}_{n-1})}, \tag{17}$$

where

$$S_n = \vec{S}_n.\vec{\sigma},\quad I = \begin{bmatrix}1 & 0\\ 0 & 1\end{bmatrix},\quad \lambda = \text{parameter},\ \vec{\sigma} = (\sigma_1,\sigma_2,\sigma_3)$$

so that

$$L_{n,t} = B_{n+1}L_n - L_nB_n$$

is equivalent to eq. (15) and the underlying nonlinear excitations are soliton modes [4].

Again considering the planar solutions of the anisotropic Heisenberg spin chain $H = \sum_{n=1}^{N-1} \vec{S}_n.\underline{J}\vec{S}_{n+1}$, $\underline{J} = \text{diag}(J_1,J_2,J_3)$ by assuming $S_n^z(t) = 0$ for all t and n, it can be easily shown that these solutions are stationary solutions [13,16] satisfying

$$J_1 S_n^y (S_{n+1}^x + S_{n-1}^x) = J_2 S_n^x (S_{n+1}^y + S_{n-1}^y) \tag{18}$$

with

$$S_n^{x^2} + S_n^{y^2} = 1.$$

A numerical study of eq. (18) shows that the solutions are aperiodic for almost all initial conditions. It is not difficult to show that after some rearrangements and iterations, eq. (18) lies on continuous invariant curves

$$S_n^x S_{n+1}^x + (J_1/J_2) S_n^y S_{n+1}^y = C, \ (C = \text{constant}) \tag{19}$$

in the phase plane (S_{n+1}^x, S_n^x). Moreover, using an elliptic substitution, eq. (18) for planar states and eq. (19) for invariant curves can be integrated completely. Additional iterations however seem to lead to chaotic planar solutions.

Finally, addition of other interactions like anisotropy and external fields lead to chaotic structures [12] in spin clusters very often.

3 Dynamics of the continuum spin chains

3.1 *Lowest order systems*

When we consider spin chains of $O(a^2)$ in the continuum limit, the elementary spin excitations are often solitons. For example, the isotropic Heisenberg chain in a constant magnetic field (which can be transformed away)

$$\vec{S}_t = \vec{S} \times \vec{S}_{xx} \tag{20}$$

can be shown to be equivalent to the integrable nonlinear Schrödinger equation (NLS) [6]. Eq. (20) is completely integrable and explicit N-soliton solutions can be obtained [15]. The unaxial anisotropic chain in a longitudinal field (which can again be transformed away)

$$\vec{S}_t = \vec{S} \times [\vec{S}_{xx} - 2AS^z\hat{m}] \tag{21}$$

is gauge equivalent to the NLS. Also, the N-soliton solutions can be constructed explicitly [1,11]. However, when the anisotropic chain is placed in a transverse magnetic field (which cannot be transformed away), the system is not known to be integrable in general [8]. On the other hand the pure biaxial anisotropic chain shows complete integrability [14].

3.2 *Higher order corrections*

3.2.1. *Invariant equations.* As a natural extension, it is interesting to study the effect of higher order corrections say up to $O(a^4)$ [9]. The corresponding equation of motion for an isotropic chain with bilinear and biquadratic exchange interactions is written from eq. (13) as

$$\vec{S}_t = \vec{S} \times [\vec{S}_{xx} + \nu\vec{S}_{xxxx} - \beta[(\vec{S}_x)^2\vec{S}_{xx} + 2(\vec{S}_x\cdot\vec{S}_{xx})\vec{S}_x]]. \qquad (22)$$

As $a \longrightarrow 0$, eq. (22) reduces to eq. (20) which represents a completely integrable dynamical system.

By associating a moving space curve in E^3 having curvature

$$\kappa(x,t) = (\vec{S}_x.\vec{S}_x)^{1/2} \qquad (23)$$

and torsion

$$\tau(x,t) = (\vec{S}_x.\vec{S}_x)^{-1}\, \vec{S}\cdot(\vec{S}_x \times \vec{S}_{xx}) \qquad (24)$$

[5] with the spin chain (22), we obtain the evolution equations

$$\kappa_t = -[2\kappa_x\tau + \kappa\tau_x + \nu(4\kappa_{xxx}\tau + 6\kappa_{xx}\tau_x + 4\kappa_x\tau_{xx} - 9\kappa^2\kappa_x\tau$$
$$- \kappa^3\tau_x + \kappa\tau_{xxx} - 4\kappa_x\tau^3 - 6\kappa\tau^2\tau_x - \beta(6\kappa^2\tau\kappa_x + \kappa^3\tau_x))], \qquad (25)$$

$$\tau_t = \{(\kappa_{xx}/\kappa) - \tau^2 + \kappa^2/2 + \nu[(\kappa_{xxxx}/\kappa) - 5\kappa_{xx}\kappa - (25/2)\kappa_x^2$$
$$- 4\tau\tau_{xx} - 6(\tau^2/\kappa)\kappa_{xx} - 3\tau_x^2 - (3/2)\kappa^4 - 12\tau\tau_x\kappa_x/\kappa$$
$$+ \tau^4 - \kappa^2\tau^2/2] + \beta(\kappa^2\tau^2 - 6\kappa_x^2 - 3\kappa\kappa_{xx} - 3\kappa^4/4)\}_x. \qquad (26)$$

Now, making the complex transformation

$$q(x,t) = \kappa(x,t)\exp\{i\int_{-\infty}^{x}\tau(x',t)dx'\} \qquad (27)$$

equations (25) and (26) take the form of a generalized nonlinear Schrödinger (GNLS) equation

$$iq_t + q_{xx} + 2|q|^2 q + \nu\, q_{xxxx} - 4(2\nu+\beta)q_{xx}^{*}q^2 - 4(3\nu+2\beta)|q|^2 q_{xx}$$
$$- 4(9\nu+4\beta)|q_x|^2 q - 4[(7/2)\nu + 2\beta]q_x^2 q^{*} - 24(\nu+\beta/2)|q|^4 q = 0. \qquad (28)$$

When a=0, eq. (28) is the NLS. In general, eq. (28) is not

integrable except for very few specific choices of the parameters involved.

3.2.2 *Integrability.* In order to identify the integrable cases of eq. (28), we carry out a Painlevé analysis [17] which ensures that the general solution of the equation is single valued around any movable singularity manifold. By denoting $q = E$ and $q^* = G$, using a Laurent type expansion

$$E = \sum_j E_j \Phi^{j+p}, \qquad G = \sum_j G_j \Phi^{j+q} \tag{29}$$

around a movable noncharacteristic singularity manifold and by equating the most dominant terms, we obtain

$$\text{Branch (i)} \;:\; p = q = -1,\ E_0 G_0 = -4\Phi_x^{\,2},$$

$$\text{Branch (ii)} \;:\; p = q = -1,\ E_0 G_0 = \Phi_x^{\,2}/[4(\nu+\beta/2)]. \tag{30}$$

By carrying out a resonance analysis, we find that the powers at which arbitrary functions can enter into the Laurent series (29) are

$$\text{Branch (i)} \;:\; j = -1,0,7,8,3/2 \pm i[399/4 + 48\beta/\nu]^{1/2},$$

$$3/2 \pm i[111/4 + 16\beta/\nu]^{1/2},$$

$$\text{Branch (ii)} \;:\; j = -1,0,3,4,7/2 \pm (1/2)^{1/2}\,[\beta/(\nu+\beta/2)$$

$$- 3/2 \pm 2i\{\nu/(\nu+\beta/2)\}]^{1/2}. \tag{31}$$

Eqs. (31) clearly show the presence of movable critical multivalued manifolds except for $\beta = (-5/2)\nu$, -4ν and $(-17/8)\nu$ when the resonances become integers. Analysing further, we find that except for $\beta = (-5/2)\nu$, for both the other two cases, logarithmic sigularity manifolds exist in the Laurent series. Thus we may conclude that for $\beta = (-5/2)\nu$ in (28), the system is integrable.

It can indeed be checked that eq. (28) corresponding to $\beta = -(5/2)\nu$ which reads

$$iq_t + q_{xx} + 2|q|^2 q + \nu\,[q_{xxxx} + 8|q|^2 q_{xx} + 2q^*_{xx} q^2$$

$$+4q|q_x|^2 + 6q_x^{\,2} q^* + 6|q|^4 q] = 0 \tag{32}$$

is completely integrable and the corresponding linearization is achieved by

$$U = A - i\lambda\sigma_3,$$

$$V = [3i\nu\lambda|q|^4 + i|q|^2 + i\nu(q^* q_{xx} + qq^*_{xx} - |q|^2) - 2\lambda\nu\ (qq^*_x - q_x q^*) - 2i\lambda^2\ (2\nu|q|^2 + 1) + 8i\lambda^4\nu]\sigma_3 - 8\lambda^3 A\nu - 4i\nu\lambda^2\sigma_3 A_x + 6i\nu A^2 A_x\sigma_3 + i\sigma_3 A_x + i\nu\sigma_3 A_{xxx} + 2\lambda(A+\nu A_{xx} - 2\nu A^3), \qquad (33)$$

where

$$A = \begin{bmatrix} 0 & q \\ -q^* & 0 \end{bmatrix},\ \sigma_3 = \begin{bmatrix} 1 & 0 \\ 0 & -1 \end{bmatrix},\ \lambda = \text{parameter}$$

so that

$U_t - V_x + [U,V] = 0$ is equivalent to (32). It is also possible to obtain the equivalent spin equation for (32) through a gauge transformation.

Defining

$$S = \vec{S}.\vec{\sigma} = g^{-1}\sigma_3 g, \qquad (34)$$

where

$$g_x = U(\lambda=0)g, \quad g_t = V(\lambda=0)g$$

we obtain a set of linear equations in terms of S

$$U = -i\lambda S,$$

$$V = (8i\nu\lambda^4 - 2i\lambda^2)\ S + \lambda(\nu\ SS_{xxx} + (5/2)\nu\ SS_x^{\ 3} + SS_x + \nu\ S_x S_{xx} + 2\nu S_{xx} S_x) - i\lambda^2\ (2\nu S_{xx} + \nu SS_x^{\ 2} - 4i\lambda\nu\ SS_x) \qquad (35)$$

which finally gives the following gauge equivalent spin system

$$\vec{S}_t = \vec{S}\ X\ [\vec{S}_{xx} + \nu\{\vec{S}_{xxxx} + (5/2)\vec{S}_x^{\ 2}\vec{S}_{xx} + 5\vec{S}_x(\vec{S}_x.\vec{S}_{xx})\}]. \qquad (36)$$

This is nothing but eq. (22) with $\beta = (-5/2)\nu$.

3.2.3 *Perturbation treatment.* Finally, the higher order effect due to discreteness on the solitons of the associated integrable system can be obtained using a multiple scaling perturbation theory [2]. For this purpose, we treat terms proportional to ν in eq. (28) as a weak perturbation and consider the envelope soliton

$$q = \hat{q}\ \exp[i\eta(x - 2\xi t - \theta_0) + i(\eta^2+\xi^2-\sigma_0)], \qquad (37)$$

where

$$\hat{q} = \eta \operatorname{sech} \eta(x - 2\xi t - \theta_0), \tag{38}$$

of the NLS as the unperturbed solution corresponding to $\nu = 0$. Treating ξ, η, θ_0 and σ_0 as functions of the new time scale $T = \nu t$, rewriting eq. (28) for $\hat{q}$, using

$$\hat{q} = \sum_{n=0} \nu^n \hat{q}_n \tag{39}$$

and solving the resultant linear differential equations successively, we obtain the perturbed solution after taking care of the secularity problem appropriately. For instance, the perturbed part of the solution of order ν corresponding to $\beta = 0$ reads as

$$\begin{aligned}\hat{q}_1 = {} & [(-1/2)(15\eta^3 - 5\xi^2\eta)\eta(\theta - \theta_0)\tanh \eta(\theta - \theta_0) - 5\xi^2\eta \\ & +15\eta^3\operatorname{sech}^3 \eta(\theta - \theta_0)+(\xi^2\eta - (45/2)\,\eta^3)\operatorname{sech}^4 \eta(\theta-\theta_0) \\ & + i\{15\xi\eta[(\eta \tanh \eta(\theta-\theta_0)-\eta(\theta-\theta_0]\}]\operatorname{sech} \eta(\theta-\theta_0),\end{aligned} \tag{40}$$

where

θ_0 and σ_0 now evolve in time as

$$\begin{aligned}\theta_{0T} &= -26\xi^2\eta^2 - 4\eta^4, \\ \sigma_{0T} &= -16\eta^4 + 37\xi^2\eta^2 + 3\xi^4.\end{aligned} \tag{41}$$

We conclude that the discrete ferromagnetic spin chain needs much more attention as the underlying nonlinear dynamics is not yet completely understood. On the other hand, in continuum chains, at the lowest order, many systems exhibit complete integrability and the associated properties. However, the addition of discreteness effect is seen to destroy in general the integrability at least in the next order in these systems.

Acknowledgement

The work reported here forms part of a Department of Science and Technology, Government of India project.

References

[1] Borovik, A.F. (1980). Exact integration of the nonlinear Landau-Lifshitz equation. *Solid State Communications* **34**, 721.

[2] Daniel, M. & Lakshmanan, M. (1983). Perturbation of solitons in the classical continuum isotropic spin system. *Physica* **A120**, 125.

[3] Herring, C. & Kittel, Ch. (1951). On the theory of spin waves in ferromagnetic media. *Phys. Rev.* **81**, 869.

[4] Ishimori, Y. (1982). An integrable spin chain. *J. Phys. Soc. Jpn.* **51**, 3417.

[5] Lakshmanan, M., Ruijgrok, Th.W. & Thompson, C.J. (1976). On the dynamics of a continuum spin chain. *Physica* **A84**, 577.

[6] Lakshmanan, M. (1977). Continuum spin system as an exactly solvable dynamical system. *Phys. Lett.* **A61**, 53.

[7] Lakshmanan, M. & Daniel, M. (1983). Comment on the classical models of electrons and nuclei and the generalizations of classical Poisson brackets to include spin. *J. Chem. Phys.* **78**, 7505.

[8] Lakshmanan, M., Daniel, M. & Nakamura, K. (1985). On the nonlinear excitations in one-dimensional unaxial anisotropic Heisenberg ferromagnetic spin chain in external magnetic fields. In *Dynamical Problems in Soliton Systems,* pp. 210-217, Takeno, S. (ed.). Springer-Verlag: Berlin.

[9] Lakshmanan, M., Porsezian, K. & Daniel, M. (1988). Effect of discreteness on the continuum limit of the Heisenberg spin chain. *Phys. Lett.* **A133**., 483.

[10] Landau, L. & Lifshitz, E. (1935). On the theory of the dispersion of magnetic permeability in ferromagnetic bodies. *Z. Physics Sowjetunion* **8**, 153.

[11] Nakamura, K. & Sasada, S. (1982). Quantum kink in the continuous one dimensional Heisenberg ferromagnet with easy plane. *J. Physics* **C15**, L1013.

[12] Nakamura, K. & Bishop, A.R. (1986). Classical dynamics and quantum spectra for a nonintegrble three spin system. *Phys. Rev.* **B33**, 1963.

[13] Roberts, J.A.G. & Thompson, C.J. (1988). Dynamics of the classical Heisenberg spin chain. *J. Physics* **A21**, 1769.

[14] Sklyanin, E.K. (1979). On complete integrability of the Landau-Lifshitz equation. *LOMI Preprint*, Leningrad.

[15] Takhtajan, L.A. (1977). Integration of the continuous Heisenberg spin chain through inverse scattering method. *Phys. Lett.* **A64**, 235.

[16] Thompson, C.J., Ross, K.A., Thompson, B.J.P. & Lakshmanan, M. (1985). Chaotic planar states of the discrete dynamical anisotropic Heisenberg spin chain. *Physica* **A133**, 330.

[17] Weiss, J., Tabor, M. & Carnevale, G. (1983). The Painlevé property for partial differential equations. *J. Math. Phys.* **24**, 522.

[18] Yang, K.H. & Hirschfelder, J.O. (1980). Generalization of classical Poisson brackets to include spin. *Phys. Rev.* **A22**, 1814.

A.R. Osborne and G. Boffetta

52. A "SUMMABLE" MULTISCALE EXPANSION FOR THE KdV EQUATION

We consider the multiscale expansion proposed by Zakharov & Kuznetsov (ZK) (1986) for deriving the nonlinear Schroedinger (NLS) equation from the Korteweg-de Vries (KdV) equation. In their derivation ZK insert a multiscale (ε) expansion into KdV and then balance the resultant equation at leading order in ε for each of the n terms in the expansion. In particular ZK considered n = 0,1,2 in their derivation of NLS from KdV. Here we show that it is relatively straightforward to evaluate recursion relations for balancing to all orders n>2. We explicitely evaluate the recursion relations for all n and write a series approximation for the original multiscale expansion. We subsequently show that this series may be summed to give a closed form solution for the "carrier" of a modulated wave train or field which may be interpreted as a small amplitude cnoidal wave. We suggest that multiscale expansions of other integrable, nonlinear wave equations may be "summable" in the same sense. We finally give an example of a special solution of an NLS wave train and its associated KdV spectral transform.

1 The Derivation

In previous papers Zakharov & Kusnetsov (1986), Tracy, *et al.* (1988, 1989a,b) have considered the derivation of the NLS equation from the KdV equation in order to study both infinite-line and periodic spectral properties of these equations. Here we investigate some of the physical consequences of this derivation and show that it may be extended to infinite order in a rather straightforward way (see also Osborne & Boffetta, 1989). The derivation given here is in dimensional form to emphasise the physics of shallow water wave propagation. The dimensional KdV equation is given by

$$\eta_t+c_0\eta_x+\alpha\eta\eta_x+\beta\eta_{xxx}= 0 \qquad (1.1)$$

where the constant coefficients are given by $c_0=(gh)^{1/2}$, $\alpha=3c_0/2h$

and $\beta=c_0h^2/6$; $\eta(x,t)$ is the wave amplitude and subscripts refer to partial derivatives with respect to space x and time t; h is the water depth and g the acceleration of gravity. The linearized KdV equation (set $\alpha=0$ in (1.1)) has the dispersion relation

$$\omega_0 = c_0k_0-\beta k_0^3 \quad . \tag{1.2}$$

After ZK we write the solution to KdV (1.1) in the following series where the coefficients are allowed to vary in space and time

$$\eta(x,t) = \sum_{n=-\infty}^{\infty} u_n(x,t)e^{in\Theta(x,t)} \tag{1.3}$$

and the phase is given by

$$\Theta(x,t) = k_0x-\omega_0t \tag{1.4}$$

The frequency ω_0 is determined from the shallow water dispersion relation $\omega_0 = \omega_0(k_0)$ as given by (1.2). The condition $u_{-n}^* = u_n$ insures real solutions for the wave amplitude $\eta(x,t)$. Slow space time scales (x',t') are introduced into the coefficients u_n by

$$u_n(x,t) = \varepsilon^{\alpha(n)} u_n'(x',t') \tag{1.5}$$

where

$$\alpha(0) = 2 \quad , \tag{1.6}$$

$$\alpha(n) = |n|, \quad |n|\geq 1 \tag{1.7}$$

$$x' = \varepsilon\left(x- \frac{\partial\omega_0}{\partial k_0} t\right) = \varepsilon[x+(3\beta k_0^2-c_0)t] \tag{1.8}$$

$$t' = \varepsilon^2\frac{\partial^2\omega_0}{\partial k^2} t = -6\beta k_0\varepsilon^2t \tag{1.9}$$

Inserting (1.3)-(1.9) into (1.1) we have

$$\left[-6\beta k_0\varepsilon^2\frac{\partial}{\partial t'} +\varepsilon(3\beta k_0^2-c_0)\frac{\partial}{\partial x'} -in\omega_0\right]\varepsilon^{\alpha(n)}u_n'$$

$$+c_0\left[\varepsilon\frac{\partial}{\partial x'} +ink_0\right]\varepsilon^{\alpha(n)}u_n'$$

$$+\frac{\alpha}{2}\left[\varepsilon\frac{\partial}{\partial x'}+ink_0\right]\sum_{m=-\infty}^{\infty}\varepsilon^{\alpha(m)+\alpha(n-m)}u_m'\,u_{n-m}'$$

$$+\beta\left[\varepsilon\frac{\partial}{\partial x'}+ink_0\right]^3\varepsilon^{\alpha(n)}u_n'=0 \qquad (1.10)$$

As in ZK we require that all terms balance at dominant order in ε for each n in (1.10). For the case n=0 ($O(\varepsilon^3)$) we find

$$u_0'=-\frac{\alpha}{3\beta k_0^2}|u_1'|^2+f_0 \qquad (1.11)$$

where f_0 is a constant of integration which physically corresponds to an arbitrary displacement of the mean surface level; we select f_0 to give a zero mean for the free surface elevation in (1.20) below. For n=1 ($O(\varepsilon^4)$) in (1.10) we have

$$3\beta k_0\left(i\frac{\partial^2 u_1'}{\partial x'^2}-2\frac{\partial u_1'}{\partial t'}\right)=i\alpha k_0(u_0'u_1'+u_2'u_{-1}') \qquad (1.12)$$

and for n=2($O(\varepsilon^2)$)

$$u_2'=\frac{\alpha}{6\beta k_0^2}u_1'^2 \qquad (1.13)$$

Balancing for $n\geq 2$ we find the following expressions (Tracy, *et al.*, 1988, 1989a,b):

$$\beta k_0^2(1-n^2)\varepsilon^{\alpha(n)}u_n'+\frac{\alpha}{2}\sum_{m=-\infty}^{\infty}\varepsilon^{\alpha(m)+\alpha(n-m)}u_m'u_{n-m}'=0 \qquad (1.14)$$

which we write as

$$u_n'=\frac{3\lambda}{k_0^2(n^2-1)}\sum_m{}' u_m'\,u_{n-m}' \qquad (1.15)$$

where $\lambda=\alpha/6\beta$; the prime on the summation implies that only the terms for which $\alpha(m)+\alpha(n-m)=\alpha(n)$ are summed over m. The expression (1.15) can be interpreted as a recursion relation for the u_n. This is because the summation involves only previous values of the indices n. With some work the recursion relation may be evaluated explicitly to find the following simple form:

$$u_n=n\left[\frac{\lambda}{2k_0^2}\right]^{n-1}u_1^n \text{ for } n\geq 2 \qquad (1.16)$$

This demonstrates that the nth harmonic u_n is locked to the first harmonic u_1 for all $n \geq 2$.

Equations (1.11)-(1.13) may be used to find the shallow water nonlinear Schroedinger equation in the "laboratory" coordinate frame

$$i\frac{\partial\psi}{\partial t} + \left(\frac{\partial\omega_0}{\partial k_0}\right)\frac{\partial\psi}{\partial x} + \frac{1}{2}\left(\frac{\partial^2\omega_0}{\partial x^2}\right)\frac{\partial^2\psi}{\partial x^2} + \left(\frac{\alpha^2}{24\beta k_0}\right)|\psi|^2\psi = 0 \tag{1.17}$$

We assume the following form for $\psi(x,t)$

$$\psi(x,t) = A(x,t)e^{i\omega' t+i\phi(x,t)} \tag{1.18}$$

where ω' is a nonlinear frequency correction given by

$$\omega' = \frac{9c_0\overline{A^2}}{16k_0h^4} \tag{1.19}$$

$A(x,t)$ is a slowly varying modulation envelope, the over bar denotes a spatial average and $\phi(x,t)$ is a slowly varying phase; the dynamical motion of A,ϕ are of course governed by the NLS equation. The free surface elevation is found from (1.11,1.13,1.16) together with (1.3) in the form of an infinite series which we write in dimensional form

$$\eta(x,t) = -\frac{3[A^2(x,t)-\overline{A^2}]}{4k_0^2h^3} + A(x,t)\sum_{n=1}^{\infty} nU^{n-1}(x,t)\cos n\theta \tag{1.20}$$

where $U(x,t)$ is a *slowly modulated* Ursell number (1953)

$$U(x,t) = \frac{3}{8}\frac{A(x,t)}{k_0^2h^3} \quad . \tag{1.21}$$

Since the spatio-temporal variations in $A(x,t)$ and $\phi(x,t)$ are assumed slow in the present theory, the coefficients in (1.20) are locally constant and consequently the series may be summed to give

$$\eta(x,t) = -\frac{3[A^2(x,t)-\overline{A^2}]}{4k_0^2h^3} + A(x,t)\left\{\frac{[1+U^2(x,t)]\cos\theta(x,t)-2U(x,t)}{[1+U^2(x,t)-2U(x,t)\cos\theta(x,t)]^2}\right\} \tag{1.22}$$

The phase $\theta(x,t)$ is given by

$$\theta = k_0x-(\omega_0+\omega')t+\phi(x,t) \tag{1.23}$$

The approximation (1.22) for the free surface elevation may be viewed as a *small amplitude, modulated cnoidal wave* whose modulation envelope $A(x,t)$ and phase $\phi(x,t)$ are found as solutions to the NLS equation through the expression (1.18, 1.19). The term proportional to A^2 in (1.22) is the *radiation stress* component associated with the free surface elevation first discussed by Longuet-Higgens & Stewart (1964). Important aspects of the above theoretical formulation are that (1) the free surface elevation has a nonlinear carrier (small amplitude cnoidal wave) rather than a sinusoidal carrier as often discussed in most derivation of NLS and that (2) radiation stress appears explicitly in the formulation.

Radiation stress is the *slowly varying mean sea level* associated with a rapidly oscillating wave field and physically results from "radiation" pressure which arises from the presence of the field (the radiation). For our purposes it is sufficient to define the presence of radiation stress in surface waves by the following space (or time) "running averages" of the free surface elevation over the space scale x or time scale t by either

$$\langle\eta(x,t)\rangle = \frac{1}{L}\int_{x}^{x+L} \eta(x',t)dx' \tag{1.24}$$

or

$$\langle\eta(x,t)\rangle = \frac{1}{T}\int_{t}^{t+T} \eta(x,t)dt \tag{1.25}$$

Here L and T are the wave length and period associated with the carrier wave: $L = 2\pi/k_0, T = 2\pi/\omega_0$. In (1.24), (1.25) we are averaging over the fast oscillations of the carrier, which leaves only the slowly varying part of the wave train due to the presence of radiation stress.

Application of either (1.24) or (1.25) to (1.22) gives the following expression for radiation stress

$$\langle\eta(x,t)\rangle = -\frac{3[A^2(x,t)-\overline{A^2}]}{4k_0^2h^3} \tag{1.26}$$

Thus a slowly modulated wave train or field depresses the mean sea level below a maximum in the modulation envelope $A(x,t)$; the depression is proportional to the envelope squared.

We now make some useful physical observations. First note that ω' may be derived directly from the NLS equation by inserting (1.18) (for $A = A_0 = \phi = \phi_0 =$ constant) into (1.19). Then by virtue of (1.23) ω' may be interpreted as a nonlinear correction to the dispersion relation (Whitham, 1974). In the presence of modulation $A(x,t)$ then ω' is proportional to the mean radiation stress. Note also that the nonlinear term in NLS (1.17) contains radiation stress. This arises from the expression

$$<\eta(x,t)> = -\frac{\alpha}{12\beta k_0^2} |\psi|^2 \tag{1.27}$$

which when inserted into NLS gives

$$i\frac{\partial\psi}{\partial t} + i\left(\frac{\partial\omega_0}{\partial k_0}\right)\frac{\partial\psi}{\partial x} + \frac{1}{2}\left(\frac{\partial^2\omega_0}{\partial x^2}\right)\frac{\partial^2\psi}{\partial x^2} - \left(\frac{\alpha}{12\beta k_0^2}\right)<\eta(x,t)>\psi = 0 \tag{1.28}$$

Hence the nonlinearity in the above evolution equation is interpretable in terms of radiation stress.

2 Comparison to the Cnoidal Wave

We view (1.22) as being small amplitude, modulated cnoidal wave. To illustrate this we write the cnoidal wave in the following dimensional form

$$\eta(x,t) = 2\eta_c \mathrm{cn}^2 \left\{\frac{K(m)}{\pi} \left[k(x-U_c t\right] \; |m\right\} - \eta_c \tag{2.1}$$

Equation (2.1) as written is an *exact nonlinear Fourier component* of the spectral transform for the KdV equation and as such has (1) a periodicity of 2π in the phase $\theta = kx-\omega_c t$ and has (2) a constant η_c subtracted from the cnoidal wave. Here the modulus m of the elliptic function is related to the amplitude η_c

$$mK^2(m) = \frac{3\pi^2}{2k_0^2h^3} \eta_c \tag{2.2}$$

and the nonlinear phase speed U_c is given by

$$U_c(k,\eta_c) = c_0\left[1+\frac{2\eta_c}{h} - \frac{2k_0^2h^2K^2(m)}{3\pi^2}\right] \tag{2.3}$$

K(m) is the complete elliptic integral of the first kind

$$K(m) = \int_0^{2\pi} (1-m \sin^2\phi)^{-1/2} d\phi \tag{2.4}$$

We give in Fig.1 a comparison of (1.22) for constant amplitude and phase with a cnoidal wave (2.1) of the same amplitude, phase and wave length (k_0=0.02683 cm^{-1} and h = 10.0 cm). The results give surprisingly good agreement for values of the modulus m up to 0.997. For values of the modulus less than about 0.9 the two curves are graphically indistinguishable. This supports our

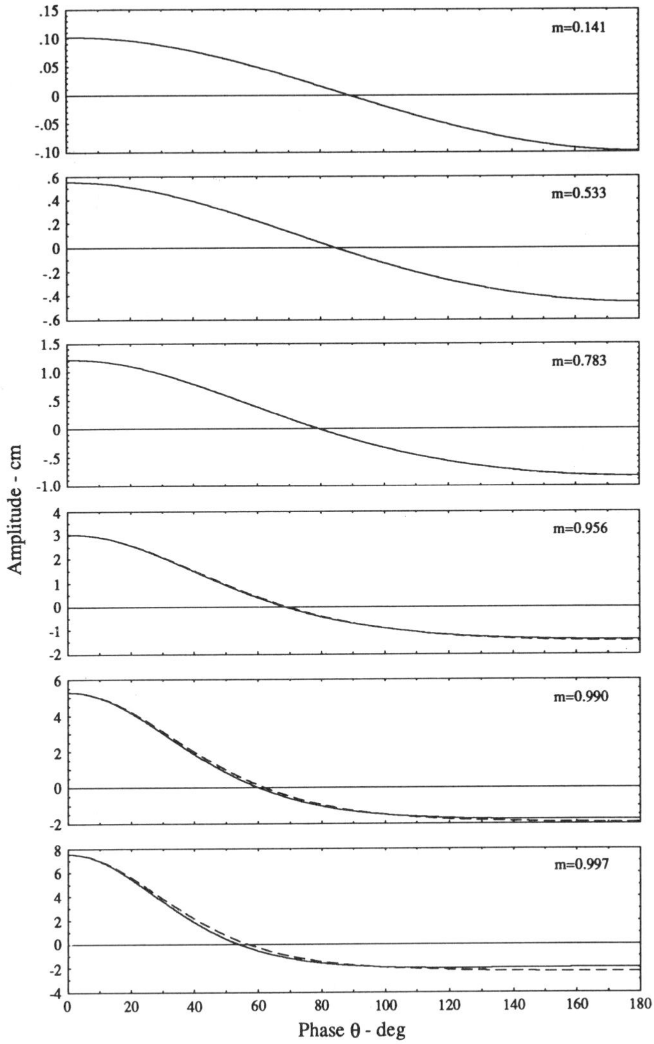

Figure 1. Comparison of the small amplitude cnoidal wave of formula (1.21) with the cnoidal wave (2.1) for the range of values of the modulus m: (a) m = 0.141, (b) m =0.533, (c) m =0.783, (d) m = 0.956, (e) m = 0.990 and (f) m =0.997.

interpretation of (1.22) as being a small amplitude cnoidal wave.

3 Special Solutions and their Spectral Transforms

Here we give two examples of wave trains governed by both the KdV and NLS equations in shallow water. The first example is shown in Figure 2 and is an unmodulated carrier given by (1.22), i.e. a

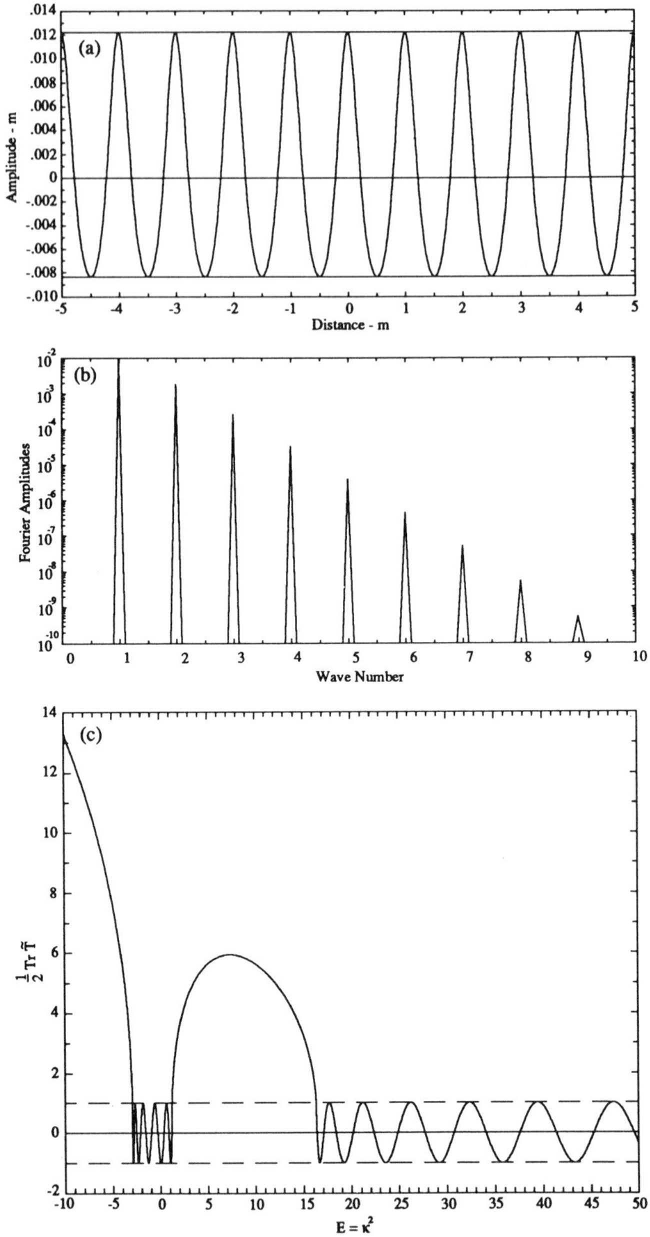

Figure 2. An unmodulated, small amplitude cnoidal wave (a), h = 10.0 cm, L = 100.0 cm, A_0 = 1.0 cm. (b) Fourier transform of wave in first panel. (c) Floquet diagram of wave in first . Note that only one degree of freedom is open since the cnoidal wave is one nonlinear Fourier component of the periodic spectral transform for KdV.

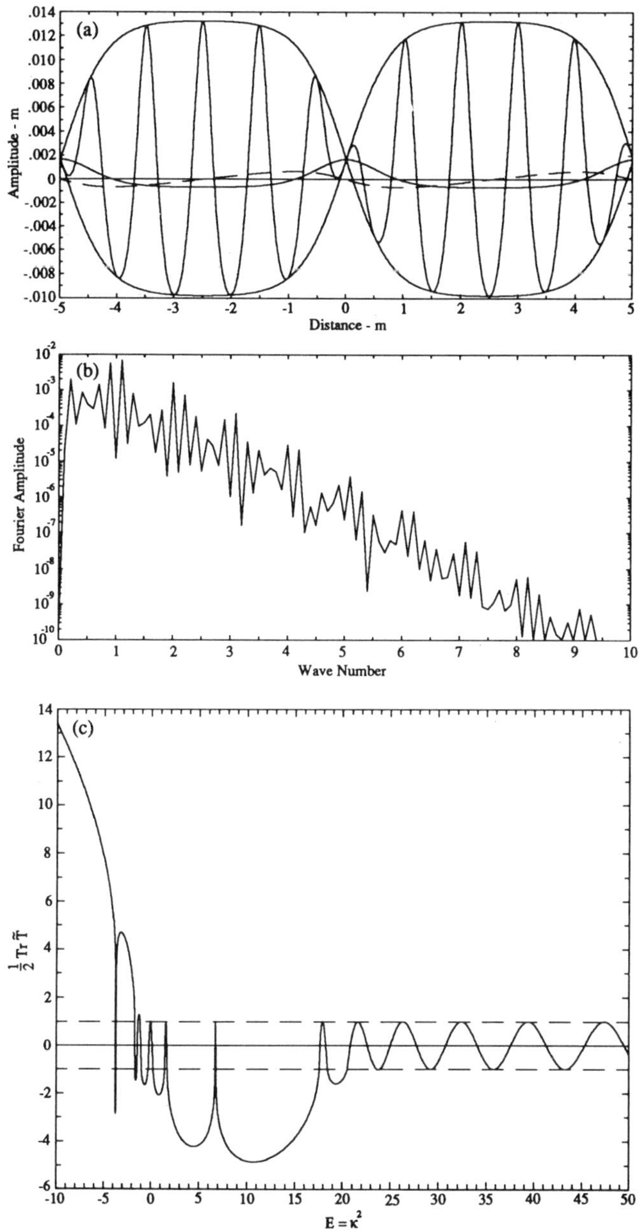

Figure 3. A small amplitude cnoidal wave carrier modulated by a soidal wave solution to the NLS equation (a). The Fourier transform of this wave is shown in (b) and the Floquet diagram in (c). Note the presence of two solitons in the radiation stress part of the spectrum and a radiation spectrum which is simple relative to (b).

small amplitude cnoidal wave of modulus m = 0.783 (see Figure 1). Note the distinct Fourier components of this wave in panel (b) of Figure 2; these are just the coefficients in the series of equation (1.20). We test how close the wave is to a cnoical wave with the spectral transform in Figure 2(c) where we show the Floquet diagram. Note that only one open band is present; this indicates that the approximation of (1.22) to a cnoidal wave is quite good. This provides an alternate verification of our contention that (1.22) is a small amplitude cnoidal wave.

The second example occurs for a modulation envelope which is the snoidal wave solution to the NLS equation (see Osborne & Boffetta, 1989). The actual modulus of the snoidal wave chosen for this example is m = 0.99 so that the envelope is very nearly a tanh function. Radiation stress is very nearly a sech^2 and is therefore graphically indistinguishable from a soliton. These can be seen in Figure 3 where in particular one observes the solitons to lie in the minima of the envelope between packets in a wave train (the dotted line is the phase from the NLS equation). The Fourier transform of this wave train is shown in Figure 3b. Note the quite complex structure which occurs both a low freqency (radiation stress) and at high frequency (higher harmonics). Relatively speaking the spectral transform is much simpler as can be seen in the Floquet diagram of Figure 3(c). The two solitons can be seen at the left; they are the only contribution to the radiation stress spectrum. The radiation spectrum (to the right in Figure 3(c) is dominated by two open bands, plus a few small adjacent bands. Note that there are no higher harmonics because the spectral transform has a much simpler perspective on the dynamics.

4 Conclusions

We have extended a multiscale expansion of the KdV equation, as discussed by Zakharov and Kusnetsov (1986), to all orders. The resultant derivation of the NLS Schroedinger equation is then associated with a "small amplitude cnoidal carrier wave," rather than the simple sinusoidal carrier which has previously been reported in the literature. There are two sources of nonlinearity in the wave motion: (1) the envelope evolves nonlinearly according to the NLS equation and (2) the carrier is a small amplitude cnoidal wave and is thus also nonlinear. The latter also gives rise to radiation stress in the low freqency part of the wave train. It may well be that multiscale expansions of other integrable, nonlinear wave equations might also be "summable" in the same sense as discussed herein. Among other predictions of the theory we find that (1) solitons reside in the radiation stress part of the spectrum and that (2) they tend to propagate in the minima between adjacent wave packets. These results are evidently generic and are expected to hold for complex wave fields as well.

Acknowledgements

Profs. C. Castagnoli and L. Bergamasco are thanked for their continuing support and valuable encouragement.

References

[1] Longuet-Higgens,M.S. & Stewart,R.W. (1964). Radiation stress in water waves; a physical discussion, with applications. Deep Sea Research, **11**, 529-562.
[2] Osborne,A.R. & Bergamasco,L. (1986). The solitons of Zabusky and Kruskal revisited: Perspective in terms of the periodic spectral transform. Physica **18D**, 26.
[3] Osborne,A.R. & Boffetta,G. (1989). Radiation stress in shallow-water surface waves: A derivation from the Korteweg-de Vries equation and nonlinear Fourier representations, submitted for publication.
[4] Tracy,E.R., Larson,J.W., Osborne,A.R. & Bergamasco,L. (1987). On the nonlinear Schrödinger equation as an averaging theory, In: Nonlinear Evolution Equations and Dynamical Systems, ed. J.Leon. World-Scientific, Singapore.
[4] Tracy,E.R., Larson,J.W., Osborne,A.R. & Bergamasco,L. (1989a). On the nonlinear Schrödinger limit of the Korteweg-de Vries equation, to appear Physica D.
[5] Tracy,E.R., Larson,J.W., Osborne,A.R. & Bergamasco,L. (1989b). The relationship between the spectral theories for the periodic Korteweg-de Vries and nonlinear Schrödinger equations, In: the Nonlinear Topics in Ocean Physics, eds. by A.R. Osborne & L. Bergamasco, Elsevier Science Publishers, Amsterdam.
[6] Ursell,F. (1953). Proceedings of the Cambridge Philosophical Society **49**, 685.
[7] Whitham,G.B., (1974). Linear and Nonlinear Waves. Interscience, New York.
[8] Zakharov,V.E. & Kusnetsov,E.A. (1986). Physica **18D**, 455.

B.M. Barbashov and A.M. Chervyakov

53. GEOMETRICAL METHOD FOR SOLVING THE BOUNDARY PROBLEM IN THE THEORY OF THE RELATIVISTIC STRING WITH MASSIVE ENDS

The differential-geometrical formulation is applied to examine the boundary-value problem in the theory of the relativistic string with massive ends in space-time E_2^1. The string world surface is described by the curvature and the torsion of the string end trajectories. The problem in question is reformulated as that for the well-known Liouville equation with appropriate boundary conditions and its solution is sketched.

The idea of strings is now used in many fields of hadronic physics [1]. In particular, massive strings provide a clear and simple picture of quark confinement [8]. Dynamics of the relativisitc string with masses m_1 and m_2 at the ends is described by the equations of motion

$$\ddot{X}_{\mu}(\tau,\sigma) - X''_{\mu}(\tau,\sigma) = 0 \tag{1}$$

and by the nonlinear boundary conditions (see, e.g. [3])

$$\begin{aligned} \frac{d}{d\tau}\left(\frac{\dot{X}_{\mu}(\tau,0)}{\sqrt{\dot{X}^2(\tau,0)}}\right) &= \frac{T}{m_1} X'_{\mu}(\tau,0), \\ \frac{d}{d\tau}\left(\frac{\dot{X}_{\mu}(\tau,\pi)}{\sqrt{\dot{X}^2(\tau,\pi)}}\right) &= -\frac{T}{m_2} X'_{\mu}(\tau,\pi). \end{aligned} \tag{2}$$

Here X^{μ} ($\mu = 0,1,2$) are coordinates of Minkowski space E_2^1 and ($u^1=\tau$, $u^2=\sigma$) are curvilinear coordinates of the pseudo-Riemannian manifold $M_1^1 \subset E_2^1$ varying on a strip $0 \le \sigma \le \pi$, $\tau_1 \le \tau \le \tau_2$, T is the string tension (kg.S^{-1}). On the string world sheet M_1^1 the local conformal coordinates were chosen to be

$$g_{ij} = \lambda(\tau,\sigma)\eta_{ij}, \quad \eta_{11} = -\eta_{22} = 1, \tag{3}$$

where g_{ij} is the metric induced on M_1^1

$$g_{ij} = (\partial x_\mu/\partial u^i)(\partial x^\mu/\partial u^j),$$

$$i,j = 1,2. \tag{4}$$

Nonlinear boundary conditions (3) represent a major problem in searching classical motions of the massive string. So far only simplest motions have been investigated for this system (see, e.g. [2]). The most familiar of them is a particular parameterization for the string world sheet $X^\mu(\tau,\sigma)$ in the form of a helicoid [3,5,6,9].

To overcome this difficulty, the differential-geometrical formation of the massive string theory has been proposed [2]. It has been shown that the minimal surface $M_1^1 \subset E_2^1$ with metric (3) is defined by the well-known Liouville equation

$$\frac{\partial^2 \log\lambda}{\partial u^+ \partial u^-} = \frac{A(u^+)A(u^-)}{2\lambda}, \tag{5}$$

where $u^\pm = \tau\pm\sigma$ and A is an arbitrary periodic function of one variable $A(\tau-\pi) = A(\tau+\pi)$, and world trajectories of the massive endpoints restricting the string sheet are asymptotic curves of a constant geodesic curvature

$$K_i(\tau) = T/m_i, \quad i = 1,2, \tag{6}$$

whereas their torsions $æ_i(\tau)$, i = 1,2 depend on the initial data. Description (2) of the trajectories of string ends in E_2^1 has been replaced by the following formulae on M_1^1

$$\frac{\partial \log\lambda}{\partial\sigma} = 2K_1\lambda^{1/2}, \quad \text{at } \sigma = 0,$$

$$\frac{\partial \log\lambda}{\partial\sigma} = -2K_2\lambda^{1/2}, \quad \text{at } \sigma = \pi. \tag{7}$$

Thus, the investigation of classical motions in the massive string theory has been reduced to the boundary value problem (7) for Liouville equation (5).

To find the minimal surface $M_1^1 \subset E_2^1$ with metric (3) restricted by two asymptotic lines of a constant geodesic curvature (6), we start with the well-known general solution to eq. (5):

$$\lambda(u^+u^-) = \frac{A(u^+)A(u^-)[f(u^+) - g(u^-)]^2}{4f'(u^+)g'(u^-)}, \tag{8}$$

with arbitrary functions $f(u^+)$ and $g(u^-)$ which are not unique: the same constant Möbius transformation

$$f \longrightarrow \frac{af + b}{cf + d}, \quad g \longrightarrow \frac{ag + b}{cg + d},$$

$$ad - cb = 1,$$

leaves (8) invariant. In what follows, to ensure the absence of superlight velocities in the string dynamics $\lambda = \dot{X}^2 > 0$, we will assume that $\varepsilon(f') = \varepsilon(g')$ where $\varepsilon(f')$ is the sign function.

Substituting (8) into the boundary conditions (7) one obtains

$$\frac{d}{d\tau} \log \frac{g'}{f'} + 2\,\frac{f' + g'}{f - g} = K_1 |A| \,\frac{|f - g|}{\sqrt{f'g'}},$$

$$f = f(\tau), \quad g = g(\tau) \tag{9}$$

$$\frac{d}{d\tau} \log \frac{g'(-)}{f'(+)} + 2\,\frac{f'(+) + g'(-)}{f(+) - g(-)} = - K_2 |A(+)| \,\frac{|f(+) - g(-)|}{\sqrt{f'(+)g'(-)}},$$

$$f(+) = f(\tau+\pi), \quad g(-) = g(\tau-\pi). \tag{10}$$

A periodic function $A(\tau)$ in (9)-(10) represents a tensor of the second quadratic form of the world sheet M_1^1 under the transformation $\tau \longrightarrow \tilde{\tau} = E(\tau)$, $E'(\tau+2\pi) = E'(\tau)$ and ensures, thereby, the invariance of eqs. (9)-(10). Depending on the values of the gauge fixing function $A(\tau)$, the solution of these equations defines local equivalent world sheets swept by the relativistic string with massive ends.

It is a generic property that set (9)-(10) has no periodic solutions of the form $f(\tau+\pi) = f(\tau-\pi)$ and $g(\tau+\pi) = g(\tau-\pi)$. In fact, using the solutions in question, one finds from (9)-(10) the condition

$$(m_1 + m_2)\left(\frac{d}{d\tau} \log \frac{g'}{f'} + \frac{f' + g'}{f - g}\right) = 0,$$

compatible with eq. (9) only if the net mass of the string is zero, $m_1 + m_2 = 0$; the latter, in turn reduces to the case of the massless string $m_1 = m_2 = 0$.

As demonstrated by Barbashov & Chervyakov [2] eqs. (9)-(10) can be put in the form

$$\frac{d}{d\tau} \log \frac{g'}{f'} + 2\,\frac{f' + g'}{f - g} = \frac{4K_1}{|æ_1(\tau)|}\,\frac{\sqrt{f'\,g'}}{|f - g|}, \tag{11}$$

$$\frac{d}{d\tau}\log\frac{g'(-)}{f'(+)} + 2\,\frac{f'(+) + g'(-)}{f(+) - g(-)} = -\,\frac{4K_2}{|æ_2(\tau)|}\,\frac{\sqrt{f'(+)\,g'(-)}}{|f(+) - g(-)|}. \tag{12}$$

Here we replace $A(\tau)$ and $A(\tau+\pi)$ by the torsions $æ_1(\tau)$ and $æ_2(\tau)$ of the trajectories of string endpoints

$$æ_1(\tau) = \frac{4f'g'}{A(f-g)^2}\,,\quad æ_2(\tau) = \frac{4f'(+)g'(-)}{A(+)(f(+)-g(-))^2}, \tag{13}$$

from which a number of restrictions follows on the world lines that $æ_i(\tau)$, $i = 1,2$ can describe

$$\frac{g'(-)}{æ_2(f(+)-g(-))^2} = \frac{g'(+)}{æ_1(+)(f(+)-g(+))^2}, \tag{14}$$

$$\frac{f'(+)g'(+)}{æ_1(+)(f(+)-g(+))^2} = \frac{f'(-)g'(-)}{æ_1(-)(f(-)-g(-))^2}. \tag{15}$$

Thus, the functions $æ_1(\tau)$ and $æ_2(\tau)$ are free parameters only inside the interval $[0,\pi]$ and can be defined there by the Cauchy data.

Let us discuss in general the simultaneous solution of set (11)-(12) with conditions (14)-(15). First of all, there is a miraculous compatibility of eqs. (11)-(12) and (14)-(15) for the case of constant torsions $æ_1(\tau) = æ_{01}$ and $æ_2(\tau) = æ_{02}$ when the string-endpoint trajectories are helices. In this case eqs. (11)-(12) coincide exactly with the boundary conditions for the Liouville equation in the Polyakov model of an open string [7] and can be resolved as follows. Multiplying eq. (11) by $|f-g|/\sqrt{f'g'}$ so that its right hand side turns into a constant and differentiating then with respect to τ one obtains

$$\left(\frac{g'''}{g'} - \frac{3}{2}\frac{g''^2}{g'^2}\right) - \left(\frac{f'''}{f'} - \frac{3}{2}\frac{f''^2}{f'^2}\right) = 0. \tag{16}$$

The expressions in parentheses are the so-called Schwarzian derivatives of the functions g and f. The above result means that g and f have the same Schwarzian derivative. As is well known, such functions are related by a constant Möbius transformation

$$g(\tau) = \frac{af(\tau) + b}{cf(\tau) + d}, \tag{17}$$

$$ad - cb = 1.$$

Substituting (17) into (11) where $æ_1(\tau) = æ_{01}$ one finally has

$$a+d = \frac{2K_1}{|æ_{01}|}\, \varepsilon(cf')\varepsilon(cf+d)\varepsilon[(cf+d)+\frac{1}{(cf+d)} - (a+d)]. \tag{18}$$

Likewise one can show for the second equation (12) with $æ_2(\tau) = æ_{02}$ that

$$g(\tau-\pi) = \frac{\alpha f(\tau+\pi) + \beta}{\gamma f(\tau+\pi) + \delta}, \qquad \alpha\delta - \gamma\beta = 1, \tag{19}$$

where

$$\alpha + \delta = - \frac{2K_2}{|æ_{02}|}\, \varepsilon(\gamma f')\varepsilon(\gamma f+\delta)\varepsilon[(\gamma f+\delta) + \frac{1}{(\gamma f+\delta)} - (\alpha+\delta)]. \tag{20}$$

From (17) and (19) one finds

$$g(\tau) = \frac{af(\tau)+b}{cf(\tau)+d} = \frac{\alpha f(\tau+2\pi)+\beta}{\gamma f(\tau+2\pi)+\delta}$$

and f (as well as g) is periodic with a period 2π up to a Möbius transformation

$$f(\tau+2\pi) = \frac{(c\beta-a\delta)f(\tau)+(d\beta-b\delta)}{(a\gamma-cd)f(\tau)+(b\gamma-d\alpha)}. \tag{21}$$

Next we have to show that conditions (14)-(15) are satisfied by the solutions (17) and (19) leading to relations between α, β, γ δ and a, b, c, d. Consider first (14). Substituting expression (17) for g(+) in terms of f(+) into the right hand side of (14) and the expression (19) for g(-) in terms of f(+) into the left hand side of (14) and then equating the coefficients of $f^2(+)$, $f^1(+)$ and $f^0(+)$ one obtains

$$(\delta-\alpha) = (\sqrt{æ_{01}}/\sqrt{æ_{02}})(d-a), \qquad \beta = (\sqrt{æ_{01}}/\sqrt{æ_{02}})b, \quad \gamma = (\sqrt{æ_{01}}/\sqrt{æ_{02}})c. \tag{22}$$

The relations (22) as well as the condition $\alpha\delta-\gamma\beta = 1$ completely define α, β, γ, δ through a, b, c, d. With (22) and ad - cb =1 it may be written as

$$(\alpha+\delta)^2/4 - æ_{01}(a+d)^2/4æ_{02} + æ_{01}/æ_{02} = 1.$$

Substitution of (18) and (20) into the latter expression yields

$$æ_{02}(1-K_2^2/æ_{02}^2) = æ_{01}(1-K_1^2/æ_{01}^2). \tag{23}$$

Equation (15) does not lead to new relations between α, β, γ, δ

and a, b, c, d apart from (22) and (23). Thus for the torsions $æ_{01}$ and $æ_{02}$ eqs. (11)-(12) and (14)-(15) have the same, up to a constant, solutions. Clearly, a world sheet restricted by helices is a helicoid (as already pointed out, such topology of the string sheet was treated in another context (see [3],[5],[6],[9])).

To solve eqs. (11)-(12) with arbitrary torsions $æ_i(\tau)$, i = 1,2, we will use the following representations

$$\frac{f'}{f-g} = \frac{f'}{f - \dfrac{af+b}{cf+d}} = \frac{cf'}{(cf+d)+(cf+d)^{-1}-(a+d)},$$

$$\frac{g'}{f-g} = \frac{cf'}{(cf+d)^2[(cf+d)+(cf+d)^{-1}-(a+d)]}.$$

Equation (11) is fulfilled for the above representations with the change of (a+d) and $f(\tau)$ to the functions $\rho(\tau)$ and $\varphi(\tau)$, respectively. As a result, the parameter representation in terms of an arbitrary function $\varphi(\tau)$ is

$$\frac{f'}{f-g} = \frac{c\varphi'}{(Z-\rho)}, \quad \frac{g}{f-g} = \frac{c\varphi'}{(c\varphi+d)^2(Z-\rho)}, \qquad (24)$$

$$Z \equiv (c\varphi+d) + (c\varphi+d)^{-1},$$

where $\rho(\tau)$ is subjected to the condition

$$\rho(\tau) = \frac{2K_1}{|æ_1(\tau)|}\,\varepsilon(c\varphi')\varepsilon(c\varphi+d)\varepsilon(Z-\rho). \qquad (25)$$

From (24), $f'(\tau), g'(\tau)$ and $f(\tau)-g(\tau)$ may be found. Extending the right hand side of the difference

$$\frac{f'-g'}{f-g} = \frac{Z'}{(Z-\rho)}$$

to the total logarithmic derivative and integrating it over τ we get the result

$$f(\tau) - g(\tau) = c^{-1}(Z-p)\exp\int^{\tau}\frac{p'}{(Z-p)}\,d\eta. \qquad (26)$$

Substitution of (26) into (24) yields

$$f'(\tau) = \varphi' \exp\int^{\tau}\frac{p'}{(Z-p)}\,d\eta,$$

$$g'(\tau) = \frac{\varphi'}{(c\varphi+d)^2}\exp\int^{\tau}\frac{p'}{(Z-p)}\,d\eta. \qquad (27)$$

It is easy to see from (27) and (25) that for the constant torsion

$æ_1(\tau) = æ_{01}$ relation (17) between $g(\tau)$ and $f(\tau)$ given here in the parametric form in terms of the function $\varphi(\tau)$ is recovered. In fact,

$$f'(\tau) = \varphi'(\tau), \quad g'(\tau) = \frac{\varphi'(\tau)}{(c\varphi+d)^2},$$

so that

$$f(\tau) = \varphi(\tau), \quad g(\tau) = - \frac{1}{c(c\varphi+d)} + \frac{a}{c} = \frac{a\varphi + b}{c\varphi + d}\ .$$

Now let us return to the second equation (12). Likewise one can show that its parameteric representation takes the form

$$f'(z+\pi) = \psi'(\tau) \exp \int^{\tau} \frac{z'}{(y - \tau)} d\eta\ ,$$

$$g'(z-\pi) = \frac{\psi'(\tau)}{(\gamma\psi+\delta)^2} \exp \int^{\tau} \frac{z'}{(y - z)} d\eta, \tag{28}$$

$$y \equiv (\phi\psi+\delta) + (\gamma\psi+\delta)^{-1},$$

where (τ) is subjected to the condition

$$(\tau) = - \frac{2K_2}{|æ_2(\tau)|} \varepsilon(\gamma\psi')\varepsilon(\gamma\psi+\delta)\varepsilon(y - z). \tag{29}$$

As it follows from (27)-(28), the parametric functions $\varphi(\tau)$ and $\psi(\tau)$ are restricted to those that obey

$$\varphi'(+)\exp \int^{\tau+\pi} \frac{p'd\eta}{(Z - p)} = \psi'(\tau) \exp \int^{\tau} \frac{'d\eta}{(y - z)}, \tag{30}$$

$$\frac{\varphi'(-)}{(c\varphi(-)+d)^2} \exp \int^{\tau-\pi} \frac{p'd\eta}{(Z - p)} = \frac{\psi'(\tau)}{(\gamma\psi+\delta)^2} \exp \int^{\tau} \frac{'d\eta}{(y - z)}\ . \tag{31}$$

Note that eqs. (30)-(31) do not impose any constraints on the choice of $æ_i(\tau)$, $i = 1,2$. For the constant torsions $æ_1(\tau) = æ_{01}$ and $æ_2(\tau) = æ_{02}$, they are satisfied by an arbitrary function which has a periodic derivative $\varphi'(\tau+\pi) = \varphi'(\tau-\pi)$. Thus, the only unsolved problem with this approach is that of finding $æ_1(\tau)$, $i = 1,2$, by using the Cauchy data on the interval $[0,\pi]$, according to (14)-(15).

References

[1] Artru, X. (1983). Classical string phenomenology. How strings work. *Physics Rep.* **97**, 147.

[2] Barbashov, B.M. & Chervyakov, A.M. (1988). Geometrical method for solving the boundary problem in theory of relativistic string with massive ends. *Theor. & Math. Phys.* **74**, 430 (in Russian).
[3] Barbashov, B.M. & Nesterenko, V.V. (1983). The exact solutions in theory of relativistic string with massive ends. In *Problems of High Energy Physics and Quantum Field Theory*, pp. 3-13. IHEP: Protvino. (In Russian.)
[4] Barbashov, B.M. & Nesterenko, V.V. (1987) *The relativistic string model in hadronic physics.* Atomizdat: Moscow.
[5] Chodos, A. & Thorn, C.B. (1974). Making the massless string massive. *Nuclear Physics* **B72**, 509.
[6] Frampton, P.H. (1975). String approaches to hadron structure. *Phys. Rev.* **D12**, 538.
[7] Gervais, J.L. & Neveu, A. (1982). The dual string spectrum in Polyakov's quantization I. *Nuclear Physics* **B199**, 59.
[8] Nambu, Y. (1974). Strings, monopoles and gauge fields. *Phys. Rev.* **D10**, 4262.
[9] Shavokhina, N.S. (1982) The relativistic two-body problem with a constant attractive force. *Reports of USSR Academy of Science* **265**, 852 (in Russian).

A.O. Barut and B.V. Baby

54. NONLINEAR DE BROGLIE WAVES AND THE RELATION BETWEEN RELATIVISTIC AND NONRELATIVISTIC SOLITONS

1. **Nonrelativistic limit**

There is no systematic method, up to now, of classifying an arbitrary nonlinear partial differential equation (NLPDE) as being C-integrable or S-integrable [10]. If a given NLPDE is integrable, then we have a number of systematic methods, such as the inverse scattering transform (IST), Bäcklund transform (BT), Darboux transform (DT), to construct exact solutions. When we are not in a position to judge the integrability then we can use other special techniques like Hirota's bilinear operator, similarity and symmetry analysis, to find the exact solutions. Most of the NLPDE are not integrable and so their solutions are difficult to find. Hirota's method will not always give exact solutions, for example, it fails for equation of motion of the $\lambda\phi^6$ Lagrangian field theory. Similarity analysis or group theoretical methods are also not always available. Even if we are able to find some symmetries, the reduced equations are nonlinear and difficult to solve in most cases.

The purpose of this work is to try to relate some nonintegrable class of equation to the well-known integrable class of NLPDEs using the nonrelativistic limiting procedure and so this method may give some information about the possible structure of the exact solutions for a nonintegrable class of equations if we know the respective integrable class of NLPDEs.

In this paper we show that the well-known envelope soliton and kink solutions of the nonlinear Schrödinger equations (NLSE) are the nonrelativistic (NR) limit of the corresponding solutions of the nonlinear Klein-Gordon equation (NLKGE). To our knowledge, the NR limit of the nonlinear field equations and their solutions have not been discussed before. As a result of this NR limiting process, we obtain new solutions for the 'double nonlinear Schrödinger equations (DNLSE)' and exact solutions for the perturbed NLSE or Generalized NLSE. We shall further show that the dispersion relations for the phase of the solution differ from the de Broglie relation by a negative internal energy ε_0 of the soliton solution. In the limit $\varepsilon_0 \longrightarrow 0$, these solutions disappear and we have the usual plane waves. Conversely, we also conjecture that corresponding to all the envelope solutions of NLSEs, there are envelope solutions of NLKGEs, such that their NR limit exactly reduces to the solutions of the NLSEs. This method provides new exact solutions for the NLKGE family. We further discuss the role of the internal degrees of freedom on the de Broglie phase.

The NR limit [4,6] or the transformation for constructing an NR Schrödinger equation from an NLKGE of motion, is defined by

$$\lim_{c \to \infty} \exp\left(\frac{imc^2 t}{\hbar}\right) \phi(\vec{x},t) \longrightarrow \psi(\vec{x},t) \tag{1}$$

where $\phi(\vec{x},t)$ is the field variable of the NLKGE.

$$\Box\phi + m^2c^4\phi - ac^2\phi|\phi|^{2\sigma} - bc^2\phi|\phi|^{4\sigma} = 0 \tag{2}$$

and σ is any real number, a, b, are real parameters and $\psi(\vec{x},t)$ is the field variable for the NLSE and

$$\Box\phi \equiv \hbar^2(\phi_{tt} - c^2\phi_{\vec{x}\vec{x}})$$

Inserting eq. (1) into eq. (2) we get

$$i\hbar\psi_t + \frac{\hbar^2}{2m}\Delta\psi + \frac{a}{2m}\psi|\psi|^{2\sigma} + \frac{b}{2m}\psi|\psi|^{4\sigma} = 0 \tag{3}$$

where $\Delta\psi = \psi_{xx} + \psi_{yy} + \psi_{zz}$. Eq. (3) is an NLSE but with two nonlinear terms, and we shall denote it as the double nonlinear Schrödinger equation (DNLSE).

Another interesting fact is that the perturbed DNLSE or Generalized nonlinear Schrodinger equation (GNLSE) [11,16,17,18,32]

$$i\hbar\psi_t + \frac{\hbar^2}{2m}\Delta\psi + \frac{a}{2m}\psi|\psi|^{2\sigma} + \frac{b}{2m}\psi|\psi|^{4\sigma} = \tilde{\beta}\psi \tag{4}$$

where $\bar{\beta}$ is the perturbation coefficient, is the NR limit of the same KGE (eq. (2)) but with a slight modification in the coefficient of linear terms

$$\Box\phi + (m^2c^4 + 2mc^2\beta)\phi + ac^2\phi|\phi|^{2\sigma} + bc^2\phi|\phi|^{4\sigma} = 0 \tag{5}$$

The above correspondence assures us that we will be able to find exact solutions of DNSLE and GNSLE if we know the solutions of NLKGEs. However it is important to note that all known solutions of NLKGEs are without an envelope phase factor except for some recent results [1,2,5], whereas the NLSE has the well-known envelope type solutions! Moreover, if we take the NR limit of the known solutions of the NLKGs we get vacuum solutions of the NLKGEs so that envelope type solutions are the relativistic form of the corresponding nonrelativistic envelope solutions of NLSEs. For this purpose we introduce a phase factor $\exp(i\theta')$ into the solutions of NLKGEs, so that when we take the modulus of the field variable the phase factor disappears. This property is analogous to L. de Broglie's earlier ideas on the 'double solutions of nonlinear equations' [5,7].

2. **Envelope solutions of nonlinear Klein-Gordon equations**

In this section we will describe the method of introducing de Broglie phase factor $\exp(i\theta')$ in the known solutions of NLKGEs. Throughout this paper our solutions will have a complex phase factor of the form

$$\exp(i\theta') = \exp\left[\frac{i\varepsilon\gamma}{\hbar}\left(\frac{\beta}{c}x - t + \delta'\right)\right] \tag{6}$$

in a relativistic quantum mechanical notation of a 'de Broglie phase'. Here $\beta = v/c$, $\gamma = (1 - v^2/c^2)^{-1/2}$ and δ' is an arbitrary parameter and if we write $\theta' = k'(x - ut) + \delta'_0$, where $k' = \dfrac{\varepsilon\beta\gamma}{\hbar c}$, then the phase velocity is u.

We introduce another exponential factor with a real exponent and with δ as constant

$$\exp(\theta) = \exp\left[\frac{\omega\gamma}{\hbar c}(x - vt) + \delta\right]. \tag{7}$$

Our soliton-like envelope solutions are expressible entirely in terms of the product of these two exponentials

$$g(x,t) = \exp(i\theta')\cdot\exp(\theta) \tag{8}$$

so that the envelope velocity is v, which is different from the phase velocity u. There is a relation between the parameter ω of the envelope with the parameter ε of the de Broglie phase

$$\omega^2 + \varepsilon^2 = m^2c^4 \tag{9}$$

where m is a parameter fixed by the nonlinear equation, whereas the parameters ε and ω do not occur in the wave equations, they characterize the solutions. The known dispersion relations are obtained by assigning $\varepsilon = 0$. We need also the combinations

$$gg^* = \exp(2\theta) \tag{10}$$

where g^* is the complex conjugate of $g(x,t)$.

The absolute valued $\gamma\phi^6$ Lagrangians gives the well-known NLKGE

$$\Box\phi + m^2c^4\phi - ac^2\phi|\phi|^{2\sigma} - bc^2\phi|\phi|^{4\sigma} = 0 \qquad (11)$$

for $\sigma = 1$.

Let $\phi(x,t)$ be a complex-valued function and assume that eq. (11) has a solution of the form

$$\phi(x,t) = \frac{h(x,t)}{\left\{\left[1 + \frac{ac^2(hh^*)^{\sigma}}{4(\sigma+1)\omega^2}\right]^2 + \frac{bc^2(hh^*)^{2\sigma}}{4(2\sigma+1)\omega^2}\right\}^{1/2\sigma}} \qquad (12)$$

where h(x,t) is also a complex valued function. Then it can be shown that h(x,t) has to satisfy the following set of equations

$$\Box h = -m^2c^4h, \qquad (\partial^{\mu}h)(\partial_{\mu}h) = -m^2c^4h^2$$

$$\Box h^* = -m^2c^4h^*, \qquad (\partial^{\mu}h^*)(\partial_{\mu}h^*) = -m^2c^4h^{*2}$$

$$(\partial^{\mu}h)(\partial_{\mu}h^*) = (-\omega^2 + \varepsilon^2)hh^*, \quad \mu = 0,1 \qquad (13)$$

where h^* is the complex conjugate of h(x,t). All solutions of eq. (13) are the solutions of eq. (11) via eq. (12). For real valued h(x,t) and $\phi(x,t)$, this result is reported earlier [9]. It can easily be verified that h(x,t) = g(x,t) is a solution of eq. (13). So the new exact envelope-type solution of the NLKGE (eq. (11)) is given by [1]

$$\phi(x,t) = \frac{e^{i\theta'}\, e^{\theta}}{\left\{\left[1 + \frac{ac^2e^{2\sigma\theta}}{4(\sigma+1)\omega^2}\right]^2 + \frac{bc^2e^{4\sigma\theta}}{4(2\sigma+1)\omega^2}\right\}^{1/2\sigma}} \qquad (14)$$

For b = 0, this solution is known [5] and for $\sigma = 1$, it reduces to the soliton-like envelope solution of the well-known $\lambda\phi^4$-Lagrangian field theory. The solution (14) is reported earlier when h(x,t) is a real valued function [9].

Of eqs. (13) the first two are linear and so we may be able to use the linear superpositions of solutions of the form

$$h(x,t) = \sum_{j=1}^{N} \exp(\theta_j + i\theta'_j) \qquad (15)$$

where

$$\theta_j = (p_{1j}x - p_{0j}t + \delta_j), \quad \theta'_j = (v_{1j}x - v_{ij}t + \delta'_j) \qquad (16)$$

then in order to find an exact solution of eq. (13) we need the following additional constraints

$$(p_{0j} + p_{0k} + v_{ok}) - c^2(p_{ij}p_{1k} + v_{ij}v_{1k}) = - \frac{m^2c^4}{\hbar 2}$$

$$p_{0j}v_{0ok} - c^2 p_{1j}q_{1k} = 0 \quad j,k = 1,2...N \tag{17}$$

But the set of eq. (17) is an overdetermined system and it is valid only for j = k = 1. In (1+1) dimensions there exists only one independent wave vector. The dimensionality of space-time (d+1) and the maximum possible number of solutions N are related by $N \le 2d - 1$. Moreover, these N-envelope solutions are highly restricted by the constraints (15) and therefore they are not of the type of the well-known N-soliton solutions of the integrable systems [20] and so we call them "constrained N-envelope soliton" like solutions.

3. Nonrelativistic reduction

In the first section we obtained, as a result of the NR limiting process, DNLSE and GNLSE from the NLKGEs. In this section we will develop exact solutions of the DNLSEs and GNLSEs using the NR limit from the solutions given in the previous sections. Before that we will find the different values of the parameter combinations and their dispersion relations under this limiting process.

Since ε has the dimension of energy it can be written in the form

$$\varepsilon = mc^2\gamma + \frac{B}{2}mv^2 + \varepsilon_{NR}, \qquad B, \varepsilon_{NR} \text{ are parameters.} \tag{18}$$

Using eq. (9), we get

$$\frac{\omega^2}{c^2} = m^2v^2 - \left\{\frac{B^2}{4c^2}m^2v^4 + \frac{Bm^2v^2}{\gamma} + \frac{Bmv^2}{c^2}\varepsilon_{NR} + \frac{2m}{\gamma}\varepsilon_{NR} + \frac{\varepsilon_{NR}}{c^2}\right\}. \tag{19}$$

We have the following limits

$$\operatorname*{lt}_{c\to\infty} \frac{\omega^2}{c^2} = (1 - B)m^2v^2 - 2m\varepsilon_{NR} = \lambda^2 \tag{20}$$

$$\operatorname*{lt}_{c\to\infty} \frac{\varepsilon\gamma\beta}{\hbar c} = \frac{mv}{\hbar} = p \tag{21}$$

$$\operatorname*{lt}_{c\to\infty} \left(\frac{\varepsilon\gamma}{\hbar} - \frac{mc^2}{\hbar}\right) = \frac{1}{\hbar}\left(\frac{B}{2}mv^2 + \varepsilon_{NR}\right) = p_0 \tag{22}$$

$$\lim_{c\to\infty} g(x,t) = g_0(x,t) \tag{23}$$

where

$$g_0(x,t) = \exp(\theta'_0 + i\theta'_0)$$

$$\theta_0 = (qx - q_0t + \delta_0), \quad \theta'_0 = (px - p_0t + \delta'_0) \tag{24}$$

$$q = \pm\lambda/\hbar, \quad q_0 = \pm\lambda v/\hbar \tag{25}$$

such that

$$pq = mq_0/\hbar \tag{26}$$

$$p^2 - q^2 = 2mq_0/\hbar, \quad q^2 = \lambda^2/\hbar^2. \tag{27}$$

Consequently by our limiting procedure we obtain the following new exact soliton like envelope solution of the DNLSE (eq. (3))

$$\psi(x,t) = \pm \frac{e^{i\theta'_o} e^{\theta_o}}{\left\{\left[1 + \frac{ae^{2\sigma\theta_0}}{4(\sigma+1)\lambda^2}\right]^2 + \frac{be^{4\sigma\theta_0}}{4(2\sigma+1)\lambda^2}\right\}^{1/2\sigma}} \tag{28}$$

For $\varepsilon_0 = \frac{-1}{2m}\lambda^2$, $b = 0$ this solution reduces to the well-known soliton envelope solutions [21,33]

$$\psi(x,t) = \pm \left[\frac{-2(\sigma+1)}{\alpha}\varepsilon_0\right]^{\frac{1}{2\sigma}} e^{\frac{1}{\hbar}[mvx-(mv^2/2+\varepsilon_0)t]} \operatorname{sech}^{\frac{1}{\sigma}}\left[\frac{\sigma}{\hbar}\sqrt{-2m\varepsilon_0}(x-vt)\right] \tag{29}$$

which satisfies the NLSE

$$i\hbar\psi_t + \frac{\hbar^2}{2m}\psi_{xx} + \frac{\alpha}{2m}\psi|\psi|^{2\sigma} = 0. \tag{30}$$

When $B = 1$ and $\varepsilon_{NR} = 0$, our solution (28) vanishes identically. Using the limiting process we can also find the exact solution of the GNLSE (eq. (4)). The new dispersion relation is now

$$\omega^2 + \varepsilon^2 = m^2c^4 + 2mc^2\tilde{\beta} \tag{31}$$

and the corresponding change of the eq. (20) becomes

$$\lim_{c\to\infty} \frac{\omega^2}{c^2} = (1 - B)m^2v^2 - 2m(\varepsilon_{NR} - \tilde{\beta}) = \lambda_0^2. \qquad (32)$$

The new exact envelope solution of the GNLSE is

$$\psi(x,t) = \frac{\pm e^{i\theta'_o} e^{\theta_o}}{\left\{\left[1 + \frac{a e^{2\sigma\theta_o}}{4(\sigma+1)\lambda_0^2}\right]^2 + \frac{b e^{4\alpha\theta_o}}{4(2\sigma+1)\lambda_0^2}\right\}^{1/2\sigma}} \qquad (33)$$

This shows that the perturbation sustains the soliton envelope solution, or any other solution of the DNLSE, obtainable from the limiting process, provided that

$$\tilde{\beta} \neq \varepsilon_{NR} = (1 - b)\,\frac{mv^2}{2}\,. \qquad (34)$$

When equality holds, the solution (eq. (33)) vanishes identically. If $\tilde{\beta}$ is sufficiently large and negative, the solution can even be purely imaginary. Another interesting fact is that the perturbation never influences the 'internal frequency' of the de Broglie phase, which is constant under all types of perturbations. When we assign b = 0 in eq. (4), we get the well-known perturbed NLSE, which has been very extensively studied using approximate solutions and Inverse Scattering transforms (IST) perturbation methods [12,19,22,24,25,26,27,28,29,32,34; see also 15]. Our results are in agreement with those earlier results. In addition to that we can find that if β is sufficiently large then $\lambda^2 < 0$ and in such situations amplitude of the soliton grows indefinitely and we get even imaginary solutions for some values of $\tilde{\beta}$ and ε_{NR}.

4. **Conclusions**

It is well-known that the NLSE in (1+1) dimensions is an integrable system and its IST is well studied, whereas, the DNLSE and GNLSE have no such property and so it is difficult to find exact solutions for those equations. The method of constructing NR limit of a relativistic equation has been used to find a number of solutions of the NR equations like DNLSE and GNLSE from the solutions of the correspondnig NLKGEs. The converse of this procedure, i.e. a systematic method of finding the solutions of relativistic equations like NLKGEs from the known solutions of NR equations like NLSEs is also interesting and will be reported separately.

For example, it is known that NLSE eq. (30) has an envelope kink solution

$$\psi(x,t) = \pm\sqrt{\frac{-2\lambda^2}{\alpha}}\; e^{i\theta'_o} \tanh\left[\pm \frac{\lambda}{\sqrt{2\hbar}} (x - vt)\right] \tag{35}$$

The respective relativistic form of the kink envelope type solutions and the $\lambda\psi^4$ Lagrangian equations are respectively

$$\phi(x,t) = \pm\sqrt{\frac{-2\omega^2}{ac^2}}\; e^{i\theta'} \tanh\left[\pm \sqrt{\frac{\omega}{2c\hbar a}} (x - vt)\right] \tag{36}$$

and

$$\Box\phi + m^2c^4\phi - ac^2\phi|\phi|^2 = 0. \tag{37}$$

Note that the known kink solution of (37) was without the de Broglie phase [8,30].

Our study reveals thus that the envelope-type solutions of NLKGEs are the natural relativistic forms of the envelope solutions of the NR equations like NLSEs. In quantum theory free particles are described by plane waves solutions of the linear Schrödinger equation with the usual probability interpretation. On the other hand, appropriate nonlinear field equations could be used to describe a localized single particle by their envelope soliton-like solutions [5,13,14,23]. The present study continues the searches in this direction [1,2,5]. The final goal, which is also related to the original thoughts of L. de Broglie [7] is to find a satisfactory description of single events in quantum phenomena.

Acknowledgements

One of the authors (B.V.B) is extremely thankful to N.C. Freeman, Martin Kruskal, A.S. Fokas, J. Hietarinta and Miki Wadati for several discussions. He would also like to thank Professor Abdus Salam, the International Atomic Energy Agency and UNESCO for hospitality at the International Centre for Theoretical Physics, Trieste.

References

[1] Baby, B.V. & Barut, A.O. (1987). On nonlinear equations admitting soliton solutions with a de Broglie phase. ICTP preprint, IC/87/294 and *Journal of Physics A: Mathematical and General* (submitted).

[2] Barut, A.O. & Baby, B.V. (1988). The relations between relativistic and nonrelativistic solitons and kinks. *Journal of Physics A: Mathematical and General* (submitted).

[3] Barut, A.O., Girardello, L. & Wyss, W. (1976). Nonlinear O(n+1) symmetric field theories, symmetry breaking and finite energy solutions. *Helv. Phys. Acta* **49**, 807.

[4] Barut, A.O. & Rączka, R. (1980). *Theory of Group Representations and Applications*, 2nd ed. (World Scientific, 1987).PWN, Polish Scientific Publishers: Warszawa.

[5] Barut, A.O & Rusu, P. (1989). On the wave-particle-like solutions of nonlinear equations. *Can. J. Physics.* **67**, 100.

[6] Barut, A.O. & Van Huele, J.F. (1985). Quantum electrodynamics based on self-energy: Lamb shift and spontaneous emission without field quantization. *Phys. Rev.* **32A**, 3187.

[7] de Broglie, L. (1960). *Nonlinear Wave Mechanics.* Elsevier Publishing Company: Paris.

[8] Bullough, R.K. & Caudrey, P.J. (1980). The soliton and its history. In *Solitons - Topics in Current Physics*, pp. 1-56. Bullough, R.K. & Caudrey, P.I. (eds.). Springer-Verlag: Berlin.

[9] Burt, P.B. (1978). Exact multiple soliton solutions of the double sine Gordon equation. *Proc. Roy. Soc. Lond.* **359A**, 479.

[10] Calogero, F. (1988). Why are certain nonlinear PDEs both widely applicable and integrable? To appear in *What is Integrability (for nonlinear PDEs)?*, Zakharov, V.E. (ed.). Springer-Verlag: Berlin.

[11] Cowan, S., Enns, R.H., Rangnekar, S.S. & Sanghera, S.S. (1986). Quasi-soliton and other behaviour of the nonlinear cubic-quintic Schrödinger equations. *Can. J. Physics* **64**, 311.

[12] Davydov, A.S. (1979). Solitons in molecular systems. *Phys. Scripta* **20**, 387.

[13] Enz, U. (1985). The Sine-Gordon breaker as a moving oscillator in the sense of de Broglie. *Physica* **17D**, 116.

[14] Enz, U. (1986). The confined breather; quantized states from classical field theory. *Physica* **21D**, 1.

[15] Fuschich, W.I. & Serov, N.I. (1987). On some exact solutions of the three dimensional nonlinear Schrödinger equations. *J. Physics A: Math. & Gen.* **20**, L929.

[16] Gagnon, L. & Winternitz, P. (1987). Lie symmetries of a generalized nonlinear Schrödinger equation. Montréal University preprint, CRM-1469.

[17] Gagnon, L. & Winternitz, P. (1988a). Exact solutions of the cubic and quintic nonlinear Schrödinger equations for a cylindrical geometry. Montréal University preprint, CRM-1528.

[18] Gagnon, L. & Winternitz, P. (1988b). Spherical quintic nonlinear Schrödinger equations. Montréal University preprint, CRM-1531.

[19] Hasse, R.W. (1980). A general method for the solution of nonlinaer soliton and kink of Schrödinger equations. *Zeit. Physik.* **37B**, 83.

[20] Hirota, R. (1972). Exact solutions of the Sine-Gordon equations for multiple collisions of solitons. *J. Phys. Soc. Jap.* **33**, 1459.
[21] Hirota, R. (1973). Exact envelop-soliton solutions of nonlinear wave equations. *J. Math. Phys.* **14**, 805.
[22] Kivshar, V.S. & Malomed, B.A. (1986). Perturbations in induced radiative losses in collision of NSE solitons. *J. Physics A: Math. & Gen.* **194**, L967.
[23] Klein, J.J. (1976). Nonlinear wave equation with intrinsic wave particle dualism. *Can. J. Physics* **54**, 1383.
[24] Kodama, Y. & Ablowitz, M.J. (1980). Transverse instability of breather in resonant media. *J. Math. Phys.* **1**, 928.
[25] Kodama, Y. & Ablowitz, M.J. (1981). Perturbations of solitons and solitary waves. *Stud. Appl. Math.* **64**, 225.
[26] Malomed, B.A. (1987a). Decay of shrinking solitons in multidimensional Sine-Gordon equations. *Physica* **24D**, 155.
[27] Malomed, B.A. (1987b). Evolution of nonsoliton and quasi-classical wave trains in nonlinear Schrödinger and Korteweg-de Vries equations with dissipative perturbations. *Physica* **29D**, 155.
[28] Pereira, N.R. & Stenflo, L. (1977). Nonlinear Schrödinger equations including growth and dampling. *The Physics of Fluids* **20**, 1733.
[29] Rabinovich, M.I. & Fabrikant, A.L. (1979). Stochastic self-modulation of waves in non equilibrium media. *Soviet Physics,* **JETP, 50,** 311.
[30] Rajaraman, R. (1975). Some nonperturbative semi-classical methods in quantum field theory. *Physics Rep.* **21**, 227.
[31] Scott, A.C. (1982). The vibrational structure of Davydov solitons. *Phys. Scripta* **25**, 651.
[32] Tuszynski, J.A., Paul, R. & Chatterjee, R. (1984). Exact solutions to the time dependent Landau-Ginsburg model of phase transitions. *Phys. Rev.* **29B**, 380.
[33] Zakharov, V.E. & Shabat, A.B. (1972). *Sov. Phys.* **JETP, 34,** 62.
[34] Zakharov, V.E. & Synakh, V.S. (1976). *Sov. Phys.* **JETP, 41,** 485.

B. Chakraborty and S. Sarkar

55. NONLINEAR ROTATION OF POLARIZATION ELLIPSE OF WAVES IN PLASMAS AND OTHER MEDIA

Methods of finding the precessional rotation of strong elliptically polarized electromagnetic waves, induced by cubic nonlinearities in plasmas and other media, are explained in this paper. These methods for secular free harmonic wave generation are important because they yield the results of the unique phenomenon of ellipse rotation of transverse waves.

1 Introduction

Nonlinear wave processes like secular free higher harmonic generation are possible by strong waves in plasma as in other media which have cubic and still higher nonlinearities. One consequence is the precessional rotation of elliptically polarized waves. This precession is also due to the nonlinearly induced birefringence of the left and right circularly polarized components (the LCP and RCP respectively) of the wave.

The ellipse rotation shows that the plasma field equations are valid in a frame which undergoes the self generated precessional rotation. This frame of reference can therefore be designated as the Principal Polarization Reference Frame, or PPRF. In addition, there is the laboratory frame, LF, of the observer, see Figs. 1 and 2.

If $(\hat{\mathbf{X}}, \hat{\mathbf{Y}}, \hat{\mathbf{Z}})$ are the unit vectors parallel to OXYZ and $(\hat{\mathbf{X}}_o, \hat{\mathbf{Y}}_o, \hat{\mathbf{Z}}_o)$ are the same of $0\ X_oY_oA_o$, then

$$\hat{\mathbf{X}} = \hat{\mathbf{X}}_o \cos \rho + \hat{\mathbf{Y}}_o \sin \rho, \qquad (1)$$

$$\hat{\mathbf{Y}} = -\hat{\mathbf{X}}_o \sin \rho + \hat{\mathbf{Y}}_o \cos \rho \qquad (2)$$

where ρ is the intensity-induced precessional angle of rotation. In temporal evolution problems ρ is a function of time, and in the spatial evolution problems ρ is a function of the Z coordinate only; see [1,3,4,5,6,7,8,9,10,11,12,13,14,16].

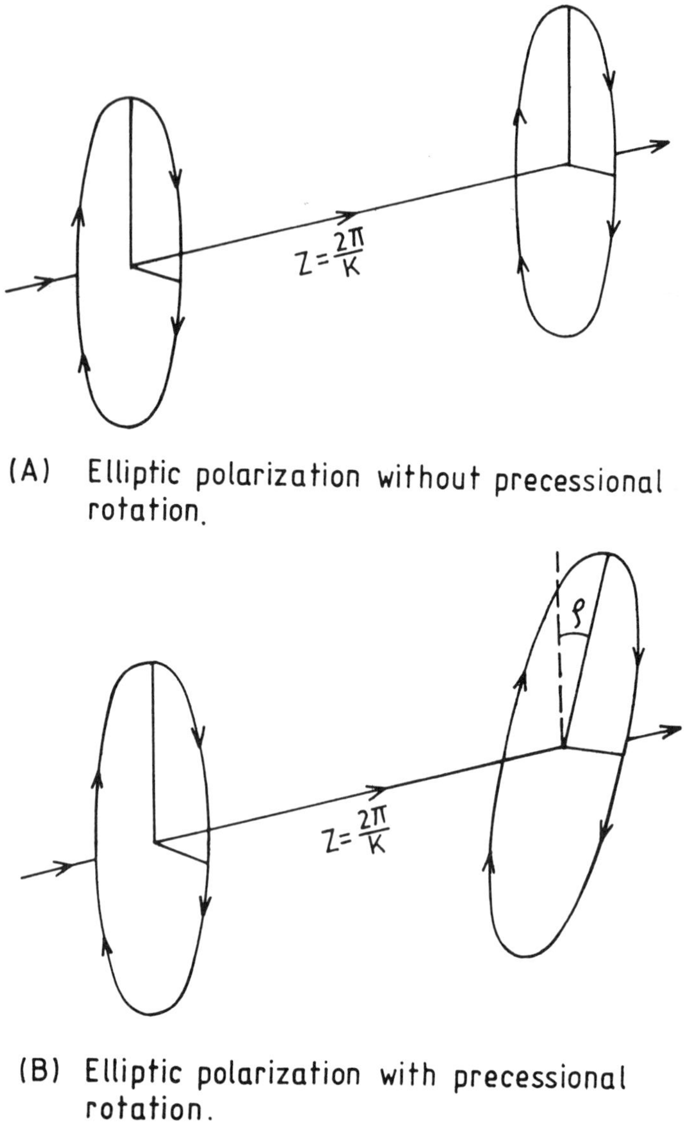

Figure 1. Precessional rotation of the polarization ellipse of a wave.

2 The mathematical model for the simplest cases of ellipse rotation

The mathematical model is the following connected pair of wave equations:

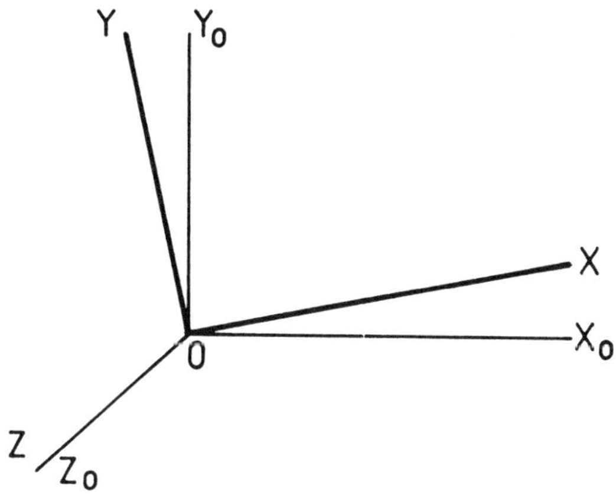

Figure 2. Shows the two frames of reference. $OX_oY_oZ_o$ *is the laboratory frame (LF) and OXYZ is the principal polarization reference frame (PPRF).*

$$\left.\begin{aligned} \Box^2\mathbf{A} &= r\ (\mathbf{A}\cdot\bar{\mathbf{A}})\mathbf{A} + S\ A^2\bar{\mathbf{A}}\ , \\ \Box^2\bar{\mathbf{A}} &= r\ (\mathbf{A}\cdot\bar{\mathbf{A}})\bar{\mathbf{A}} + S\ \bar{A}^2\mathbf{A} \end{aligned}\right\} \qquad (3)$$

where the Klein-Gordon (K-G) wave operator

$$\Box^2 = \frac{\partial^2}{\partial x^2} + \frac{\partial^2}{\partial y^2} + \frac{\partial^2}{\partial z^2} - \frac{1}{c^2}\frac{\partial^2}{\partial t^2} - \frac{w_p^2}{c^2} \qquad (4)$$

r and s are constants, $\bar{\mathbf{A}}$ is the complex conjugate (CC) of the complex Vector **A**. For a secular-free plane wave train solution in a harmonic series, let

$$\begin{aligned} \mathbf{A} + \bar{\mathbf{A}} &= \mathbf{a}e^{i\theta} + \bar{\mathbf{a}}e^{-i\theta} \\ &= \hat{\mathbf{X}}\ \alpha\ \cos\theta + \hat{\mathbf{Y}}\beta\ \sin\theta \end{aligned} \qquad (5)$$

with

$$\theta = kz - wt,\ \frac{\partial}{\partial x} = 0,\ \frac{\partial}{\partial y} = 0 \qquad (6)$$

α, β and the complex vectors **a** and $\bar{\mathbf{a}}$ are constants. Since the direction of wave propagation is the Z axis, in view of (6),

$$\Box^2 = \frac{\partial^2}{\partial z^2} - \frac{1}{c^2}\frac{\partial^2}{\partial t^2} - \frac{w_p^2}{c^2} \qquad (7)$$

Moreover, the fields being transverse, **a** and $\bar{\mathbf{a}}$ are normal to OZ. The first and third harmonics are those of transverse waves; the second harmonic effect, being dependent on a nonlinear source, proportional to $\mathbf{a} \times \bar{\mathbf{a}}$, is that of longitudinal waves.

In place of the connected pair of equations (3) we can also use the following model set of connected inhomogeneous K-G equations:

$$\left.\begin{aligned} \Box^2(\mathbf{a}e^{i\theta}) &= [r(\mathbf{a}\cdot\bar{\mathbf{a}})\mathbf{a} + S\,\mathbf{a}^2\bar{\mathbf{a}}]e^{i\theta}, \\ \Box^2(\bar{\mathbf{a}}\,e^{-i\theta}) &= [r(\mathbf{a}\cdot\bar{\mathbf{a}})\bar{\mathbf{a}} + S\,\bar{\mathbf{a}}^2\mathbf{a}]e^{-i\theta} \end{aligned}\right\} \tag{8}$$

Some methods of obtaining secular free solution are the following:

a) Method of co-ordinate stretching of Lindstedt and others [2];

b) Method of expansion of the wave parameter of Bogoliubov, Krilov & Mitropolsky [15];

c) A direct method of successive approximation, yielding the dispersion relation nonlinearly correct up to the third order of small quantities (Section 4);

d) A method involving complex combination of the two transverse components of each of the wave fields, depending on the nonlinearly induced birefringence (Section 5);

e) A normalization technique, developed by Arons & Max [1].

3 The basic field equations and solutions up to the second order

The basic field equations are those of a cold, unmagnetized electron plasma. These are

$$m\frac{d}{dt}\left(\frac{\mathbf{u}}{\sqrt{1-\frac{u^2}{c^2}}}\right) = -e\mathbf{E} - \frac{e}{c}(\mathbf{u}\times\mathbf{H}). \tag{9}$$

$$\frac{\partial N}{\partial t} + \nabla\cdot(N\mathbf{u}) = 0, \tag{10}$$

$$\nabla\times\mathbf{E} = -\frac{1}{c}\frac{\partial\mathbf{H}}{\partial t}, \tag{11}$$

$$\nabla\times\mathbf{H} = \frac{1}{c}\frac{\partial\mathbf{E}}{\partial t} - \frac{4\pi e}{c}N\mathbf{u}, \tag{12}$$

$$\nabla\cdot\mathbf{H} = 0,\ \nabla\cdot\mathbf{E} = 4\pi e(N_o - N) \tag{13}$$

with

$$\frac{d}{dt} = \frac{\partial}{\partial t} + (\mathbf{u}\cdot\nabla) \tag{14}$$

$-e$ is the charge and m mass of an electron, $\mathbf{u}$ is its velocity, N is its number density, $\mathbf{E}$ is the electric field vector and $\mathbf{H}$ is the magnetic field vector.

The first order field equations are

$$\frac{\partial\mathbf{u}_1}{\partial t} = -\frac{e}{m}\mathbf{E}_1 \tag{15}$$

$$\frac{\partial N_1}{\partial t} + N_o(\nabla \cdot \mathbf{u}_1) = 0, \tag{16}$$

$$\nabla \times \mathbf{E}_1 = - \frac{1}{c} \frac{\partial \mathbf{H}_1}{\partial t}, \tag{17}$$

$$\nabla \times \mathbf{H}_1 = \frac{1}{c} \frac{\partial \mathbf{E}_1}{\partial t} - \frac{4\pi e}{c} N_o \mathbf{u}_1, \tag{18}$$

$$(\nabla \cdot \mathbf{E}_1) = -4\pi e N_1, \tag{19}$$

$$\nabla \cdot \mathbf{H}_1 = 0 \tag{20}$$

For an elliptically polarized transverse wave we write

$$\mathbf{E}_1 = (a \cos \theta,\ b \sin \theta,\ 0),\ \theta = Kz - wt \tag{21}$$

$$(\nabla \cdot \mathbf{E}_1) = 0,\ N_1 = 0 \tag{22}$$

The dispersion relation is

$$K^2 c^2 = w^2 - w_p^2 \tag{23}$$

with

$$w_p^2 = \frac{4\pi N_o e^2}{m} \tag{24}$$

Then evidently

$$\mathbf{u}_1 = \frac{e}{mw} (a \sin \theta,\ - b \cos \theta,\ 0), \tag{25}$$

$$\mathbf{H}_1 = - \frac{kc}{w} (b \sin \theta,\ - a \cos \theta,\ 0). \tag{26}$$

The second order equations are

$$\frac{\partial \mathbf{u}_2}{\partial t} = \frac{e}{m} \mathbf{E}_2 - \frac{e}{mc} (\mathbf{u}_1 \times \mathbf{H}_1), \tag{27}$$

$$\frac{\partial N_2}{\partial t} = - N_o(\nabla \cdot \mathbf{u}_2) \tag{28}$$

$$(\nabla \cdot \mathbf{E}_2) = -4\pi e N_2,\ \nabla \cdot \mathbf{H}_2 = 0 \tag{29}$$

$$\nabla \times \mathbf{E}_2 = - \frac{1}{c} \frac{\partial \mathbf{H}_2}{\partial t} \tag{30}$$

$$\nabla \times \mathbf{H}_2 = \frac{1}{c}\frac{\partial \mathbf{E}_2}{\partial t} - \frac{4\pi e}{c} N_o \mathbf{u}_z \qquad (31)$$

The only nonlinear source of excitation of the second order effects is the last term of equation (27). Putting the values of $\mathbf{u}_1$ and $\mathbf{H}_1$ from (25) and (26) respectively, it becomes

$$-\frac{e}{mc}(\mathbf{u}_1 \times \mathbf{H}_1) = \left[0,\ 0,\ -\frac{ke^2(a^2-b^2)}{2m^2w^2}\sin 2\theta\right] \qquad (32)$$

The second harmonic wave fields, excited by the nonlinearities, being purely longitudinal in nature, we write

$$\mathbf{E}_{2\perp} = 0, \quad \mathbf{H}_2 = 0 \qquad (33)$$

and obtain

$$\mathbf{u}_2 = \left[0,\ 0,\ -\frac{kew_p^2(a^2-b^2)\cos 2\theta}{m^2w(4w^2-w_p^2)}\right] \qquad (34)$$

$$\mathbf{E}_2 = \left[0,\ 0,\ \frac{kew_p^2(a^2-b^2)\cos 2\theta}{2mw^2(4w^2-w_p^2)}\right] \qquad (35)$$

$$N_2 = -\frac{k^2N_oe^2(a^2-b^2)\cos 2\theta}{m^2w^2(4w^2-w_p^2)} \qquad (36)$$

4 The third order fields and equations, and secular free solution correct up to that order

The third order fields are linear combinations of cos $(\theta \pm \theta \pm \theta)$ and sin $(\theta \pm \theta \pm \theta)$, so these are linear combinations of cos θ, sin θ, cos 3θ, sin 3θ. The first harmonic fields, correct up to that order of approximation, exist together with the third harmonic fields which start appearing only from the third order approximation. We therefore drop the second order fields from the field equations for the nonlinear evolution of the first harmonic field, but not the nonlinearly induced second order effect terms which interact with the first order fields for the stated purpose. So the effective third order field equations are

$$\frac{\partial \mathbf{u}_1}{\partial t} + u_2 \cdot \frac{\partial \mathbf{u}_1}{\partial z} + \frac{\partial}{\partial t}\left(\frac{u_1^2\mathbf{u}_1}{2c^2}\right) = -\frac{e}{m}(\mathbf{E}_1+\mathbf{E}_3) - \frac{e}{mc}(\mathbf{u}_2 \times \mathbf{H}_1) \qquad (37)$$

$$\nabla^2(\mathbf{E}_1+\mathbf{E}_3) - \frac{1}{c^2}\frac{\partial^2}{\partial t^2}(\mathbf{E}_1+\mathbf{E}_3) = -\frac{4\pi e}{c^2}\frac{\partial}{\partial t}(N_o\mathbf{u}_1 + N_2\mathbf{u}_1) \qquad (38)$$

We have eliminated $\mathbf{H}_1 + \mathbf{H}_3$ by taking curl of curl $\mathbf{E}$ and using (12). Moreover, since all the four cubic nonlinear sources,

$$u_2 \frac{\partial \mathbf{u}_1}{\partial z}, \ \frac{\partial}{\partial t}\left(\frac{u_1^2 \mathbf{u}_1}{2c^2}\right), \ -\frac{e}{mc}(\mathbf{u}_2 x \mathbf{H}_1), \ -\frac{4\pi e}{c^2}\frac{\partial}{\partial t}(N_2 \mathbf{u}_1) \qquad (39)$$

are absolutely transverse in nature, we assume that $\mathbf{E}_3$, like $\mathbf{E}_1$, is absolutely transverse, so use the condition

$$(\nabla \cdot \mathbf{E}_3) = 0 \qquad (40)$$

like $(\nabla \cdot \mathbf{E}_1) = 0$. With the help of (37) equation (38) becomes

$$\left(\frac{\partial^2}{\partial z^2} - \frac{1}{c^2}\frac{\partial^2}{\partial t^2}\right)(\mathbf{E}_1 + \mathbf{E}_3) = -\frac{4\pi N_o e}{c^2}\left[-u_2 \frac{\partial \mathbf{u}_1}{\partial z} - \frac{\partial}{\partial t}\left(\frac{u_1^2 \mathbf{u}_1}{2c^2}\right) - \right.$$

$$\left. - \frac{e}{m}(\mathbf{E}_1 + \mathbf{E}_3) - \frac{e}{mc}(\mathbf{u}_2 \times \mathbf{H}_1) + \frac{1}{N_o}\frac{\partial}{\partial t}(N_2 \mathbf{u}_1)\right]$$

$$= \frac{w_p^2}{c^2}(\mathbf{E}_1 + \mathbf{E}_3) + \frac{4\pi N_o e}{c^2}\left[u_2 \frac{\partial \mathbf{u}_1}{\partial z} + \frac{\partial}{\partial t}\left(\frac{u_1^2 \mathbf{u}_1}{2c^2}\right) + \right.$$

$$\left. + \frac{e}{mc}(\mathbf{u}_2 \times \mathbf{H}_1) - \frac{1}{N_o}\frac{\partial}{\partial t}(N_2 \mathbf{u}_1)\right] \qquad (41)$$

It can be expressed as the inhomogeneous Klein-Gordon equation

$$\left[\frac{\partial^2}{\partial z^2} - \frac{1}{c^2}\frac{\partial^2}{\partial t^2} - \frac{w_p^2}{c^2}\right](\mathbf{E}_1 + \mathbf{E}_3) = \square^2(\mathbf{E}_1 + \mathbf{E}_3)$$

$$= \frac{4\pi N_o e}{c^2}\left[u_2 \frac{\partial \mathbf{u}_1}{\partial t} + \frac{\partial}{\partial t}\left(\frac{u_1^2 \mathbf{u}_1}{2c^2}\right) + \frac{e}{mc}(\mathbf{u}_2 \times \mathbf{H}_1) - \frac{1}{N_o}\frac{\partial}{\partial t}(N_2 \mathbf{u}_1)\right]$$

$$= -\frac{w_p^2}{w}\left[-\frac{w}{8c^2}\{-\hat{\mathbf{x}}\, a\,(3\alpha^2+\beta^2)\cos\theta - \hat{\mathbf{Y}}\, b\,(\alpha^2+3\beta^2)\sin\theta + \right.$$

$$+ 3(\hat{\mathbf{x}}\, a \cos 3\theta + \hat{\mathbf{Y}}\, b \sin 3\theta)(\alpha^2-\beta^2)\} +$$

$$\left. + \frac{k^2(\alpha^2-\beta^2)}{2w(4-x)}\{-\hat{\mathbf{x}}\, a\,(\cos\theta - 3\cos 3\theta) + \hat{\mathbf{Y}}\, b\,(\sin\theta + 3\sin 3\theta)\}\right] \qquad (42)$$

where we have used the dimensionless symbols x, α, β defined by

$$x = \frac{w_p^2}{w^2}, \quad \alpha = \frac{ea}{mwc}, \quad \beta = \frac{eb}{mwc} \tag{43}$$

Equating from both sides the first harmonic and third harmonic fields we obtain the following two equations

$$\left(\frac{\partial^2}{\partial z^2} - \frac{1}{c^2}\frac{\partial^2}{\partial t^2} - \frac{w_p^2}{c^2}\right)\mathbf{E}_1 =$$

$$= x\left[\frac{w^2}{8c^2}\{-\hat{\mathbf{x}}\ a\ (3\alpha^2+\beta^2)\cos\theta - \mathbf{Y}\ b(\alpha^2+3\beta^2)\sin\theta\}\right]$$

$$+ \frac{k^2(\alpha^2-\beta^2)}{2(4-x)}(\hat{\mathbf{x}}\ a\cos\theta - \hat{\mathbf{Y}}\ b\sin\theta), \tag{44}$$

$$\left(\frac{\partial^2}{\partial z^2} - \frac{1}{c^2}\frac{\partial^2}{\partial t^2} - \frac{w_p^2}{c^2}\right)\mathbf{E}_3 =$$

$$= x\left[\frac{3w^2}{8c^2}(\alpha^2-\beta^2)(\hat{\mathbf{x}}\ a\cos 3\theta + \hat{\mathbf{Y}}\ b\sin 3\theta) - \right.$$

$$\left. - \frac{3k^2(\alpha^2-\beta^2)}{2(4-x)}(-\hat{\mathbf{x}}\ a\cos 3\theta - \hat{\mathbf{Y}}\ b\sin 3\theta)\right]$$

$$= \frac{3x}{2}(\alpha^2-\beta^2)\left(\frac{w^2}{4c^2} + \frac{k^2}{4-x}\right)(\hat{\mathbf{x}}\ a\cos 3\theta + \hat{\mathbf{Y}}\ b\sin 3\theta) \tag{45}$$

4.1 *Temporal evolution of the self-induced rotation*

In the temporal evolution problem the wave precession angle ρ of equations (1) and (2) is a function of time only. So, differentiating the unitvectors of these with respect to t, we obtain

$$\dot{\hat{\mathbf{X}}} = (\dot{\rho} \times \hat{\mathbf{X}}), \quad \dot{\hat{\mathbf{Y}}} = (\dot{\rho} \times \hat{\mathbf{Y}}) \tag{46}$$

where $\dot{\rho}$ is the precession frequency. Hence evidently

$$\frac{\partial^2 \mathbf{E}_1}{\partial t^2} = -w^2\mathbf{E}_1 + 2w\dot{\rho}\hat{\mathbf{z}}x(\hat{\mathbf{X}}\ a\sin\theta - \hat{\mathbf{Y}}\ b\cos\theta) \tag{47}$$

where terms of $\ddot{\rho}$ and $\dot{\rho}^2$ are neglected because these are of still higher order, and since

$$\frac{\partial^2}{\partial z^2} = -k^2,$$

equation (45) gives

$$\left(-k^2 + \frac{w^2 - w_p^2}{c^2}\right) (\hat{\mathbf{X}}\ a \cos\theta + \hat{\mathbf{Y}}\ b \sin\theta) -$$

$$- \frac{2w}{c^2} \dot{\rho}\ (\hat{\mathbf{Y}}\ a \sin\theta + \hat{\mathbf{X}}\ b \cos\theta)$$

$$= \frac{x}{2} \left[\frac{k^2}{4c^2} \{-\hat{\mathbf{X}}\ a(3\alpha^2+\beta^2) \cos\theta - \hat{\mathbf{Y}}\ b(\alpha^2+3\beta^2) \sin\theta\} + \right.$$

$$\left. + \frac{k^2}{4-x} (\alpha^2-\beta^2)(\hat{\mathbf{X}}\ a \cos\theta - \hat{\mathbf{Y}}\ b \sin\theta)\right] \qquad (49)$$

Equating coefficients of $\hat{\mathbf{X}}$ a cos θ and $\hat{\mathbf{Y}}$ b sin θ from both sides gives the nonlinear evolution equations

$$(-k^2 + \frac{w^2-w_p^2}{c^2})\ a - \frac{2w}{c^2} \dot{\rho}\ b$$

$$= \frac{x}{2}\ a \left[\frac{w^2}{4c^2} (3\alpha^2+\beta^2) + \frac{k^2(\alpha^2-\beta^2)}{4-x}\right], \qquad (50)$$

$$\left(-k^2 + \frac{w^2-w_p^2}{c^2}\right) b - \frac{2w}{c^2} \dot{\rho}\ a$$

$$= \frac{x}{2}\ b \left[-\frac{w^2}{4c^2} (\alpha^2+2\beta^2) - \frac{k^2(\alpha^2-\beta^2)}{4-x}\right] \qquad (51)$$

Solving for $(-k^2 + \frac{w^2-w_p^2}{c^2})$ and $\dot{\rho}$ we obtain

$$k^2c^2 - (w^2-w_p^2) = \frac{(\alpha^2+\beta^2)w_p^2}{2} \left(\frac{3}{4} - \frac{k^2c^2}{4w^2-w_p^2}\right) \qquad (52)$$

$$\dot{\rho} = \frac{\alpha\beta}{2}\ x\ w \left[\frac{k^2c^2}{w^2(4-x)} - \frac{1}{4}\right] \qquad (53)$$

Equation (52) is the field intensity-dependent dispersion relation correct up to the third order. Equation (53) gives the nonlinearly induced wave precession rate; integrating it with respect to t we obtain

Rotation of the wave polarization ellipse

$$\rho = \frac{\alpha\beta \; x \; wt}{2} \left[\frac{k^2 c^2}{w^2 (4-x)} - \frac{1}{4} \right] \tag{54}$$

Putting $\rho = 2\pi$, $t = T_p$, where T_p is the time period for the precessional rotation, gives

$$T_p = \frac{2T}{\alpha\beta x} \Big/ \left(\frac{1-x}{4-x} - \frac{1}{4} \right) \tag{55}$$

where $T = 2\pi/w$; T is the wave time period.

4.2 *Spatial evolution of wave precession*

In this case the wave precession angle ρ (of Fig. 2) and the relations (1) and (2)) is a function of Z only. Denoting derivative with respect to Z by a prime, we find that

$$\hat{\mathbf{X}}' = (\rho' \times \hat{\mathbf{X}}) \;, \quad \hat{\mathbf{Y}} = (\rho' \times \hat{\mathbf{Y}}) \tag{56}$$

$$\frac{\partial^2 \mathbf{E}_1}{\partial z^2} = -k^2 \mathbf{E}_1 - 2k\rho' \mathbf{Z} \times (\hat{\mathbf{X}} \; a \sin\theta - \hat{\mathbf{Y}} \; b \cos\theta) \tag{57}$$

$$\frac{\partial^2 \mathbf{E}_1}{\partial t^2} = -w^2 \mathbf{E}_1 \tag{58}$$

Hence, in place of (50) and (51), we now have

$$\left(-k^2 + \frac{w^2 - w_p^2}{c^2} \right) a - 2k\rho' b$$

$$= \frac{x}{2} \left[- \frac{w^2}{4c} \; a \; (3\alpha^2 + \beta^2) + \frac{k^2}{4-x} \; a \; (\alpha^2 - \beta^2) \right], \tag{59}$$

$$\left(-k^2 + \frac{w^2 - w_p^2}{c^2} \right) b - 2k\rho' a$$

$$= \frac{x}{2} \left[- \frac{w^2}{4c^2} \; b(\alpha^2 + 3\beta^2) - \frac{k^2}{4-x} \; b(\alpha^2 - \beta^2) \right] \tag{60}$$

Solving for $-K^2 + W^2 - W_p^2/C^2$ we recover (52); however

$$\rho' = \frac{\alpha\beta X W^2}{2KC^2} \left[\frac{K^2 C^2}{W^2 (4-x)} - \frac{1}{4} \right] \tag{61}$$

Hence, in place of (55), we now have

$$\lambda_p = \frac{2K^2C^2}{\alpha\beta XW^2} \Big/ \left(\frac{1-x}{4-x} - \frac{1}{4}\right) \tag{62}$$

where $Z = \lambda_p$ when $\rho = 2\pi$, and $\lambda = 2\pi/K$. So λ_p is the precessional rotation period length and λ is the wave length of the wave.

4.3 *Nonlinearly induced frequency shift and wave number shift*

The nonlinear dispersion relation (52), common both for temporal and spatial evolution problems, gives the nonlinearly induced frequency shift δw for temporal problems and the wave number shift δk for spatial problems. We first replace w by $w+\delta w$ in the left hand side and retain only terms up to the first power of δw. Then, with the help of (23), we find that

$$\frac{\delta w}{w} = -\frac{(\alpha^2+\beta^2)}{4w^2} w_p^2 \left(\frac{3}{4} - \frac{1-x}{4-x}\right) \tag{63}$$

Similarly

$$\frac{\delta k}{k} = \frac{(\alpha^2+\beta^2)}{4k^2c^2} w_p^2 \left(\frac{3}{4} - \frac{1-x}{4-x}\right)$$

5 Wave precession from the nonlinearly induced birefringence

The first harmonic elliptically polarized electric field,

$$\mathbf{E}_1 = \frac{1}{2}\hat{\mathbf{X}}\, a(e^{i\theta}+e^{-i\theta}) - \frac{i}{2}\hat{\mathbf{Y}}\, b(e^{i\theta}-e^{-i\theta}) \tag{64}$$

is replaced by the complex combinations

$$\begin{aligned} E_{1x} \pm iE_{1y} &= \frac{a}{2}(e^{i\theta}+e^{-i\theta}) \pm \frac{b}{2}(e^{i\theta}-e^{-i\theta}) \\ &= (pe^{i\theta}+qe^{-i\theta};\ qe^{i\theta}+pe^{-i\theta}) \end{aligned} \tag{65}$$

where $p = \frac{a+b}{2}$, $q = \frac{a-b}{2}$. (66)

Then the first order Maxwell equations (15)- (20) take the form

$$\dot{U}_{1\pm} = -\frac{e}{m} E_{1\pm}, \tag{67}$$

$$\dot{H}_{1\pm} = \mp ieE'_{1\pm} \tag{68}$$

$$\pm iCH'_{1\pm} = E'_{1\pm} - 4\pi N_o eU_{1\pm} \tag{69}$$

$$N_1 = 0, \quad U_{1Z} = 0, \quad E_{1Z} = 0 \tag{70}$$

where by prime derivative with respect to Z is denoted, and dot means the time derivative. For $E_{1\pm}$ the homogeneous K-G equation is

$$C^2E''_{1\pm} - \ddot{E}_{1\pm} - W_p^2E_{1\pm} = 0 \tag{71}$$

It yields the dispersion relation (23), and gives

$$H_{1+} = \frac{ikc}{w}(pe^{i\theta}+qe^{-i\theta}),\ H_{1-} = -\frac{ikc}{w}(qe^{i\theta}+pe^{-i\theta}), \tag{72}$$

$$U_{1+} = -\frac{ie}{mw}(pe^{i\theta}-qe^{-i\theta}),\ U_{1-} = -\frac{ie}{mw}(qe^{i\theta}-pe^{-i\theta}) \tag{73}$$

The second order equations are

$$\left.\begin{aligned} \dot{U}_{2Z} &= -\frac{3}{m}E_{2Z} - \frac{ie}{2mc}(U_{1-}H_{1+} - U_{1+}H_{1-}), \\ \dot{N}_2 &= -N_oU'_{2Z},\ E'_{2Z} = -4\pi eN_2,\ \dot{E}_{2Z} = 4\pi eN_oU_{2Z} \end{aligned}\right\} \tag{74}$$

Hence, if n is the refractive index of the plasma,

$$\begin{aligned} \ddot{U}_{2Z} + W_p^2U_{2Z} &= \frac{ie}{2mc}\frac{\partial}{\partial t}(U_{1-}H_{1+} - U_{1+}H_{1-}) \\ &= -\frac{2e^2n}{m^2c}pq(e^{2i\theta}+e^{-2i\theta}) \end{aligned} \tag{75}$$

$$n = \frac{kc}{w} \tag{76}$$

The solution is

$$U_{2Z} = \frac{2nc}{4-x}p_oq_o(e^{2i\theta}+e^{-2i\theta}) \tag{77}$$

where

$$p_o = \frac{ep}{mwc},\quad q_o = \frac{eq}{mwc}, \tag{78}$$

So p_o and q_o are dimensionless amplitude-like quantities. Also

$$N_2 = \frac{2N_on^2}{4-x}p_oq_o(e^{2i\theta}-e^{-2i\theta}) \tag{79}$$

$$E_{2Z} = \frac{i4\pi N_oenc}{w(4-x)}p_oq_o(e^{2i\theta}-e^{-2i\theta}) \tag{80}$$

The first harmonic and third harmonic electric fields, correct up to the third order, are given by

$$= \frac{4\pi N_o e}{c^2}\left[U_{2Z}\frac{\partial u_{1\pm}}{\partial z} + \frac{\partial}{\partial t}\left(\frac{u_{1+}u_{1-}u_{1\pm}}{2c^2}\right) \pm \frac{ie}{mc}U_{2Z}U_{1\pm} - \frac{1}{N_o}\frac{\partial}{\partial t}(N_2U_{1\pm})\right] \tag{81}$$

Since

$$U_{2Z}\left(\frac{\partial u_{1\pm}}{\partial z} \pm \frac{ie}{mc}H_{1\pm}\right) = 0, \tag{82}$$

this equation simplifies to

$$\left(-k_\pm^2 + \frac{w_\pm^2-w_p^2}{c^2}\right)E_\pm = \frac{4\pi N_o e}{c^2}\left[\frac{\partial}{\partial t}\left(\frac{u_{1+}u_{1-}u_{1+}}{2c^2}\right) - \frac{1}{N_o}\frac{\partial}{\partial t}(N_2u_{1\pm})\right] \tag{83}$$

where in the left hand side only the first harmonic solution

$$E_\pm = E_\pm^o\, e^{i(k_\pm z - w_\pm t)} \tag{84}$$

has been retained. So, in the right hand side, only the first harmonic contribution from the nonlinearities is to be retained in (83). Since

$$\frac{\partial}{\partial t}\left(\frac{u_{1+}u_{1-}u_{1+}}{2c^2}\right) = \frac{e}{2m}\left[-p(p_o^2+2q_o^2)e^{i\theta} - q(q_o^2+2p_o^2)e^{-i\theta}\right], \tag{85}$$

$$-\frac{1}{N_o}\frac{\partial}{\partial t}(N_2u_{1+}) = \frac{2n^2e}{m(4-x)}p_oq_o(qe^{i\theta}+pe^{-i\theta}), \tag{86}$$

the equation for E_+, obtained from (83), reads

$$\left(-K_+^2 + \frac{w_+^2-w_p^2}{c^2}\right) = \frac{2w_p^2}{c^2}\left[\{-\frac{1}{4}(p_o^2+2q_o^2) + \frac{n^2q_o^2}{4-x}\}\, pe^{i\theta} - \right.$$

$$\left. - \{\frac{1}{4}(q_o^2+2p_o^2) - \frac{n^2}{4-x}p_o^2\}\, qe^{-i\theta}\right] \tag{87}$$

Equating the coefficients of $pe^{i\theta}$ from both sides of this relation we obtain the intensity dependent dispersion relation for the RCP correct up to the third order:

$$-K_+^2 + \frac{w_+^2-w_p^2}{c^2} = \frac{2w_p^2}{c^2}\left[-\frac{1}{4}(p_o^2+2q_o^2) + \frac{n^2}{4-x}q_o^2\right] \tag{88}$$

Similarly, equating coefficients of $qe^{i\theta}$, from both sides of the equation for E_-, the dispersion relation for the LCP wave is obtained; the same is also obtained from (88) by interchanging the roles of p and q. Thus,

$$-K_-^2 + \frac{w_-^2-w_p^2}{c^2} = \frac{2w_p^2}{c^2}\left[-\frac{1}{4}(2p_o^2+q_o^2) + \frac{n^2p_o^2}{4-x}\right] \tag{89}$$

In the temporal evolution problem the corresponding intensity induced frequency shifts δw_+ and δw_- are given by

$$\frac{\delta w_+}{w} = x\left[-\frac{1}{4}(p_o^2+2q_o^2) + \frac{n^2q_o^2}{4-x}\right] \tag{90}$$

$$\frac{\delta w_-}{w} = x\left[-\frac{1}{4}(2p_o^2+q_o^2) + \frac{n^2p_o^2}{4-x}\right] \tag{91}$$

The average of these two frequency shifts is

$$\frac{\delta w}{w} = \frac{1}{2w}(\delta w_+ + \delta w_-) = \frac{x}{2}(p_o^2+q_o^2)\left[-\frac{3}{4} + \frac{n^2}{4-x}\right] \tag{92}$$

The difference is given by

$$\frac{\delta w_- - \delta w_+}{2w} = -\frac{x}{2}(p_o^2-q_o^2)\left(\frac{1}{4} - \frac{n^2}{4-x}\right) \tag{93}$$

Changing to the original dimensionless amplitude parameters α and β, with the help of (43), (66) and (78), we find that

$$\frac{1}{2w}(\delta w_+ + \delta w_-) = \frac{x}{4}(\alpha^2+\beta^2)\left(\frac{n^2}{4-x} - \frac{3}{4}\right), \tag{94}$$

$$\frac{1}{2w}(\delta w_- - \delta w_+) = \frac{x}{2}\,\alpha\beta\left(\frac{n^2}{4-x} - \frac{1}{4}\right) = \frac{\dot{\rho}}{w} \tag{95}$$

where $\dot{\rho}$ is the precession frequency evaluated in (53).

Acknowledgements

The authors are thankful to Dr. Manoranjan Khan, Secretary, Science Faculty, Jadavpur University and Dr. B. Bhattacharyya, Dept. of Mathematics, North Bengal University for useful comments and suggestions. This work is supported by the DAE projects "Energy transport in laser irradiated targets" and "Magnetic field generation and transport inhibition in laser-plasma interaction" and the UGC FAT project 1594, entitled "Electrodynamics and Mechanics in Rotating Frames".

References

[1] Arons, J. & Max, C.E. (1974). Self-precession and frequency shift for electromagnetic waves in homogeneous plasmas. *Physics of Fluids* **17**, 1983.

[2] Bellman, R. (1964). *Perturbation Techniques in Mathematics, Physics and Engineering.* Holt, Rinehart and Winston: New York.

[3] Bhattacharyya, B. & Chakraborty, B. (1979). Nonlinear self-precession and wave number shift of electromagnetic waves under resonance and Alfven waves in plasmas. *Physica (Utrecht)* **97C**, 257.

[4] Chakraborty, B. (1977). Parametric rotation of principal polarization axes and other aspects due to four transverse waves in a plasma. *Phys. Rev.* **A16**, 1297.

[5] Chakraborty, B. & Chandra, S.K. (1978). Self-precession and frequency shift for two waves in a plasma. *Physics of Fluids* **21**, 2336.

[6] Chakraborty, B., Khan Manoranjan & Bhattacharyya, B. (1981). Impact of strong radiation on neutral atoms in a plasma background. *Phys. Rev.* **23A**, 344.

[7] Chakraborty, B. & Paul, S.N. (1983). (i) Wave-precession as a consequence of the nonlinearly developed birefringence in plasmas. *Physics of Fluids* **26**, 2193; (ii) Method of averaged Lagrangian for evaluating precessional rotation and other nonlinear effects in plasmas. *Austral. J. Physics* **36** 867.

[8] Chakraborty, B., Paul, S.N., Khan Manoranjan & Bhattacharyya, B. (1984). Wave-precession and related nonlinear effects in plasmas. *Phys. Rep.* **114**, 181.

[9] Chiao, R.Y. & Godine, J.Jr. (1969). Polarization dependence of stimulated Rayleigh-Wing scattering and the optimal frequency Kerr-effect. *Phys. Rev.* **185**, 430.

[10] Katz, I., Parks, D.E. & Rotenberg, M. (1973). Precession of high intensity elliptically polarized light in a plasma. *Physics of Fluids* **18**, 895.

[11] Khan, M.R. & Chakraborty, B. (1979). Precessional rotation and other nonlinear effects due to standing waves and other two-wave resonant interactions in a magnetized plasma. *J. Appl. Physics* **50**, 7979.

[12] Lai, C.S. & Wonnacott, E. (1976). Frequency shift and precession of elliptically polarized electromagnetic waves in plasmas. *Can. J. Physics* **54**, 962.

[13] Maker, P.D. & Terhune, R.w. (1965). Study of optical effects due to an induced polarization of third order in the electric field strength. *Phys. Rev.* **A137**, 801.

[14] Maker, P.D., Terhune, R.W. & Savage, C.M. (1964). Intensity-dependent change in the refractive index of liquids. *Phys. Lett.* **12**, 507.

[15] Minorsky, N. (1962). *Nonlinear Oscillations.* D. Van Nostrand: Princeton.

[16] Paul, S.N. & Chakraborty, B. (1983). Some applications of transformation to the space independent frame in the processes of strong radiation in plasmas. *Austral. J. Physics* **36**, 67.

R.K. Bullough, J. Timonen, Yu-zhong Chen, Yi Cheng and M. Stirland

56. QUANTUM AND CLASSICAL STATISTICAL MECHANICS OF THE INTEGRABLE MODELS

1 Introduction

This paper is concerned with the following system of nonlinear integral equations

$$\begin{aligned}\epsilon(k) = \omega(k) + \nu_1(2\pi\beta)^{-1}\int_{\alpha_1}^{\alpha_2} dk' \frac{d}{dk}\Delta(k,k')\ln(\beta\epsilon(k')) \\ - \nu_2(2\pi\beta)^{-1}\int_{\delta}^{\infty} dp \frac{d}{dk}\Delta_s(k,p)e^{-\beta E_s(p)}\end{aligned} \tag{1a}$$

$$\begin{aligned}E_s(p) = E(p) - \nu_1(2\pi\beta)^{-1}\int_{\alpha_1}^{\alpha_2} dk \frac{d}{dp}\Delta_s(p,k)\ln(\beta\epsilon(k)) \\ -\nu_2(2\pi\beta)^{-1}\int_{\delta}^{\infty} dp' \frac{d}{dp}\Delta_{ss}(p,p')e^{-\beta E_s(p')},\end{aligned} \tag{1b}$$

in which ν_1, ν_2 are integers (1 or 2). The equations are physically based and indeed $\beta^{-1} = k_B T$ where T is temperature and k_B is Boltzmann's constant. The functions $\omega(k)$ and E(p) are already determined dispersion relations. The functions $\Delta(k,k')$, $\Delta_s(k,p)$ and $\Delta_{ss}(p,p')$ are already determined phase shifts.

The equations arise in the calculation of the **classical** partition function Z, or the free energy $F=-\beta^{-1}\ln Z$, of many (with appropriate generalizations perhaps "all") of the integrable models in 1+1 dimensions [1]. Once the system (1) is solved for the excitation energies $\epsilon(k)$, $E_s(p)$ the free energy is found from

$$\lim_{L\to\infty} Fl^{-1} = \nu_1(2\pi\beta)^{-1}\int_{\alpha_1}^{\alpha_2} \ell n(\beta\epsilon(k))dk - \nu_2(2\pi\beta)^{-1}\int_{\delta}^{\infty} e^{-\beta E_s(p)} dp, \quad (2)$$

in which L is a length (a period under periodic boundary conditions). F is thus per unit length and the limit is a **finite density** thermodynamic limit [1].

The integrable models in 1+1 dimensions include the sine-Gordon (s-G) model

$$\phi_{xx} - \phi_{tt} = m^2 \sin\phi, \quad (3)$$

the sinh-Gordon (sinh-G) model

$$\phi_{xx} - \phi_{tt} = m^2 \sinh\phi, \quad (4)$$

the two nonlinear Schrödinger (NLS) models

$$-i\phi_t = \phi_{xx} - 2c\phi|\phi|^2 \quad (5)$$

($c>0$ is repulsive NLS, $c<0$ is attractive), the Landau-Lifshitz (L-L′) model

$$\vec{S}_t = \vec{S} \times \vec{S}_{xx} + \vec{S} \times J \cdot \vec{S}, \quad J = \mathrm{diag}\,(J_1, J_2, J_3) \quad (6)$$

the Heisenberg ferromagnet ($J_1 = J_2 = J_3$), and many integrable lattices. The lattice of interest in this paper is the now famous Toda lattice [2,3,4,5] which has equations of motion

$$\ddot{Q}_n = -V'(Q_n - Q_{n-1}) + V'(Q_{n+1} - Q_n) \quad (7a)$$

where

$$V(r) = \frac{a}{b}[e^{-\sqrt{b/a}\, r} + \sqrt{b/a}\, r]. \quad (7b)$$

The dimensionless quantity b/a can be chosen as the coupling constant. For $b/a \to 0$ one has the linear limit $V(r) = \frac{1}{2}r^2$. For $b/a \to \infty$ one has the hard core limit $V(r)=0, r>0$; $V(r)=\infty, r<0$.

These integrable models have various applications in physics. Very much in particular the classical sine-Gordon model has been used as a model of the thermal excitations in one dimensional magnets [6,7]. It also arises as a model of the biological molecule of DNA [8]. In the spirit adopted for the applications of the other models circular DNA [9] is evidently modelled by a Toda lattice under periodic boundary conditions of **finite** period. A question of some interest is then whether at body temperatures a significant number of soliton-like excitations are excited on DNA

[10,11]. Once the energies $\epsilon(k)$, $E_s(p)$ are found from equations (1) one can compute particle **densities** from further nonlinear integral equations [12]

$$\rho(k)\beta\epsilon(k) = \frac{1}{2\pi} + \frac{\nu_1}{2\pi}\int_{\alpha_1}^{\alpha_2} dk' \frac{d}{dk}\Delta(k,k')\rho(k') - \frac{\nu_2}{2\pi}\int_{\delta}^{\infty} dp\frac{d}{dk}\Delta_s(k,p)\rho_s(p) \tag{8a}$$

$$\rho_s(p)e^{\beta E_s(p)} = \frac{1}{2\pi} + \frac{\nu_1}{2\pi}\int_{\alpha_1}^{\alpha_2} dk\frac{d}{dp}\Delta_s(p,k)\rho(k) + \frac{\nu_2}{2\pi}\int_{\delta}^{\infty} dp'\frac{d}{dp}\Delta_{ss}(p,p')\rho_s(p'). \tag{8b}$$

The density $\rho(k)$ is the "phonon" density; $\rho_s(p)$ is the "soliton" density. We have not reported these integral equations (8a,b) as such before. However, the equations (1) have been reported for the s-G, sinh-G and repulsive NLS models: e.g. for s-G $\nu_1=1$, $\nu_2=2$, $\alpha_2 = \infty$, $\alpha_1 = \delta = -\infty$ [13]. Moreover, equations (1), or equivalents, appear for s-G in [14,15] as the classical limit of comparable integral equations for the quantum s-G model. We have also found examples of them for the L-L′ and ferromagnet models [12,16], and we shall see they apply, with qualifications, to the Toda lattice.

The dispersion relations $\omega(k)$,$E(p)$ play crucial roles in both the systems (1) and (8). For example for classical s-G our procedure [13] replaces the usual $\omega(k) = (k^2 + m^2)^{1/2}$ by the dispersion relation of some linear lattice of s-G type such as [13]

$$\omega(k) = [m^2 + (\sin(\tfrac{1}{2}ka_0)/\tfrac{1}{2}a_0)^2]^{1/2}, \tag{9a}$$

$a_0>0$ the lattice spacing. With (9a) the phonon term in (2) is then

$$\beta a_0^{-1}[\ell n(\beta a_0^{-1}) + \tfrac{1}{2}ma_0] \equiv F_{KG}. \tag{9b}$$

The limit $a_0 \to 0$ cannot be taken in (9b) and this is the **classical** ultraviolet divergence. The precise dependence of F_{KG} on a_0 is determined by the actual lattice model employed (in the work of [13] we really derive the linear lattice from an integrable nonlinear lattice). However, it is evident that Z is to be renormalized: $Z \to Z/Z_{KG}$, which means that F_{KG} is simply subtracted from F. **It is a remarkable fact that one can set $a_0 \to 0$ in the**

remaining part of F and obtain exact results (Section 3).

An opposite example is the Toda lattice: for this in simplest form (Section 5) $\alpha_1 = \delta = 0$, $\alpha_2 = \pi$, $\nu_1 = \nu_2 = 2$ and

$$\omega(k) = \sin(\tfrac{1}{2}ka_0)/\tfrac{1}{2}a_0, \qquad (10)$$

and we choose $a_0=1$. Then the phonon term in (2) converges at $\alpha_2 = \pi$ without difficulty, and converges at the lower limit as $\alpha_1 \ln\alpha_1$, $\alpha_1 \to 0$. This is an **infra-red** "divergence". Evidently the soliton number 2N, at temperature β^{-1} is (with L=2N+1)

$$\lim_{N\to\infty} \frac{2N_s}{2N+1} = \nu_2 \int_\delta^\infty \rho_s(p)dp, \qquad (11)$$

and there is a corresponding expression for the phonon number density. The latter diverges as $\ell n N$ (i.e. there is no thermodynamic limit for this quantity), and this is apparently the only significant manifestation of the infra-red divergence.

We have derived [1,13,15,16,17] the integral equations (1) and (8) by working in thermodynamic limit at finite density. For this we have used the action-angle variables of an **integrable** nonlinear lattice [1,13] under periodic boundary conditions of period L with $L \to \infty$. But we are then able to show that as $N \to \infty$ ($L \to \infty$) one can relate the action-angle variables to ones defined on the real line $\mathcal{R}$ (or infinite lattice) with decaying boundary conditions at infinity. Thus far our only way to solve (1) and so (8) has been to iterate these equations. So it is a question of some interest whether the equations iterate or not. The answer depends on the model dependent functions and constants in these equations, and below we give that answer for some particular chosen models.

2 Results for the sinh-Gordon model

The classical statistical mechanics of this model is the easiest of the models treated here. Both its classical and **quantum** statistical mechanics are given in [17]. The constants to be used in (1) and (8) are now $\nu_1 = 1$, $\alpha_1 = -\infty$, $\alpha_2 = \infty$. There are no soliton solutions of sinh-G and eqns. (1b),(8b) are irrelevant. The phonon dispersion relation is in principle that of the integrable lattice we use, but in leading order we can use (9a). The phonon phase shifts are

$$\Delta(k,k') = -\tfrac{1}{4}\gamma_0 m^2[k\omega(k') - k'\omega(k)]^{-1} \qquad (12)$$

with $\omega(k) = (k^2 + m^2)^{1/2}$. The dimensionless number $\gamma_0>0$ in (12) is the coupling constant which appears also in the sinh-G Hamiltonian [17]

$$H[\phi] = \gamma_0^{-1} \int_{-\infty}^{\infty} [\tfrac{1}{2}\gamma_0^2\Pi^2 + \tfrac{1}{2}\phi_x^2 + m^2(\cosh\phi - 1)]dx. \tag{13}$$

The model is a field model and the classical partition function Z is the functional integral

$$Z = \int D\Pi D\phi e^{-\beta H[\phi]}. \tag{14}$$

Our mode of calculation transforms to action-angle variables on the real line, and (14) becomes [17]

$$\int D\mu e^{-\beta H[p]} \tag{15}$$

with

$$H[p] = \int_{-\infty}^{\infty} \omega(k)P(k)dk \leftrightarrow \sum_{n=-N}^{N-1} \omega(\tilde{k}_n)P_n. \tag{16}$$

The $D\mu$ is an appropriate measure [17] while in H[p] $0 \le P(k) < \infty$. However, periodic boundary conditions induce a constraint on the allowed $\tilde{k}_n$ which must satisfy

$$k_n = k_n - L^{-1} \sum_{n' \neq n} \Delta(\tilde{k}_n, \tilde{k}_{n'})P_{n'}, \tag{17}$$

and this with (15) leads to the integral equation (1a) with $\Delta_s=0$ and the expressions (2) for F with $\nu_1=1$, $\nu_2=0$. The resultant system (1a) with (2) then explicitly iterates: all the integrals arising can be done, and the final strictly **asymptotic** development takes the form [17] (with $t \equiv M\beta^{-1}$ and $M \equiv 8m\gamma_0^{-1}$)

$$\lim_{L\to\infty} FL^{-1} = m\beta^{-1}\Big[\frac{1}{4}t - \frac{1}{8}t^2 + \frac{3}{16}t^3 - \frac{53}{128}t^4 + \frac{594}{2^9}t^5 - \frac{7922}{2^{11}}t^6 + \frac{121454}{2^{13}}t^7 - \dots\Big] + F_{KG}. \tag{18}$$

All the considerations noted in Section 1 surround this result: the limit $a_0 \to 0$ can be taken in, but only in, $FL^{-1} - F_{KG}$. It is actually possible to calculate exactly the same free energy (18) by the transfer integral method [17,18,19].

The quantum theory is also fairly easy [17]. We find, in analogy to (1), that

$$\epsilon(k) = \omega(k) + (2\pi\beta)^{-1} \int_{-\infty}^{\infty} dk' \frac{d}{dk}\Delta_b(k,k')\ell n(1 - e^{-\beta e(k')}) \tag{19a}$$

$$\lim_{L\to\infty} FL^{-1} = (2\pi\beta)^{-1} \int_{-\infty}^{\infty} \ell n(1 - e^{-\beta e(k)})dk, \qquad (19b)$$

where

$$\Delta_b = \Delta_f + 2\pi\theta(k' - k), \qquad (20a)$$

and $\theta(x)$ is the unit step function. The "fermion" phase shift Δ_f is the smooth branch $-2\pi<\Delta_f<0$ of

$$\Delta_f(k,k') = -2\tan^{-1}\left[\frac{m^2\sin(\frac{1}{8}\gamma_0'')}{k\omega(k') - k'\omega(k)}\right] \qquad (20b)$$

in which $\gamma_0'' \equiv \gamma_0/(1 + \gamma_0/8\pi)$ and $\omega(k) = (k^2 + m^2)^{1/2}$. It is noteworthy that there is bose-fermi equivalence [1,17] and eqns. (19) transform to a fermion form in terms of Δ_f rather than Δ_b.

3 Results for the sine-Gordon model

These are reported in [13]. An important point is that s-G derives from sinh-G by analytical continuation in the coupling constant $\gamma_0 \to -\gamma_0$. Thus the Hamiltonian of s-G is (13) with potential $m^2\gamma_0^{-1}(1 - \cos\phi)$, $\gamma_0>0$.

This system has soliton solutions on the real line $x \in \mathcal{R}$ and these show up in the integral equations (1). For these $\nu_1=1$, $\nu_2=2$, $\alpha_1 = \delta = -\infty$ and $\alpha_2 = \infty$. However, it is convenient to work with rapidities [13] so that ($p \to M\sinh y, k \to m\sinh x$)

$$\Delta(x,x') = \tfrac{1}{4}\gamma_0(\sinh(x - x'))^{-1}, \qquad (21a)$$

$$\Delta_s(x,y) = -2\tan^{-1}[(\sinh(x - y))^{-1}], \qquad (21b)$$

$$\frac{\partial}{\partial y}\Delta_{ss}(y,y') = 8\gamma_0^{-1}\ell n\left[\frac{\cosh(y - y') - 1}{\cosh(y - y') + 1}\right]. \qquad (21c)$$

The system (1) with (2) now iterates. Again all the integrals can be done and the result is [13]

$$\lim_{L\to\infty} FL^{-1} = F^{(1)} + F^{(2)} + F^{(3)} + \ldots + F_{KG}; \qquad (22a)$$

$$F^{(1)} = -m\beta^{-1}(8/\pi t)^{1/2}e^{-1/t}[1 - \tfrac{7}{8}t - \tfrac{59}{128}t^2 - \tfrac{897}{1024}t^3 - \qquad (22b)$$

$$- \tfrac{75005}{32768}t^4 - \ldots] - m\beta^{-1}[\tfrac{1}{4}t + \tfrac{1}{8}t^2 + \tfrac{3}{16}t^3 + \tfrac{53}{128}t^4 + \ldots];$$

$$F^{(2)} = \frac{8m}{\pi}Me^{-2/t}\{\ell n(\frac{4C}{t}) - \frac{5}{4}t[\ell n(\frac{4C}{t}) + 1] - \frac{t^2}{32}[13\ell n(\frac{4C}{t}) + 2] + \dots\}; \tag{22c}$$

$F^{(j)} = \mathcal{O}(e^{-j/t})$, $j = 3,4,\dots$; F_{KG} is as before; $\ell nC = 0.5771\dots$ (Euler's constant). It is evident that (22) is the analytic continuation in γ_0 of (18). Because there are now soliton solutions the $F^{(j)}$ have the j-kink-plus-antikink contributions depending on $e^{-j/t}$. It is relevant to the analysis of the Toda lattice, which has **zero** mass corresponding to m, that the development in multisoliton contributions is not possible and a different development arises. Note that it is the finiteness of the mass M which controls the convergence of (22b,c,...) as $T \to 0$.

The robust features of both (18) and (22) seem to us quite extraordinary given the apparent delicacy of the means of finding these asymptotic expansions either by our methods or by the transfer integral method. Still more so do we find it extraordinary that these asymptotic expansions can sum to the integral equations of type (1) with (2), as they do, given the particular problem associated with F_{KG}, the first iterate of these systems.

A number of people have given integral equations comparable with the classical equations (1) with (2) for the quantum s-G model. Our own result which is **not** restricted to very particular values of the coupling constant γ_0 is given in [16]. We cannot give the details of the quantum problem here; it suffices to say that in classical limit it actually **reduces** to (1) with (2) together with the phase shifts (21) [15].

4 Results for the repulsive NLS

For this we get for equations (1)

$$\epsilon(k) = \omega(k) + (2\pi\beta)^{-1}\int_{-\infty}^{\infty} dk'\frac{d}{dk}\Delta(k - k')\ln(\beta\epsilon(k')) \tag{23a}$$

with $\Delta(k - k') = -2c(k - k')^{-1}$, $c > 0$ is the coupling constant and $\omega(k) = k^2$. The free energy (2) is

$$\lim_{L\to\infty} FL^{-1} = (2\pi\beta)^{-1}\int_{-\infty}^{\infty} \ell n(\beta\epsilon(k))dk. \tag{23b}$$

This system (23) does **not** iterate. However, we found [20] from equations (23) and also by applying the transfer integral method to the corresponding functional integral similar to (14) that

$$\lim_{L\to\infty} FL^{-1} = (\beta a_0)^{-1}\ln(\beta a_0^{-2}) + Kc^{1/3}\beta^{-4/3}, \tag{24}$$

where a_0 is the lattice spacing and a_0^{-2} is an energy in our units. The constant K is determined by a numerical solution for the lowest eigenvalue of the radial Schrödinger equation (derived through the transfer integral method) as $K = 1.49 \times 2^{-2/3}$ [20]. Reference to (8) shows that if both (8) and (1) iterate and there are solitons, the leading order in the soliton density $\lim_{L\to\infty} 2N_s L^{-1}$ is $-\beta$ times the leading order in $\lim_{L\to\infty} FL^{-1}$. Remarkably enough [16] we also find (24) for the **attractive** NLS which does have soliton solutions: in this case $K \longrightarrow -K$. The attractive NLS has soliton solutions but it is open at present whether $|K|(c/\beta)^{1/3} \propto T^{1/3}$ forms an estimate of the soliton number density. The attractive NLS in any case needs special discussion [16].

5 Results for the Toda lattice

Our results for this model are not complete so our remarks take the form of a comment contrasting the Toda lattice model with the s-G model.

The parameters a,b are the parameters of Toda's potential in usual dimensioned form

$$V(r) = ab^{-1}e^{-br} + ar. \tag{25}$$

By choosing suitable units of energy and time (with $\hbar=1$), ba^{-1} is dimensionless and acts as the coupling constant in

$$H[Q_n] = ab^{-1} \sum_{n=-N}^{N} [\tfrac{1}{2}b^2a^{-2}p_n^2 + e^{-(Q_n-Q_{n-1})} - 1] \tag{26}$$

with $\{P_n,Q_{n'}\} = \delta_{nn'}$. There is a canonical transformation to new variables with the same bracket in which V(r) is given by (7b). But note that periodic boundary conditions are assumed and there are 2N + 1 points in each period L while the energy zero is chosen as H = 0 for all $P_n,Q_{n'}=0$. We also fix $Q_{-N-1}=Q_N$ (periodicity) - set $Q_{-N-1} = Q_N = 0$ for convenience - corresponding to translations of the whole lattice. There are therefore 2N degrees of freedom. Note that the ar terms in (25) cancel, i.e. can be dropped, under periodic boundary conditions.

The 2N coordinates $r_n \equiv Q_n - Q_{n-1}$ can be used as canonical coordinate variables [2]: they satisfy $\Sigma_{n=-N+1}^{N} r_n = Q_{-N}$. The classical partition function Z is just

$$Z = (2\pi)^{-2N} \int \prod_{n=-N+1}^{N} dP_n dQ_n \, e^{-\beta H[Q_n]}, \tag{27}$$

and this apparently compares favourably with a functional integral like (14) for the continuum models.

Under the chosen periodic boundary conditions

$$Z = Z_{kin} \cdot Z_{pot}; \quad Z_{kin} = (2\pi b a^{-1}\beta)^{-N}, \tag{28a}$$

$$Z_{pot} = e^{(2N+1)ab^{-1}\beta} \left(\prod_{j=1}^{2N} \int_0^\infty \frac{dy_j}{y_j} \right) e^{-ab^{-1}\beta\left(\sum_{j=1}^{2N} y_j + \prod_{j=1}^{2N} y_j^{-1}\right)}. \tag{28b}$$

The 2N-fold integral for Z_{pot} is not too far away from the product of 2N identical Γ-functions $\Gamma(ab^{-1}\beta)$, and it actually takes this form if the boundary conditions are that **one** point of a Toda **chain** is fixed, $Q_{-N-1}=0$. In this case the volume is not fixed. However, we have not reduced (28) to any combination of tabled integrals. Nevertheless, for large enough $ab^{-1}\beta$ we find from it the asymptotic expansion

$$\lim_{N\to\infty} F(2N+1)^{-1} = \beta^{-1}\ell n\beta - \beta^{-1}\ell n[1 + \mathcal{O}(ba^{-1}\beta^{-1})] + \mathcal{O}\left(\frac{\ell n 2N}{2N+1}\right) + \mathcal{O}\left(\frac{1}{2N+1}\right). \tag{29}$$

Compared with s-G or sinh-G the new feature is the error term $\mathcal{O}(\ell n2N/(2N + 1))$: this is a consequence of the infra-red "divergence" induced by the phonon dispersion $\omega(0)=0$. In fact for the **linearized** problem with dispersion (10)

$$N_{ph}(2N+1)^{-1} = \ell n 2N - \ell n(\pi/2). \tag{30}$$

There is therefore no thermodynamic limit for these phonons. However, the free energy $-\beta^{-1}\ell nZ$ converges as is also evident from (29).

To get the contribution of solitons to the free energy and to get the soliton number density, we must work with appropriate "soliton coordinates". The energy of the single soliton solution [2] for the infinite Toda lattice is $ab^{-1}[\sinh 2\kappa - 2\kappa]$. From this we infer that the action-angle variables are

$$0 \le P(k) < \infty, \quad 0 \le Q(k) < 2\pi; \quad -\infty < p_j, q_j < \infty \tag{31a}$$

with brackets

$$\{P(k), Q(k')\} = \delta(k - k'); \quad \{p_i, q_j\} = \delta_{ij} \tag{31b}$$

in the usual way [19], while the energy will be [16]

$$H[p] = 2\int_0^{\pi} dk(2\sin(\tfrac{1}{2}k))P(k) + ab^{-1}\sum_{j=1}^{2N_s}[\sinh(2ba^{-1}p_j) - 2ba^{-1}p_j] \tag{32}$$

(with $p_j > 0$). We shall finally be obliged to amend (31b) - see [5] and (37) below. With the results (31) and (32) our methods [12,16] yield the integral equations (1) with $\nu_1 = \varphi_2 = 2$, $\alpha_2 = \pi$, $\delta = 0$, $\omega(k) = 2\sin(\frac{1}{2}k)$ and $E(p) = ab^{-1}[\sinh(2ba^{-1}p) - 2ba^{-1}p]$. The lower limit α_1 in the k-space is kept to control the $k \to 0$ limit. We here quote only the phase shifts [16]

$$\Delta_s = -2\tan^{-1}[\cot(\tfrac{1}{2}k)\tanh(\tfrac{1}{2}ba^{-1}p)], \tag{33a}$$

$$\Delta_{ss} = -4\tanh^{-1}[\tanh(\tfrac{1}{2}ba^{-1}p)\coth(\tfrac{1}{2}ba^{-1}p)]. \tag{33b}$$

The phonon-phonon shift Δ is calculated in [16]. It is important that there are actually two phase shifts Δ_s, and two Δ_{ss}'s depending on whether pairs of excitations move in the same or opposite directions [2,16]. This already means we must **extend** the integral equations (1) and the expression (2) for F. We have not yet solved the extended integral equations for the Toda lattice, and we therefore complete this section with several remarks.

We note that (32) shows a simple soliton energy $E(p) \sim p^3$ for small p. It is then easy to show that, without the interactions described by the integral equations (1), the soliton contribution to F is $\sim T^{4/3}$. This compares with the results (24) for the repulsive NLS (and see also [11] for the soliton number density).

We have not demonstrated whether the system of integral equations (1) with (2) (or their extended forms) for the Toda lattice iterates or not. The "exact" asymptotic expansion (29) suggests it does and that the small parameter in the expansion is $ba^{-1}\beta^{-1}$. To sketch how the expansion could be found in terms of the present formulation, we first move the linear term $-2ba^{-1}p$ out of the soliton energy: $E(p) \longrightarrow E_0(p) = ab^{-1}\sinh(2ba^{-1}p)$. Note that this can actually be done ([5] and see below).

We then calculate $E_s(p)$ from (1b) in the form

$$-\beta E_s(p) = -\beta E_0(p) + \ell n\beta + I_1 + I_2, \tag{34}$$

where I_1, I_2 are the remaining integrals in (1b). The soliton-soliton interaction I_2 produces a prefactor $e^{-(ab^{-1}\beta)^{\frac{1}{2}}}$ in the soliton contribution to the free energy. Despite the square root this factor controls the equivalent of the s-G multisoliton contributions in the Toda problem. However, we shall not pursue I_2 further here.

The phonon-soliton contribution I_1 can be calculated in leading order in β^{-1} and p to produce

$$-\beta E_s(p) = -\beta p + \ell n(ba^{-1}\beta p) + \ldots \tag{35}$$

There is now a steepest descents development in the soliton contribution $-(\pi\beta)^{-1} \int_0^\infty e^{-\beta E_s(p)} dp$ to F, and the higher powers in (35) will produce a series in $ba^{-1}\beta^{-1}$. We expect to find the whole expression (29) in this way.

We have to explain the means of shifting E(p) to $E_0(p)$: this was done to make (35) have a saddle point analytic in $ab^{-1}\beta$. Now the action variables P(k) and p_j are **not independent** [5]. They satisfy the condition

$$-\int_0^{2\pi} (\omega(k))^{-1} P(k)dk + \sum_{j=1}^{2N_s} 2p_j = 0. \tag{36}$$

There is a corresponding change in the brackets (31b) which become

$$\{P(k),Q(k')\} = \delta(k - k') - \frac{\sin \frac{1}{2}k}{\sin \frac{1}{2}k'} (\delta(k) + \delta(k - 2\pi)),$$

$$\{P(k),\tilde{q}_j\} = -\frac{4ba^{-1} \sin(\frac{1}{2}k)e^{-ba^{-1}p_j}}{ab^{-1}e^{-2ba^{-1}p_j} - 1}(\delta(k) + \delta(k - 2\pi)), \tag{37}$$

with $\{\tilde{p}_i,\tilde{q}_j\} = \delta_{ij}$ and e.g. $\tilde{p}_i = 2ba^{-1} \cosh(ba^{-1}p_j)$ with $-\infty<\tilde{q}_j<\infty$. Although (36) means soliton contributions in H[p] can be moved to the phonons, significant problems with the phonons are introduced this way. Moreover (36) remains as a constraint on the problem. All of these ramifications are still being explored. Note that to find a quantum theory of Bethe Ansatz type for the Toda lattice all such problems must recur [1]. The same is true of any calculation of the soliton number.

Acknowledgement

One of us (RKB) is grateful to MIDIT, DTH and NORDITA; Copenhagen, for financial support during a vist when the work on the Toda lattice was begun.

References

[1] Bullough, R.K., Pilling, D.J. & Timonen, J. (1988). In *Solitons*, Lakshmanan, M. (ed.). Springer-Verlag: Heidelberg; and references therein.
[2] Toda, M. (1981). *Theory of Nonlinear Lattices.* Springer-Verlag: Berlin.
[3] Flaschka, H. & McLaughlin, D.W. (1976). *Prog. Theor. Phys.* **55**, 438.
[4] Bullough, R.K. & Caudrey, P.J. (eds.) (1980). *Solitons.* Springer-Verlag: Heidelberg.
[5] Faddeev, L.D. & Takhtajan, L.A. (1987). *Hamiltonian Methods in the Theory of Solitons.* Springer-Verlag: Berlin.
[6] See e.g. Lovesey, S.W., Balucani, U., Borsa, F. & Tognetti, V. (eds.) (1984). *Magnetic Excitations and Fluctuations.* Springer-Verlag: Heidelberg.
[7] Timonen, J. & Bullough, R.K. (1981). In *Inverse Problems - Nonlinear Evolution,* Sabatier, P.C. (ed.). Editions du CNRS, Paris, 1980; (1981). *Phys. Lett.* **82A**, 183.
[8] Yomosa, S. (1983). *Phys. Rev.* **A27**, 2120; Englander, S.W., Kallenbach, N.R., Heeger, A.J., Krumhansl, J.A. & Litwin, S. (1980). *Proc. Nat. Acad. Sci. USA* **77**, 7222; Banerjee, A. & Sobell, H.M. (1984). In *Nonlinear Electrodynamics in Biological Systems,* Adey, W.R. & Lawrence, A.F. (eds.). Plenum: New York.
[9] Stryer, L. (1981). *Biochemistry* (pp. 573-4). W.H. Freeman & Co.: San Francisco.
[10] Muto, V. (1988) *Nonlinear Models for DNA Dynamics.* Ph.D. Thesis, Laboratory of Applied Mathematical Physics, The Technical University of Denmark, DK-2800 Lyngby, Denmark, October 1988.
[11] Muto, V. (1989). In *Nonlinear Coherent Structures in Physics, Mechanics and Biological Systems,* Pouget, J. Remoissenet, N. & Ribotta, R. (eds.), Les Editions de Physique: Paris; Bullough, R.K., Pilling, D.J., Cheng Yi, Chen, Y.Z. & Timonen, J. (1989). Ibid.
[12] Bullough, R.K., Timonen, J. & Chen, Y.Z., to be published.
[13] Timonen, J., Stirland, M., Pilling, D.J., Cheng, Yi & Bullough, R.K. (1986). *Phys. Rev. Lett.* **56**, 2233.
[14] Chen, N.N., Johnson, M.D. & Fowler, M. (1986). *Phys. Rev. Lett.* **56**, 907.
[15] Timonen, J., Bullough, R.K. & Pilling, D.J. (1986). *Phys. Rev.* **B34**, 6525.
[16] Chen, Y.Z., Ph.D. Thesis, Department of Mathematics, University of Manchester, July 1989.
[17] Bullough, R.K., Pilling, D.J. & Timonen, J. (1986). *J. Phys.* **A19**, L955.
[18] Stone, M. & Reeve, J. (1986). *Phys. Rev.* **D18**, 4746; Dingle, R.B. & Müller, H.J.W. (1962). *J. Reine und Angew Math.* **211**, 11.
[19] Bullough, R.K., Pilling, D.J. & Timonen, J. (1985). In *Nonlinear Phenomena in Physics,* Claro, F. (ed.). Springer-Verlag: Heidelberg.

[20] Bullough, R.K., Pilling, D.J. & Timonen, J. (1985). In *Dynamical Problems in Soliton Systems,* Takeno, S. (ed.). Springer-Verlag: Berlin.
[21] Yi Cheng (1987). Ph.D. Thesis, Department of Mathematics, University of Manchester, February 1987.